Resilience and Sustainability of Civil Infrastructures under Extreme Loads

Resilience and Sustainability of Civil Infrastructures under Extreme Loads

Special Issue Editors

Zheng Lu
Ying Zhou
Tony Yang
Angeliki Papalou

MDPI • Basel • Beijing • Wuhan • Barcelona • Belgrade

Special Issue Editors

Zheng Lu
Department of Disaster Mitigation for
Structures, College of Civil Engineering,
Tongji University
China

Ying Zhou
Department of Disaster Mitigation for
Structures, College of Civil Engineering,
Tongji University
China

Tony Yang
Department of Civil Engineering,
The University of British Columbia
Canada

Angeliki Papalou
Department of Civil Engineering,
University of Peloponnese
Greece

Editorial Office
MDPI
St. Alban-Anlage 66
4052 Basel, Switzerland

This is a reprint of articles from the Special Issue published online in the open access journal *Sustainability* (ISSN 2071-1050) from 2018 to 2019 (available at: https://www.mdpi.com/journal/ sustainability/special_issues/Resilience_Infrastructures_Sustainability)

For citation purposes, cite each article independently as indicated on the article page online and as indicated below:

LastName, A.A.; LastName, B.B.; LastName, C.C. Article Title. *Journal Name* **Year**, *Article Number*, Page Range.

ISBN 978-3-03921-401-3 (Pbk)
ISBN 978-3-03921-402-0 (PDF)

Contents

About the Special Issue Editors . . . . **vii**

Preface to "Resilience and Sustainability of Civil Infrastructures under Extreme Loads" . . . **ix**

Zheng Lu, Ying Zhou, Tony Yang and Angeliki Papalou
Special Issue: Resilience and Sustainability of Civil Infrastructures under Extreme Loads
Reprinted from: *Sustainability* **2019**, *11*, 3292, doi:10.3390/su11123292 . . . . **1**

Cheng-Yu Yang, Xue-Song Cai, Yong Yuan and Yuan-Chi Ma
Hybrid Simulation of Soil Station System Response to Two-Dimensional Earthquake Excitation
Reprinted from: *Sustainability* **2019**, *11*, 2582, doi:10.3390/su11092582 . . . . **4**

Yongmei Zhai, Xuxia Fu, Yihui Chen and Wei Hu
Study on Column-Top Seismic Isolation of Single-Layer Latticed Domes
Reprinted from: *Sustainability* **2019**, *11*, 936, doi:10.3390/su11030936 . . . . **19**

Shuo Jiang and Yimin Wang
Long-Term Ground Settlements over Mined-Out Region Induced by Railway Construction
and Operation
Reprinted from: *Sustainability* **2019**, *11*, 875, doi:10.3390/su11030875 . . . . **28**

Yun Chen, Junzuo Li and Zheng Lu
Experimental Study and Numerical Simulation on Hybrid Coupled Shear Wall with
Replaceable Coupling Beams
Reprinted from: *Sustainability* **2019**, *11*, 867, doi:10.3390/su11030867 . . . . **47**

**Xiong-Fei Ye, Kai-Chun Chang, Chul-Woo Kim, Harutoshi Ogai, Yoshinobu Oshima and
O.S. Luna Vera**
Flow Analysis and Damage Assessment for Concrete Box Girder Based on Flow Characteristics
Reprinted from: *Sustainability* **2019**, *11*, 710, doi:10.3390/su11030710 . . . . **69**

Dongwang Tao, Jiali Lin and Zheng Lu
Time-Frequency Energy Distribution of Ground Motion and Its Effect on the Dynamic Response
of Nonlinear Structures
Reprinted from: *Sustainability* **2019**, *11*, 702, doi:10.3390/su11030702 . . . . **101**

Hyunjin Ju, Sun-Jin Han, Hyo-Eun Joo, Hae-Chang Cho, Kang Su Kim and Young-Hun Oh
Shear Performance of Optimized-Section Precast Slab with Tapered Cross Section
Reprinted from: *Sustainability* **2019**, *11*, 163, doi:10.3390/su11010163 . . . . **118**

Nima Pirhadi, Xiaowei Tang, Qing Yang and Fei Kang
A New Equation to Evaluate Liquefaction Triggering Using the Response Surface Method and
Parametric Sensitivity Analysis
Reprinted from: *Sustainability* **2019**, *11*, 112, doi:10.3390/su11010112 . . . . **136**

Yabin Chen, Longjun Xu, Xingji Zhu and Hao Liu
A Multi-Objective Ground Motion Selection Approach Matching the Acceleration and
Displacement Response Spectra
Reprinted from: *Sustainability* **2018**, *10*, 4659, doi:10.3390/su10124659 . . . . **160**

Jianwei Zhang and Yulong Chen
Experimental Study on Mitigations of Seismic Settlement and Tilting of Structures by Adopting
Improved Soil Slab and Soil Mixing Walls
Reprinted from: *Sustainability* **2018**, *10*, 4069, doi:10.3390/su10114069 . . . . **185**

Sang-Yun Lee, Sam-Young Noh and Dongkeun Lee
Evaluation of Progressive Collapse Resistance of Steel Moment Frames Designed with Different
Connection Details Using Energy-Based Approximate Analysis
Reprinted from: *Sustainability* **2018**, *10*, 3797, doi:10.3390/su10103797 . . . . **202**

Jeongwook Choi, Do Guen Yoo and Doosun Kang
Post-Earthquake Restoration Simulation Model for Water Supply Networks
Reprinted from: *Sustainability* **2018**, *10*, 3618, doi:10.3390/su10103618 . . . . **230**

Jihong Ye and Liqiang Jiang
Simplified Analytical Model and Shaking Table Test Validation for Seismic Analysis of Mid-Rise
Cold-Formed Steel Composite Shear Wall Building
Reprinted from: *Sustainability* **2018**, *10*, 3188, doi:10.3390/su10093188 . . . . **247**

Xingji Zhu, Zaixian Chen, Hao Wang, Yabin Chen and Longjun Xu
Probabilistic Generalization of a Comprehensive Model for the Deterioration Prediction of RC
Structure under Extreme Corrosion Environments
Reprinted from: *Sustainability* **2018**, *10*, 3051, doi:10.3390/su10093051 . . . . **265**

Zhiguang Zhou, Liuyun Xu, Chaoxin Sun and Songtao Xue
Brazier Effect of Thin Angle-Section Beams under Bending
Reprinted from: *Sustainability* **2018**, *10*, 3047, doi:10.3390/su10093047 . . . . **282**

Bo Fu, Huanjun Jiang and Tao Wu
Nonlinear Error Propagation Analysis of a New Family of Model-Based Integration Algorithm
for Pseudodynamic Tests
Reprinted from: *Sustainability* **2018**, *10*, 2846, doi:10.3390/su10082846 . . . . **293**

Zaixian Chen, Xueyuan Yan, Hao Wang, Xingji Zhu and Billie F. Spencer
Substructure Hybrid Simulation Boundary Technique Based on Beam/Column Inflection Points
Reprinted from: *Sustainability* **2018**, *10*, 2655, doi:10.3390/su10082655 . . . . **308**

Jinping Yang, Peizhen Li and Zheng Lu
Numerical Simulation and In-Situ Measurement of Ground-Borne Vibration Due to
Subway System
Reprinted from: *Sustainability* **2018**, *10*, 2439, doi:10.3390/su10072439 . . . . **328**

Zheng Lu, Junzuo Li and Chuanguo Jia
Studies on Energy Dissipation Mechanism of an Innovative Viscous Damper Filled with Oil
and Silt
Reprinted from: *Sustainability* **2018**, *10*, 1777, doi:10.3390/su10061777 . . . . **348**

Rui Fukumoto, Yuji Genda and Mikiko Ishikawa
Characteristics of Corporate Contributions to the Recovery of Regional Society from the Great
East Japan Earthquake Disaster
Reprinted from: *Sustainability* **2018**, *10*, 1717, doi:10.3390/su10061717 . . . . **361**

About the Special Issue Editors

Zheng Lu (College of Civil Engineering, Tongji University) has more than ten years of experience working in structural vibration control, earthquake resistance of engineering structures, and particle damping technology. The studies he has carried out focus on developing particle-damping-based new dissipative devices and their corresponding fundamental applications. He directed two projects subsidized by the Natural Science Foundation of China and received an award from the Shanghai "Chen Guang" project, which is given to young scholars below the age of 30. He has published more than 60 first-authored or corresponding-authored SCI-indexed papers, all of which are widely approved. His h-index is 16 in Web of Science. He has published review papers pertaining to nonlinear dissipative dampers in two high-quality SCI-indexed journals, namely, Structural Control and Health Monitoring and the Journal of Sound and Vibration.

Ying Zhou (College of Civil Engineering, Tongji University) focuses on the seismic performance of complex tall buildings, performance-based seismic design, structural performance of composite structures and hybrid structures, and methodology and technology for the structural dynamic test. In 2008, she received the Outstanding Paper Award for Young Experts in the 14th World Conference on Earthquake Engineering. She received an award from the National Science Fund for Excellent Young Scholars (2014–2016) for her project titled "Seismic Resistance of High-Rise Buildings.".

Tony Yang (Department of Civil Engineering, The University of British Columbia) focuses on improving structural performance through advanced analytical simulation and experimental testing. He has developed the next-generation performance-based design guidelines (adopted by the Applied Technology Council, the ATC-58 research team) in the United States and has developed advanced experimental testing technologies, such as hybrid simulation and nonlinear control of shake table, to evaluate structural response under extreme loading conditions.

Angeliki Papalou (Department of Civil Engineering, University of Peloponnese, Greece) focuses on the seismic protection of structures, structural control, and health monitoring in the seismic protection and restoration of historic structures and in the analysis and design of structures. She was the principal investigator of a funded research project and was a member of research groups for five other funded research projects. She is a reviewer for more than ten international scientific journals and is member of the American Society of Civil Engineers (ACSE), the Technical Chamber of Greece (TEE) and the American Mechanics Institute.

Preface to "Resilience and Sustainability of Civil Infrastructures under Extreme Loads"

Resilience and Sustainability of Civil Infrastructures under Extreme Loads contains papers that focus on cutting-edge approaches to enhancing the resilience and sustainability of structures under extreme loading events such as theoretical investigation, numerical simulation, and experimental study. Earthquakes, which are detrimental to the safety of civil structures, have been occurring more frequently in many regions and nations in recent years, and these seismic excitations can generate severe and unexpected damages to civil infrastructures. For example, seismic excitations usually contain multiple frequency components. The plastic deformation of the structure caused by them would continuously accumulate in a certain direction, resulting in a large nonlinear displacement. Moreover, earthquakes can cause liquefaction, which in turn can cause seismic settlement and tilting of structures. Even worse, the progressive collapse of buildings is more likely to happen if structural redundancy suddenly decreases because of unexpected abnormal loads. In light of that, there is an urgent need to explore innovative methods that can improve the resilience and sustainability of civil infrastructures under extreme loading events.

The contents of this book cover a wide variety of topics divided into these three categories: The theoretical study of structural resilience and sustainability under specific loading events (academic fundamental phase); new simulation tools for evaluating structural damages under extreme loading events and new experimental methods for seismic analysis of civil infrastructures (current cutting-edge research); and finally, new energy dissipative devices and resilient structural forms (industrial application phase). To broaden the scope of structural resilience and sustainability, this book also addresses the concept of resilient communities, specifically with the idea that communities in devastated areas can develop resilience that will enable them to autonomously and efficiently recover from natural disasters through the help of corporations. Additionally, this book also introduces research related to post-earthquake restoration simulation models for water supply networks. These models help formulate prompt system recovery plans and restoration priorities in the case of an actual seismic hazard that may occur in water supply networks.

This book contains 20 very high quality papers. The author groups represent currently active researchers in the field of resilient and sustainable structures. The topics not only provide current and cutting-edge research but are also of great academic (fundamental phase) and industrial (applied phase) interest. Readers will observe that current technology utilized to enhance structural resilience and sustainability mainly concentrates on theoretical investigation, numerical simulation, and experimental study. Hence, rich ground for further development is freely available in the applied phase field of resilient and sustainable structures.

Zheng Lu, Ying Zhou, Tony Yang, Angeliki Papalou
Special Issue Editors

Editorial

Special Issue: Resilience and Sustainability of Civil Infrastructures under Extreme Loads

Zheng Lu [1,*] , **Ying Zhou** [1], **Tony Yang** [2] **and Angeliki Papalou** [3]

[1] Department of Disaster Mitigation for Structures, College of Civil Engineering, Tongji University, Shanghai 200092, China; yingzhou@tongji.edu.cn
[2] Department of Civil Engineering, University of British Columbia, Vancouver, BC V6T 1Z4, Canada; yang@civil.ubc.ca
[3] Department of Civil Engineering, University of Peloponnese, 26334 Patra, Greece; papalou@teiwest.gr
* Correspondence: luzheng111@tongji.edu.cn; Tel.: +86-21-6598-6186

Received: 6 June 2019; Accepted: 14 June 2019; Published: 14 June 2019

Abstract: The special issue entitled "Resilience and Sustainability of Civil Infrastructures under Extreme Loads" updates the state of the art and perspectives focused on cutting-edge approaches to enhance structures' resilience and sustainability under extreme loading events, including theoretical investigation, numerical simulation, and experimental study, keeping an eye on the seismic performance of civil structures. Notably, some innovative energy dissipative devices and resilient structural forms, which are encompassed in this special issue, would provide valuable references for the engineering application of resilient and sustainable civil infrastructures in the near future.

Keywords: resilience; sustainability; civil infrastructures; extreme loads; natural hazards; earthquakes; seismic performance; energy dissipative devices

---

1. Rationale

Natural hazards, such as earthquakes, that are detrimental to the safety of civil structures, have appeared more frequently in many regions and nations in recent years. Seismic excitations generate many unexpected damages to civil infrastructures. For example, seismic excitations usually contain multiple frequency components, and therefore the plastic deformation of the structure continuously accumulates in a certain direction, resulting in a large nonlinear displacement [1,2]. Moreover, liquefaction may occur after earthquakes, as it is worth mentioning that liquefaction can cause seismic settlement and tilting of structures [3,4]. Even worse, the progressive collapse of buildings is more likely to happen if structural redundancy suddenly decreases because of unexpected abnormal loads [5,6]. In light of that, there is an urgent need to exploring innovative methods to improve the resilience and sustainability of civil infrastructures under extreme loading events.

The Special Issue on Resilience and Sustainability of Civil Infrastructures under Extreme Loads contains papers that focus on the cutting-edge approaches to enhance structures' resilience and sustainability under extreme loading events, including theoretical investigation, numerical simulation, and experimental study. The contents of this special issue cover a wide variety of topics, which can be mainly divided into three categories, namely, (i) academic fundamental phase—the theoretical study of structural resilience and sustainability under specific loading events (contributions 6, 7, 9, 15); (ii) current cutting-edge research—new simulation tools for evaluating structural damage under extreme loading events (contributions 3, 5, 8, 11, 14, 18), and new experimental methods for seismic analysis of civil infrastructures (contributions 1, 4, 10, 13, 16, 17); and (iii) industrial application phase—new energy dissipative devices and resilient structural forms (contributions 2, 19). It is worth mentioning that to broaden the scope of structural resilience and sustainability, the concept of resilient communities is also encompassed in this book. To be specific, through corporations, communities in devastated areas can be resilient in order to autonomously and efficiently recover from natural disasters (contribution 20). Additionally, this special issue also introduces the research related to the post-earthquake restoration simulation modal for water supply networks, which is conducive to determining prompt system recovery plans and restoration priorities in the case of an actual seismic hazard that may occur in water supply networks (contribution 12).

2. List of Contributions

The Special Issue contains 20 very high-quality papers. The author groups represent currently active researchers in the field of resilient and sustainable structures. The topics are not only current (cutting-edge research) but also of great academic (fundamental phase) and industrial (applied phase) interest. The readers will observe that the current technology utilized to enhance structural resilience and sustainability are mainly concentrated in the theoretical investigation, numerical simulation, and experimental study. Hence, further effort should be made on the applied phase of resilient and sustainable structures. It is worthy of mention that the development of efficient energy dissipative devices plays an important part in the actual implementation of resilient and sustainable structures. In fact, apart from innovative viscous damper filled with oil and silt, as introduced in the Special Issue, nonlinear energy dissipative devices (e.g., particle impact dampers) could also contribute a lot to improving structural resilience and sustainability [7]. The contributions of this Special Issue are as listed below:

1. Hybrid Simulation of Soil Station System Response to Two-Dimensional Earthquake Excitation, by Yang et al.
2. Study on Column-Top Seismic Isolation of Single-Layer Latticed Domes, by Zhai et al.
3. Long-Term Ground Settlements over Mined-Out Region Induced by Railway Construction and Operation, by Jiang and Wang.
4. Experimental Study and Numerical Simulation on Hybrid Coupled Shear Wall with Replaceable Coupling Beams, by Chen et al.
5. Flow Analysis and Damage Assessment for Concrete Box Girder Based on Flow Characteristics, by Ye et al.
6. Time-Frequency Energy Distribution of Ground Motion and Its Effect on the Dynamic Response of Nonlinear Structures, by Tao et al.
7. Shear Performance of Optimized-Section Precast Slab with Tapered Cross Section, by Ju et al.
8. A New Equation to Evaluate Liquefaction Triggering Using the Response Surface Method and Parametric Sensitivity Analysis, by Pirhadi et al.
9. A Multi-Objective Ground Motion Selection Approach Matching the Acceleration and Displacement Response Spectra, by Chen et al.
10. Experimental Study on Mitigations of Seismic Settlement and Tilting of Structures by Adopting Improved Soil Slab and Soil Mixing Walls, by Zhang and Chen.
11. Evaluation of Progressive Collapse Resistance of Steel Moment Frames Designed with Different Connection Details Using Energy-Based Approximate Analysis, by Lee et al.
12. Post-Earthquake Restoration Simulation Model for Water Supply Networks, by Choi et al.
13. Simplified Analytical Model and Shaking Table Test Validation for Seismic Analysis of Mid-Rise Cold-Formed Steel Composite Shear Wall Building, by Ye and Jiang.
14. Probabilistic Generalization of a Comprehensive Model for the Deterioration Prediction of RC Structure under Extreme Corrosion Environments, by Zhu et al.
15. Brazier Effect of Thin Angle-Section Beams under Bending, by Zhou et al.
16. Nonlinear Error Propagation Analysis of a New Family of Model-Based Integration Algorithm for Pseudodynamic Tests, by Fu et al.

17. Substructure Hybrid Simulation Boundary Technique Based on Beam/Column Inflection Points, by Chen et al.
18. Numerical Simulation and In-Situ Measurement of Ground-Borne Vibration Due to Subway System, by Yang et al.
19. Studies on Energy Dissipation Mechanism of an Innovative Viscous Damper Filled with Oil and Silt, by Lu et al.
20. Characteristics of Corporate Contributions to the Recovery of Regional Society from the Great East Japan Earthquake Disaster, by Fukumoto et al.

Author Contributions: All authors have conceived the special volume and the Preface, and have worked on that.

Conflicts of Interest: The authors declare no conflict of interest.

References

1. Cao, H.; Friswell, M.I. The effect of energy concentration of earthquake ground motions on the nonlinear response of rc structures. *Soil Dyn. Earthq. Eng.* **2009**, *29*, 292–299. [CrossRef]
2. Yaghmaei-Sabegh, S. Time-frequency analysis of the 2012 double earthquakes records in north-west of iran. *Bull. Earthq. Eng.* **2014**, *12*, 585–606. [CrossRef]
3. Rasouli, R.; Towhata, I.; Rattez, H.; Vonaesch, R. Mitigation of nonuniform settlement of structures due to seismic liquefaction. *J. Geotech. Geoenviron. Eng.* **2018**, *144*. [CrossRef]
4. Smyrou, E.; Bal, I.E.; Tasiopoulou, P.; Gazetas, G. Wavelet analysis for relating soil amplification and liquefaction effects with seismic performance of precast structures. *Earthq. Eng. Struct. Dyn.* **2016**, *45*, 1169–1183. [CrossRef]
5. Alashker, Y.; El-Tawil, S.; Sadek, F. Progressive collapse resistance of steel-concrete composite floors. *J. Struct. Eng.* **2010**, *136*, 1187–1196. [CrossRef]
6. Lu, Z.; Chen, X.; Lu, X.; Yang, Z. Shaking table test and numerical simulation of an RC frame-core tube structure for earthquake-induced collapse. *Earthq. Eng. Struct. Dyn.* **2016**, *45*, 1537–1556. [CrossRef]
7. Lu, Z.; Wang, Z.X.; Zhou, Y.; Lu, X.L. Nonlinear dissipative devices in structural vibration control: A review. *J. Sound Vib.* **2018**, *423*, 18–49. [CrossRef]

*sustainability*

MDPI

Article

Hybrid Simulation of Soil Station System Response to Two-Dimensional Earthquake Excitation

Cheng-Yu Yang [1], Xue-Song Cai [2,*], Yong Yuan [1] and Yuan-Chi Ma [3]

1 State Key Laboratory of Disaster Reduction in Civil Engineering, Tongji University, Shanghai 200092, China; seismicyang@tongji.edu.cn (C.-Y.Y.); yuany@tongji.edu.cn (Y.Y.)
2 Department of Geotechnical Engineering, Tongji University, Shanghai 200092, China
3 Department of Bridge Engineering, Tongji University, Shanghai 200092, China; 1630427@tongji.edu.cn
* Correspondence: caixiaoshuai@tongji.edu.cn

Received: 30 March 2019; Accepted: 28 April 2019; Published: 5 May 2019

Abstract: Soil station system seismic issues have been highly valued in recent years. In order to investigate the dynamic seismic behaviors of the intermediate column in soil station systems, a hybrid test of a soil station system was conducted. The soil station model was performed with OpenSees. Virtual hybrid simulation was fulfilled with adapter elements. A hybrid model, composed of the steel column specimen and the remainder numerical model, was assembled using the OpenFresco framework. An intermediate column was treated as the physical substructure, while the rest of the soil station system was treated as the numerical substructure in a hybrid simulation. The hybrid test results are compared with the analytical results. The data obtained from such tests show that the system can accurately reflect the mechanical properties of intermediate columns in soil station systems. A hybrid simulation would be a proper way to assess the seismic performance of a soil station system.

Keywords: hybrid simulation; intermediate column; subway station; OpenFresco; OpenSees

---

1. Introduction

The uninterrupted service of transportation infrastructure after an earthquake is of great importance for the immediate recovery and long-term economic sustainability of the impacted region. Subway stations and railway systems, essential components of the transportation infrastructure, are designed to provide continued function after both frequent and rare seismic events [1]. It is generally believed that the seismic performance of underground structures is superior to superstructure since underground structures are surrounded and restrained by soil and rock. However, several catastrophic earthquake events have happened in recent years such as the Kobe earthquake, Chi-Chi earthquake, and Wenchuan earthquake, and thus the seismic performance of an underground structure is of concern to many researchers [2–6]. Especially in 1995, the Daikai subway station suffered serious damage under the Hyogoken-Nanbu earthquake, including shear failure of the intermediate column, which had never been observed in the past. The failure positions of the intermediate column are mostly located at the base and top part of the intermediate column [7]. Moreover, a large number of cracks have caused serious damage to underground engineering facilities. Insufficient horizontal shear resistance of the intermediate column is the main reason why the Daikai station failed during the seismic event [8]. Most research has conducted dynamic analyses to study the failure mechanism of the subway station, few studies have adopted seismic testing methods. Seismic testing methods based on intermediate columns are necessary and significant ways to evaluate the seismic performance of a soil station system.

There are several seismic testing methods, including the pseudo-static test (PST), shaking table test (STT) and pseudo-dynamic (PSD) test. The PST method is to apply prescribed cyclic displacement or cyclic force history on a structure component with actuators at a low speed. The shortcoming of PST method is that the predefined load and displacement cannot reflect real seismic responses of structures [9]. In the STT method, structure specimens are fixed on a shaking table. The shaking table provides acceleration boundary conditions at the bottom of structure specimens. When it comes to performing the shaking table test on underground structures, such as tunnel or subway station specimens, there exist two main issues that are often improperly considered. The first one is that the size and payload of the tested structure is greatly limited to the size and payload of the table used. The second one is the gravity distortion effect.

The hybrid simulation (HS) testing method, initially named the pseudo-dynamic testing method, was proposed in 1969 [10]. This method combines the physical test of the nonlinear components of a structure and the numerical simulation of the remainder [11]. Without the size and payload limitation, hybrid simulation can easily enlarge the size of the specimen so that test results would not be impacted by scale factors. Hybrid simulation development trends are composed of two aspects. One aspect involves developing next-generation hybrid simulation methods that will provide more realistic structural responses, including robust numerical integration techniques [12–15] and the loading control method [16–19]. The other aspect involves providing validated general hybrid simulation procedure suitable to various projects and testing facilities to promote its awareness and broader applications, including the geographically distributed hybrid simulation, and developing a benchmark hybrid simulation [20–22]. Based on the experience of hybrid projects in the George E. Brown, Jr. Network for Earthquake Engineering Simulation (NEES), hybrid simulation, and especially displacement-based PSD simulation, is a viable approach to generate reliable structural seismic responses [23]. For instance, Tessari utilizes both shaking tables and dynamic actuators as an ideal experimental method to investigate soil structure interaction issues, where a massive test specimen of soil can be accommodated by the shaking table and the interactions between the soil and structure can be simulated by the actuators [24]. Hybrid simulation is a unique way to experimentally judge the seismic performance of underground structures. With the hybrid simulation test method, a reasonable full-scale model of the intermediate column in a subway station could be studied when an earthquake excitation proceeds.

A study of the dynamic responses of the intermediate column in underground structures by applying the hybrid simulation method is presented in this paper. Both virtual hybrid simulation and physical hybrid simulation are performed in this paper. A virtual hybrid simulation is conducted before the implementation of an actual hybrid test, which works as a preparatory simulation in order to check all the setting/adjustments, make sure the numerical calculator is working fine and there is no data exchange problem between numerical and physical subparts of the structure. A novel steel specimen is designed for the physical hybrid simulation. According to different study needs, the stiffness of the steel specimen could be easily changed by a simple calculation and replacement of some screws before the test. Hybrid simulation with such steel specimens allow for repetitive tests with low costs. Thus, hybrid simulation emerges as a best way to experimentally assess the performance of underground structures under earthquake excitation.

2. Theory of Hybrid Simulation

In traditional PSD testing, the dynamic equation is solved to get displacement response during test process. In Equation (1), M is the mass matrix, C is the damping matrix, $R(u_i)$ is the resisting force, $(\ddot{u}_i, \dot{u}_i, u_i)$ is the acceleration vector, velocity vector and displacement vector of numerical model, and $a_{g,i}$ is the input motion.

$$M\ddot{u}_i + C\dot{u}_i + R(u_i) = -Ma_{g,i} \tag{1}$$

$$M_N\ddot{u}_N + C_N\dot{u}_N + R_N(u_N) + R_E(u_E) = -Ma_{g,i} \tag{2}$$

In Equation (2), M is the mass matrix of whole structure, M_N is the mass matrix of numerical substructure, C_N is the damping matrix of numerical substructure, R_N is the resisting force of numerical substructure, R_E is the resisting force of experimental substructure, $(\ddot{u}_N, \dot{u}_N, u_N)$ is the acceleration vector, velocity vector and displacement vector of numerical substructure, u_E is the displacement vector of experimental substructure, and $a_{g,i}$ is the input motion [25]. As shown in Figure 1, PSD substructure testing theory is adopted as analysis method for realizing hybrid simulation of underground structures in this paper. The implementation procedures are as follows: treat structure response $(\ddot{u}_i, \dot{u}_i, u_i)$ as initial conditions at time t_i, measure the resisting force of experimental substructure $R_{E(i)}$, obtain the solution of structure response $(\ddot{u}_{i+1}, \dot{u}_{i+1}, u_{i+1})$ at time t_{i+1}; treat structure response $(\ddot{u}_{i+1}, \dot{u}_{i+1}, u_{i+1})$ as initial conditions at time t_{i+1}, measure the resisting force of experimental substructure $R_{E(i+1)}$, obtain the solution of structure response $(\ddot{u}_{i+2}, \dot{u}_{i+2}, u_{i+2})$ at time t_{i+2}. Such a calculation cycle should be repeated until the hybrid simulation process is over.

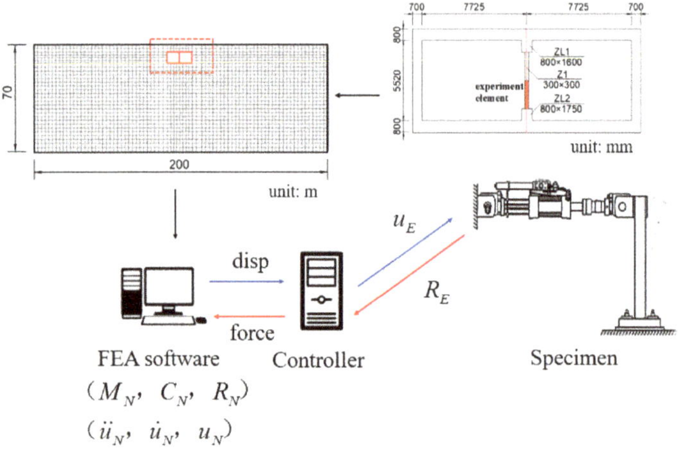

Figure 1. Key components of a hybrid simulation model.

3. Numerical Substructure

3.1. Input Motions

The Shanghai artificial wave, El Centro wave and Kobe wave are adopted as the input earthquake motions. Figure 2 shows the seismic acceleration time histories and spectra. It should be noted that the Shanghai artificial wave is a kind of synthetic earthquake motion developed to predict bedrock movements under the specific construction site [26]. The frequency content of the Shanghai artificial motion is mainly below 10 Hz and the dominate frequency ranges from to 1.7 to 3.6 Hz based on the results of a Fourier transform [27]. The peak base acceleration (PBA) of such three motions are scaled to 0.12 g in this paper.

It is assumed that the bedrock is rigid, the bottom of the model is fixed in such a way that no movement is allowed on the vertical and horizontal direction. Moreover, the lateral boundary is set as the equal displacement and impermeable boundary [28]. In this way, vertically polarized shear waves (SV seismic waves) can be properly considered in the seismic issue of the soil station system.

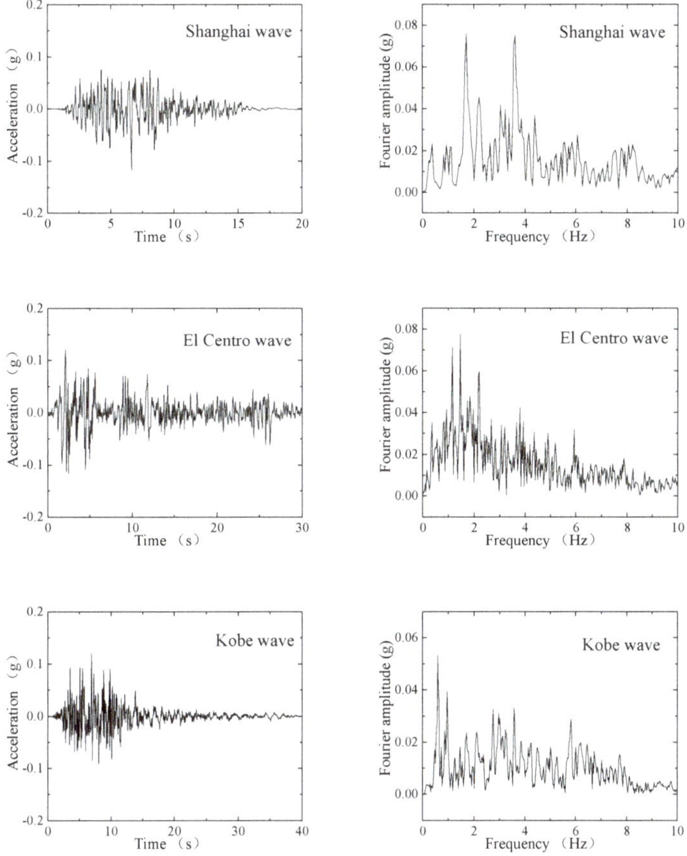

Figure 2. Input earthquake motions.

3.2. Soil Modeling

The analysis is carried out under plane strain assumptions [29]. Note that the seismic response of the soil-station system is a 3D issue, a 2D analysis model is carried out by assuming plane strain condition. Linear elastic material is adopted in the model for simulating the soil layer. In addition, the soil layer is simulated with total stress analysis. Moreover, the soil layer in finite element analysis (FEA) model is homogeneous, and the properties of soil are constant along the depth in a vertical direction. The dimension of the soil domain is 200 m long and 70 m deep (bedrock level). The soil parameters applied in the FEA model are listed in Table 1. Quad elements are employed to model the soil layer. Four-noded quad elements can be used to perform drained analysis, total stress analysis, and undrained analysis.

Table 1. Properties of soil.

Variables	Parameters
Density	1.48 ton/m^3
Elastic Modulus	86.7 MPa
Poisson ratio	0.3
Shear velocity	150 m/s
Shear Modulus	33.4 MPa

3.3. Structure Modeling

A typical rectangular subway station is adopted as the computational model. Figures 3 and 4 show the FEA model in detail. The FEA model is taken sufficiently long so that lateral boundaries would not influence the seismic response of soil layer. The dimensions of the soil domain are 200 m long and 70 m deep. The rectangular subway station section is 17 m × 7.2 m, the station is embedded 4.8m deep below ground surface. It should be emphasized that the bedrock is 70 m away from the ground surface. The soil elements are 3.5 m in thickness based on the column spacing. In addition, the top slab is 0.8 m thick, the bottom slab is 0.85 m thick and the side wall is 0.7 m thick. Further, the intermediate column section is 0.3m × 0.3m.

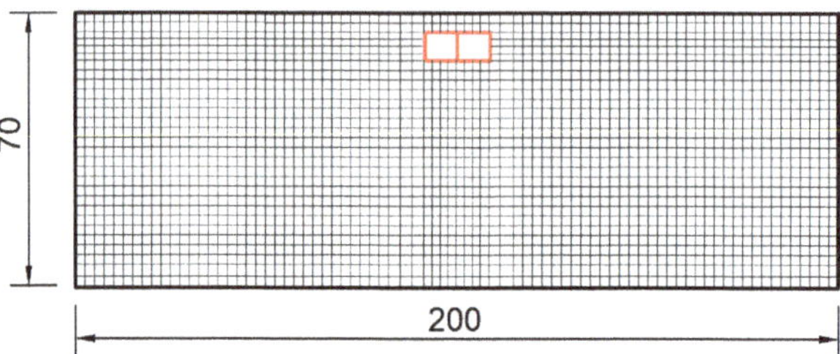

Figure 3. Cross section of soil station model (unit: m).

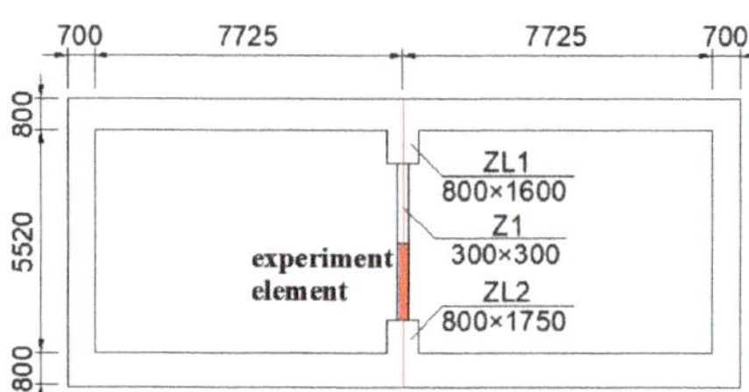

Figure 4. Cross section of station (unit: mm).

In the FEA model, the concrete compressive strength is 39.8 MPa, and the concrete strain at maximum strength is 0.002. The concrete crushing strength is 20 MPa, and the concrete strain at crushing strength is 0.004. Nonlinear beam column elements are adopted to simulate the slabs, walls and intermediate columns. Concrete 01 material is selected in OpenSees. Such material is used for constructing a uniaxial Kent-Scott-Park concrete material object with a degraded linear unloading/reloading stiffness. No tensile strength is considered in Concrete 01 material [30].

The soil and structure are directly bonded together using the equalDOF command. This is a simplified algorithm and the slippage of soil near the structure could not be considered. In the refined analysis model, the above factors need to be considered, and the contact element can be adopted for the simulation. A one-dimensional nonlinear spring and a tangentially nonlinear spring could simulate the behaviour of the slippage of soil near the structure [31]. A section of 2 × 2 m soil elements are adopted for the reason that the soil element length should be less than approximately one-tenth to one-eighth of the wavelength associated with the maximum frequency f_{max} component of the input seismic motion. For the case of a 70 m deposit excited by the Shanghai artificial wave, a target damping ratio of 5% is calculated by calibrating parameters in Equations (3)–(5). The first mode of the site and five times this frequency for f_m and f_n are selected as parameter for the target damping ratio [32].

$$C = \alpha M + \beta K \tag{3}$$

$$\left\{ \begin{array}{c} \alpha \\ \beta \end{array} \right\} = \frac{2\zeta}{\omega_i + \omega_j} \left\{ \begin{array}{c} \omega_i \omega_j \\ 1 \end{array} \right\} \tag{4}$$

$$\omega_m = 2\pi f_m \tag{5}$$

4. Hybrid Simulation Procedure

4.1. Virtual Hybrid Simulation Procedure

A virtual hybrid simulation is performed based on the FEA model mentioned before. In physical hybrid simulation case, middleware platform OpenFresco provides a bridge between a standard finite element analysis program and laboratory control data-acquisition systems. In virtual hybrid simulation, one OpenSees process stands for numerical substructure, the other OpenSees process stands for experimental substructure. Adapter elements are utilized to manage communication between two OpenSees processes with OpenFresco [33,34]. This approach provides an important advantage that all of the connected codes run continuously without need to shut down and restart, hence reducing analysis time consumptions significantly. In this way, virtual hybrid simulation can be realized with two OpenSees processes connected by adapter elements on OpenFresco platform.

When coupling two or more finite element programs, the structure is generally divided into two parts: master and slave. The master is the main part which solves the equation of motion of the whole structure, while the slave represents different substructures and is responsible for the modelling and analysis of them [35]. As shown in Figure 5, the half intermediate column was analyzed in the slave program, the rest of the structure's components and soil were analyzed in the master program. The adapter element, which connected the slave program and master program on the interface node, ensured that the two OpenSees shared same displacement order during virtual hybrid test process. In the slave program, the base of the column is restrained at all three degrees of freedom. The initial stiffness matrix of the column element needs to be specified. Such a matrix can be determined from geometric condition.

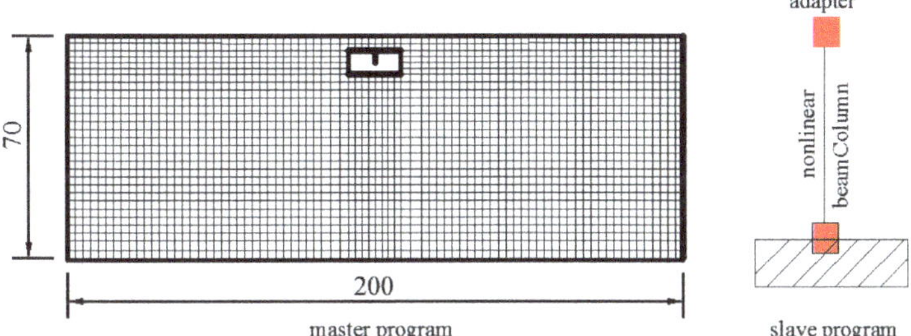

Figure 5. Master program and slave program in soil station model (unit: m).

In the slave program, the substructure conforms to Equation (6):

$$M\ddot{U}(t) + P_r(U(t), \dot{U}(t)) = P(t) - P_0(t) \tag{6}$$

where M represents the global mass matrix; P_r represents the global resisting force vector; P represents seismic load vector; P_0 represents the global element load vector.

$$p_{0,adpt} = -k_{adpt} u_{imp}(t)$$
$$p_{adpt} = p_{r,adpt} + p_{0,adpt} = k_{adpt} u_{adpt}(t) - k_{adpt} u_{imp}(t) \tag{7}$$

In Equation (7), $p_{0,adpt}$ represents externally applied load vector due to the imposed displacements, k_{adpt} represents the stiffness of the adapter element, $u_{imp}(t)$ represents imposed displacements, p_{adpt}

represents nodal force vector of the adapter element, $p_{r,adpt}$ represents load vector due to deformations, $u_{adpt}(t)$ represents the adapter element deformations. According to Equation (7), the element force of the adapter element is composed of two parts, one part is from the nodal force caused by the imposed displacement from the master program, and the other part is from the nodal force caused by the deformation of the adapter element itself. When a calculation step begins, the displacement order is transmitted to the slave program from the master program. When a calculation step finishes in the slave program, unbalanced force feedback is transmitted to the main program. The flow chart of data interaction between master program and slave program is shown in Figure 6. When the first analysis step begins, target displacement is calculated by the master program. It then sends these displacements using a TCP/IP socket to the OpenFresco simulation application server.

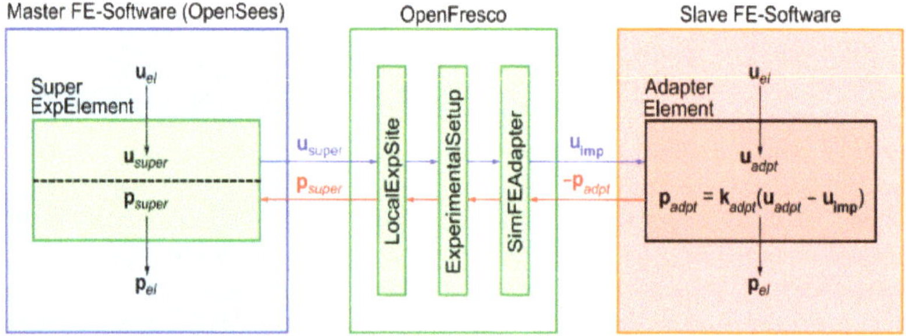

Figure 6. Sequence of operations and data exchange [34].

On OpenFresco platform, the experimental site module and the experimental setup module are responsible for storing, transmitting and converting displacement signals and force signals. The displacement order is transformed on OpenFresco platform by setting related parameters of experimental site module and experimental setup module. The experiment beam column element is presented in this case. Degrees of freedom (DOF) transformation for the experiment beam column element is shown in Figure 7. The adapter element then combines the received displacements $u_{imp}(t)$ with its own element displacements $u_{adpt}(t)$. Once the equilibrium solution process of the slave program has converged, the negative of the element force vector p_{adpt} is returned to the SimFEAdapter experimental control object across the TCP/IP socket. Finally, the element force p_{adpt} is transmitted to the master program, which is then capable to determine the new trial displacements and proceed to the next time step.

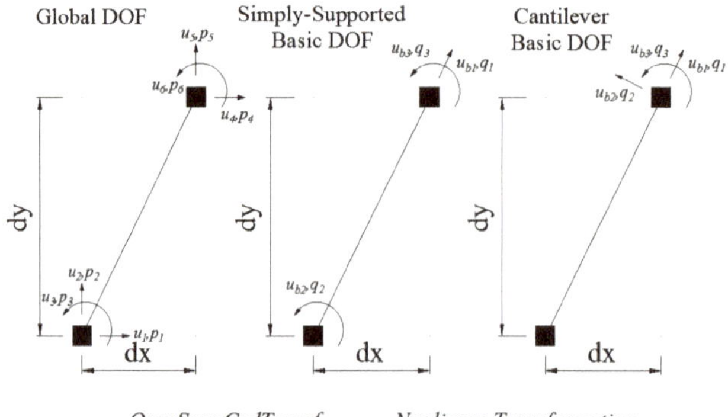

Figure 7. Transformation in beam column experiment element [34].

It should be emphasized that the integration methods of the finite element calculation should be adapted to each other in the master program and the slave program. In this case, the static integrator is applied in the slave program for the reason that the influence of inertial force and damping is not considered in the slave program. The Hilber-Hughes-Taylor integration method is applied in the master program. In order that the command displacement at the interface point is consistent with the top point displacement of column in slave program, it should be ensured that the stiffness of the adapter element is much larger than the lateral stiffness of the center column in slave program. In this case, the stiffness of the adapter element is selected as 2500 kN/mm.

In order to verify the accuracy of the virtual hybrid test, a numerical model with the same material parameters and geometric parameters is established on OpenSees platform. The drift responses of the intermediate column in the soil station system are selected as a calibration index to verify the accuracy of the virtual hybrid test. As shown in Figures 8–10, The drift responses of the virtual hybrid test (labeled as "Coupled") and the complete numerical model (labeled as "Complete") match well during the whole seismic process, which illustrates that the adapter elements work well during virtual hybrid simulation process.

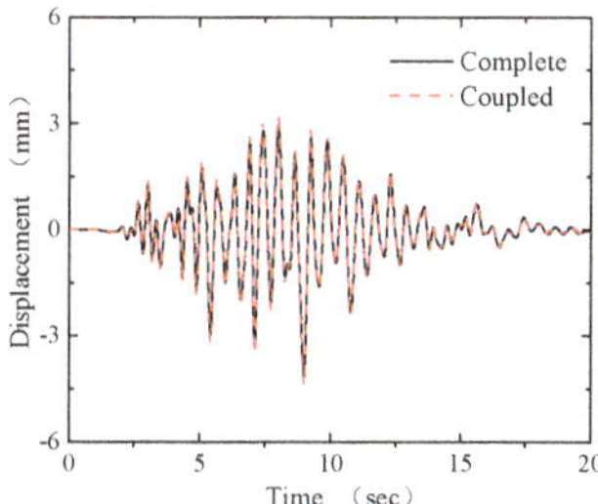

Figure 8. Drift response for Shanghai wave case.

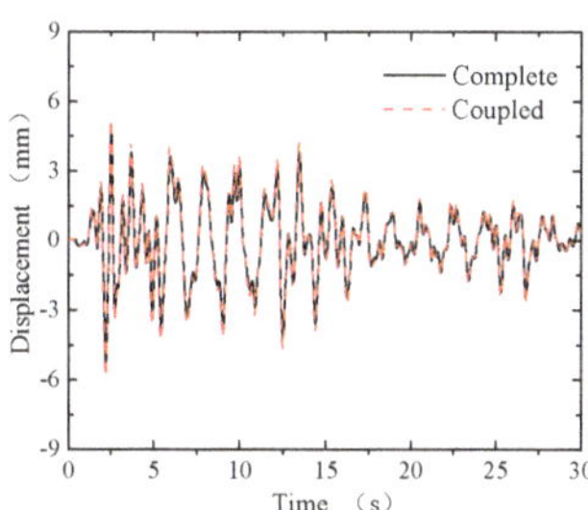

Figure 9. Drift response for El Centro wave case.

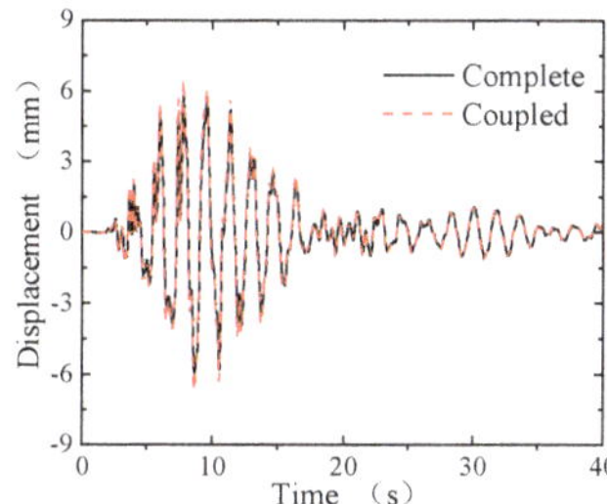

Figure 10. Drift response for Kobe wave case.

As shown in Figure 8, peak drift in complete numerical model is −4.14 mm and peak drift in virtual hybrid simulation is −4.40 mm when the Shanghai wave is applied as input motion. A complete numerical response could be regarded as standard solution, the test error of virtual hybrid simulation is 6.4%. As shown in Figure 9, peak drift in complete numerical model is −5.30 mm and peak drift in virtual hybrid simulation is −5.69 mm when El Centro wave is applied as input motion. The test error is 7.3%. As shown in Figure 10, peak drift in complete numerical model is −5.94 mm and peak drift in virtual hybrid simulation is −6.40 mm when Kobe wave is applied as input motion. The test error is 7.8%.

In short, the virtual hybrid test process is a feasible hybrid simulation method. Two OpenSees programs run simultaneously with the adapter elements. The virtual hybrid test provides a theoretical basis for the physical hybrid test and verifies the feasibility of the FEA model.

4.2. Physical Hybrid Simulation Procedure

An intermediate column in the subway station is adopted as an experimental physical model, while the remainder of the subway station and geotechnical medium is classified as an analytical numerical model. In the experimental physical model, horizontal force is applied on the top of intermediate column by actuator in order to consider the horizontal degree of freedom. The bottom of the intermediate column is fixed to a strong floor to satisfy the boundary conditions in the analytical model. Experimental element, experimental site, experimental setup and experimental control should be defined to connect the experimental physical model and the analytical model through Openfresco. The purpose of OpenFresco middleware is to mediate the transactions between the computational driver and the physical transfer system. With a computational driver, the physical transfer system and middleware hybrid simulation framework, the hybrid simulation system is established.

The MTS-CSI control method is adopted as an experimental control method with the MTS loading system. MTS-CSI defines a control point that manages the displacement command of the test interface node at a specified degree of freedom. Moreover, this point can feed back the displacement and load responses of the interface node from experiment substructure. In such a hybrid test, the control point is selected as the inflection point of the intermediate column element. A horizontal actuator is connected with steel specimen during the loading process. In this case, the specimen was loaded with one actuator in horizontal direction. The signal of experiment substructure is converted by defining experimental setup setting in OpenFresco. The numerical substructure and the experimental substructure are both in the same laboratory. The communication is carried out in the local area network. The type of hybrid test is a local hybrid test. In OpenSees FEA model, nonlinear beam column elements are selected to simulate the seismic behaviors of intermediate column, side wall, station roof and station floor. On the OpenFresco platform, an experiment beam column element with initial stiffness matrix is adopted to represent the steel specimen in hybrid test. Since the numerical model is a 2D analytical model, the experiment element is also a 2D experimental beam column element. The experimental beam column element has two end nodes, each of which has three degrees of freedom in the axial, tangential, and rotational directions. As shown in Equation (8), the initial stiffness matrix of the experiment element can be obtained by theoretical calculation. According to the geometric information and material properties of the hybrid test specimen, the initial parameters of the experiment element are calculated. In this way, the experiment beam column element is defined in OpenFresco.

$$K_{\exp} = \begin{bmatrix} \frac{EA}{l} & 0 & 0 \\ 0 & \frac{12EI}{l^3} & -\frac{6EI}{l^2} \\ 0 & -\frac{6EI}{l^2} & \frac{4EI}{l} \end{bmatrix} \tag{8}$$

5. Experimental Setup and Test Program

The hybrid simulation system is composed of the FEA platform OpenSees, middleware platform OpenFresco and the MTS loading system. The physical hybrid simulation was carried out in the multi-functional shaking tables lab of Tongji University. OpenFresco built a bridge between the numerical substructure and experiment substructure. Data communication and control coordination were performed on the OpenFresco platform. In addition, in order to investigate the experimental error of the hybrid simulation process, the corresponding FEA model of the soil station system was carried out by OpenSees software for each working condition. The solution of the numerical model could be regarded as a target solution for the corresponding hybrid simulation process. In the physical hybrid test, the actuator is fixed at the top of the column specimen to simulate the horizontal displacement under earthquake excitation. The top point of column specimen is the interface between numerical substructure and physical substructure. As shown in Figure 11, the top of the column and the bottom of the column are hinged. A special design is adopted for the bottom part of the column specimen. Four screws are installed to connect steel column specimen and bottom base. The bending stiffness of steel column specimen could be changed by replacing the screws at the column foot. In this way, the bending stiffness could be specified according to the test needs. Since the bending stiffness of the column itself is much larger than that of the screws, the failure mode of such tests would be the bucking failure of screws. Such an experiment could be easily repeated by replacing screws so that theupper column specimen would not be damaged during test process. In the seismic event of soil station system, numerical results demonstrated that the collapse of the structure was caused by the poor ductility of the intermediate columns. The intermediate column is the key component of seismic design in underground structures. Therefore, a cantilevered steel column is designed for hybrid simulation. Furthermore, the cantilever steel column represents the lower half of the intermediate column of the station. The lateral stiffness of steel column specimen can be obtained from Equation (9).

$$
\begin{aligned}
I_1 &= \frac{b_1 h_1{}^3}{12} - \frac{b_2 h_2{}^3}{12} \\
k_m &= \frac{M}{\varphi} = \frac{\pi d^2 a^2 E_2}{4 l_2} \\
\delta &= \frac{l_1{}^3}{3 E_1 I_1} + \frac{l_1{}^2}{k_m} 0.5 e^{0.073 d}
\end{aligned}
\tag{9}
$$

where δ is flexibility coefficient; l_1 is column height; E_1 is column elastic modulus; I_1 is moment of inertia; k_m is rotation stiffness; b_1 is outer surface width; h_1 is outer surface height; b_2 is inner surface width; h_2 is inner surface height; d is screw diameter; a is screw space; E_2 is screw elastic modulus; l_2 is screw length.

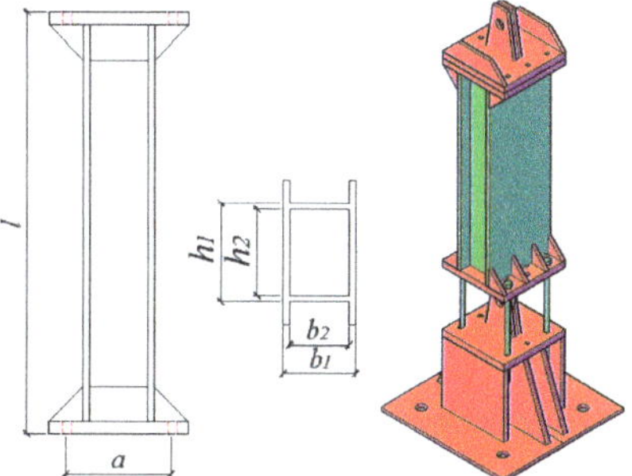

Figure 11. Design drawing of steel column specimen.

The cantilever steel column specimen is the physical substructure of the soil station system hybrid test system, and the remaining parts of the FEA model are treated as numerical substructure. After each integration step, the numerical substructure in the soil station system calculates the displacement response at the control point, horizontal load is applied to physical substructure by loading system according to displacement order. After the displacement command is applied to the physical substructure, the restoring force signal of steel column specimen is measured by the actuator and transmitted back to the data interaction system. On the OpenFresco platform, restoring force response order and displacement response order would be exchanged at each integration step. Such orders would be converted according to geometric transformations by setting experiment site, experiment element, experiment control and experiment actuator. After the numerical substructure obtains a restoring force signal from the physical substructure, such a signal would be transmitted back to dynamic equations to complete the calculation of the next time step in the numerical substructure. The displacement command for next time step would be calculated on basis of restoring force signal accepted. Such a cycle would be repeated until the test is over. Based on the results of the virtual hybrid test procedure and the analytical model, the physical substructure specimens were designed according to the principle of equivalent lateral stiffness. The lateral stiffness of the test substructure was adjusted by changing the screws which were installed at the bottom of steel column. In this case, four vertical screws with a diameter of 15 mm were placed at the bottom of the column, and the lateral stiffness of the physical substructure was 2.10 kN/mm measured by loading system. The lateral stiffness of the intermediate column element is extracted in OpenSees, which is 2.06 kN/mm correspondingly. During the hybrid simulation process, a stiffness matrix of the physical substructure is updated according to the feedback force signal.

6. Overview of Experimental Observation

6.1. Phase 1: Pseudo-Static Tests

In order to investigate the plastic behaviors and energy dissipation capacity of a steel specimen during a real seismic event, a pseudo-static test of the steel specimen was conducted before a hybrid simulation test. The bottom of the column was fixed on the strong floor and the column end was connected to a hydraulic actuator which permitted free horizontal degree of freedom of the column. Each test specimen was subjected to a pseudo-static reversed cyclic simulated earthquake loading as shown in Figure 12. In order to investigate the plastic behaviors and energy dissipation capacity of steel specimen during a real seismic event, the loading sequence adopted in this testing program followed the typical pseudo-static test sequence. In the current study, it is believed that the probable seismic resistance of a sub-component evaluated following the simple testing sequence is able to provide a satisfactory behavior during a real seismic event [34]. The horizontal cyclic loading applied to the specimens throughout the test was displacement controlled. The horizontal loading was applied to the top of the column using a 500kN capacity hydraulic actuator. No axial load was applied to the test specimens.

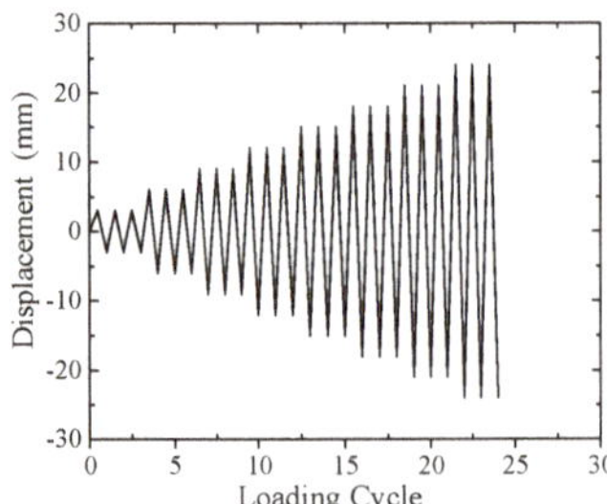

Figure 12. Loading scheme.

As shown in Figure 13, shear force versus horizontal displacement relationship illustrates loading capacity and mechanical behaviors of test specimen. Steel column specimen reached the yield load (26.5 kN) when horizontal displacement reached 13.2 mm. Steel column specimen reached the ultimate load (32.3 kN) when horizontal displacement reached 17.4 mm. The hysteretic response was mainly characterized by the stiffness degradation and the strength degradation. As shown in Figure 14, the most important type of failure mode in the bolt rod area was mainly screw yielding failure.

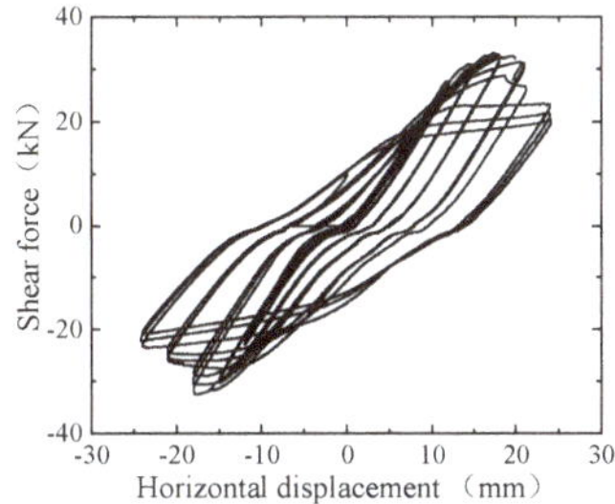

Figure 13. Hysteresis curve of steel specimen.

Figure 14. Bolt failure pattern.

6.2. Phase 2: Hybrid Simulation Tests

With the previously mentioned method, a series of hybrid simulation tests were carried out on the test model under the following base accelerations: Shanghai artificial wave, El Centro wave and Kobe wave. The peak acceleration of three waves is 0.12g. Figure 15 shows lateral displacement and shear force time histories of control point in Shanghai artificial wave case. Figure 16 shows the lateral displacement and shear force time histories of control point in the El Centro wave case. Figure 17 shows lateral displacement and shear force time histories of control point in Kobe wave case. It should be emphasized that the shear force test results in the three working conditions were larger than the analytical results for the reason that the actual stiffness of the intermediate column specimen was greater than the theoretical stiffness in the analytical model. Based on the pseudo-static test results above, it should be noted that the specimen stayed in elastic stage during the three working conditions. Overall, such a specimen applied in hybrid simulation tests was reasonable in this case.

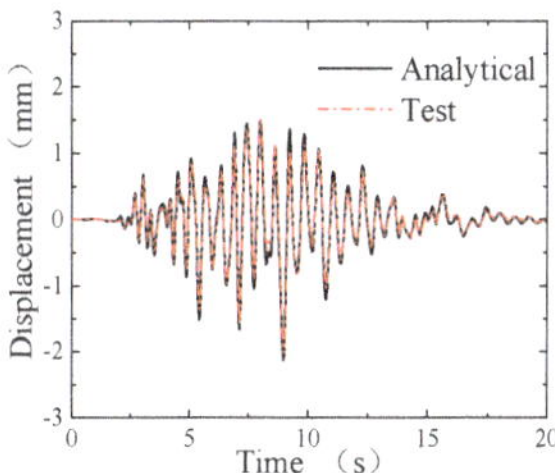
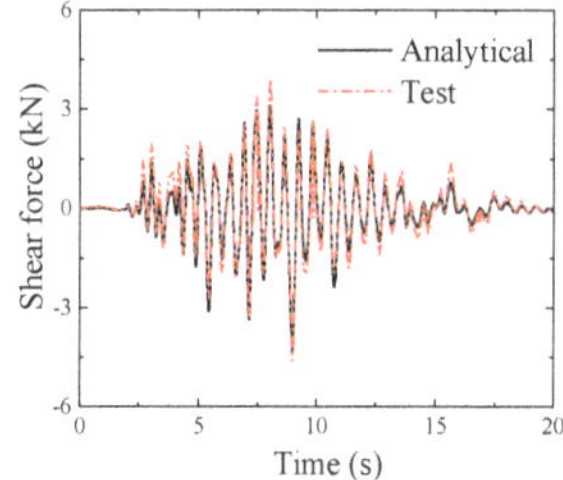

Figure 15. Displacement and force response for Shanhai case.

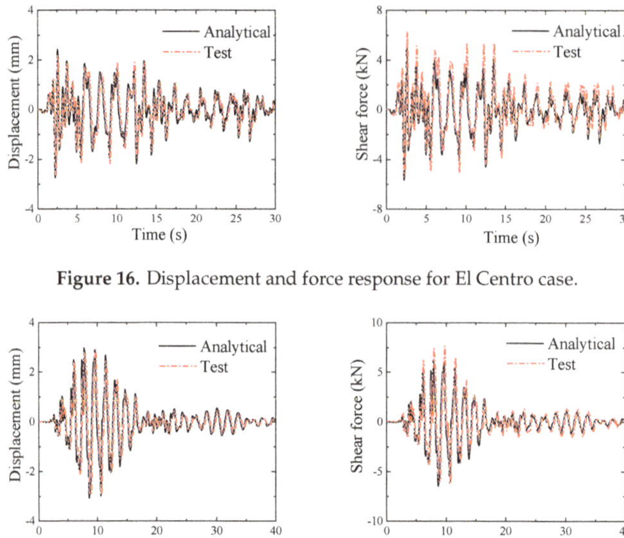

Figure 16. Displacement and force response for El Centro case.

Figure 17. Displacement and force response for Kobe case.

In order to calibrate the accuracy of the hybrid test, shear force and horizontal displacement response of the numerical model were plotted in the same figure. It can be seen that the numerical simulation results matched well with the hybrid simulation test results under earthquake excitation. Thus, the feasibility and accuracy of the physical hybrid simulation test could be verified. The good correspondence between the hybrid simulation and analytically obtained response data validated the analytical model of soil station system. The comparison suggested a good agreement, which indicated the proposed hybrid simulation composed by OpenSees, OpenFresco and MTS performed well in the soil station system hybrid simulation.

7. Conclusions

A study of the two-dimensional seismic response of a soil station system by applying hybrid simulation is reported in this paper. Firstly, the soil station system FEA model is developed for the subsequent hybrid simulation process. The virtual hybrid simulation method is implemented on bases of OpenSees and OpenFresco with an adapter element. The adapter element is then used to connect two FEA programs to implement a virtual hybrid test process. The virtual hybrid simulation provides a theoretical basis for the following physical hybrid test. Moreover, the virtual hybrid simulation also helps to verify the feasibility of a developed numerical substructure in the hybrid test. According to the principle of equivalent lateral stiffness, a novel steel specimen is designed to physically model the intermediate column in this paper. The hybrid test results are compared with the analytical results, to validate the effectiveness and accuracy of the proposed hybrid simulation method. The following conclusions can be drawn:

1. The implementation of a virtual hybrid simulation with two OpenSees processes based on adapter elements is a good way to verify the effectiveness and robustness of the hybrid simulation system. The results of the virtual hybrid simulation are in agreement with the numerical simulation results, which verify the correctness of the established hybrid test system.

2. The variable stiffness steel column specimens applied in the hybrid test were designed according to the corresponding stiffness in the numerical substructure. The lateral stiffness of the steel

specimen is changed by replacing the screws. The stiffness could be obtained from design parameters before testing.

3. The mechanical behavior of the intermediate column under earthquake excitation could be reproduced by the proposed hybrid simulation system composed of OpenSees, OpenFresco, MTS-CSI and loading system.

Author Contributions: Conceptualization, C.-Y.Y. and Y.Y.; Data curation, Y.-C.M.; Investigation, X.-S.C.; Methodology, C.-Y.Y. and X.-S.C.; Software, X.-S.C. and Y.-C.M.; Supervision, Y.Y.; Writing—original draft, X.-S.C.; Writing—review & editing, C.-Y.Y. and Y.Y.

Funding: This research was funded by Shanghai Committee of Science and Technology (16DZ1200302).

Conflicts of Interest: The authors declare no conflict of interest.

References

1. Terzic, V.; Stojadinovic, B. Hybrid Simulation of Bridge Response to Three-Dimensional Earthquake Excitation Followed by Truck Load. *Struct. Eng.* **2014**, *140*, 1–11. [CrossRef]
2. Iida, H.; Hiroto, T.; Yoshida, N. Damage to Daikai subway station. *Spec. Issue Soils Found.* **1996**, *1*, 283–300. [CrossRef]
3. Lu, C.; Hwang, J. Nonlinear collapse simulation of Daikai Subway in the 1995 Kobe earthquake. *Tunn. Undergr. Space Technol.* **2019**, *87*, 78–90. [CrossRef]
4. Cilingir, U.; Madabhushi, S.P.G. A model study on the effects of input motion on the seismic behavior of tunnels. *Soil Dyn. Earthq. Eng.* **2011**, *31*, 452–462. [CrossRef]
5. Zhuang, H.; Hu, Z.; Wang, X. Seismic responses of a large underground structure in liquefied soils by FEM numerical modelling. *Bull. Earthq. Eng.* **2015**, *13*, 3645–3668. [CrossRef]
6. Tsinidis, G. Response characteristics of rectangular tunnels in soft soil subjected to transversal ground shaking. *Tunn. Undergr. Space Technol.* **2017**, *62*, 1–22. [CrossRef]
7. Ma, C.; Lu, D.; Du, X. Seismic performance upgrading for underground structures by introducing sliding isolation bearings. *Tunn. Undergr. Space Technol.* **2018**, *74*, 1–9. [CrossRef]
8. Huo, H.; Bobet, A.; Fernández, G. Load Transfer Mechanisms between Underground Structure and Surrounding Ground: Evaluation of the Failure of the Daikai Station. *J. Geotech. Geoenviron. Eng.* **2005**, *131*, 1522–1533. [CrossRef]
9. Williams, M.S.; Blakeborough, A. Laboratory Testing of Structures under Dynamic Loads: An Introductory Review. *Philos. Trans. Math. Phys. Eng. Sci.* **2001**, *359*, 1651–1669. [CrossRef]
10. Hakuno, M.; Shidawara, M.; Hara, T. Dynamic Destructive Test of a Cantilever Beam Controlled by an Analog-computer. *Proc. Jpn. Soc. Civ. Eng.* **1969**, *1969*, 1–9. [CrossRef]
11. Chen, Z.; Wang, H.; Wang, H. Application of the Hybrid Simulation Method for the Full-Scale Precast Reinforced Concrete Shear Wall Structure. *Appl. Sci.* **2018**, *8*, 252. [CrossRef]
12. Phillips, B.M.; Spencer, B.F. Model-Based Feedforward-Feedback Actuator Control for Real-Time Hybrid Simulation. *J. Struct. Eng.* **2013**, *139*, 1205–1214. [CrossRef]
13. Chen, C.; Ricles, J.M.; Marullo, T.M. Real-time hybrid testing using the unconditionally stable explicit cr integration algorithm. *Earthq. Eng. Struct. Dyn.* **2010**, *38*, 23–44. [CrossRef]
14. Kolay, C.; Ricles, J.M.; Marullo, T.M. Implementation and application of the unconditionally stable explicit parametrically dissipative KR-method for real-time hybrid simulation. *Earthq. Eng. Struct. Dyn.* **2015**, *44*, 735–755. [CrossRef]
15. Kolay, C.; Ricles, J.M. Force-Based Frame Element Implementation for Real-Time Hybrid Simulation Using Explicit Direct Integration Algorithms. *Struct. Eng.* **2018**, *144*. [CrossRef]
16. Pan, P.; Nakashima, M.; Tomofuji, H. Online test using displacement-force mixed control. *Earthq. Eng. Struct. Dyn.* **2005**, *34*, 869–888. [CrossRef]
17. Günay, S.; Mosalam, K.M. Enhancement of real-time hybrid simulation on a shaking table configuration with implementation of an advanced control method. *Earthq. Eng. Struct. Dyn.* **2015**, *44*, 657–675. [CrossRef]
18. Yang, T.Y.; Tung, D.P.; Li, Y. Theory and implementation of switch-based hybrid simulation technology for earthquake engineering applications. *Earthq. Eng. Struct. Dyn.* **2017**, *46*, 2603–2617. [CrossRef]

19. Wu, B.; Ning, X.; Xu, G. Online numerical simulation: A hybrid simulation method for in complete boundary conditions. *Earthq. Eng. Struct. Dyn.* **2018**, *47*, 889–905. [CrossRef]

20. Spencer, B.F.; Carrion, J.E.; Phillips, B.M. Real-time hybrid testing of semi-actively controlled structure with MR damper. In Proceedings of the Second International Conference on Advances in Experimental Structural Engineering, St. Louis, MO, USA, 10–12 June 2009.

21. Kim, S.; Christenson, R.; Phillips, B.; Spencer, B.F. Geographically distributed real-time hybrid simulation of MR dampers for seismic hazard mitigation. In Proceedings of the 20th Analysis and Computation Specialty Conference, Chicago, IL, USA, 29–31 March 2012; ASCE: Reston, VA, USA, 2012.

22. Li, X.; Ozdagli, A.I.; Dyke, S. Development and Verification of Distributed Real-Time Hybrid Simulation Methods. *Comput. Civ. Eng.* **2017**, *31*. [CrossRef]

23. Shao, X.; Griffith, C. An overview of hybrid simulation implementations in NEES projects. *Eng. Struct.* **2013**, *56*, 1439–1451. [CrossRef]

24. Stefanaki, A.; Sivaselvan, M.V.; Tessari, A. Soil-foundation-structure interaction using hybrid simulation. In Proceedings of the International Conference on Structural Mechanics in Reactor Technology, Manchester, UK, 10–14 August 2015.

25. Mahin, S.A.; Shing, P.B. Pseudodynamic Method for Seismic Testing. *J. Struct. Eng.* **1985**, *111*, 1482–1503. [CrossRef]

26. Shanghai Tunnel Engineering & Rail Transit Design Research Institute. *Preliminary Design of Shanghai Riverine Tunnel*; Book 1, General Specification; Shanghai Tunnel Engineering & Rail Transit Design Research Institute: Shanghai, China, 2013. (In Chinese)

27. Tao, D.; Lin, J.; Lu, Z. Time-Frequency Energy Distribution of Ground Motion and Its Effect on the Dynamic Response of Nonlinear Structures. *Sustainability* **2019**, *11*, 702. [CrossRef]

28. Zienkiewicz, O.C.; Bicanic, N.; Shen, F.Q. Earthquake Input Definition and the Trasmitting Boundary Conditions. In *Computational Nonlinear Mechanics*; Springer: Vienna, Austria, 1989; pp. 109–138.

29. Bao, X.; Xia, Z.; Ye, G. Numerical analysis on the seismic behavior of a large metro subway tunnel in liquefiable ground. *Tunn. Undergr. Space Technol.* **2017**, *66*, 91–106. [CrossRef]

30. McKenna, F.; Fenves, G.L. *The OpenSees Command Language Manual*; Version 1.2; University of California: Berkeley, CA, USA, 2001.

31. Zhang, Y.; Conte, J.P.; Yang, Z.; Elgamal, A.; Bielak, J.; Acero, G. Two-Dimensional Nonlinear Earthquake Response Analysis of a Bridge-Foundation-Ground System. *Earthq. Spectra* **2008**, *24*, 343–386. [CrossRef]

32. Amorosi, A.; Boldini, D.; Elia, G. Parametric study on seismic ground response by finite element modelling. *Comput. Geotech.* **2010**, *37*, 515–528. [CrossRef]

33. Schellenberg, A.; Huang, Y.; Mahin, S.A. Structural FE-software coupling through the experimental software framework. In Proceedings of the 14th World Conference on Earthquake Engineering, Beijing, China, 12–17 October 2008.

34. Schellenberg, A.; Kim, H.K.; Takahashi, Y. *OpenFresco Command Language Manual*; University of California: Berkeley, CA, USA, 2009.

35. Ghaffary, A.; Karami Mohammadi, R. Framework for virtual hybrid simulation of TADAS frames using opensees and abaqus. *J. Vib. Control* **2016**, *12*. [CrossRef]

Article

Study on Column-Top Seismic Isolation of Single-Layer Latticed Domes

Yongmei Zhai [1],*, Xuxia Fu [2] , Yihui Chen [3] and Wei Hu [4]

[1]   Shanghai Institute of Disaster Prevention and Relief, Tongji University, Shanghai 200082, China
[2]   Department of Civil Engineering, Tongji University, Shanghai 200082, China; 1732347@tongji.edu.cn
[3]   ARTS GROUP Co., Ltd., Suzhou 215000, China; chenyihui@artsgroup.cn
[4]   College of Architecture and Urban Planning, Tongji University, Shanghai 200082, China;
    1990huwei@sina.com
*   Correspondence: zymww@tongji.edu.cn

Received: 27 December 2018; Accepted: 7 February 2019; Published: 12 February 2019

Abstract: In this paper, a single-layer lamella reticulated dome with reinforced concrete bearings was studied, and a method of column-top isolation was proposed to improve the seismic performance of the whole structure, thereby avoiding too large support stiffness in engineering practice. A nonlinear time-history analysis showed that lead rubber bearings (LRB) can reduce the support reaction to a certain extent and make it distribute uniformly, reducing the support design requirements under frequent earthquakes. During rare earthquakes, the LRB was basically in the plastic state and the support reaction remained near the yield force, which was reduced greatly compared with that of the original structure. The bearing hysteresis curve was full, while the plasticity development degree of the upper reticulated dome was greatly reduced and the elasticity was basically maintained, thus achieving a good damping effect.

Keywords: column-top isolation; single-layer reticulated dome; nonlinear time-history analysis; damping effect

---

1. Introduction

A single-layer reticulated dome is an important form of space structure in China that combines the characteristics of skeletal structures and thin-shell structures with beautiful shapes, reasonable forces and simple structures that are widely used in various large and medium span buildings [1]. However, China is among the world's most earthquake-prone countries, and some spatial structures in the Wenchuan earthquake demonstrated various degrees of earthquake disaster phenomena [2,3]. Common failure forms include lower support failure, bearing connection damage, excessive plastic deformation of the upper steel roof and overall collapse of the structure [4]. Therefore, it is very important to study the seismic behavior of reticulated domes and seismic isolation techniques. Isolation technology can effectively block seismic action and reduce the seismic response of a structure. It reduces the earthquake effect by prolonging the structure period and dissipating energy through additional damping, so that the structure displacement is controlled within the allowable range [5–7]. Early isolation technology has been primarily applied to support multi-story building foundations and bridge structures. With the development of structural vibration control technology and the increasing demand for seismic performance in large-scale engineering projects, this method has also gradually been applied to large span spatial structures. The roof of the Shanghai International Speedway News Center adopts a high isolation bearing—a new composite isolation bearing consisting of one basin bearing and four common rubber bearings—which effectively releases temperature stress and significantly reduces the seismic response of the structure [7]. The pyramid-shaped roof of the Ataturk Airport terminal in Turkey uses a column-top friction-pendulum isolation scheme [8].

To simplify the calculation, the roof structure is analyzed separately by the simply supported boundary, disregarding the dynamic interaction between the roof and the lower supporting structure in the engineering design. In fact, the existence and stiffness of the lower supporting structure will have an important influence on the seismic performance of the integral structure, especially when the stiffness of the supporting structure is much stronger than that of the upper roof. In this case, the seismic amplification effect of the structure is very significant, and the failure limit load of the superstructure can be greatly reduced. To meet the requirements of the construction in engineering practice when the supporting structure introduces more space frames, slabs, or shear walls, its support stiffness is generally large. To improve the seismic performance of this kind of "strong support" structure, this paper takes a single-layer spherical reticulated dome with a concrete bearing under the surface as its research subject. The common spherical steel bearing at the top of the column is replaced by an isolation lead rubber bearing in this study, and the seismic behavior of the structure before and after the isolation is compared and analyzed.

2. Integral Model of Dome and Support Structure

The roof structure analyzed in this paper is a single-layer spherical reticulated dome (Figure 1) with a span of 40 m, a vector span ratio of 1/5, a round steel pipe as the reticulated dome rod, a material of q235b steel, a circular members section of φ132 × 4, and radial and oblique rod sections of φ116 × 4. The lower support structure adopts the cylindrical-ring beam system. The column height is 10 m. To simulate the "strong support rigidity" condition, the column section diameter is set to 1.5 m, and the ring beam covers a 0.6 m × 0.6 m rectangular section. The whole structure contains 24 pillars, and the concrete strength grade of the beam and column is C30. Through refinement of finite element analysis in Abaqus, it is shown that when the diameter of the column cross section is 1.5 m, the damage factor of the lower supporting structure concrete is very small. The reticulated shell, the beam and column adopt B31 beam units based on the fiber bundle model. Steel adopts an ideal elastoplastic model. In order to achieve a certain calculation accuracy, all beam elements are controlled to a length of about 1 m during meshing. Material nonlinearity and geometric nonlinearity are considered simultaneously. A welded ball joint is used to connect the members of the upper reticulated dome rigidly, and a common spherical steel bearing or an isolation bearing is used to connect the ring beam between the reticulated shell and the lower support structure at the top of each column. The integral structural model is shown in Figure 1b.

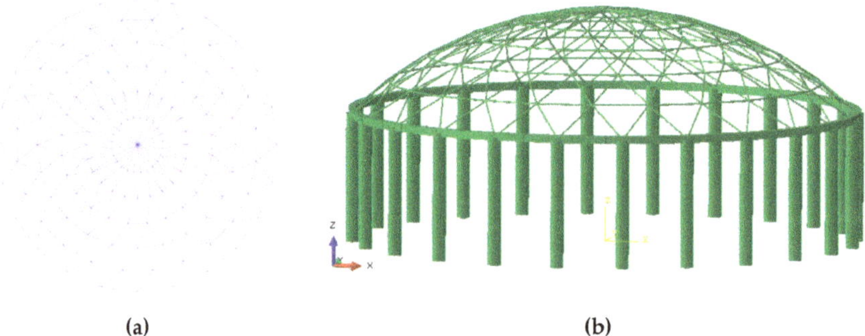

(a)	**(b)**

Figure 1. Layout of dome (a) single lamella spherical dome model; (b) model of integral structure.

3. Selection of a Seismic Isolation Device

In this paper, a lead rubber bearing is chosen as the isolation device. Lead has good mechanical properties, as its yield shear stress is relatively low at only approximately 10 MPa, the initial shear stiffness is higher, the shear modulus is approximately 130 MPa, and it is also the ideal elastoplastic body. Lead has good fatigue resistance for the plastic cycling load, and high-purity Pb (99.99%) is easy to obtain, which makes its mechanical properties more reliable. This article selects the Fuyo lead rubber bearing series product LRB600, produced by the Wuxi construction new material Limited Company [9]. Its mechanical performance parameters are shown in Table 1.

Table 1. Mechanical performance parameters of the bearing.

Type	Base Level Pressure (MPa)	Long-Term Load (kN)	Vertical Stiffness (kN/mm)	Horizontal Performance of 100% Deformation				
				Initial Stiffness (kN/mm)	Post-Yield Stiffness (kN/mm)	Yield Force (kN)	Equivalent Stiffness (kN/mm)	Equivalent Damping (%)
LRB600	6	1178	1513	6.27	0.483	51	0.878	23.9

Therefore, this paper discusses two structural models: The original structure, whose column top adopts the common steel bearing, namely, the hinged connection, and the isolation structure whose column top is connected by the LRB.

The analytical model of the LRB adopts the rubber isolator unit in SAP2000 [10], as shown in Figure 2, which consists of six internal "springs" representing the components of axial, shearing, bending, and torsion. For each degree of freedom of the deformation, linear or nonlinear behavior can be specified independently, and the plasticity properties are based on the hysteresis behavior proposed by Wen (1976) [11] and Park, Wen, and Ang (1986) [12].

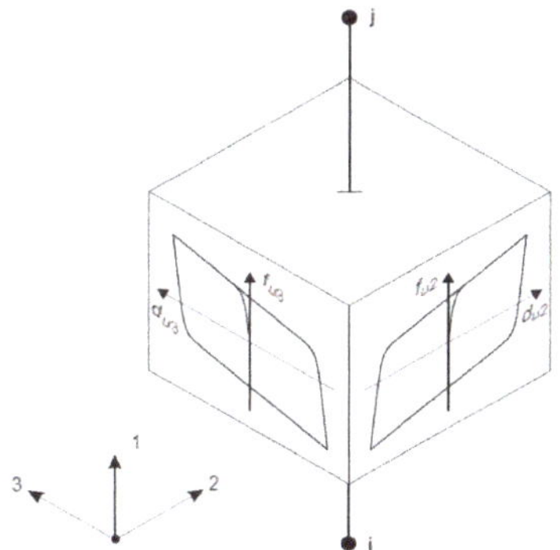

Figure 2. Mechanical model of a lead rubber bearing.

4. Results

First, the natural frequency of the structure before and after isolation was compared. Since the LRB is a nonlinear element, its stiffness is a function of the bearing deformation, and modal analysis is a linear perturbation analysis. To objectively evaluate the impact of the isolation device on the structural dynamic characteristics, the parameters of the LRB, including the equivalent stiffness and the equivalent damping ratio in the modal analysis, are shown in Table 1. As shown in Table 2, the natural frequency of the isolated structure was significantly lower than that of the original structure.

Table 2. The comparison of natural frequency for original and seismically-isolated structures.

Mode Number	Natural Frequency (Hz)		Percentage Scatter
	Original Structure	Seismic Isolation Structure	
1	3.80	1.6207	57%
2	3.80	1.6208	57%
3	4.22	2.3588	44%
4	4.22	3.2405	23%
5	4.50	3.8370	15%
6	4.5091	3.8371	15%
7	4.6865	4.0540	13%
8	4.6867	4.0544	13%
9	4.6920	4.1578	11%
10	4.6921	4.1580	11%
11	4.6976	4.3140	8%
12	4.7029	4.3140	8%
13	4.7032	4.4963	4%
14	4.8545	4.4974	7%
15	4.8545	4.6357	5%

5. Analysis of Frequent Earthquake Response

The acceleration peaks of the El Centro wave and Taft wave adjusted to 70 gal under frequent earthquakes were input into the structure in three directions, and the nonlinear time-history response was analyzed. The distribution curve of the reaction force envelope value of all 24 bearings at the top of the original structure and the isolated structure column are given in Figure 3. It was apparent that the distribution of the maximum bearing reaction force was very uneven for the original structure because of the uncertain direction of the action of the earthquake. The actual design was based on taking the most unfavorable bearing force, so the design requirements of the bearing were often relatively high and could easily act as a weak link during an earthquake. The bearing force distribution was relatively flat for the isolated structure, and the value of the force was lower than that of the original structure, which was beneficial to the design of the bearing and reduced the seismic effect of the transmission to the upper reticulated dome. A typical bearing in the sixth support position (Figure 3) is taken as an example.

Figure 3. Support position.

As shown in Figures 4 and 5, under the El Centro and Taft waves, the maximum bearing reaction force of the original structure in the X direction was 32.11 kN and 29.01 kN, respectively, and the maximum bearing force of the isolation structure in the X direction was reduced to 18.45 kN and 18.59 kN, a decrease of 42.5% and 35.9%, respectively. The comparison for the time-history curve of the bearing No.6 between the original structure and the isolation structure is given in Figure 6; the structural reaction apparently decreased with the isolation bearing.

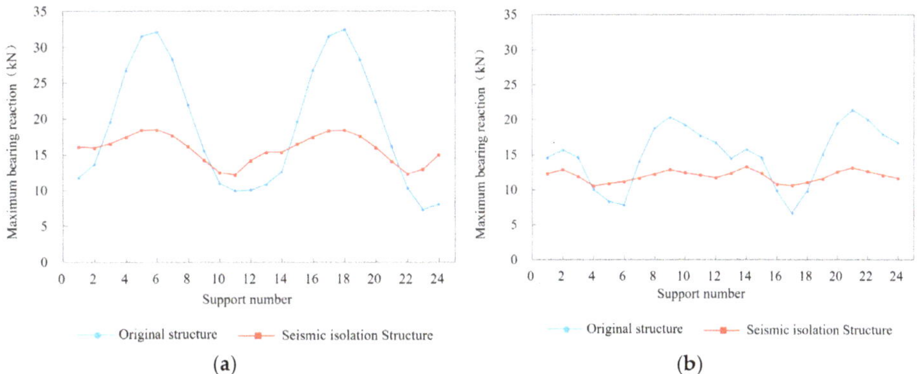

Figure 4. Comparison of maximum inverse force distribution direction between original structure and seismic isolation structure under the El Centro wave in frequent earthquakes, (**a**) in the X direction, (**b**) in the Y direction.

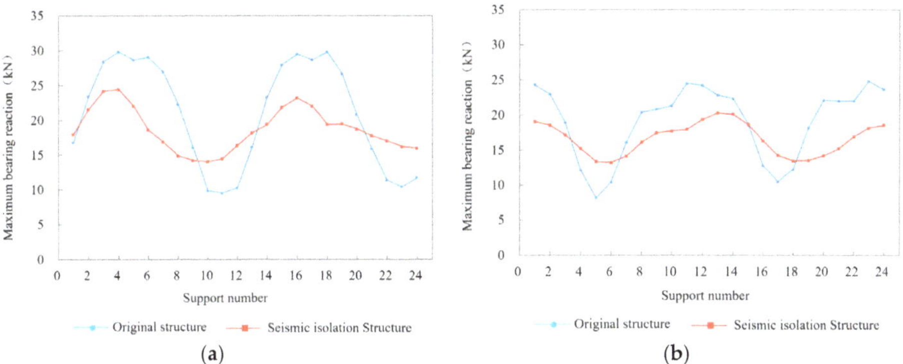

Figure 5. Comparison of maximum inverse force distribution between original structure and seismic isolation structure under the Taft wave in frequent earthquakes, (**a**) in the X direction, (**b**) in the Y direction.

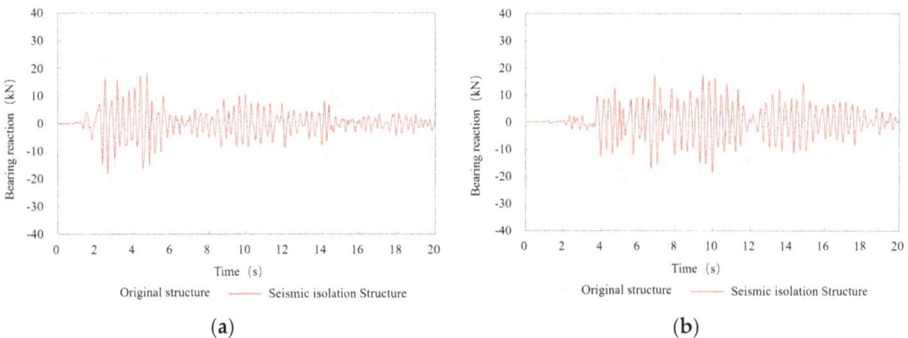

Figure 6. Comparison of time-history curve of the reaction force of bearing No. 6 in the X direction between original structure and seismic isolation structure in frequent earthquakes, (**a**) under the El Centro wave, (**b**) under the Taft wave.

Under frequent earthquakes, all LRB supports did not reach their yield force, so the bearings were still in the initial elastic state. The typical hysteresis curve of the No. 6 bearing is given in Figure 7, which shows that the LRB did not dissipate any energy under frequent earthquakes and that the isolation effect was realized by prolonging the structure period.

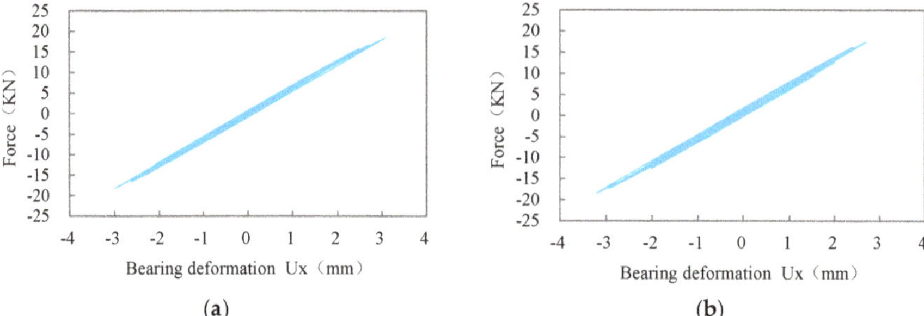

Figure 7. Hysteretic curve in the X direction in frequent earthquakes, (**a**) under the El Centro wave, (**b**) under the Taft wave.

The maximum displacement envelope value of all supports under the El Centro wave and Taft wave was 3.07 mm and 3.26 mm, respectively, which is within the elastic deformation range and can guarantee the normal use of the structure under earthquake or wind load.

6. Analysis of Rare Earthquake Response

The acceleration peak of the seismic wave was adjusted to the value of rare earthquakes stipulated in the specification, 400 gal, and then input into the structure basement in three directions in order to analyze the nonlinear time-history reaction. Compared with frequent earthquakes, the response of structures under rare earthquakes was very similar, but because of the plastic energy dissipation caused by most of the bearing in the yield state, the seismic isolation structure under rare earthquakes showed a better damping effect. Figures 8 and 9 give the distribution curve of the reaction force envelope value of all 24 bearings at the top of the original structure and the isolated structure column. Note that the bearing reaction of the original structure under rare earthquakes is quite large, and the distribution is very uneven. Taking bearing No. 6 as an example, the resultant force of the two horizontal forces under the El Centro wave acted up to 188 kN, making the integral structure prone to fail early under the strong earthquake due to the support force. The bearing force of the isolated structure was basically located near the yield force, which greatly reduced the seismic action transmitted to the upper reticulated dome. The comparison of the time-history curve of bearing No. 6 in the X direction between the original structure and the isolation structure is given in Figure 10, showing that the structural reaction apparently decreased with the isolation bearing.

The typical hysteresis curve of the No. 6 bearing in the X direction is given in Figure 11, which shows that the hysteretic curve of the LRB under rare earthquakes was full, and the energy dissipation effect was significant. The maximum displacement of the bearing in the X direction under the El Centro wave and the Taft wave was 20.5 mm and 18.0 mm, respectively, which is less than the maximum allowable deformation of the bearing (according to the product specification gauge deformation of 400%).

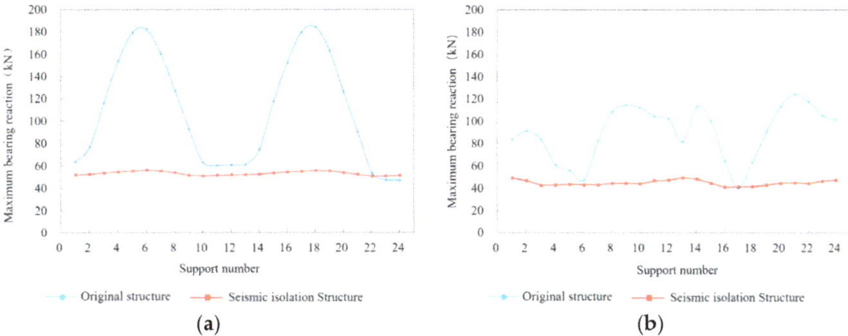

Figure 8. Comparison of maximum inverse force distribution between original structure and seismic isolation structure under the El Centro wave in rare earthquakes, (**a**) in the X direction, (**b**) in the Y direction.

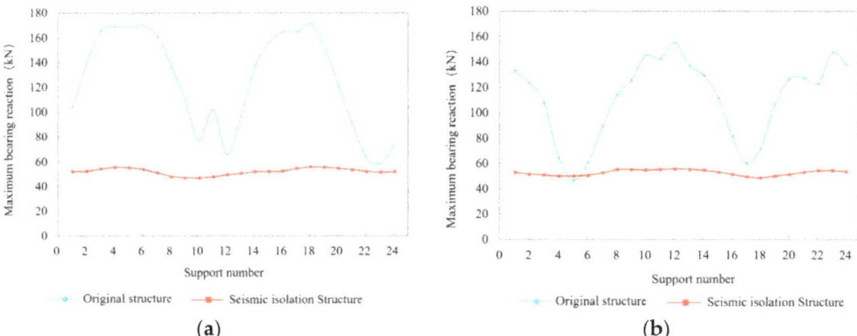

Figure 9. Comparison of maximum inverse force distribution between original structure and seismic isolation structure under the Taft wave in rare earthquakes, (**a**) in the X direction, (**b**) in the Y direction.

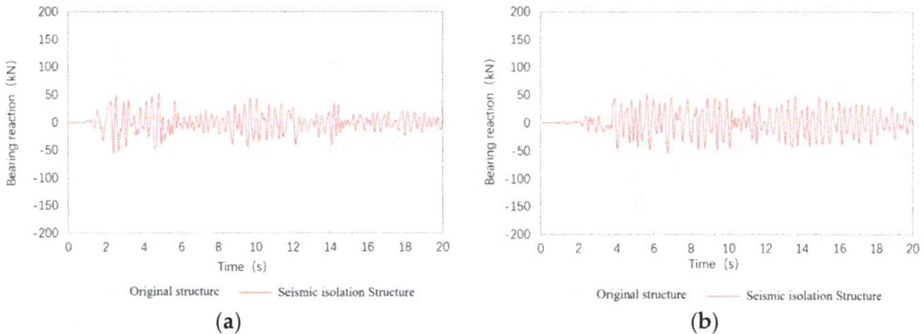

Figure 10. Comparison of time-history curve of the reaction force of bearing No. 6 between original structure and seismic isolation structure in rare earthquakes, (**a**) under the El Centro wave, (**b**) under the Taft wave.

A comparison of the plastic development of the upper reticulated dome before and after the isolation is given in Figures 12 and 13, which indicates more plastic development of the original structure under rare earthquakes, while the isolation structure was basically in an elastic state. The plastic deformation only occurs at the outer part of the lateral beam under the Taft wave.

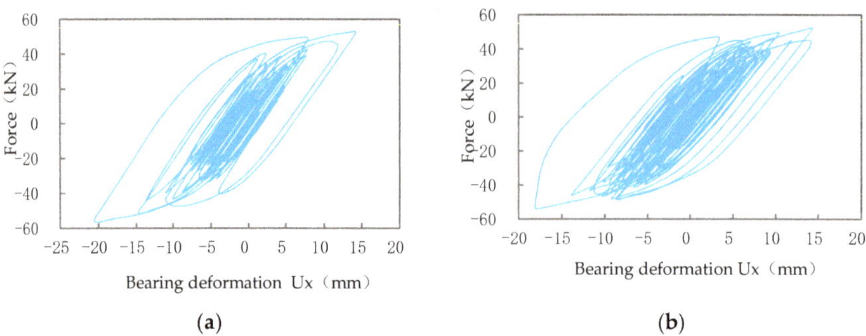

Figure 11. Hysteretic curve in the X direction under rare earthquakes, (**a**) under the El Centro wave, (**b**) under the Taft wave.

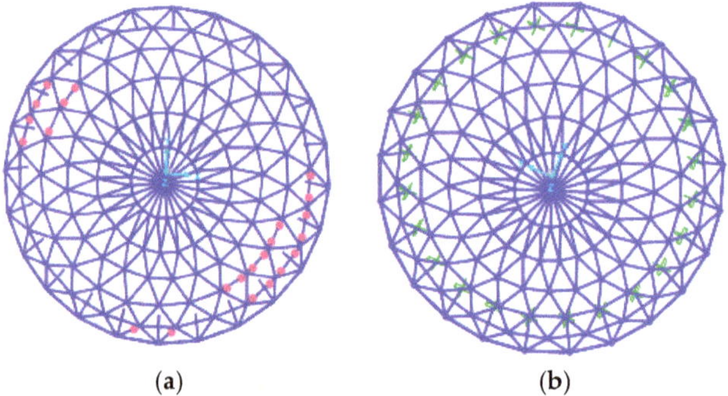

Figure 12. Comparison of plastic distribution of the original upper structure under the El Centro wave, (**a**) original structure, (**b**) seismic isolation structure.

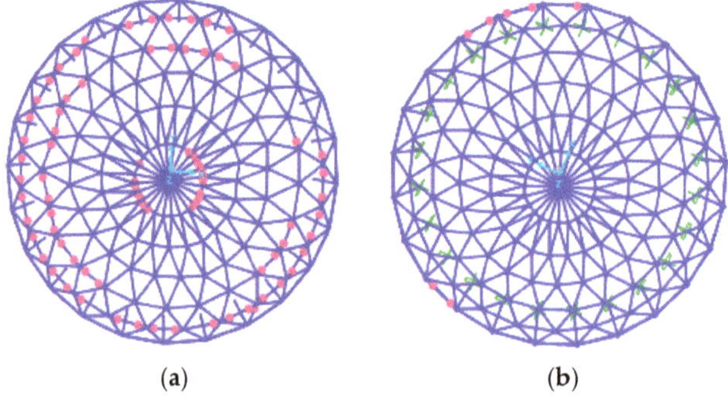

Figure 13. Comparison of plastic distribution of the seismic upper structure under the Taft wave, (**a**) original structure, (**b**) seismic isolation structure.

7. Conclusions

The modal analysis of the structure before and after the isolation showed that the LRB can prolong the natural period of the structure effectively [13–15], thus avoiding the excellent period of earthquake motion and reducing the seismic action of the superstructure. Time-history analysis showed that the LRB can reduce the bearing reaction force to a certain extent under frequent earthquakes, the maximum reduction of which under the El Centro wave and Taft wave was up to 42.5% and 35.9%, respectively. The distribution of the bearing force tended to be uniform, which reduced the requirement of the bearing design. Under rare earthquakes, the LRB bearing was basically in a plastic state, and the bearing reaction force was maintained near the yield force, which decreased greatly compared with that of the original structure. The bearing hysteresis curve was full, the energy dissipation effect was significant, and the plasticity development degree of the upper reticulated shell was greatly reduced so that it was basically elastic.

Overall, for the strong support structure, the column-top seismic isolation method used in this paper is a new idea that effectively improves the seismic performance of the whole structure, from "resisting" to "eliminating".

Author Contributions: Conceptualization, Y.Z.; methodology, Y.Z.; software, Y.C.; validation, X.F.; formal analysis, Y.C.; investigation, W.H.; resources, Y.C.; data curation, Y.C.; writing—original draft preparation, Y.C.; writing—review and editing, X.F.; visualization, X.F.; supervision, X.F.; project administration, Y.Z.; funding acquisition, Y.Z.

Funding: This research was funded by NATIONAL KEY R&D PROJECTS IN 13TH FIVE-YEAR OF CHINA, grant number 2016YFC0800209, 2017YFC0803300.

Conflicts of Interest: The authors declare no conflict of interest.

References

1. Ming, G.; Tao, W. Application of reticulated shell structure. *Archicreation* **2000**, *2*, 48–51.
2. Liu, W.; Deng, K. Preliminary investigation on earthquake damage of "5.12" Wenchuan in spatial structure. *Sichuan Archit.* **2009**, *29*, 112–113.
3. Deng, K. Investigation on earthquake damage of Wenchuan large earthquake. *China Steel Struct. Assoc. Spat. Struct. Bull.* **2008**, *3*, 64–68.
4. Qin, S.; Lu, J.; Dai, C. Design and stability analysis of large-span single-layer spherical reticulated shells. *Spec. Struct.* **2012**, *28*, 9–12.
5. Feng, Y. Discussion on Seismic Mitigation and Isolation Schemes for High-speed Railway Large-span Continuous Girder Bridges. *Earthq. Res.* **2015**, *38*, 167–172.
6. Liu, S.; Wei, X.; Zhang, C.; Ma, F. Analysis on Anti-mining Deformation and Anti-seismic Protection of Bridge Structure based on Isolation Technology. *Earthq. Res.* **2014**, *37*, 86–97.
7. Zhuang, P.; Xue, S.; Li, B. Seismic Isolation Analysis of SMA-rubber Bearings in a Single-layer Spherical Lattice Shell Structure. *Build. Struct.* **2006**, *11*, 103–106.
8. Shi, W.; Sun, H.; Li, Z.; Ding, M. High-Position Seismic Isolation of Press Center at Shanghai International Circuit. *J. Tongji Univ. Nat. Sci. Ed.* **2005**, *33*, 1576–1580.
9. Constantinou, M.; Whittaker, A.S.; Velivasakis, E. Seismic evaluation and retrofit of the Ataturk international airport terminal building. In Proceedings of the 2001 Structural Congress and Exposition, Washington, DC, USA, 21–23 May 2001.
10. Wuxi Construction Materials Co., Ltd. Fuyo Rubber Isolation Bearing Standard Product Design Data. Available online: http://fuyotech.com/Product.aspx?ptid=2 (accessed on 1 February 2019).
11. Wen, Y.K. Method for random vibration of hysteretic system. *J. Eng. Mech. Div.-ASCE* **1976**, *102*, 249–263.
12. Park, Y.J.; Wen, Y.K.; Ang, A.H.-S. Random vibration of hysteretic systems under bi-directional ground motions. *Earthq. Eng. Struct. Dyn.* **1986**, *14*, 543–557. [CrossRef]
13. Beijing Golden Civil Engineering software Technology Co., Ltd. *SAP2000 Chinese Version (Second Edition)*; People's Transportation Press: Beijing, China, 2012; Available online: https://max.book118.com/html/2018/0428/163589441.shtm (accessed on 1 February 2019).
14. Chiara, B.; Antonino, M. Dynamic testing and parameter identification of a base-isolated bridge. *Eng. Struct.* **2014**, *60*, 85–99.
15. Corrado, C.; Chiara, B.; Claudio, A. Dynamic and static identification of base-isolated bridges using Genetic Algorithms. *Eng. Struct.* **2015**, *102*, 80–92.

 *sustainability*

Article

Long-Term Ground Settlements over Mined-Out Region Induced by Railway Construction and Operation

Shuo Jiang * and Yimin Wang

School of Civil Engineering and Transportation, South China University of Technology, Guangzhou 510641, China; ctymwang@scut.edu.cn
* Correspondence: 201410101336@mail.scut.edu.cn

Received: 20 December 2018; Accepted: 3 February 2019; Published: 8 February 2019

Abstract: With the rapid development of railway construction and the massive exploitation of mineral resources, many railway projects have had to cross mining areas and their caverns. However, the settlement of the ground surface may cause severe damage to human-built structures and lead to the loss of human lives. The research on ground deformation monitoring over caverns is undoubtedly important and has a guiding role in railway design. Settlement observation points were set up around the mine, establishing a ground subsidence monitoring level network that has been in operation for 11 years. The ground settlement and lateral displacement along the designed railway were studied. A finite element model was established to predict the long-term ground settlements over the mined-out region induced by designed railway embankment construction and train operation. The results show that the predicted ground settlement induced by railway embankment construction is smaller than the ground settlement induced by the mined-out cavity. One train pass-by has an insignificant impact on the safety of train operation. However, when the number of train pass-bys increases to 10,000,000 times and 20,000,000 times, the cumulative deformations of the ground at different depths are quite large, which may affect the safety of the railway operation. Thus, it is necessary to deal with settlement issues when designing railway construction.

Keywords: settlement; mined-out region; railway construction; dynamic model

---

1. Introduction

In China, railway is an important national infrastructure and a popular means of transportation, which plays a key role in China's comprehensive transportation system. China has a vast territory, a large population, and an uneven distribution of resources. Therefore, economical and fast railways generally have greater advantages and become a widely used mode of transportation. The same thing is happening around the world. The reliable, efficient, safe, and environmentally sound inter-urban long distance rail freight is needed to support economic growth and maintain the quality of life [1].

However, there are many problems in major countries related to railways crossing mineral caves. In China, the line selection principle of railway engineering above the mined-out region is mainly to avoid. Even though the railway needs to cross the mining areas, it should also retain a certain width of safe pillars to ensure the safety of railway operation. In recent years, with the rapid development of railway construction and the massive exploitation of mineral resources, many railway projects have to cross the mining area and its cavern. The original survey design specifications and related manuals cannot meet the requirements of the design standards for the evaluation and treatment of the cavern. The settlement of the ground surface, especially when it occurs in a rapid and differential manner, may cause severe damage to human-built structures and the loss of human lives [2]. Therefore, the research on long-term ground settlements over mined-out regions has undoubtedly played an important guiding role in railway designs.

Salt mine subsidence is more likely to be manifested as ground collapse, cracking, squirting, and surface subsidence, which makes predictions very difficult [3,4]. In order to mine salt mines more effectively and reduce the damage and loss caused by ground settlement, many scholars have done relevant research regarding the establishment of a salt mine subsidence evaluation system in recent decades [5–7]. Most of these studies focus on salt mine water-soluble mining, a summarization of accident causes, determination of reasonable roof span evaluation methods, the establishment of relevant mechanical models based on the stratigraphic characteristics of salt mines, modification and improvement of probabilistic integration methods in salt mine water-soluble mining, and the prediction of water-soluble mining surface failure and subsidence caused by various mathematical systems and computer methods [8–12].

Salt mine mining subsidence prediction is a project that is closely related to practical experience. There are many influencing factors, and each influencing factor needs to accumulate a large amount of field experiences [13,14]. There have been many research results in the summary of the scene accidents of surface subsidence in salt mining areas. For example, Cai and Li [15] obtained the rock stratum and surface movement after actual observation and theoretical analysis according to the surface collapse of the Yingcheng salt mine; the general law presented in time and space. Yu et al. [16] carried out the roadway deformation measurement to observe and analyze the ground trend of change after mining in the Qiaohou Salt Mine in Yunnan. Wang et al. [17] used the engineering geological evaluation method to predict and evaluate the surface subsidence in view of the geological conditions and mining characteristics of the Jintan Salt Mine in Jiangsu Province of China.

The evaluation of the ground and building foundation deformation and stability is a very complicated problem. It is difficult to fully reflect the various factors affecting the deformation and stability by using on-site monitoring, statistical analysis, and other theoretical methods. Numerical analysis can take all the influencing factors into consideration, which saves time and effort, as well as money. Therefore, using numerical analysis to analyze the ground and building foundation deformation and stability over mined-out regions becomes an effective and applicable method. In the late 1990s, Yao [18] successively used the boundary element method and the finite element method to study the stability of the foundation over the mined-out regions, and then, the numerical method was widely used in the evaluation of the stability. A two-dimensional numerical model was used to simulate the effects on surface subsidence due to the discontinuity of overly hard rock formations of longwall mining. Diaz-Fernandez used the influence function to automatically predict the surface subsidence [19]. Alexander [20] proposed that geological structures (such as fault location and joint orientation) play an extremely important role in affecting the surface subsidence over the mined-out regions.

Fan and Li [21] simulated the water-soluble mining subsidence of a salt rock mine by finite element analysis software, and revealed the law of water-soil mining subsidence, and analyzed the ground deformation under the action of a multi-solution cavity. The simulation results showed that the internal water pressure played a controlling role for the stability of the cavity. Increasing the internal water pressure can support the roof of the cavity. Liu et al. [22] used the ADINA nonlinear finite element program to simulate the stability of the thin-layer salt rock water-soluble cavity. The effects of the cavity span, height, salt angle, and cavity geometry on the stability of the cavity were analyzed. Li [23] calculated the influence degree of cavity stability on surface subsidence through engineering examples, and summarized the degree of surface subsidence caused by the instability or even collapse of the cavity. The boundary element method is used to simulate and analyze the stability of the cavity in the Yipinglang salt mine in Yunnan by Yu et al. [24]. According to the difference in the lateral pressure coefficient, different cavity modes should be selected. Ren et al. [25] used the cusp catastrophe theory to study the critical conditions of the instability of the pillars in the Yanyan well group, and analyzed the release mechanism of the column's collapse instability.

The prediction of surface settlement over the mined-out regions can also be predicted by fuzzy mathematics. Li et al. [26] used the fuzzy mathematics to analyze the comprehensive influence matrix of various factors of salt mining. However, a large amount of actual observation data is required in the prediction process. As for the mining depth of minerals, due to the restriction of production conditions and technical factors at that time, the mining depth is generally not more than 500 m, and few minerals can reach 1000 m, as shown in Table 1. Moreover, the study of railway crossing mine is rare.

Table 1. Statistics of mining depth.

Name	Depth/m	Name	Depth/m
Catalina mine	280	Nantong mine	281
Poblek mine	441	Panzhihua mine	56–102
Sverdlov mine	360	Denghe mine	305–325
Southlowitz mine	250	Zibo mine	173
Yorkshire ore	90	Ganshan mine	100
North England mine	550	Weishan mine	213
Hucknall mine	176	Shengli mine	506
Lancashire mine	640	Linyi mine	112

Recently, a railway was designed inevitably to cross a mine area. The construction of this railway plays an important role in improving the regional railway network structure, improving the radiation capacity of the railway network, improving the inter-city transport capacity of the Pearl River Delta, and optimizing the transport organization. It also develops the railway system as a ring junction pattern with goods outside and passengers inside, and realizes the separate operation of passenger and cargo lines. It has the advantage of greatly improving the line capacity, and solving the cross-interference problem between passenger and cargo trains. In order to obtain sufficient monitoring data to ensure the safety of railway embankment construction, the settlement observation was set up on the ground over the nitrate mine, and the ground subsidence monitoring level network was established for 11 years to study the ground settlement and lateral displacement along the designed railway. A finite element model was established to predict the long-term ground settlements over the mined-out region induced by railway embankment construction and train operation. The safety of the railway construction was comprehensively evaluated to explore the feasibility of railway construction over the mined-out region.

2. Statement of the Problem

The Guangzhou Railway Northeast Line Project is located in the Pearl River Delta region, within the Huadu District, Baiyun District, Luogang District. Some part of the designed railway with the distance of 2400 m is inevitable to cross a mine area, which is shown in Figure 1. The Nitrate Mine is located in the northern part of Guangzhou city with an area of 46.86 km^2, which was mined since 1994 and stopped in 2017. The products of the mine are raw material brine with the depth of 500–600 m underground. The ground surface of the mining area has a river, and the terrain is relatively flat and open, and there also built some residential buildings, fish ponds, and farmlands.

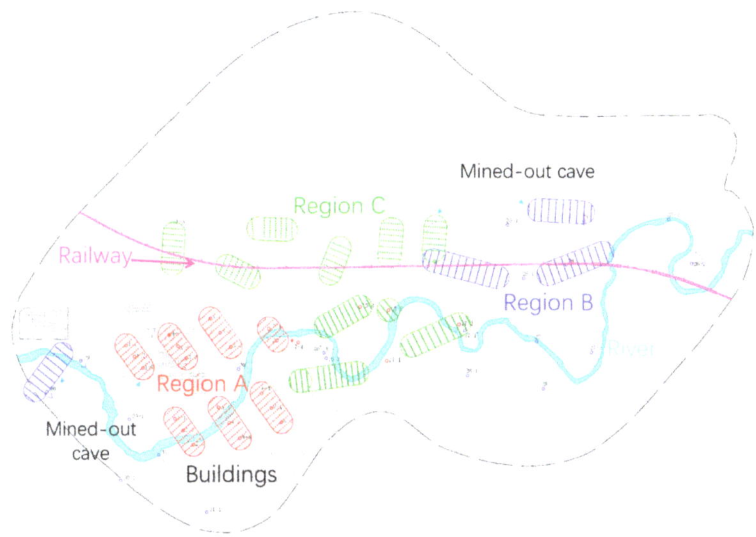

Figure 1. Plan view of railway and mined-out cave.

As shown in Figure 1, the mining area is divided into three regions, including region A, region B, and region C, which correspond to different mining times. The mined-out cave of three regions are represented by different colors. Region A (red) consisted of 21 wells, which went into operation in 1994. Region B (blue) opened in 1998 and consisted of 22 wells, and region C (green) had 31 wells and use began in 2013. The wells can be seen in Figure 1, which was represented by small circle of number. By the end of 2017, the mining area had consumed more than 40 million tons of stone salt ore. The mining area has formed an underground mined area of 1.066 km^2, of which several cavities were distributed under and near the proposed railway. Most of the upper salt layers in this mining area are calcareous argillaceous rocks, which are easy to be softened. Under the disturbance of the water-soluble mining method, the natural balance is easy to be destroyed, which may cause ground fractures and ground subsidence as well as seriously threaten the safety of the designed railway construction and operation [27]. For the mechanized excavation of the mine, a long-term settlement observation system was established to monitor the ground deformation.

3. Ground Deformation Monitoring

The mine has set up observation points around all the wells, and established a ground deformation monitoring level network, as shown in Figure 2. The observation points are named as CJ with numbers in the figure, and a total of 89 observation points were set. However, due to the destruction of some observation points in the construction of residential buildings, factory buildings, and agricultural production, the settlement observations could not be continuously tracked and measured, and some observation points even lost their usefulness. Through statistics on effective deformation data for 11 years from 2007 to 2017, the ground and building surface cracking, ground settlement, and lateral displacement were analyzed.

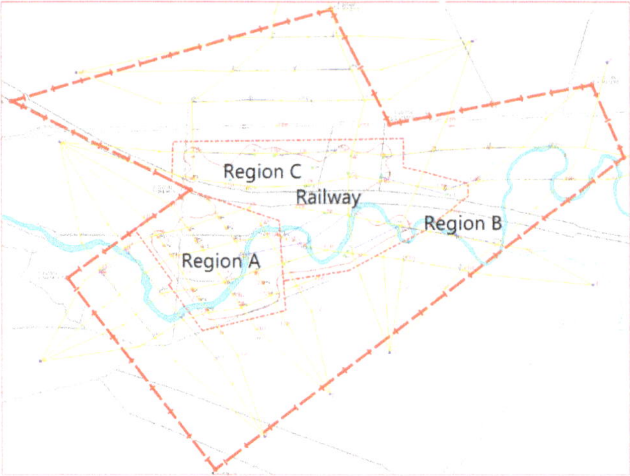

Figure 2. Ground deformation monitoring level network.

3.1. Ground Surface Cracking

Through field investigation, in some regions of the mining area, there have been unfavorable phenomena such as ground and building surface cracking. More serious is that large ground settlement has been observed the roads in the eastern part of region A. In addition, the slate joints laid in front of a six-story residential building in the south of region A were burst and sounded strongly, and some villagers felt shaking. The two adjacent residential buildings showed cracks of up to about one cm. The ground cracking pictures taken at the mine area were shown in Figure 3. Most of the cracks are one to four meters in length, with a maximum length of eight meters, and 0.5–30-mm wide.

Figure 3. Ground crack.

According to the investigation, ground deformation monitoring and other relevant data, combined with the comprehensive factors such as the mining horizon, mining process, and influence range of Nitrate Mine, it is believed that the main reason for the cracking of some building surface in this area was due to the different ground settlement rates induced by the top part settlement of the cavern. However, the superposition of the natural compaction of the Quaternary soil layer is not ruled out. The range of settlement is relatively large, and the settlement rate is relatively uniform, and the influence on the ground structures is slow.

3.2. Ground Settlement

If the ground settlement analysis of the monitoring points only focuses on a single point, the actual settlement of the mining area cannot be explained. Therefore, the accumulated settlement of the monitoring points should be calculated, and the monitoring data of the regions A, B, and C will be effectively superimposed. Figure 4 shows a contour map of the ground settlement over the mines from January 2007, and the period of monitoring is 11 years. In order to avoid misunderstanding, all the time uses months to the present, in which January 2007 is the first month.

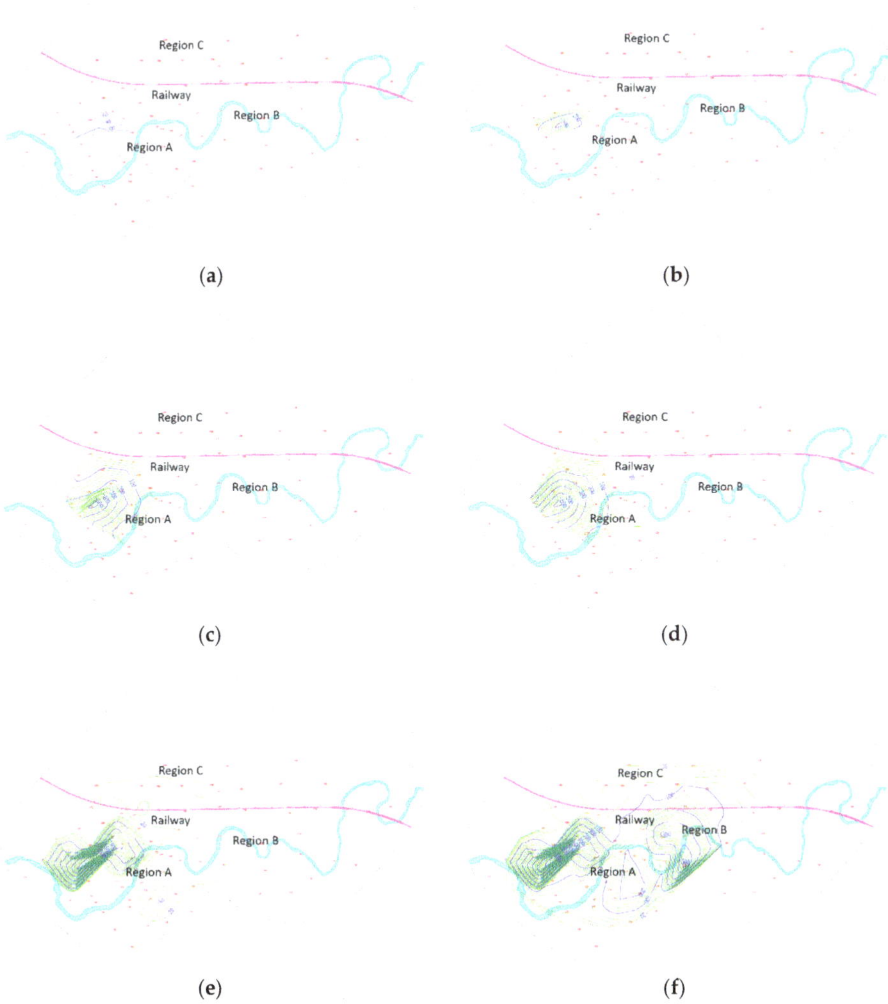

Figure 4. Settlement contour map of mine area after (**a**) 12 months; (**b**) 36 months; (**c**) 60 months; (**d**) 84 months; (**e**) 108 months; and (**f**) 132 months.

The settlement of region A was small for 12 months from the beginning of monitoring, and then accelerated subsidence began to occur, and the maximum cumulative settlement at 36 months reached 323 mm. With the mining progress of region B and region C, the settlement of region A did not stop, and the settlement continued at a large rate. The region A formed a settlement area with a length of about 600 m and a width of about 400 m with the distributed of the northeastward ellipse. The subsidence area is about 0.32 km^2. The cumulative maximum settlement is more than 750 mm. The average annual settlement rate in 132 months was 37.3 mm, of which the central settlement rate of 62.7 mm per year in region A was the largest.

Region B was started at 48 months in Figure 4. In 108 months, the settlement rate of region B increased, and forming a settlement area with a length of about 1500 m and a width of about 700 m. The settlement area is about 1.0 km^2. In 132 months, the average annual settlement rate of region B was 63.91 mm. Compared with the region A, the settlement rate of region B was significantly larger, and the settlement area gradually increased.

Mining began at region C at 108 months. At present, the settlement in region C is also being developed. It can be seen from the contour map that the settlement range is expanding, and the settlement value is increasing. The settlement of region A has always been a center of settlement, which is showed as a funnel-shaped feature. The data of the comprehensive three region settlement observation shows that the settlement rate has a significant increase after a period of mining. At present, the ground settlement rate is increasing from west to east, and the settlement area is still expanding at a relatively rapid rate. The settlement range is gradually expanding to the north and east.

After surveying the four km^2 area of the mining area and surrounding areas, 36 ground settlement locations were found, mainly including ground and pavement cracking, wall cracks, etc. The ground settlement area was about 1.3 km^2, which was similar to the mineral burial area. The village buildings and factories in the mining area is relatively dense. Due to the regional differences of settlement in the mining area, the building structure of the small area has not yet been significantly impacted.

Generally, the salt caverns formed by salt mining generally have the changes related to stress, the roof deformation of the cavern, roof collapse, and ground subsidence. Especially after the mined-out area is formed, the salt mine will still dissolve and soften further, and the roof may collapse. The collapsed mudstone will soften in water and it is unable to support the upper rock. The dome formed by collapse gradually rises, the bedrock surface flexes and sinks, and land subsidence occurs.

The ground subsidence caused by salt mining can be divided into three stages: the stable subsidence stage, the accelerated subsidence stage, and the collapse stage. If no large-scale underground mined-out area communication is formed after mining, the stability of the mine will be maintained for a long time after the end of mining, and the subsidence will gradually disappear. Most salt mines in the world maintain in the stable settlement stage. In the 19th century, the Winsford Meadowbank salt mines in England adopted the chamber and pillar method to extract 152 m to 213 m underground of salt with a recovery rate of 65% to 75%, and no surface subsidence appeared. In the 1930s, controlled salt extraction was adopted near Holford, Scotland, and no subsidence occurred. In China, the control measures such as brine filling and drilling sealing were adopted at the mining area of Xiangheng salt mine. After 15 years of mining, no surface subsidence has occurred. Due to the deep mining location of the salt mine in Shalongda, China, the water-soluble depth is below 2000 m underground. In addition, with good water-isolating strata and good water source pressure supply around, the ground subsidence is extremely small. This shows that the possibility of surface subsidence can be reduced or even avoided by setting a reasonable mining plan according to the geological conditions of the salt mine itself. From the monitoring data of ground subsidence in mining areas region A and region B, it can be found that the subsidence has accelerated, and there is no sign of convergence, indicating that the region A and region B mining areas are still in the phase of accelerated subsidence. By the end of 2017, the maximum settlement along the designed railway appeared near the center of the region B mining area, with the settlement of about 140 mm.

3.3. Ground Lateral Displacement

The cumulative horizontal displacement vector obtained in 132 months is shown in Figure 5. The horizontal displacement is relatively small, and the maximum horizontal displacement is less than 20 mm. It is mainly due to the horizontal movement caused by uneven settlement.

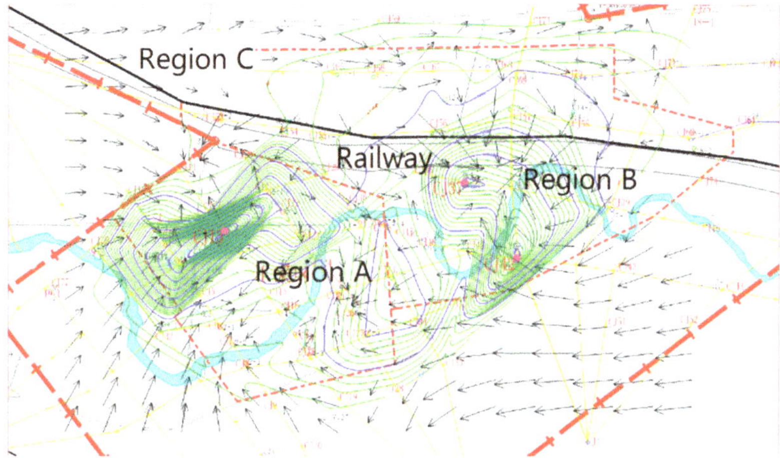

Figure 5. Lateral displacement vector.

4. Ground Settlements over Mined-Out Region Induced by Railway Construction

In order to ensure the safety of railway embankment construction and operation, it is necessary to predict long-term ground settlement after construction. The numerical simulation method has unique advantages. It can simulate the influence of complex surface and special geological structures, study the surface deformation law, and save time and effort in analysis. This section uses the finite element method to construct a three-dimensional geological model to study the impact of mechanized excavation on ground and building settlements.

According to the actual mining situation, the model length is 2500 m, the width is 2000 m, and the depth is 600 m, as shown in Figure 6. The soil layers are divided into five types from top to bottom, including a quaternary soil layer, silty mudstone, mudstone, halite, and interlayer, as listed in Table 2. The physical and mechanical parameters are based on an engineering geological survey report, which is shown in Table 3.

Table 2. Soil layer distribution.

Depth (m)	Soil Layer
25.5	Quaternary soil layer
324.5	Silty mudstone
412.5	Mudstone
481.52	Mudstone
491.49	Halite
494.72	Interlayer
497.88	Halite
505.13	Interlayer
509.91	Halite
512.79	Interlayer
516.22	Halite
600	Mudstone

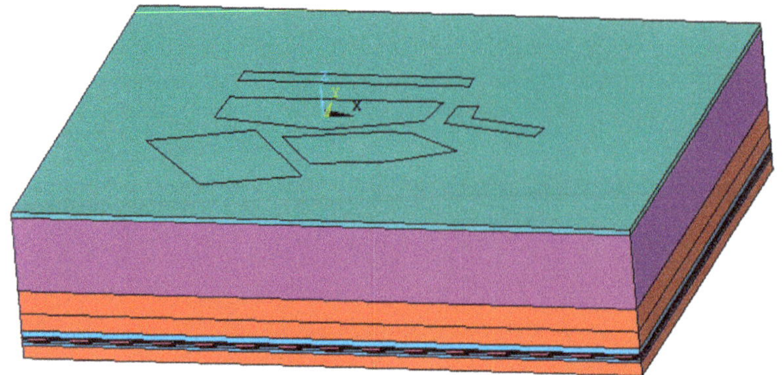

Figure 6. Finite element model.

Table 3. Physical and mechanical properties of soil layers.

Soil Layer	Density (kg/m³)	Elasticity Modulus (GPa)	Poisson's Ratio	Cohesion (MPa)	Inner Friction Angle (°)
Quaternary soil layer	1800	0.01	0.3	/	/
Silty mudstone	2200	0.9	0.27	0.5	35
Mudstone	2400	0.9	0.27	0.5	35
Halite	2200	1.2	0.3	0.5	30
Interlayer	2200	0.4	0.2	0.5	30
Railway embankment	2000	0.015	0.27	/	/

The bottom of the model is fixed with constraints, and the four sides are hinged to the hinge. Soil is assumed as ideal elastoplastic model, which compliance with the Mohr–Coulomb failure criterion.

The new railway embankment is built on the ground according to the design location in Figure 1. A triangular arch is set on the railway embankment surface, and a 4% drainage slope is provided on both sides from the center of the railway line. On the embankment section, a concrete shoulder with a top width of 0.4 m is built. The drawing of the railway embankment is shown in Figure 7. The height of the embankment is eight meters, which uses stratified paving and the height of each layer is based on the designed guideline. A settlement observation section is set at different depths below the embankment, including one meter, 4 m, 10 m, 16 m, and 24 m below the ground surface.

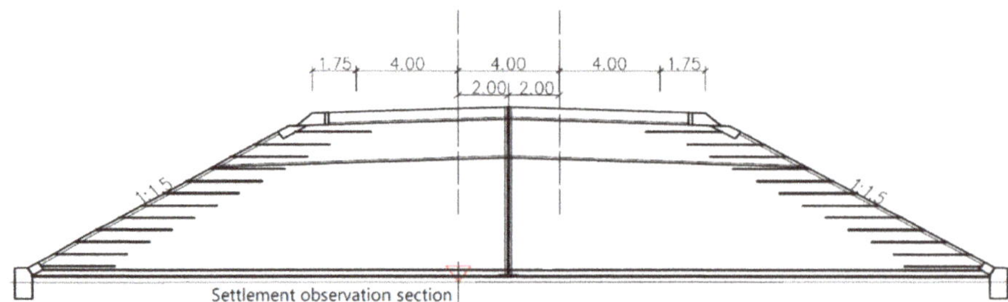

Figure 7. Railway embankment.

For the convenience of the research, this model has the following assumptions:

(a) Due to the complexity of the rock mass and its structure, it is impossible to take all the factors into consideration when performing large-scale calculations. Therefore, the same stratum is treated according to the isotropic medium.

(b)   The formation of the soil and rock layer is affected by many factors in the formation process; its shape cannot be regular, and the inclination and thickness of the rock layer are constantly changing. In order to facilitate the establishment of models and calculations, the interface of the rock layer is usually not considered.

(c)   The tectonic stress is ignored; the self-heavy stress is regarded as the ground stress.

(d)   Due to the complexity of the problem and the unpredictability of the size of the cavity, it is only possible to model the entire mining zone and simplify the cavity. The similar cavities are spliced into a large mined-out cavity. Therefore, the cavities are finally reduced to five areas; the finite element model is shown in Figure 8. The order of salt mine excavation is in accordance with the actual project.

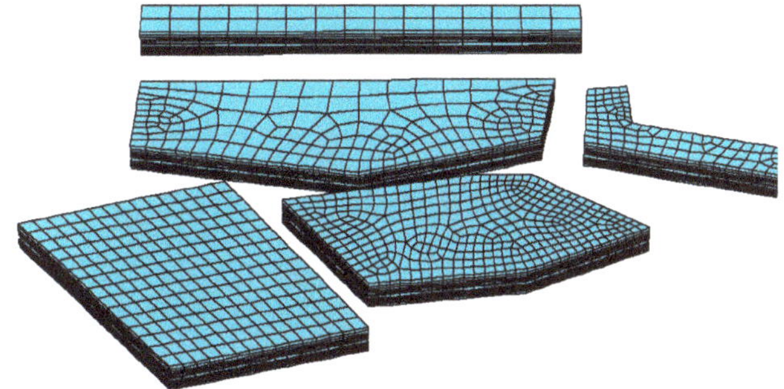

Figure 8. Finite element model of mined-out cavities.

In this model simulation, firstly, the validity of the model is verified by the monitoring settlement data. Then, settlement at different depths from the ground was calculated, and the long-term ground surface settlement was predicted.

4.1. Model Validation

In order to verify the accuracy of the model, five monitoring points—CJ53, CJ54, CJ55, CJ56, and CJ58—along the railway are selected to compare the predicted settlement and measured settlement. The comparison is shown in Figure 9.

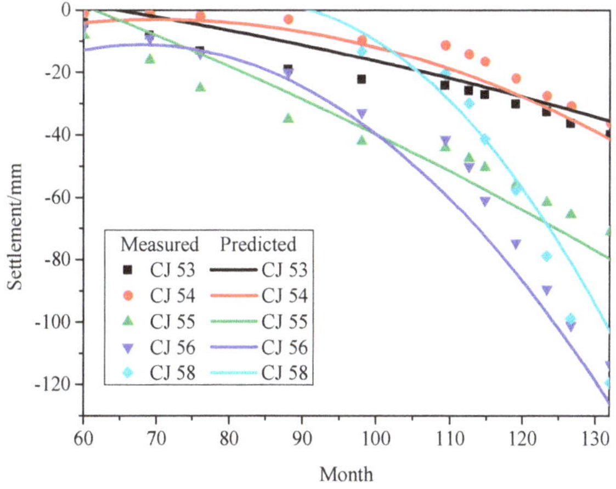

Figure 9. Comparison of predicted settlement and measured settlement.

As shown in Figure 4, the selected monitoring points along the railway show obvious settlement after 60 months. Therefore, the comparison of predicted settlement and measured settlement is from 60 months to 132 months. In Figure 8, a good match can be seen between the measured settlement and the predicted settlement. It can be considered that the finite element model can effectively predict the ground settlement.

4.2. Soil Settlement below Railway Embankment

In order to obtain the soil settlement at different depths below the railway embankment, a section near monitoring point CJ56 is selected due to the settlement around CJ56 being the largest along the railway. However, the total settlement consists of a consolidation settlement induced by the mined-out cavity and an additional load induced by embankment soil. Therefore, the consolidation settlement induced by the mined-out cavity should be removed to only consider the settlement induced by additional embankment load. Figure 10 shows the settlement of zero meters, 4 m, 10 m, 16 m, and 24 m below the railway embankment change with the different embankment heights.

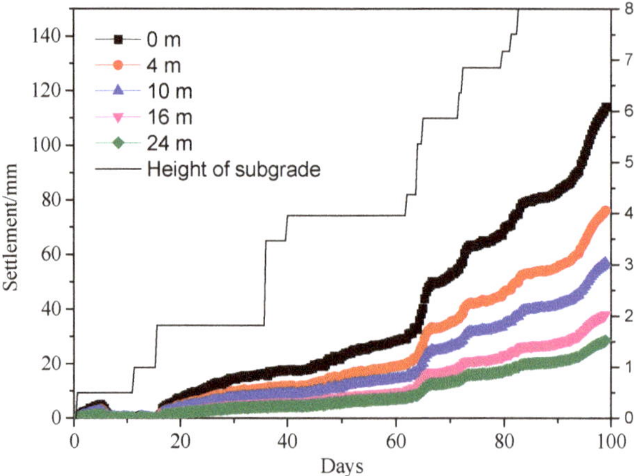

Figure 10. Soil settlement at different depths below the railway embankment change with different embankment heights.

As shown in Figure 10, the settlement of soil layers at different depths has a significant increasing trend with the increase of embankment height. When the embankment height is small, the settlement of the soil layer at different depths is also small. With the increase of the embankment height, the settlement of the soil layer at different depths also increases accordingly.

During the embankment filling period, the settlement of the soil layer at different depths increases rapidly, which is reflected in the curve as the slope becomes larger. During the intermittent period of filling, the settlement of the soil layer at different depths will fluctuate up or down. This is also in line with the general rule, because in the intermittent period of filling, the base soil layer is filled with load pressure from the top, which leads to the compression deformation of the soil.

The settlement of the soil layer at different depths is also different. The depth of the soil layer is inversely proportional to the settlement. The deeper the soil layer, the smaller the settlement. On the contrary, the shallower the soil layer, the larger the settlement. It means the settlement of the deep soil layer is less than the settlement of the shallow soil layer.

4.3. Long-Term Ground Surface Settlement

Long-term settlement is a matter of great concern. It is related to whether the train can operate safely, smoothly, and comfortably. The Figure 11 shows the different periods of settlement along the railway surface of the embankment from the left side to the right side of the mining area with the distance of 2400 m.

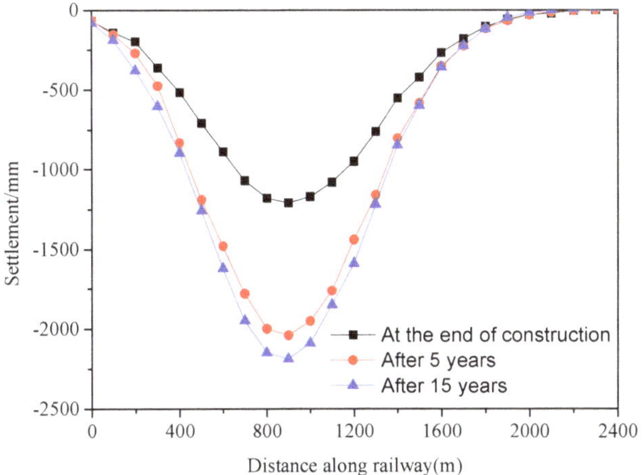

Figure 11. Predicted long-term settlement along the railway.

It can be seen from the figure that ground settlement along the railway is basically the same at different times. The settlement along the railway surface of the embankment can be divided into two stages. The first stage is the linearization of settlement, and the settlement increased quickly, which is from about 1.2 m at the end of the construction to about 2.0 m after five years. The second stage is that the settlement value tends to be stable from five years to 15 years. The location of maximum settlement is close to region B. Compared to the settlement induced by the railway embankment construction, the total settlement is much larger. It indicates that the main cause of settlement is not railway embankment construction, but rather a mined-out cavity. In addition, the creep properties of the salt rock are not considered. After the end of mining, the salt rock around the mined-out cavity is in a state of partial stress, and the salt rock will continuously shrink to the mined-out cavity, which will continue to cause ground settlement.

Due to the settlement becoming stable after five years, the predicted settlement for the whole mining area is shown for five years, as shown in Figure 12. The maximum settlements of three regions are between two and three meters. For the residential buildings in the south of region A, the settlements are between one and 1.5 m, which will increase the damage of the buildings. It needs to reinforce the mined-out area urgently to avoid bigger losses.

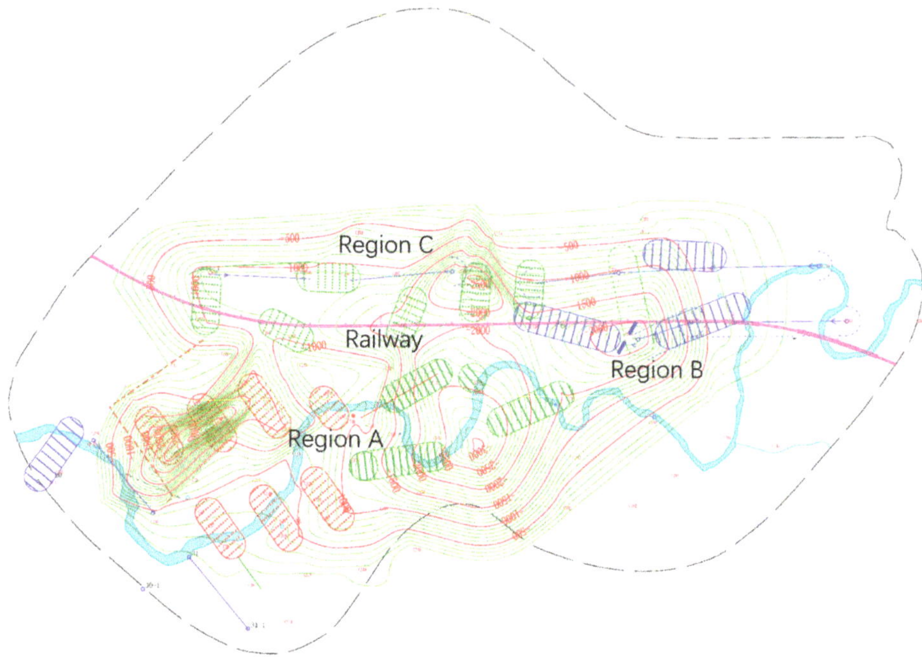

Figure 12. Predicted settlement contour map of the mine area after five years of railway construction.

5. Ground Settlements over Mined-Out Region Induced by Train Operation

In order to analyze the dynamic deformation mechanism caused by moving train loads, this section establishes the train track coupling dynamic model, fully considering the influence of track irregularity, and calculating the wheel–rail interaction when the train runs on the track structure in the time domain. On this basis, the finite element model of the track structure and ground is built based on the finite element model of Section 4 and excitation by moving train loads, and the dynamic response characteristics of the ground under moving load are obtained.

5.1. Dynamics Model of Train

A vertical dynamics model with 10 degrees of freedom (DOFs) is built for the train based on the theory of the multi-body system dynamics. Figure 13 illustrates the schematic drawing of the vertical dynamics model of a train. In this mode, the car body, two bogies, and four wheelsets are represented using mass blocks. All of these mass blocks can move up and down along the vertical direction. Also, the car body and two bogies can rotate in the vertical plane. The spring-dampers are used to connect each mass block to simulate the primary and secondary suspensions in the train.

According to Newton's second law, the dynamics equations for the train model can be expressed as:

$$M\ddot{x} + C\dot{x} + Kx = F \tag{1}$$

where x means the vector of the 10 DOFs in the vehicle; and M, C, and K represents the mass, damping, and stiffness matrix of the multi-body system. The symbol F is the load vector applied to the vehicle, which is actually the contribution of the dynamic interaction forces between the wheelsets and the rail. The wheel/rail forces can be calculated according to the nonlinear Hertz contact theory, in which the effect of the rail irregularity is also taken into account. The dynamics parameters of train are listed in Table 4.

Table 4. Dynamics parameters of the train.

Distance between two bogies	12.6 m	Wheelset mass	1800 kg
Distance between two axles of bogie	2.0 m	Wheel rolling radius	0.42 m
Car mass	50,320 kg	Primary suspension stiffness	2.45×10^6 kN/m
Car body inertia	1.513×10^6 kg·m^2	Secondary suspension stiffness	1.04×10^6 kN/m
Bogie mass	3000 kg	Primary suspension damping	8×10^4 kN·s/m
Bogie inertia	2100 kg·m^2	Secondary suspension damping	6×10^4 kN·s/m

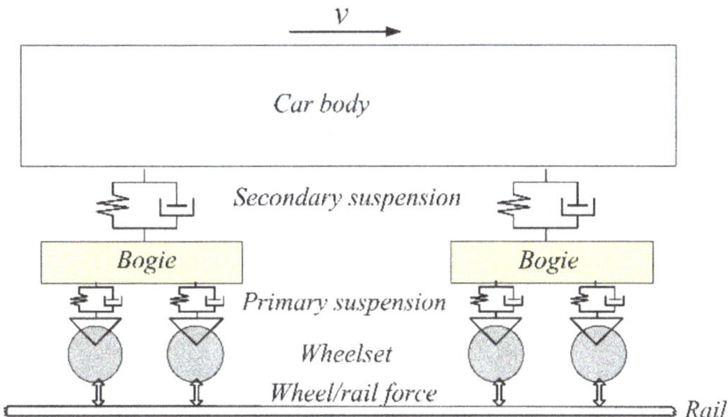

Figure 13. Schematic drawing of the dynamics model of a train.

5.2. Dynamics Model of Track Structure

A finite element model is established for the track structure. Figure 14 shows the schematic drawing of the finite element model of the track structure. In this model, the rails and sleepers with real geometry sizes are all modeled using hexahedral solid elements. The rail pads under the rail are modeled using the spring-dampers. The mechanical parameters of each component in the finite element model such as the density, the Young's modulus, and damping are chosen according to the material properties of the field track structures, as shown in Table 5. The track adopts 60 kg/m, U75V hot-rolled steel rail, and the distance between two rails is 1.435 m.

Figure 14. Schematic drawing of the dynamics model of track structure.

Table 5. Track structure parameters.

Rail elastic modulus	N/m^2	2.06×10^{11}	Rail mass	kg/m	60.64
Rail cross-sectional area	m^2	7.745×10^{-3}	Rail section moment of inertia	m^4	3.217×10^{-5}
Rail density	kg/m^3	7830	Rail elastic pad damping	N·s/m	5×10^4
Rail elastic pad stiffness	MN/m	100	Sleeper mass	kg	251
Sleeper spacing	m	0.6	Ballast elastic modulus	Pa	0.8×10^8
Sleeper elastic modulus	MPa	30,000	Ballast elastic modulus	MPa	300
Ballast mass	kg	630	Embankment elastic modulus	Pa/m	1.3×10^8
Ballast damping	N·s/m	1.6×10^5	Embankment damping	N·s/m	6.32×10^4

Track irregularity is the main reason of dynamic response caused by train operation, and it is the excitation source of the wheel–rail system. Considering that the established train track dynamic model only considers the vertical dynamic response, the track irregularity is based on the spectrum used by Lombaert [28]. The simulation is as follows:

$$S(k_1) = S(k_{1,0})\left(\frac{k_1}{k_{1,0}}\right)^{-w} \tag{2}$$

where $k_{1,0} = 1\,\mathrm{rad/s}$, $S(k_{1,0}) = 1 \times 10^{-8}\,\mathrm{m^3}$, $w = 3.5$.

For the train load, this paper simulates moving train loads by applying wheel–rail contact force, which was obtained from the train track dynamic model to the rail surface. The train speed is 160 km/h for the boundary condition, which was obtained by coupling the finite element and the infinite element to simulate infinite ground [29]. It has wide applicability in the problem of simulating the infinite domain without losing the accuracy of the calculation.

5.3. Train-Induced Ground Settlement

This section mainly studies the settlement of train dynamic load on the ground above the mined-out cavity. Figure 15 shows the process of simulating the train passing through the mined-out cavity. At this time, the train is running in the center of the mined-out cavity.

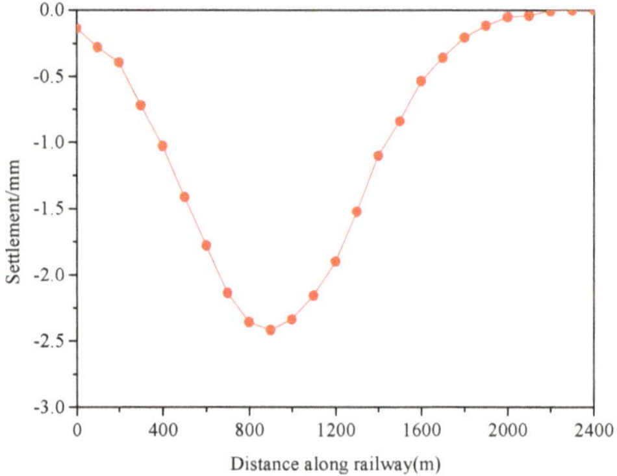

Figure 15. Ground settlement along the railway during the train runs in the central area of the mined-out cavity.

The settlement induced by the dynamic load of the train has time-shift characteristics; the settlement changed with different time of the train operation. It can be seen from Figure 15 that the maximum settlement induced by the train operation is about 2.5 mm. When the train runs to the non-mining area, the ground settlement is less than the settlement induced by train runs inside the mining area. The settlement is obvious when the train runs at the center of the mining area. Beyond the area of the mined-out cavity, the dynamic load has less impact on the ground settlement.

At the end of the train operation, the maximum residual settlement of 1.4 mm occurred on the ground surface. For the evaluation of the stability of the railway embankment, it can analyze the residual deformation of the ground surface caused by the additional dynamic load. Therefore, the maximum residual settlement of 1.4 mm caused by train operation has less impact on the safety of train operation.

Figure 16 shows the soil settlement at different depths below the ground surface. The impact depth of about 16 m induced by train operation can be found, which indicates that the influence of one pass-by train is small.

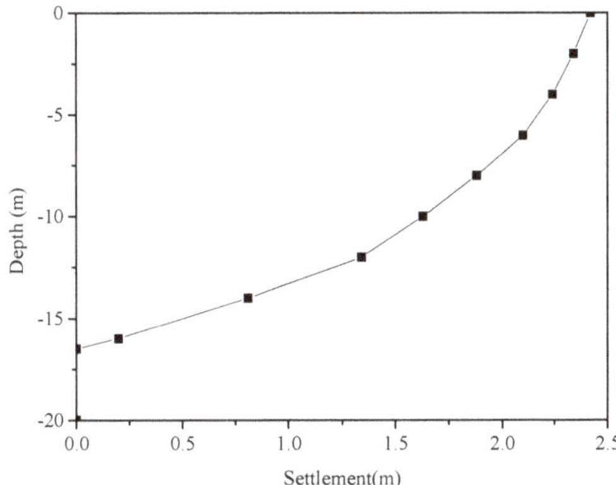

Figure 16. Soil settlement at different depths below the ground surface.

In order to predict the cumulative deformation of the ground under the moving train dynamic load, the empirical formula that was proposed by Chai and Miura [30] was used. The formula was established by the soil cumulative plastic strain results based on the indoor dynamic triaxial test, which is defined as:

$$\varepsilon_p = a \left(\frac{q_d}{q_f} \right)^m \left(1 + \frac{q_s}{q_f} \right)^n N^b \tag{3}$$

where ε_p is the cumulative plastic strain, N is the number of cyclic loadings, q_d is the dynamic deviatoric stress, q_s is the initial static deviatoric stress, and q_f is the static strength of the soil. When the soil is undrained, $q_f = 2C_u$, C_u can be obtained by the shear test.

The specific analysis process is:

(a) Dynamic analysis of the three-dimensional track ground finite element model. The dynamic moving load is the wheel–rail contact force obtained by the train track dynamic model.
(b) The dynamic analysis is carried out to obtain the dynamic response of the soil after one single train pass-by. The maximum dynamic deviator stress q_d of the soil can be obtained.
(c) The static analysis of the model can obtain the initial static deviator stress q_s of the soil.
(d) Use the empirical formula Equation (3) to calculate the plastic cumulative strain of the ground under any vehicle load times, and calculate the cumulative deformation of the ground.

The integral deformation of the soil can be obtained by integrating the plastic strain in the depth direction. It can be seen from Figure 17 that the cumulative deformation of the ground increases with the increase of the number of times of loading. The deformation rate is the largest in the early stage of loading, and as the number of loads increases, the growth rate of cumulative deformation gradually decreases, and 500,000 times is the inflection point of the curve. Before 500,000 train operations, the cumulative deformations of the soil at the ground surface, one meter under the ground surface, and three meters under the ground surface are close to each other. However, when the number of loading increases to 10,000,000 times and 20,000,000 times, the cumulative deformations of the ground at different depths are quite different. This may affect the safety of the railway operation; thus, foundation reinforcement measures ought to be considered during construction.

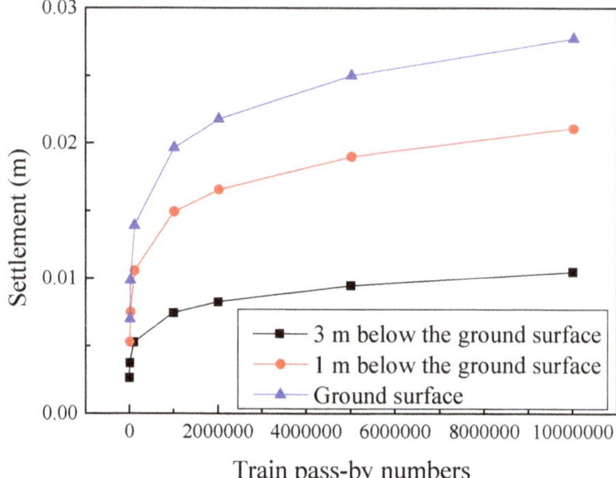

Figure 17. Cumulative soil settlement changed with different train pass-by numbers.

According to the above analysis, the ground settlement is caused by mining, rather than embankment construction and train operation. The mined-out region is too large and difficult to manage. The salt mine should stop mining immediately to avoid the mined-out region expanding further and thus increasing processing difficulty. For abandoned wells in the mining area, it is necessary to carry out grouting sealing in the hole and to reinforce the weak broken zone by grouting. On the one hand, it can block the loss path of quaternary sand and soil particles, thus preventing further loss and the tunneling of quaternary strata. On the other hand, this method can seal the brine in the cave and make use of the good tightness of salt rock and cover mudstone, so that the brine pressure will increase with the shrinkage of the cave and increase the brine pressure's supporting force on the roof, and further inhibit the subsidence of the overlying strata and ground subsidence.

6. Conclusions

In this paper, the settlement observation points were set up around the salt mine, and the ground settlement monitoring level network was established. The ground settlement and lateral displacement along the railway were studied last for 11 years. A finite element model was established to predicted long-term ground settlements over the mined-out region induced by designed railway construction and train operation. The conclusions are as follows:

1. The finite element model can effectively predict the ground settlement induced by mechanical excavation of the salt mine and railway embankment construction as well as train operation. The cause of settlement is mainly due to the impact of the mined-out cavity, rather than railway construction and operation. For the design of the railway crossing the mining area, it is necessary to avoid crossing the settlement center area as much as possible, and establish a long-term settlement observation system.

2. The mechanical excavations of the salt mine mainly lead to large ground settlement, and the horizontal displacement is relatively small.

3. One train pass-by has less impact on the ground settlement. However, when the number of train pass-by increases to 10,000,000 times and 20,000,000 times, the cumulative deformations of the ground at different depth are quite large, which may affect the safety of the railway operation.

Author Contributions: Conceptualization, Y.W.; methodology, Y.W.; software, S.J.; validation, S.J. and Y.W.; data curation, S.J.; writing—original draft preparation, S.J.; writing—review and editing, Y.W.

Funding: This research was funded by China Railway Guangzhou Bureau Group Co., Ltd.

Conflicts of Interest: The authors declare no conflict of interest.

References

1. Aditjandra, P.T.; Zunder, T.H.; Islam, D.M.Z.; Palacin, R. Green Rail Transportation: Improving Rail Freight to Support Green Corridors. In *Green Transportation Logistics*; Springer: Cham, Switzerland, 2016; pp. 413–454.
2. Desir, G.; Gutiérrez, F.; Merino, J.; Carbonel, D.; Benito-Calvo, A.; Guerrero, J.; Fabregat, I. Rapid subsidence in damaging sinkholes: Measurement by high-precision leveling and the role of salt dissolution. *Geomorphology* **2018**, *303*, 393–409. [CrossRef]
3. Chengzong, C.; Guangzhong, C.; Peisong, D. A study on the stability of railway subgrade in a salt lake. *Proc. Int. Symp. Eng. Complex Rock Form.* **1988**, 607–614. [CrossRef]
4. Yang, C.; Jing, W.; Daemen, J.J.K.; Zhang, G.; Du, C. Analysis of major risks associated with hydrocarbon storage caverns in bedded salt rock. *Reliab. Eng. Syst. Saf.* **2013**, *113*, 94–111. [CrossRef]
5. Einstein, H.H.; Vick, S.G. Geological model for a tunnel cost model. In Proceedings of the Rapid Excavation and Tunneling Conference, San Francisco, CA, USA, 24–27 June 1974; pp. 1701–1720.
6. Einstein, H.H. Risk and risk analysis in rock engineering. *Tunn. Undergr. Space Technol.* **1996**, *11*, 141–155. [CrossRef]
7. Diamantidis, D.; Zuccarelli, F.; Westhäuser, A. Safety of long railway tunnels. *Reliab. Eng. Syst. Saf.* **2000**, *67*, 135–145. [CrossRef]
8. Cagno, E.; De Ambroggi, M.; Grande, O.; Trucco, P. Risk analysis of underground infrastructures in urban areas. *Reliab. Eng. Syst. Saf.* **2011**, *96*, 139–148. [CrossRef]
9. Bérest, P.; Brouard, B. Safety of salt caverns used for underground storage blow out; mechanical instability; seepage; cavern abandonment. *Oil Gas Sci. Technol.* **2003**, *58*, 361–384. [CrossRef]
10. Yang, C.; Wang, T.; Li, Y.; Yang, H.; Li, J.; Qu, D.; Xu, B.; Yang, Y.; Daemen, J.J.K. Feasibility analysis of using abandoned salt caverns for large-scale underground energy storage in China. *Appl. Energy* **2015**, *137*, 467–481. [CrossRef]
11. Chan, K.S.; Bodner, S.R.; Fossum, A.F.; Munson, D.E. A damage mechanics treatment of creep failure in rock salt. *Int. J. Damage Mech.* **1997**, *6*, 121–152. [CrossRef]
12. Li, Y.; Yang, C. On fracture saturation in layered rocks. *Int. J. Rock Mech. Min. Sci.* **2007**, *6*, 936–941. [CrossRef]
13. Heusermann, S.; Rolfs, O.; Schmidt, U. Nonlinear finite-element analysis of solution mined storage caverns in rock salt using the LUBBY2 constitutive model. *Comput. Struct.* **2003**, *81*, 629–638. [CrossRef]
14. Zhang, N.; Ma, L.; Wang, M.; Zhang, Q.; Li, J.; Fan, P. Comprehensive risk evaluation of underground energy storage caverns in bedded rock salt. *J. Loss Prev. Process Ind.* **2017**, *45*, 264–276. [CrossRef]
15. Cai, K.; Li, Y. The theory and practice of infrared radiation surveying technology of geological structure in pit. *J. Xiangtan Min. Inst.* **2002**, *17*, 5–8. [CrossRef]
16. Yu, X.D.; Yang, Z. Environmental Impacts of the Mining Activities in the High Mountain Regions of Yunnan Province, China. In Proceedings of the ISRM International Symposium—EUROCK 2002, Madeira, Portugal, 25–27 November 2002; International Society for Rock Mechanics and Rock Engineering: Salzburg, Austria, 2002.
17. Guangya, W.; Bin, S.; Zulin, Q.; Shiliang, W. Investigation on Hetang earth fissure in south Jiangyin. *Geotech. Investig. Surv.* **2009**, *4*, 2.
18. Yao, X.L.; Reddish, D.J.; Whittake, B.N.R. Non-linear finite element analysis of surface subsidence arising from inclined seam extraction. *Int. J. Rock Mech. Min. Sci. Geomech.* **1993**, *30*, 4. [CrossRef]
19. O'connor, K.M.; Dowding, C.H. Distinct element modeling and analysis of mining-induced subsidence. *Rock Mech. Rock Eng.* **1992**, *25*, 1–24. [CrossRef]
20. Díaz-Fernández, M.E.; Álvarez-Fernández, M.I.; Álvarez-Vigil, A.E. Computation of influence functions for automatic mining subsidence prediction. *Comput. Geosci.* **2010**, *14*, 83–103. [CrossRef]
21. Fan, Y.; Li, X. Study of 3D finite-element modeling on the law of solution mining subsidence. *Ind. Miner. Process.* **2008**, *9*, 6.

22. Liu, C.; Xu, L.; Xian, X. Fractal-like kinetic characteristics of rock salt dissolution in water. *Colloids Surf. A Physicochem. Eng. Asp.* **2002**, *201*, 231–235. [CrossRef]

23. Li, X.; Wang, S.; Malekian, R.; Hao, S.; Li, Z. Numerical Simulation of Rock Breakage Modes under Confining Pressures in Deep Mining: An Experimental Investigation. *IEEE Access* **2016**, *4*, 5710–5720. [CrossRef]

24. Yu, X.B.; Xie, Q.; Li, X.Y.; Na, Y.K.; Song, Z.P. Cycle loading tests of rock samples under direct tension and compression and bi-modular constitutive model. *Chin. J. Geotech. Eng.* **2005**, *27*, 988.

25. Ren, S.; Jiang, D.Y.; Yang, C.H.; Jiang, Z.W. Study on a new probability integral 3D model for forecasting solution mining subsidence of rock salt. *Rock Soil Mech.* **2007**, *28*, 133–138.

26. Wenxiu, L. Fuzzy mathematics models on rockmass displacements due to open-underground combined mining for thick ore body with steep dip angle. *Chin. J. Rock Mech. Eng.* **2004**, *23*, 572–577.

27. Zou, C.; Wang, Y.; Moore, J.A.; Sanayei, M. Train-induced field vibration measurements of ground and over-track buildings. *Sci. Total Environ.* **2017**, *575*, 1339–1351. [CrossRef] [PubMed]

28. Lombaert, G.; Degrande, G. Ground-borne vibration due to static and dynamic axle loads of InterCity and high-speed trains. *J. Sound Vib.* **2009**, *319*, 1036–1066. [CrossRef]

29. Yang, J.; Li, P.; Lu, Z. Numerical simulation and in-situ measurement of ground-borne vibration due to subway system. *Sustainability* **2018**, *10*, 2439. [CrossRef]

30. Chai, J.C.; Miura, N. Traffic-load-induced permanent deformation of road on soft subsoil. *J. Geotech. Geo-Environ. Eng.* **2002**, *10*, 907–916. [CrossRef]

 *sustainability*

Article

Experimental Study and Numerical Simulation on Hybrid Coupled Shear Wall with Replaceable Coupling Beams

Yun Chen [1,2], Junzuo Li [2] and Zheng Lu [2,3,*] 

[1] College of Civil Engineering and Architecture, Hainan University, Haikou 570228, China; chenyunhappy@163.com
[2] Research Institute of Structural Engineering and Disaster Reduction, Tongji University, Shanghai 200092, China; 1732580@tongji.edu.cn
[3] State Key Laboratory of Disaster Reduction in Civil Engineering, Tongji University, Shanghai 200092, China
[*] Correspondence: luzheng111@tongji.edu.cn; Tel.: +86-21-659-86-186

Received: 26 December 2018; Accepted: 1 February 2019; Published: 7 February 2019

Abstract: The coupled shear wall with replaceable coupling beams is a current research hotspot, while still lacking comprehensive studies that combine both experimental and numerical approaches to describe the global performance of the structural system. In this paper, hybrid coupled shear walls (HSWs) with replaceable coupling beams (RCBs) are studied. The middle part of the coupling beam is replaced with a replaceable "fuse". Four $\frac{1}{2}$-scale coupled shear wall specimens including a conventional reinforced concrete shear wall (CSW) and three HSWs (F1SW/F2SW/F3SW) with different kinds of replaceable "fuses" (Fuse 1/Fuse 2/Fuse 3) are tested through cyclic loading. Fuse 1 is an I-shape steel with a rhombic opening at the web; Fuse 2 is a double-web I-shape steel with lead filled in the gap between the two webs; Fuse 3 consists of two parallel steel tubes filled by lead. The comparison of seismic properties of the four shear walls in terms of failure mechanism, hysteretic response, strength degradation, stiffness degradation, energy consumption, and strain response is presented. The nonlinear finite element analysis of four shear walls is conducted by ABAQUS software. The deformation process, yielding sequence of components, skeleton curves, and damage distribution of the walls are simulated and agree well with the experimental results. The primary benefit of HSWs is that the damage of the coupling beam is concentrated at the replaceable "fuse", while other parts remain intact. Besides, because the "fuse" can dissipate much energy, the damage of the wall-piers is also alleviated. In addition, among the three HSWs, F1SW possesses the best ductility and load retention capacity while F2SW possesses the best energy dissipation capacity. Based on this comprehensive study, some suggestions for the conceptual design of HSWs are further proposed.

Keywords: replaceable coupling beam; beam; shear wall; cyclic reversal test; seismic behavior

1. Introduction

Based on the working mechanism of structures under seismic loads and previous earthquake damage surveys, it is found that in many cases, structures fail only because of some specific components' failure, not all of them [1–3]. The repair work of damaged components can be very costly and even impractical, especially for cast-in-place reinforced concrete structure. Coupled shear wall system is an efficient structural system to resist lateral forces, which is widely used in high-rise buildings [4,5]. Under large lateral forces, plastic hinges are designed to appear at the beam ends to dissipate energy [6], and then lead to failure. To improve the reliability of the coupling beam, many scholars are committed to reduce the damage of the beam mainly in two ways. Traditional designers propose special reinforcement layouts or constructional measures to enhance the energy dissipation capacity and ductility of coupling beams [7–10]. However, those methods do not change the failure mechanism of the coupled wall that the beam ends are most vulnerable to damage, and repairing the coupling beam is very difficult.

In recent years, the design of replaceable components has been proposed [11]. By consciously making the risky parts of the structure into a replaceable form, the replaceable parts can dissipate energy by plastic deformation to protect the other components from damage. The replaceable components can be called "fuses" [12,13]. As for the coupling beams in coupled shear walls, Fortney et al. first put forward the concept of replaceable coupling beams (RCBs) [14]. They divide the coupling beam into 3 parts. The middle part which is named the "fuse" is intentionally weakened to dissipate energy by shear yielding. A design methodology is presented in great detail, which will ensure that the outer sections remain elastic through the inelastic range of the fuse section. Cyclic reversal test of a steel coupling beam with a replaceable "fuse" shows that the damage of the beam is concentrated at the middle part, which is easy to be replaced. Li et al. develop a replaceable coupling beam which is designed as a steel truss with a replaceable buckling restrained steel web. The buckling restrained web can dissipate energy by shear deformation and is confined by two precast reinforced concrete panels to avoid out-of-plane buckling. Cyclic tests show that the "fuse" has desirable deformation and energy absorption capacities and inelastic deformation is concentrated at steel webs. Besides, they also developed a modified strip model to predict the bearing capacity of the beam [15]. Recently, Ji et al. [16] developed a type of replaceable steel coupling beam which consists of a central "fuse" shear link connecting to steel beam segments at its two ends. Large-scale specimens of the beams are tested, showing that all specimens have fully developed the shear strength and shown large inelastic rotation capacities of no less than 0.06 rad. Besides, different types of the connection between the fuse and the beam segments were compared in the test, showing that the end plate connection is most convenient for the post-damage repair whereas the bolted web connection can bear the largest residual deformation. These experimental studies demonstrate the advantages of RCBs compared to conventional coupling beams. However, previous experimental studies have focused on component performance without a comprehensive discussion of global performance of the structural system.

On the other hand, much research uses numerical methods to study the global seismic performance of new shear walls. Wang et al. develops a type of RCB in which the middle part of the coupling beam is replaced by a shape memory alloy (SMA) damper [17]. Two groups of SMA wires which can dissipate energy by inelastic tensional deformation are assembled on the damper. The relative flexural deformation of the wall limbs is transferred to the ends of coupling beams and then to the SMA dampers. After earthquakes, the deformation of the damper can recover automatically because of the pseudo elasticity of austenite SMA material. Besides, an elastic-plastic time history analysis of a 12-story frame-shear wall structure with SMA damper is carried out to illustrate the performance of the proposed coupling beam. Kurama proposes to connect the steel coupling beam and the wall-piers with prestressed steel strand and install steel angles at the joints [18,19]. Nonlinear behavior of the proposed coupled wall is studied through numerical simulation, proving that when there is relative deformation between the wall-piers and the beam, the steel angles which can be replaced after the earthquake, will suffer severe plastic deformation to dissipate energy. In addition, a nonlinear numerical analysis of a hybrid coupled wall system (HCW) with replaceable steel coupling beams versus traditional RC coupling beams under seismic excitations is carried out by Ji et al. [20], showing that HCW can adequately meet code defined objectives in terms of global and component behavior under service level earthquakes. Under extreme events, the interstory drifts and beam rotations of HCW are much lower than the conventional coupled walls. Recently, Shahrooz et al. [21] have conducted nonlinear static and dynamic analyses of a 20-story prototype building with RCBs. The results show that: (1) the mid-span fuses are the primary energy-dissipating components and yield first; (2) the wall-piers experience little or no damage under design ground motions; (3) residual deformations are small which facilitates the replacement of the fuses. These numerical studies illustrate the benefits of the coupled shear wall with RCBs over the conventional coupled wall system, but lack the support of experimental data.

Although previous studies have done a lot of work on RCBs, there are few experimental studies of the structural system to validate the global performance of HSW. In this paper, the cyclic reversal tests and elaborate finite element simulation of three types of HSWs (F1SW/F2SW/F3SW) and a conventional coupled shear wall (CSW) are conducted. The global seismic behaviors of HSWs and CSW are compared. The authors must mention that in a former article, Lu et al. [22] have presented the study of F1SW. The practical design method of RCB and the hysteretic performance of Fuse 1 are first investigated. Then, the cyclic reversal test of F1SW and the numerical simulation by OpenSEES software are presented to show the global performance of the wall. In addition, the comparative study of F1SW and CSW are also briefly presented in review articles [8,23]. However, this paper presents the experimental and numerical study of other two types of HSWs (F2SW/F3SW), which have different RCBs, to illustrate the performance of HSWs more comprehensively. Furthermore, the concrete damage is taken into account in the finite element simulation conducted by ABAQUS software, to reveal the damage of HSWs. This paper also compares the similarities and differences between the three HSWs in seismic performance. The issues on how different configurations of "fuses" influence the global structural system's performance are discussed and suggestions for conceptual design are put forward.

2. Experimental Setup and Instrumentation

2.1. Experimental Specimens

Specimens include a conventional reinforced concrete coupled shear wall (CSW) and three hybrid coupled shear walls (HSWs) with different RCBs (F1SW/F2SW/F3SW), which are designed based on Chinese design codes [24]. The scale of the specimens is 1/2. The dimensions and reinforcement layouts of the four specimens are the same except the coupling beams (shown in Figure 1). The concrete grade of the shear wall is C25; the longitudinal reinforcements of the concealed column adopt HRB335 grade steel bars; the longitudinal reinforcements of the coupling beams, the stirrups of the beams and concealed columns, and the distributing bars of the shear wall adopt HPB235 grade steel bars. The steel and concrete grades are regulated by Chinese code [25].

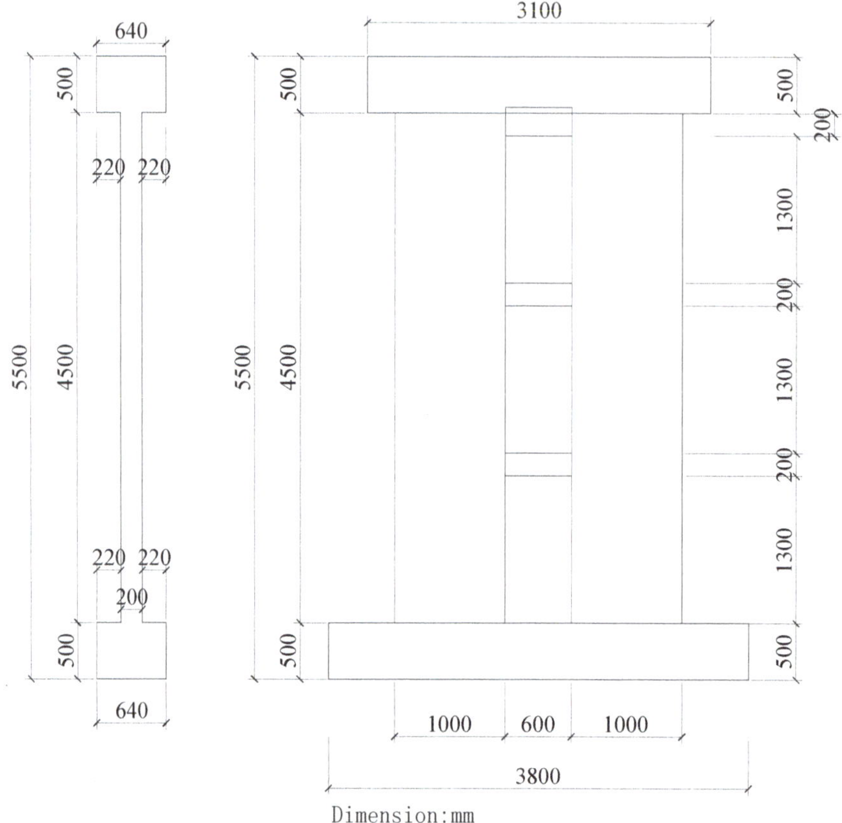

Figure 1. Dimensions of the specimens.

The RCB consists of two non-yield sections embedded in the wall-piers and a replaceable "fuse". To make the "fuse" replaceable, the "fuse" is connected to the embedded steel by end plates and bolts, shown in Figure 2b. The design method of the RCB is discussed in former study [22]. The bearing capacities of the beams are presented in Table 1 [22]. Fuse 1 is an I-shape steel with a rhombic opening at the web. The opening can enlarge the yield range of the web to enhance the energy dissipation capacity. Fuse 2 is a double-web I-shape steel with lead filled in the gap between the two webs. Fuse 3 consists of two parallel steel tubes filled by lead. After being installed at the mid-span of the coupling beam, Fuse 1 and Fuse 2 are designed to suffer shear failure while Fuse 3 are designed to suffer bending failure. The lead in Fuse 2 and Fuse 3 can dissipate energy and prevent the steel plates or tubes from buckling. The details of the coupling beams are presented in Figure 2.

Table 1. Design bearing capacity of the specimens.

	Design Shear Force	Design Bending Moment
Conventional coupling beam	16.053 kN	4.816 kN · m
Replaceable coupling beam	16.053 kN	1.926 kN · m

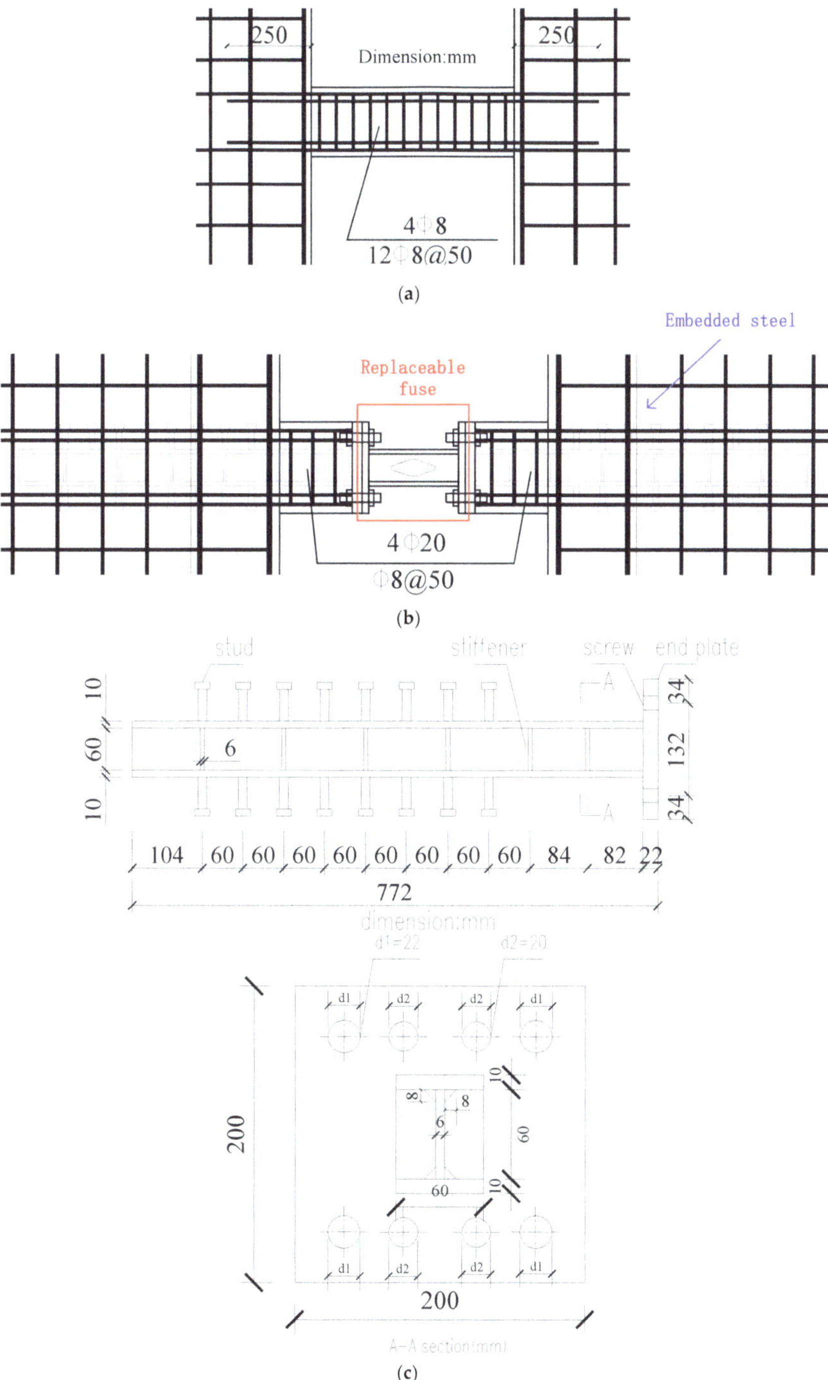

Figure 2. *Cont.*

(**d**)

Figure 2. Details of the beams: (**a**) conventional coupling beam; (**b**) replaceable coupling beam; (**c**) embedded steel; (**d**) pictures of three types of fuses.

2.2. Experimental Setup

The experimental setup is shown in Figure 3. The bottom of the shear wall is fixed on the foundation while the top is free. The vertical force is preloaded by prestresses pull rods and the lateral force is loaded by the actuator. The prestressed pull rod is anchored by bolts and steel plates at the base, so the prestressed rods can be rotated slightly during the repeated loading process. The preloaded vertical load is 1200 kN. The loading process of the lateral force is controlled by load before yielding. The initial loading value is 25% of the estimated yield load. When reaching 75% of the estimated yield load, the gap between each load level is reduced. Please note that each level of load cycles once. After yielding, the process is controlled by displacement. The displacement increases step by step, and the step length is equal to the final displacement of the load control period. Each level of displacement cycles 3 times until the horizontal bearing capacity drops to 85% of the maximum value or the specimen cannot withstand the predetermined axial pressure. The specific loading history is shown in Figure 3b.

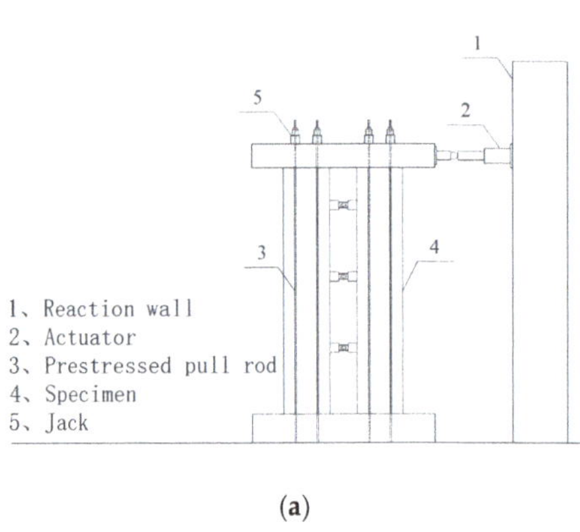

(**a**)

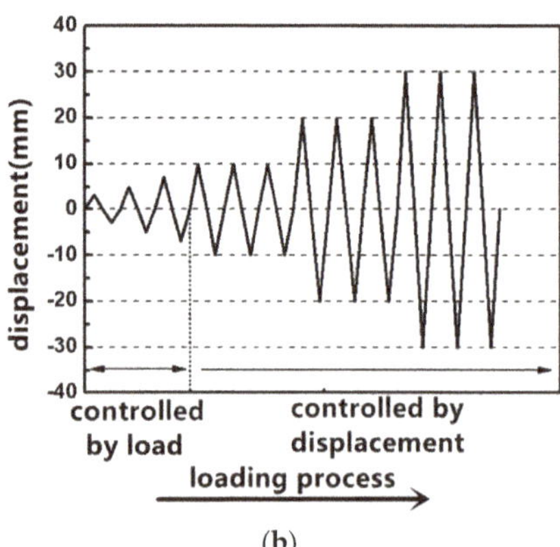

(**b**)

Figure 3. *Cont.*

(c)

Figure 3. Experimental setup: (**a**) loading device; (**b**) loading history; (**c**) test configuration.

3. Experimental Results and Discussion

3.1. Experimental Observations

The failure modes of the coupling beams and the wall-piers under cyclic reversal forces were introduced as below.

3.1.1. Failure Mode of Coupling Beams

As for CSW, when the lateral force achieved the calculated yield force, vertical bending cracks formed at the beam ends of the first and second floor at the same time. The cracks continued to extend, and then connected. Furthermore, the concrete of the beam ends was crushed on the first floor. Ultimately, except the beam ends, the coupling beam had almost no damage. The failure mode was typical bending failure with plastic hinges, shown in Figure 4a. As for RCBs, the coupling beams of F1SW, F2SW, and F3SW had quite similar failure modes. Some micro cracks firstly appeared at the non-yield sections and beam ends. However, those cracks did not extend visibly any more. The ultimate failure was caused by the destruction of the "fuses". Taking F1SW as an example, the final cracking status is shown in Figure 4b. In general, coupled walls with RCBs achieved the design intention. The non-yield sections and beam ends were generally not-damaged, with very small residual deformations. The plastic deformation was concentrated at the "fuse", which was easy to be removed after loading, as shown in Figure 4c.

Probably due to eccentrically loading caused by installation and manufacturing error, F3SW appeared out-of-plane buckling during the loading process. To avoid further buckling, after that F3SW was loaded in one direction, which means the specimen was repeatedly loaded to the maximum displacement and then unloaded until it failed. While the other specimens were repeatedly loaded to the maximum displacement, then unloaded, and then reversely loaded to the maximum displacement.

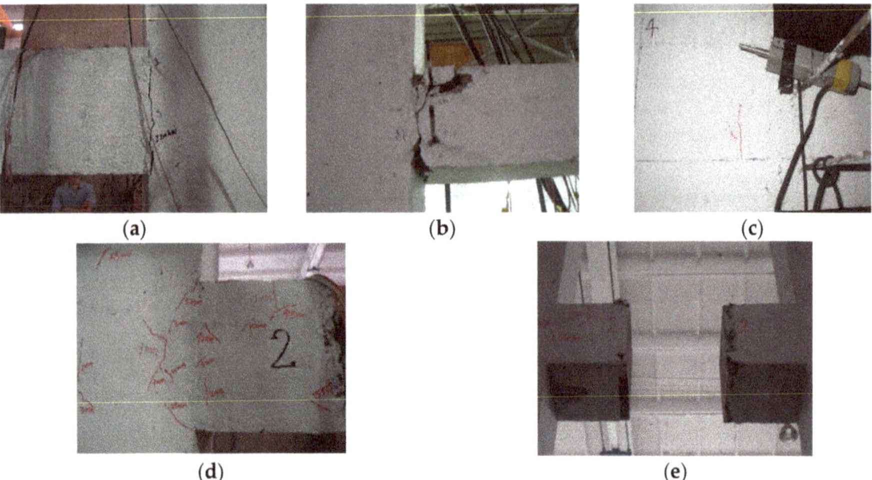

(a) (b) (c)

(d) (e)

Figure 4. Failure mode of coupling beams: (**a**) initial cracking of CSW; (**b**) ultimate state of CSW; (**c**) initial cracking of F1SW; (**d**) ultimate state of F1SW; (**e**) remove the "fuse".

Although the failure modes of the three RCBs were similar, the yielding mechanisms of the three "fuses" were different, shown in Figure 5. The cracks of the fuses almost occurred at the place where the stress was concentrated, such as the corner of the opening and the weld seam. As for Fuse 1, when the top displacement reached 60mm, cracks formed at the web around the rhombic opening. The cracks first appeared at the corners of the opening due to stress concentration. Then, the cracks continued to extend to the flanges until the web was torn. As for Fuse 2, when the top displacement reached 80mm, cracks first formed at the intersections of the stiffeners and the web probably because of the stress concentration induced by welding. With the increase in displacement, the web near the stiffeners was finally torn. As for Fuse 3, cracks appeared at the tube ends and led to the failure. The difference is that Fuse 3 suffered typical bending failure while Fuse 1 and Fuse 2 suffered shear failure.

(a)

Figure 5. *Cont.*

(b)

(c)

Figure 5. Failure mode of "fuses": (**a**) Fuse 1; (**b**) Fuse 2; (**c**) Fuse 3.

3.1.2. Failure Mode of Wall-Piers

The failure modes of the wall-piers are basically the same, hence the following description is commonly applicable for all specimens. The damages of four specimens are shown in Figure 6. To better describe the experimental phenomena, the wall is divided into the outer side and the inner side as shown in Figure 6a.

At the beginning, small horizontal cracks appeared at the bottom of the outer side, and then those cracks extended obliquely downward. With the increase in loading displacement, some parallel horizontal cracks gradually formed at the outer side. After the coupling beams yielded, horizontal cracks also started to form at the inner side. Those cracks extended obliquely downward as well, and then intersected with the cracks from the outer side. As for CSW, eventually, a concealed column was crushed. Two longitudinal bars of the concealed column were broken while the other two were completely buckled. In addition, the four longitudinal distributing bars near the concealed column were also completely buckled and the concrete was severely crushed. The horizontal length of the crushed part was about 1/2 of the wall-pier's width. Compared to the outer side, concrete at the inner side was crushed very slightly and the steel bars were not exposed. As for HSWs, the wall-piers had quite similar phenomena to that of CSW, with a ductile failure mode. However, compared to CSW, the damages of HSWs were much lighter.

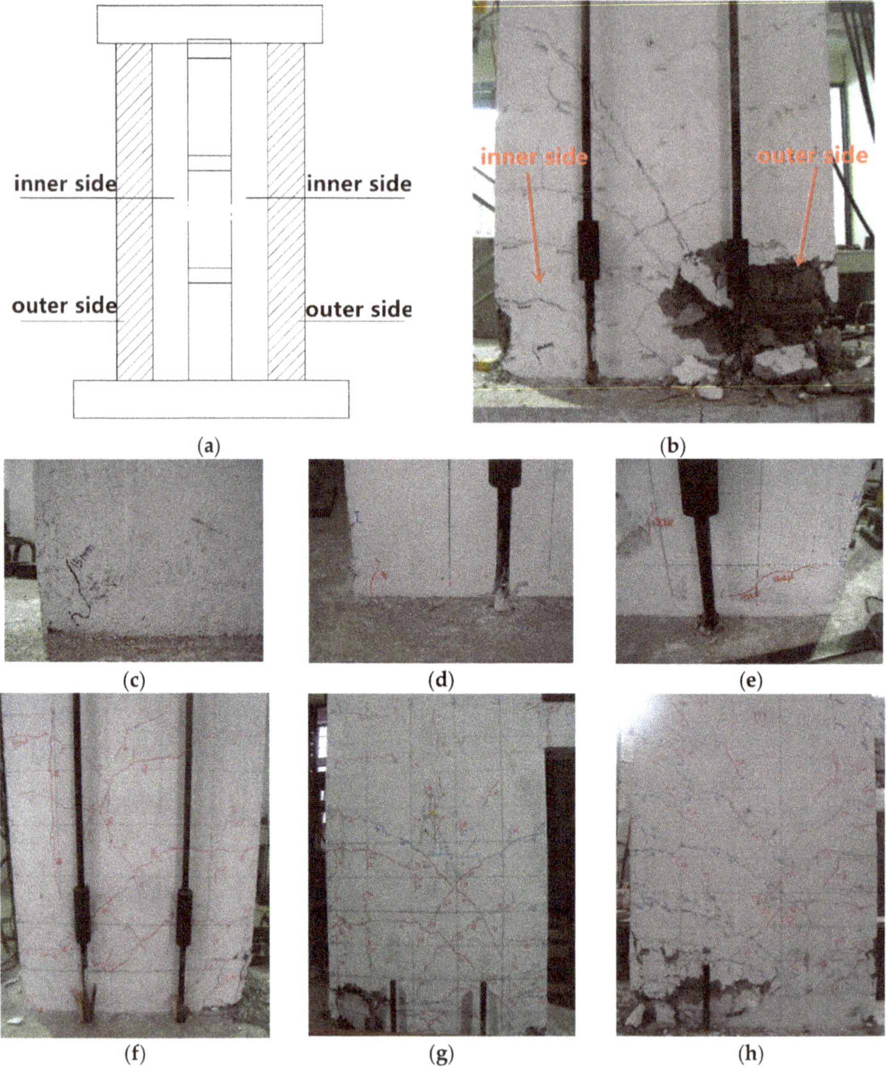

Figure 6. Failure mode of wall-piers: (**a**) two sides of the wall-pier; (**b**) CSW's failure mode; (**c**) initial cracking of F1SW; (**d**) initial cracking of F2SW; (**e**) initial cracking of F3SW; (**f**) F1SW's failure mode; (**g**) F2SW's failure mode; (**h**) F3SW's failure mode.

3.2. Hysteretic Curve and Skeleton Curve

The lateral force and the displacement of the loading point are measured. The hysteretic curves and skeleton curves of the four specimens are drawn in Figures 7 and 8. Because the experiment of CSW is first carried out and the anti-slip measures on the base are not taken in this loading process, the hysteretic curve and skeleton curve of CSW are not symmetrical. After that, anti-slip measures are adopted for the other remaining tests. As for F3SW, Figure 7d depicts the curve before the out-of-plane buckling.

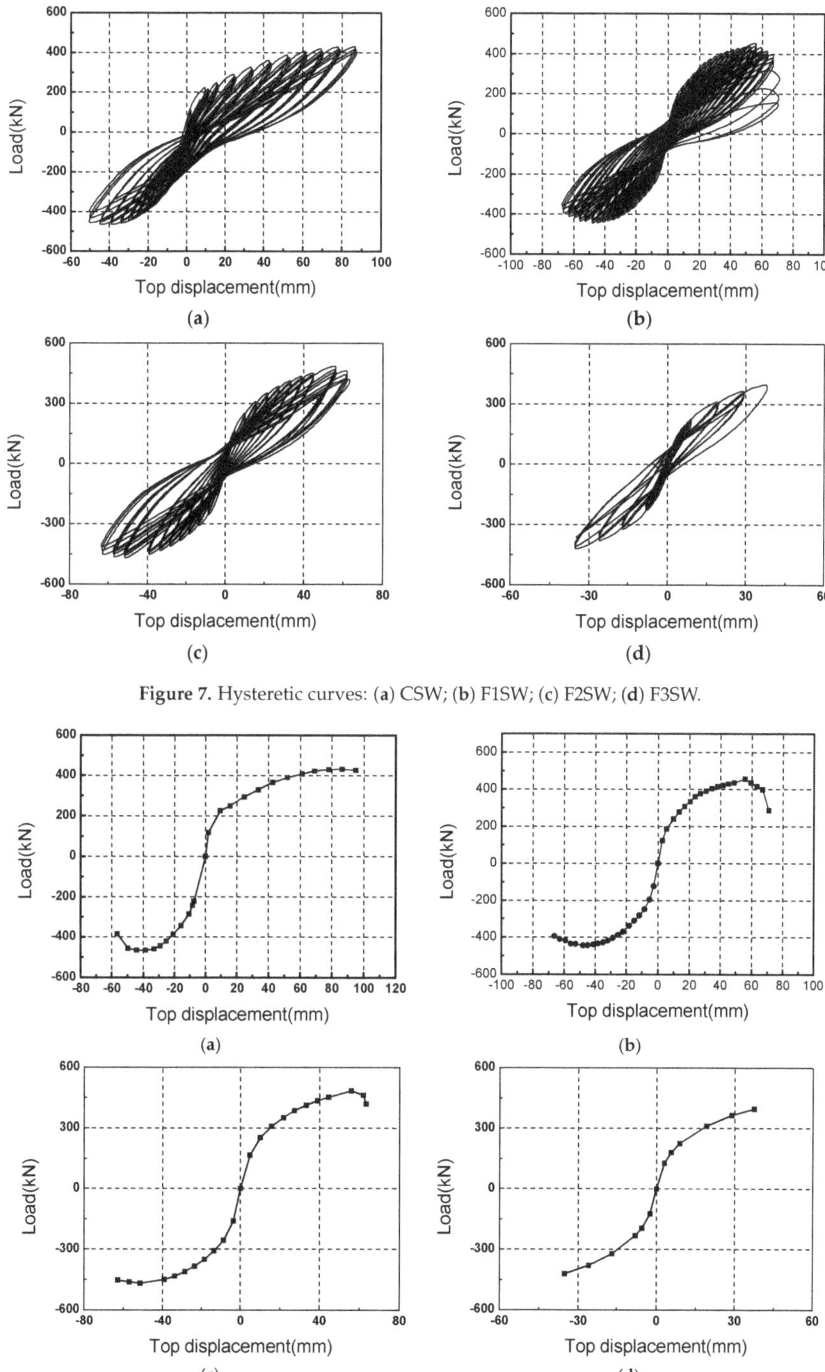

Figure 7. Hysteretic curves: (**a**) CSW; (**b**) F1SW; (**c**) F2SW; (**d**) F3SW.

Figure 8. Skeleton curves: (**a**) CSW; (**b**) F1SW; (**c**) F2SW; (**d**) F3SW.

To make a comprehensive comparison of four specimens in terms of bearing capacity, yield displacement, ultimate displacement, etc., Table 2 summarizes the relevant experimental results. The yield displacement is determined by the method put forward by Park [26]. The ultimate bearing capacity equals the damage load or the load when the maximum bearing capacity decreases by 15%. The ductility ratio is the ratio of ultimate deformation to initial yield deformation.

Table 2. Experimental data.

Specimen	CSW		F1SW		F2SW		F3SW	
Loading direction (+: Positive; −: Negative)	+	−	+	−	+	−	+	−
Yield displacement (mm)	34.07	−24.73	26.36	−20.97	28.57	−22.41	20.30	−23.70
Yield load (kN)	351.69	−376.30	374.16	−358.65	393.71	−378.88	317.12	−362.70
Peak displacement (mm)	73.75	−55.10	55.75	−48.06	55.84	−51.63	—	−54.27
Peak load (kN)	431.46	−463.82	455.06	−444.27	484.04	−468.54	—	−453.03
Ultimate displacement (mm)	81.80	−77.43	67.29	−66.46	63.27	−62.80	—	−81.73
Ultimate load (kN)	426.74	−388.99	386.80	−393.71	419.32	−453.03	—	−429.44
Ductility ratio	2.38	3.13	2.55	3.17	2.21	2.80	—	3.45

All specimens have similar strength. The yield loads of F1SW, F2SW, and F3SW differ from that of CSW by 0.66%, 6.13% and −6.59%, respectively. At the same time, the peak loads of F1SW, F2SW, and F3SW differ from that of CSW by 0.45%, 6.4% and 1.2%, respectively. In general, the bearing capacities of HSWs are similar to that of CSW.

The ductility ratios of four specimens are also calculated, shown in Table 2. Considering the average ratios of positive and negative directions, the ductility ratio of CSW is 2.75 while those of F1SW, F2SW, and F3SW are 2.86, 2.51 and 3.45. The ratios of F1SW, F2SW, and F3SW differ from that of CSW by 4.00%, −8.73% and 25.45%, respectively. In general, HSW also has desirable ductility, which is very important for earthquake resilience. Among the three HSWs, F3SW possesses the best ductility. However, because F3SW is only loaded in the negative direction after the out-of-plane buckling, the experimental results of F3SW are not sufficiently comparable to that of other specimens. In addition to F3SW, the ductility of F1SW is also satisfactory. The reason is that the rhombic opening at the web of Fuse 1 can enhance the deformability of the fuse by enlarging the yield range of the web.

3.3. Strength Degradation and Stiffness Degradation

In the experiment, the actuator cycles 3 times at each displacement level. To fully evaluate the load retention capacity of the HSWs, the strength degradation ratio is calculated. The degradation ratio is equal to the ratio of the peak load of the third cycle to that of the first cycle. The secant stiffness is used to calculate the stiffness degradation ratio. The degradation curves of the four specimens are shown in Figures 9 and 10.

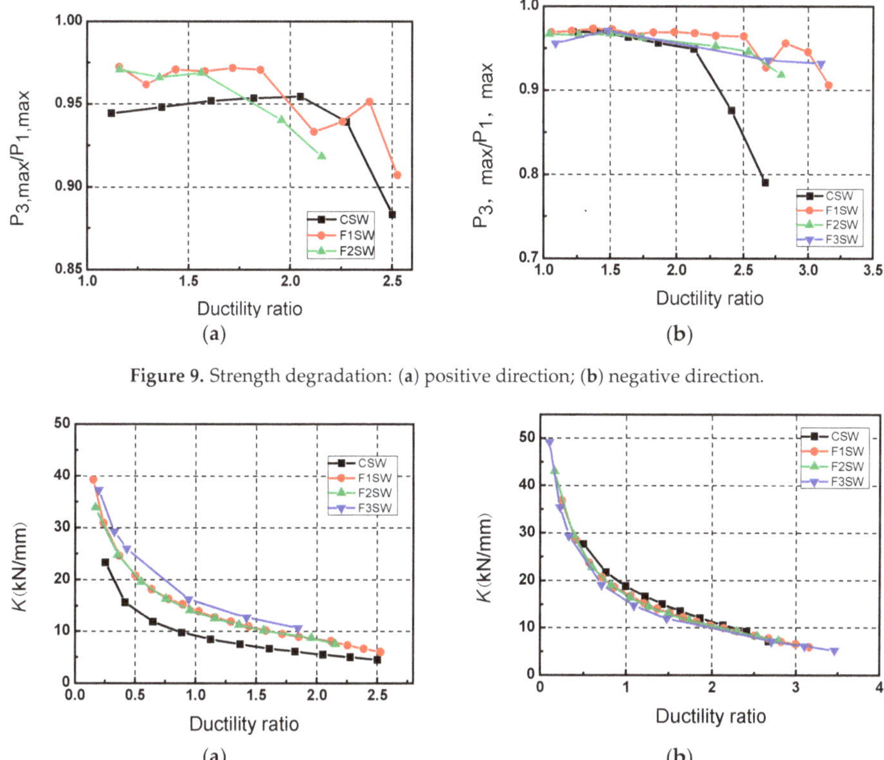

Figure 9. Strength degradation: (**a**) positive direction; (**b**) negative direction.

Figure 10. Stiffness degradation: (**a**) positive direction; (**b**) negative direction.

In the comparison of strength degradation in positive direction, F3SW is not considered because it is only tested in the negative direction after out-of-plane buckling. From Figure 9a, in the early stage, the strength degradation of F1SW and F2SW are lower than that of CSW. After that, the degradation of F1SW and F2SW are accelerated and exceed CSW temporarily. At last, the strength of CSW degrades rapidly and the degradation ratio eventually reaches 12% while those of F1SW and F2SW are 8.9% and 8%, respectively. When loading in the negative direction, the four curves are very close before the ductility ratio (displacement/yield displacement) reaches 2.14, with the degradation ratios less than 5%. However, CSW's strength degradation becomes severer after that, and ultimately reaches 21%. Compared to CSW, the degradation curves of HSWs are gentler and the ultimate degradation ratio are less than 10%. In general, the strength retention capacities of HSWs are close or even better than that of CSW. Among the three HSWs, it can be seen that the strength retention capacity of F1SW is relatively better. As for stiffness degradation, the degradation trends of the four specimens are generally the same. However, due to the foundation slip, CSW has some differences in initial stiffness.

3.4. Energy Dissipation Capacity

Equivalent viscous damping coefficient is adopted to evaluate the energy dissipation of the specimens. The calculation of the coefficient is shown in Figure 11a and Equation (1). The viscous damping coefficients of the four specimens are shown in Figure 11b. It can be seen that the equivalent viscous damping coefficients of the specimens are generally very close. Because the lead core in Fuse 2 can dissipate much energy and prevent the webs from buckling, the energy dissipation capacity of

F2SW is better. As for F3SW, the coefficient curve is incomplete because of the out-of-plane buckling. In general, HSWs have desirable energy dissipation capacities.

$$h_e = \frac{S_{ABC} + S_{CDA}}{2\pi(S_{OBE} + S_{ODF})} \tag{1}$$

where $S_{ABC} + S_{CDA}$ represents the area of the oval; S_{OBE} and S_{ODF} represent the area of the triangle *OBE* and *ODE*, respectively.

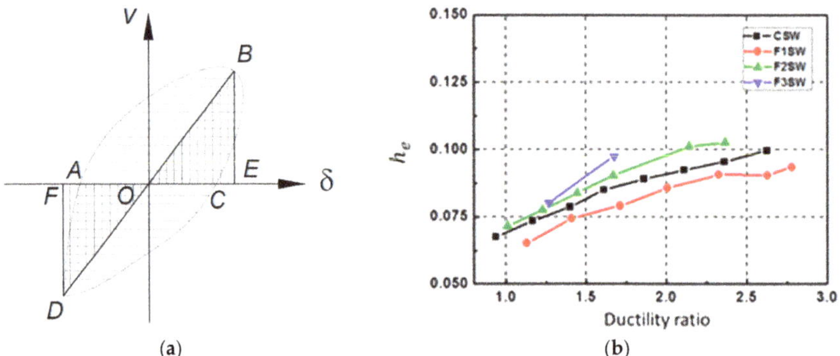

Figure 11. Energy dissipation capacity: (**a**) graphics for Equation (1); (**b**) equivalent viscous damping coefficient.

3.5. Strain Analysis

According to the design purpose of the RCB, the non-yield section of the beam should remain intact during the deformation process of the shear wall. The strain analysis can reveal the damage of the coupling shear walls. Hence, strain gauges are embedded in the concrete to examine the strains at critical locations. By analyzing the strain of the embedded steel and reinforcements, it is shown that the non-yield section does not yield although there are some micro cracks in the test. The strains of the longitudinal bars, the stirrups, and the embedded steel (flange and web) are quite small. The absolute values of the strains for F1SW, F2SW, and F3SW are less than 500×10^{-6}, 1200×10^{-6} and 300×10^{-6} respectively, which are much lower than the yield strains.

In addition, the strains of the longitudinal reinforcements of the concealed columns are also measured. All specimens have plastic hinges at the wall foots. As for CSW, the plastic area extends to the second floor. However, as for HSWs, the plastic area is only within the first floor and the longitudinal reinforcements on the second floor do not yield. Because the "fuses" can dissipate much energy, the wall foots of HSWs do not dissipate as much energy as CSW does. The strain analysis shows that the damage area of wall-piers for HSWs is smaller than that of CSW, which is consistent with the experimental phenomena.

3.6. Discussions and Suggestions

The following discussions are based on the experimental phenomena and test data. In addition, some preliminary suggestions for the conceptual design of HSWs with RCBs are proposed. In general, this experiment proves that although the three fuses are different in working mechanism, the three HSWs exhibit similar global performance in terms of failure mode, hysteretic behavior, etc. However, the three HSWS still have some differences in specific properties.

As for F1SW, Fuse 1 is an I-shape steel with a rhombic opening at the web. Compared to CSW, although the equivalent viscous damping coefficient of F1SW is slightly smaller, the coupling beams of F1SW are still considered to be good at dissipating energy considering that the wall-piers of F1SW dissipate much less energy. In addition, the ductility and load retention capacity of F1SW are also desirable.

Considering the simple construction of Fuse 1, this type of RCBs can be widely used in engineering. Besides, if the opening at the web is more rounded to reduce the stress concentration, the anti-cracking ability of Fuse 1 will be enhanced.

As for F2SW, Fuse 2 is a double-web I-shape steel and the gap between the two webs is filled by lead to dissipate more energy. Compared to CSW, the stress of the longitudinal reinforcement of the concealed column is smaller and the plastic area is also smaller, which indicates that the energy dissipated by the wall-piers of F2SW is less. However, the equivalent viscous damping coefficient of F2SW is even larger, which proves that the RCBs of F2SW can dissipate much energy. This proves that installing components with high energy dissipation capacity at mid-span of the coupling beam is practical to improve the energy dissipation capacity of the coupled wall system. In addition, because the cracks of Fuse 2 almost formed at the intersections of the stiffeners and web due to stress concentration, improving the welding quality of the stiffeners might enhance the ductility of F2SW.

As for F3SW, due to the out-of-plane buckling, the experimental data is incomplete in positive direction. Although the remaining data is not sufficiently comparable to that of other specimens, it still reflects the performance of F3SW. The failure mode is similar to that of F1SW and F2SW, and the ductility and load retention ability in the negative direction are satisfactory. In addition, F3SW possesses the best energy dissipation capacity before the out-of-plane buckling. The experiment indicates F3SW still has a good researching prospect. In the further research, improvements such as adding longitudinal stiffeners at the fuse should be taken to prevent out-of-plane buckling.

4. Finite Element Implementation

To further study the seismic behavior of HSWs with RCBs, finite element models are established by ABAQUS software. The same loading conditions as the experiment are applied. The deformations and damage distribution are properly simulated. The skeleton curves of the four specimens are also obtained, which are compared to the experimental results.

4.1. Modeling Method

The wall-piers and coupling beams are modeled by 8-node reduced integrated solid element C3D8R [27]. C3D8R has one less integration point in each direction compared to the fully integrated element. Hence, even if torsional deformation appears, the accuracy of the analysis will not be greatly affected. In addition, this element is not prone to shear locking under bending moment. The three "fuses" and the embedded steel are also modeled by C3D8R. The slip between the lead core and steel plate in Fuse 2 and Fuse 3 is neglected. Furthermore, the reinforcements are modeled by the truss element T3D2. The reinforcements and embedded steel members are embedded into the concrete elements using the command *Embedded Element, neglecting the slip between steel and concrete.

In the simulation, the strengths of the steel and concrete are consistent with the measured values in the experiment, shown in Table 3. The constitutive model of the steel is isotropic hardening model [28] while that of the concrete is plastic-damage model [29]. In ABAQUS software, Concrete Damaged Plasticity model is chosen. Command *Concrete Damaged Plasticity defines some critical parameters of the concrete model shown in Table 4.

Table 3. Strengths of the steel and concrete.

Steel Grade	Diameter (mm)	Yield Strength (MPa)	Ultimate Strength (Mpa)
HPB235	6	330	405
HPB235	8	340	400
HRB335	12	400	530
HRB335	20	355	500

Concrete Grade	Cubic Compressive Strength f_{cu} (Mpa)	Elastic Modulus E (Mpa)
C25	21.16	21375

Note: The steel and concrete grades are regulated by Chinese code [25].

Table 4. Parameters of the concrete model

$\psi(°)$	$\in$	α_f	K_c	μ
32	0.1	1.16	0.6667	0.003

Notes: ψ: dilatancy angle; $\in$: plastic potential eccentricity; μ: viscosity coefficient; α_f: the ratio of biaxial compressive strength to uniaxial compressive strength; K_c: the ratio of the second stress invariant on tensile meridian plane to that on compressive meridian plane;

The uniaxial compressive stress-strain relationship of the concrete is defined by Hognestad model [30]. Furthermore, the tensile stress-strain relationship of the concrete is assumed to be linear elastic before the peak stress. After that, post-peak tensile model given by Reinhardt and Cornelissen is applied [31]:

$$\frac{\sigma}{f_t} = \left[1 + c_1 \left(\frac{\varepsilon}{\varepsilon_{tu}}\right)^4\right] e^{-c_2 \frac{\varepsilon}{\varepsilon_{tu}}} \tag{2}$$

where f_t represents the tensile strength; ε_{tu} represents the ultimate tensile strain; $c_1 = 9$ and $c_2 = 5$.

According to previous research [32], the compressive and tensile damage factors are obtained by equations as follows:

$$\beta_c = \varepsilon_c^{pl} / \varepsilon_c^{in} \tag{3}$$

$$\varepsilon_c^{in} = \varepsilon_c - \sigma_c E_0^{-1} \tag{4}$$

$$d_c = \frac{(1 - \beta_c)\varepsilon_c^{in} E_0}{\sigma_c + (1 - \beta_c)\varepsilon_c^{in} E_0} \tag{5}$$

$$\beta_t = \varepsilon_t^{pl} / \varepsilon_t^{in} \tag{6}$$

$$\varepsilon_t^{in} = \varepsilon_t - \sigma_t E_0^{-1} \tag{7}$$

$$d_t = \frac{(1 - \beta_t)\varepsilon_t^{in} E_0}{\sigma_t + (1 - \beta_t)\varepsilon_t^{in} E_0} \tag{8}$$

where ε_c and ε_t represent the compressive and tensile strain while σ_c and σ_t represent the corresponding stress; ε_c^{pl} and ε_t^{pl} represent the compressive and tensile plastic strain; ε_c^{in} and ε_t^{in} represent the compressive and tensile inelastic strain (considering concrete damage, $\varepsilon^{pl} \neq \varepsilon^{in}$); E_0 represents the initial elastic modulus; d_c and d_t represent the compressive and tensile damage factors, respectively.

4.2. Simulated Deformation

The simulated deformations of the four specimens are obtained. The coupling beams of CSW suffer typical bending deformation, which is identical to the experimental phenomenon that plastic hinges form at the ends of the coupling beam. As for F1SW, F2SW, and F3SW, the deformations of the coupling beams are concentrated at the "fuses". The non-yield sections have almost no deformation and remain vertical to the wall-piers. The deformations are shown in Figure 12. For more obvious demonstration, the deformations are magnified three times in this figure.

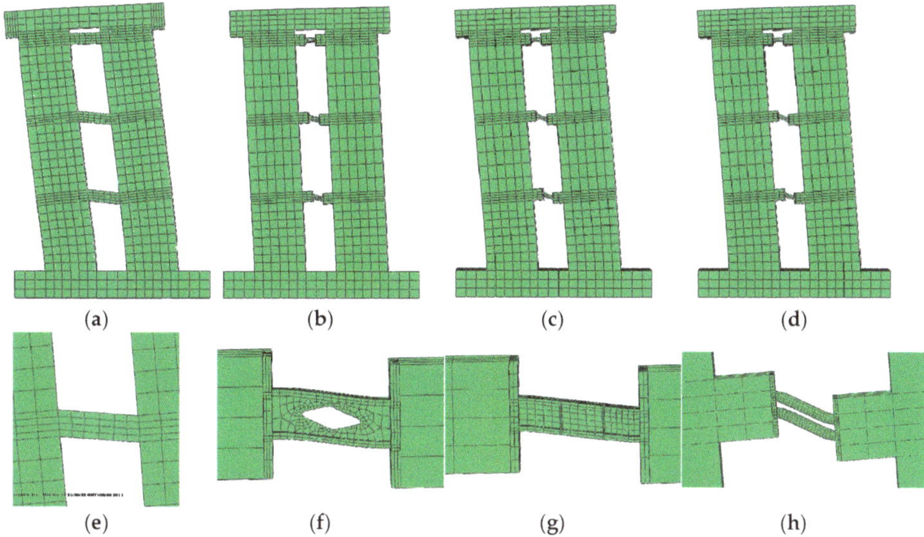

Figure 12. Simulated deformations for: (**a**) CSW; (**b**) F1SW; (**c**) F2SW; (**d**) F3SW; (**e**) the coupling beam of CSW; (**f**) the "fuse" of F1SW; (**g**) the "fuse" of F2SW; (**h**) the "fuse" of F3SW.

4.3. Yield Sequence

By observing the equivalent plastic strains of the reinforcements and fuses at different loading steps, the yield sequence of different parts of the shear wall is determined. As for F1SW, the loading process includes 307 steps. At step 7, the "fuses" of the first and second floor first yield while the longitudinal reinforcements at wall foots remain elastic. At step 20, the longitudinal reinforcements at the wall foots start to yield. As for F2SW and F3SW, the yield sequences of different parts are the same as F1SW in that the "fuses" yield first and then the longitudinal reinforcements inside the wall yield. The ideal yield sequence is achieved. The final equivalent plastic strain diagrams of the RCBs are shown in Figure 13. The plastic strains of the "fuses" are much larger than those of the longitudinal reinforcement, stirrups and embedded steels, which are consistent with the experimental phenomena.

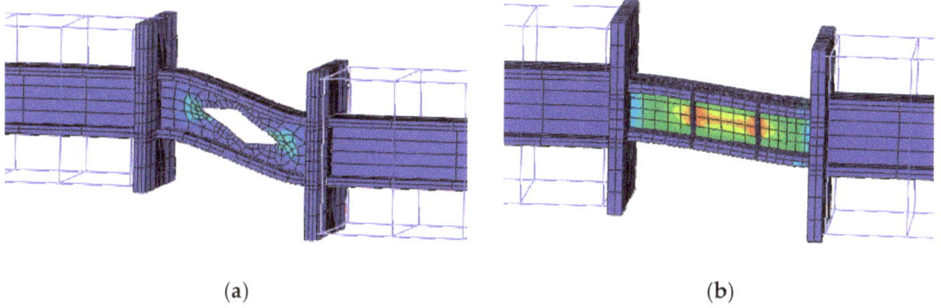

Figure 13. *Cont.*

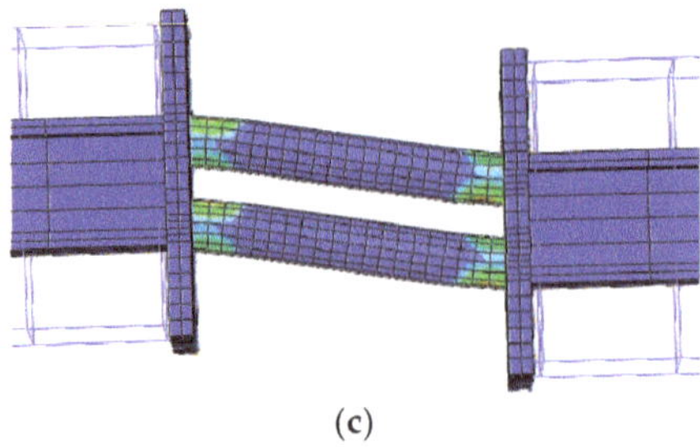

(c)

Figure 13. Final equivalent plastic strain: (**a**) F1SW; (**b**) F2SW; (**c**) F3SW.

4.4. Concrete Damage Analysis

By extracting the tensile damage factor d_t, the cracking of the concrete can be visually reflected, shown in Figure 14.

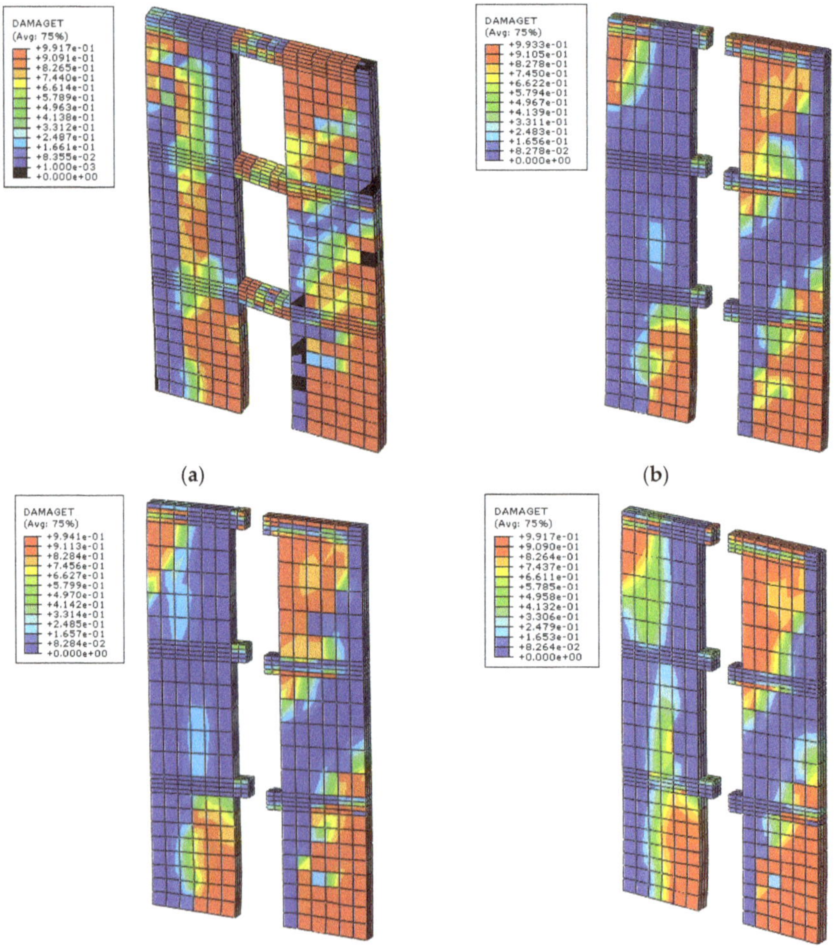

(a)

(b)

(c)

(d)

Figure 14. Damage condition of: (**a**) CSW; (**b**) F1SW; (**c**) F2SW; (**d**) F3SW.

From Figure 14a, it can be seen that the tensile parts of the wall-piers and coupling beams are suffering severe damages. The damage area has extended to the second floor and the top of the wall also has severe damages due to stress concentration. The damage condition is consistent with the experimental phenomenon. Compared to CSW, HSWs are much less damaged at the coupling beams. Although the tensile parts of the wall-piers also have visible damages, the damage areas of the three HSWs are smaller and do not extend to the second floor. In general, the damage distribution of the finite element simulation agrees well with the experimental results discussed in Section 3.5.

4.5. Skeleton Curve

The simulated and experimental skeleton curves of the four specimens are shown in Figure 15. It can be seen that the simulated peak bearing capacities of the four specimens all agree well with the experimental results, with the maximum deviation less than 5%. In addition, the simulated curves and the experimental curves are very consistent in terms of initial stiffness as well. However, some differences between the simulated and experimental yield bearing capacities are still noticed. As for CSW, the simulated yield bearing capacity is visibly larger than the experimental result especially in the positive direction, which is probably due to the base slip in the experiment. As for F1SW and F2SW, the simulated yield bearing capacities are also slightly larger than experimental results, but the deviations are acceptable. As for F3SW, the skeleton curve of is incomplete in positive direction due to out-of-plane buckling but the simulated curve is close to the experimental curve in negative direction.

In conclusion, the skeleton curves of HSWs can be well simulated by ABAQUS software. However, the descending stage of the curve is hard to simulate. It is very difficult to simulate the hysteretic response of reinforced concrete low shear wall with ABAQUS. Firstly, it is difficult to converge, and secondly, the pinching effect of hysteretic curve is hard to simulate. Hence, in Ref [22], the authors apply OpenSEES software to simulate the hysteretic curves, which are in good agreement with the experimental curves.

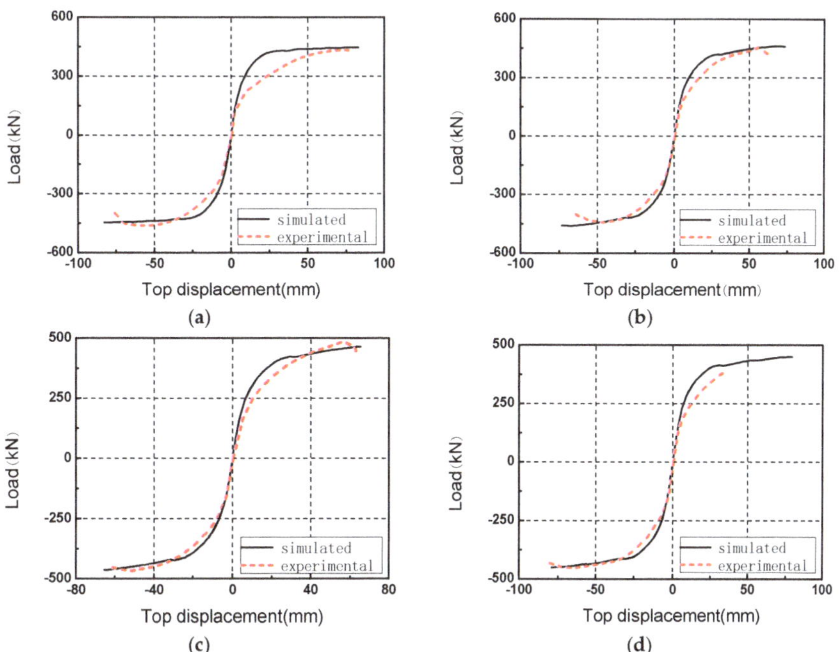

Figure 15. Simulated and experimental skeleton curves. (**a**) CSW; (**b**) F1SW; (**c**) F2SW; (**d**) F3SW.

5. Conclusions

This paper compares the seismic performance of HSWs with RCBs and a concrete shear wall (CSW) in terms of bearing capacity, deformability, energy consumption, and damage distribution, etc. Based on the results of cyclic reversal test and finite element analysis, the following conclusions are drawn.

(1)　Although the three fuses are different in working mechanism, the three HSWs exhibit similar global seismic performance. Under cyclic lateral forces, the plastic deformations of conventional coupling beams are concentrated at the beam ends, which are also the most vulnerable to damage. As for the three HSWs, the deformations of the beams are concentrated at the "fuses", which can be easily replaced. The wall foots of CSW are severely crushed while those of HSWs are not.

(2)　It is proved that HSWs meet the design requirements as CSW, with many desirable properties. The strength, ductility, energy dissipation capacity, and stiffness degradation rules of HSWs are similar to those of CSW. At the same time, the load retention capacity of HSWs are also close or even better than that of CSW.

(3)　Strain analysis shows that the plastic area of the wall-piers of CSW extends to the second floor, causing widespread damage. As for HSW, the plastic area is only within the first floor. At the same time, the embedded steel, and longitudinal steel bars of the non-yield sections are far from yielding.

(4)　The deformation, yield sequence, hysteretic behavior, and damage distribution of HSWs can be well simulated in ABAQUS software. In addition, nonlinear finite element analysis proves that HSWs have ideal yield sequence that the "fuses" yield earlier than the wall foots. More importantly, numerical simulation verifies that the damage of the RCBs is concentrated at the "fuses" and the concrete damage of HSWs is visibly alleviated.

(5)　Among the three HSWs, F1SW is better in ductility and load retention ability, while F2SW is better in energy consumption. In general, both F1SW and F2SW have good application prospects, with desirable seismic properties. The cracks of Fuse 1 and 2 first appear at the place where the stress is concentrated. If the stress concentration can be alleviated by optimizing the opening angle and improving the welding quality, the anti-cracking capacity of Fuse 1 and 2 may be improved. As for F3SW, despite the out-of-plane buckling, it still has ideal failure mode and good seismic performance. In the further research, improvements such as adding longitudinal stiffeners at the fuse should be taken to prevent out-of-plane buckling. In the authors' opinion, F1SW is the easiest one to be designed and fabricated in engineering applications because the configuration of Fuse 1 is relatively simpler, hence engineers can easily simulate its mechanical behavior and further obtain the practical design methods. Compared to Fuse 1, the configurations of Fuse 2 and 3 are more complicated. The interaction mechanism between lead core and steel plate is complicated, which will need more works to solve it.

Author Contributions: Funding acquisition, Z.L.; Investigation, Y.C.; Project administration, Y.C. and Z.L.; Software, J.L. and Z.L.; Writing—original draft, J.L.; Writing—review and editing, Z.L.

Funding: Financial supports from the National Key Research and Development Program of China (2017YFC1500701) and the Program of Shanghai Academic Research Leader (18XD1403900) are highly appreciated. The financial supports from the National Natural Science Foundation of China (51408170), the Academic Innovation Plan of Hainan Science and Technology Association for Young Scientists (201601), Hainan Key R & D Program (ZDYF2016151), and the Midwest Key Areas Construction Project Plan of Hainan University under (ZXBJH-XK011) are also gratefully appreciated.

Conflicts of Interest: The authors declare no conflict of interest.

References

1.　Ang, H.S. Seismic Damage Analysis of Reinforced Concrete Buildings. *J. Struct. Eng.* **1985**, *111*, 740–757.

2.　Takeda, T. Reinforced Concrete response to simulated earthquakes. *J. Struct. Div. Proc. Am. Soc. Civil Eng.* **1970**, *96*, 2557–2573.

3. Lu, Z.; Chen, X.; Lu, X.; Yang, Z. Shaking table test and numerical simulation of an RC frame-core tube structure for earthquake-induced collapse. *Earthq. Eng. Struct. Dyn.* **2016**, *45*, 1537–1556. [CrossRef]

4. El-Tawil, S.; Kuenzli, C.; Hassan, M.; Kunnath, S. Inelastic Behavior of Hybrid Coupled Walls. *J. Struct. Eng.* **2004**, *130*, 285–296.

5. Aristizabal-Ocfaoa, J.D. Seismic Behavior of Slender Coupled Wall Systems. *J. Struct. Eng.* **1987**, *113*, 2221–2234. [CrossRef]

6. Paulay, T. Coupling Beams of Reinforced Concrete Shear Walls. *J. Struct. Div.* **1971**, *97*, 843–862.

7. Hindi, R.A.; Hassan, M.A. Shear capacity of diagonally reinforced coupling beams. *Eng. Struct.* **2004**, *26*, 1437–1446. [CrossRef]

8. Galano, L.; Vignoli, A. Seismic behavior of short coupling beams with different reinforcement layouts. *Struct. J.* **2000**, *97*, 876–885.

9. Ding, D.; Cao, Z.; Zhang, S. Experimental studies of new ductile coupling beams and multi-storey shear walls. *Mater. Struct.* **1997**, *30*, 566–573.

10. Parramontesinos, G.J.; Canbolat, B.A. Experimental Study on Seismic Behavior of High-Performance Fiber-Reinforced Cement Composite Coupling Beams. *Struct. J.* **2005**, *102*, 159–166.

11. Lu, X.; Mao, Y.; Chen, Y.; Liu, J.; Zhou, Y. New structural system for earthquake resilient design. *J. Earthq. Tsunami* **2013**, *7*, 1350013. [CrossRef]

12. Vargas, R.; Bruneau, M. Experimental Response of Buildings Designed with Metallic Structural Fuses. II. *J. Struct. Eng.* **2009**, *135*, 394–403. [CrossRef]

13. Vargas, R.; Bruneau, M. Analytical Response and Design of Buildings with Metallic Structural Fuses. I. *J. Struct. Eng.* **2009**, *135*, 386–393. [CrossRef]

14. Fortney, P.J.; Shahrooz, B.M.; Rassati, G.A. Large-Scale Testing of a Replaceable "Fuse" Steel Coupling Beam. *J. Struct. Eng.* **2007**, *133*, 1801–1807. [CrossRef]

15. Li, X.; Lv, H.L.; Zhang, G.C.; Ding, B.-D. Seismic behavior of replaceable steel truss coupling beams with buckling restrained webs. *J. Constr. Steel Res.* **2015**, *104*, 167–176. [CrossRef]

16. Ji, X.; Liu, D.; Sun, Y.; Hutt, C.M. Cyclic Behavior of Replaceable Steel Coupling Beams. *Earthq. Eng. Struct. Dyn.* **2017**, *46*, 517–535. [CrossRef]

17. Wang, L.; Mao, C.; Dong, J. Seismic performance of shear wall structure with shape memory alloy dampers in coupling beams. *World Earthq. Eng.* **2011**, *27*, 101–107.

18. Kurama, Y.C.; Shen, Q. Posttensioned Hybrid Coupled Walls under Lateral Loads. *J. Struct. Eng.* **2004**, *130*, 297–309. [CrossRef]

19. Kurama, Y.C.; Shen, Q. Nonlinear Behavior of Posttensioned Hybrid Coupled Wall Subassemblages. *J. Struct. Eng.* **2002**, *128*, 1290–1300.

20. Ji, X.; Liu, D.; Sun, Y.; Hutt, C.M. Seismic performance assessment of a hybrid coupled wall system with replaceable steel coupling beams versus traditional RC coupling beams. *Earthq. Eng. Struct. Dyn.* **2017**, *46*, 517–535. [CrossRef]

21. Shahrooz, B.M.; Fortney, P.J.; Harries, K.A. Steel Coupling Beams with a Replaceable Fuse. *J. Struct. Eng.* **2018**, *144*, 04017210. [CrossRef]

22. Lu, X.; Chen, Y.; Jiang, H. Earthquake resilience of reinforced concrete structural walls with replaceable "fuses". *J. Earthq. Eng.* **2016**, *22*, 801–825. [CrossRef]

23. Lu, X.; Jiang, H. Recent progress of seismic research on tall buildings in China Mainland. *Earthq. Eng. Eng. Vib.* **2014**, *13*, 47–61. [CrossRef]

24. GB50011. *Code for Seismic Design of Buildings in China*; China Architecture & Building Press: Beijing, China, 2010.

25. GB50010. *Code for Design of Concrete Structures*; China Architecture & Building Press: Beijing, China, 2015.

26. Park, R.; Paulay, T. *Reinforced Concrete Structures*; John Wiley & Sons.: New York, NY, USA, 1975.

27. Hibbit, Karlsson & Sorensen, Inc. *ABAQUS User's Manual*; Hibbit, Karlsson & Sorensen, Inc.: Johnston, RI, USA, 2018.

28. Håkansson, P.; Wallin, M.; Ristinmaa, M. Comparison of isotropic hardening and kinematic hardening in thermoplasticity. *Int. J. Plast.* **2005**, *21*, 1435–1460. [CrossRef]

29. Lubliner, J.; Oliver, J.; Oller, S.; Oñate, E. A plastic-damage model for concrete. *Int. J. Solids Struct.* **1989**, *25*, 299–326. [CrossRef]

30. Wight, J.K.; Macgregor, J.G. *Reinforced Concrete-Mechanics and Design*; Pearson Prentice Hall: Upper Saddle River, NJ, USA, 2009.

31. Reinhardt, H.W.; Cornelissen, H.A.W. Post-peak cyclic behaviour of concrete in uniaxial tensile and alternating tensile and compressive loading. *Cem. Concr. Res.* **1984**, *14*, 263–270. [CrossRef]
32. Zhang, J.; Wang, Q.; Hu, S. Parameters Verification of Concrete Damaged Plastic Model of ABAQUS. *Build. Struct.* **2008**, *38*, 127–130.

 *sustainability*

Article

Flow Analysis and Damage Assessment for Concrete Box Girder Based on Flow Characteristics

Xiong-Fei Ye [1,*], Kai-Chun Chang [2], Chul-Woo Kim [2,*], Harutoshi Ogai [1], Yoshinobu Oshima [3] and O.S. Luna Vera [2]

[1] Faculty of Science and Engineering, Waseda University, Tokyo 169-8050, Japan; ogai@waseda.jp
[2] Graduate School of Engineering, Kyoto University, Kyoto 615-8540, Japan;
    chang.kaichun.4z@kyoto-u.ac.jp (K.-C.C.); os.luna.vera@gmail.com (O.S.L.V.)
[3] Public Works Research Institute, Tsukuba 305-8516, Japan; y-ooshima@pwri.go.jp
* Correspondence: yexiongfei@gmail.com (X.-F.Y.); kim.chulwoo.5u@kyoto-u.ac.jp (C.-W.K.)

Received: 26 December 2018; Accepted: 21 January 2019; Published: 29 January 2019

Abstract: For a system such as the concrete structure, flow can be the dynamic field to describe the motion, interactions, or both in dynamic or static (Eulerian description) states. Further, various kinds of flow propagate through it from the very start to the end of its lifecycle (Lagrangian description) accompanied by rains, winds, earthquakes, and so forth. Meanwhile, damage may occur inside the structure synchronously, developing from micro- to macro-scale damage, and eventually destroy the structure. This study was conducted to clarify the content of flow which has been implicitly used in the damage detection, and to propose a flow analysis framework based on the combination data space and the theory of dissipative structure theory specifically for nondestructive examination in structural damage detection, which can theoretically standardize the mechanism by which flow characteristics vary, the motion of the structure, or the swarm behavior of substructures in engineering. In this paper, a destructive experiment (static loading experiment) and a following nondestructive experiment (impact hammer experiment) were conducted. According to the experimental data analysis, the changing of flow characteristics shows high sensitivity and efficient precision to distinguish the damage exacerbations in a structure. According to different levels of interaction (intensity) with the structure, the information flow can be divided into two categories: Destructive flow and nondestructive flow. The method used in this research is named as a method of "flow analysis based on flow characteristics", i.e., "FC-based flow analysis".

Keywords: flow; analysis; concrete; girder; damage; NDE

---

1. Introduction

From some damage surveys [1–3] in Japan and the United States of America, bridges are deteriorated or damaged by material aging, overload, fatigue, temperature, corrosion, rust, and disasters such as earthquakes, hurricanes, and so forth. In addition to the natural factors, various kinds of damage may occur for the following reasons: (a) Nonconservative design, (b) severe service environment, and (c) improper construction or operation management maintenance. No matter whether the damage is caused by nonhuman factors or human factors, the damage usually initiates at a microscopic level, grows and extends to a macroscopic level, and eventually causes the collapse of a structure when the damage is not detected in time. The detection and evaluation of the potential damage and minimization of the probability of structural failures are keen issues for many countries in the world.

There are many damage detection methods. They can be majorly categorized into the destructive examination and nondestructive examination (NDE) [4]. NDE is a group of damage detection methods that do not affect or harm the test material, component, or system, such as ultrasonic testing (UT), radiographic testing (RT), infrared thermography (IT), and acoustic emission (AE), among others. In the laboratory experiment, the NDE is often conducted to test the damage after some static loading experiment (one kind of destructive experiments, which would cause damage to the structure). If we view the above methods from a general point, many of the NDE methods share a common concept to show the change of structural status: "dynamic field [5]" or otherwise named as "flow". The dynamic field (flow) is a term defining a field with smooth uninterrupted movement or progress in physics, very commonly used in fluid mechanics nowadays, and also implicitly used in NDE methods; for example, the electromagnetic field in RT and the sound field in AE. Simultaneously, in the static loading experiment, there is a stress field [6] and crack-tip stress field [7] in the static stress–strain response.

According to Newton's third law, in the research of damage detection, a global system consists of three parts: The target structure, the environment (referring to everything outside the target structure with a limit domain), and interactions (often described as one kind of flow). In the common sense, an equilibrium state is the state in which the system state variables remain unchanged, as originated from thermodynamics and applied in various departments. The equilibrium state is also commonly studied in dissipative structure theory [8] (one special kind of system theory for but not only restricted to nonequilibrium thermodynamics describing the system "far from equilibrium state") and general system theory (the general system theory often concerns the system near "equilibrium state"), and exists in the water flow on dams and bridges, wind on buildings, rain on steel plants, and radiation in nuclear reactors, and so on; even for vibration in an ordinary concrete beam-like structure. For these systems, which can realize their new balance in self-organization [9] (a system can continuously reduce its entropy and improve its order by exchanging substance, energy, and information with the outside world), if the structures are nonlinear and nonstationary, there may be many equilibrium shifts (the process from one equilibrium to another, i.e., "the process from an equilibrium state to a nonequilibrium state, and then to another equilibrium state", or "from an equilibrium state to another equilibrium state directly in a short time"). In the global systems, the state variables describing the flow characteristics in multiple equilibrium shifts may dynamically fluctuate. For the reason that it is almost impossible to describe the evolutions of equilibrium and nonequilibrium states chemically and physically for these highly complex structures [10], it is difficult to obtain an analytical solution for the problems within the dynamic field (for example, turbulent flow). Then, to understand the flow in terms of structural damage detection, the black-box methods involving only inputs and outputs or grey-box methods involving inputs, outputs, and certain structural features could be used. Moreover, in solid mechanics, the flow can be described as vibration by the Eulerian description, which concerns the change of the whole dynamic field and can be modeled as a function of time; also, the flow can be described as a wave by the Lagrangian description, which concerns the difference between one place and another when the wave is propagating and can be modeled as a function of space. Though time is concerned in the Lagrangian description, the main issue we are concerned with is more about the space.

In recent years, to identify the potential damage in civil structures, the information model of the structure [11] for health monitoring concerns many kinds of structural characteristics as damage indicators, including flexibility and stiffness [12], frequency [13], damping ratio change [14], and mode shape [15], among others, in a variety of works through modal analysis, hazard analysis [16], and so forth. However, for comprehensively understanding the structural system in decision-making, more attention should be paid to the global system's characteristics rather than solely to the structure characteristics, in acknowledging that all structures are more or less interacting with their environment. Therefore, the characteristics of the structural environment [17,18], as well as kinds of characteristics to describe the flow (flow characteristics), should be considered.

In this research, two kinds of flow are defined and classified in this research: Physical flow and information flow. Concerning the information flow, it has four classes of basic characteristics describing the spread of flow, the channel for flow, the flow amount in the channel, and the expansion of flow in the system. When the structure has damage, the equilibrium of swarm behavior [19] of the flow may immediately be broken and its characteristics changed at the same time. By analyzing these changing but distinguishable characteristics (such as via AE [20] and local wavenumber technique [21]), it is possible to evaluate whether the structure is damaged or not. Further, measuring data used in flow analysis can be directly inherited from general static and dynamic damage detection experiments. In this paper, an offline case of global damage evaluation for a structure (girder with artificial damage) is discussed in flow analysis by conducting a static experiment (static loading experiment and tendon-cut) and dynamic experiment (impact hammer experiment) to simulate the accelerated destruction and nondestructive examination for flow analysis to evaluate and diagnose the existing structures.

The paper is organized as follows: Following this introduction, the basic concept of the present FC-based method is given in Section 2. A laboratory experiment is conducted on two nearly full-scaled box girders to test the present method and its details are given in Section 3. In Section 4, experiment results are presented and discussed, including quantitative study, qualitative study, comparative analysis, error analysis, and possible applications. Finally, several concluding remarks and future works are summarized.

2. Basic Concepts of Flow Analysis

2.1. Definition and Classification of Flow

Flow is often referred to via simulating the phenomena which are similar to the motion of fluid [22] with a continuous tendency of points in time or space. From the perspective of mathematics, flow can be described as the dynamic field as a set of changes over time or space [23], i.e., flow is described as a group action of the real number on a set; for instance, a vector flow is determined by a vector field. A flow can be modeled as the function of time or space (t) as a group action of the additive real numbers $\mathbb{R}$ on X [24]. More explicitly, a flow is a mapping of φ [25] in time or space t using the iterated function [26]:

$$\varphi(t) : X \times \mathbb{R} \to X \tag{1}$$

The mathematical definition of flow is used widely in computational fluid mechanics and may be mainly used in the experimental data processing later.

In engineering, according to the observation of flow, a flow system can be divided into two categories: Incoming flow and outgoing flow. Since the flow is described as the function of the time or space, the flow in the previous space (s_a) or previous time (t_a) is the incoming flow (or referred to the input of flow and denoted as flow$_{in}$), and the flow in the later space ($s_b = s_a + \triangle s$) or later time ($t_b = t_a + \triangle t$) is the outgoing flow (or referred to the output of flow and denoted as flow$_{out}$). Sometimes a third category is considered: The cycling flow between incoming and outgoing flow within $\triangle s$ or during $\triangle t$.

In our classification, there are two kinds of flow in mechanical engineering. The first kind is the flow whose carrier or intermedium moves along with it, such as the water flow in the river (hydromechanics); the second kind is the flow whose carrier or intermedium does not move along with it, only using the dynamic field of some information (such as "force, energy, momentum", etc.) to describe the integral motion of some kind of space (with the function of time) in the Eulerian description, or the information field describing the motion passing through the space in the Lagrangian description (in mechanics, it means the transferring of interaction between structure and environment, or among substructures, from one point to another, represented as the function of space). In this paper, we name the first kind of flow as the physical flow and the second kind as the information flow (the flow describing the physical information of the system or the structure). In the following chapters, the information flow will be mainly discussed, and both the Eulerian description and Lagrangian description will be used when conducting the experimental analysis. Further, there are some other methods to classify the information flow, such as destructive flow or nondestructive flow, according to the influence on the structure; the information flow can be of either scalar, vector, or tensor nature.

Furthermore, for a measurement of a system, there are three kinds of views: microscopic, mesoscopic, and macroscopic [27]; so, in deeply understanding these, the research of flow can have three corresponding views:

- In the microscopic view, it mainly focuses on the phonon (a phonon is a quantum mechanical description of an elementary vibrational motion in which a lattice of atoms or molecules uniformly oscillates at a single frequency) and the formation mechanism of flow (here, wave and flow have the same meaning).
- In the mesoscopic view, it mainly concentrates on the evolution mechanism of basic flow, and how the basic flow (wave) joins to a swarm (here, the flow is treated as the swarm behavior of simple harmonic waves).
- In the macroscopic view, the research is mainly for statistical mechanism and pattern recognition for the flow (usually, the flow is of turbulence, which is hard to describe using the analytical wave functions).

In this paper, the research aspects for the evaluation of a global system can be about structure, environment, and flow. For a specific environment in the measurement, suppose the data in any survey (recorded as time series, etc.) contain the information (i.e., some characteristics; according to the definition proposed by Wiener [28], information is a set of marks of a thing's attributes) of the structure, environment, and flow simultaneously, more or less. This paper is trying to propose a new kind of damage analysis within a unified concept of flow.

In some sense, the interaction between environment and structure can be treated as the interactions between structure and flow, as well as between flow and the environment; the interactions among substructures can be converted as the interaction between substructures and flow. Analogous to the wave–particle duality [29] in physics, where the wave and particle are coexisting integrally, structure and flow are coexisting too, despite the description that particles show the system's framework updating and the description that waves may reveal the system dynamically changing as well. Further, in some cases of statistical mechanics, it is more probable to know the behavior of the swarm rather than that of any individual; the statistical grouping parameters [30] of the motion of the waves can also be described as flow to show the trend of motion shown as the Lattice Boltzmann method (LBM) simulation in Figure A1 (Appendix A). To study the flow in the structure, in dynamics, flow analysis is to investigate the swarm behavior of the waves or vibrations (here, flow is equivalent to the swarm of waves in the structure from the view of space. Further, flow is equivalent to the swarm of vibrations in the structure from the view of time. Using different measuring equipment, the recorded data may be a time series of displacement, acceleration, velocity, etc.). In statics, flow analysis is used to investigate the field of force (here, flow is equivalent to a varying force field) or field of displacement (here, flow is equivalent to a varying displacement field), etc. However, in traditional structural mechanics for engineering, the structure itself (the particle perspective) is more likely to be concerned, such as for some structural characteristics in modal analysis, and the perspective of flow is often selectively ignored, and it is obviously insufficient. Moreover, if we consider the problem in damage analysis in the view of system, the structure and its environment are interacting (described by a dynamic field, i.e., flow) with each other continuously.

2.2. Flow Characteristics-Based (FC-based) Method

We often define the system Φ as a black box in many cases of damage detection. In the Eulerian description, for a field Θ existing in Φ and bounded by Φ, its status is denoted as $\varphi(t, \Theta_{t_0})$: A status of field depending on the time (t) and the initial condition $\Theta = \Theta_{t_0}$ [31]. If $\varphi(t, \Theta_{t_0})$ is a loop function (or its differential equals 0), the field may have a dynamic equilibrium corresponding to time. The interaction can described by field x, where $x \in \left\{ \mu, \dot{\mu}, \ddot{\mu}, m\dot{\mu}, m\ddot{\mu}, 1/2m\dot{\mu}^2 \right\}$, in which μ is the displacement and m is the mass, and its status is $\varphi(t, \Theta_{t_0})$; this description of flow can be named as x-flow. Usually, the function of $\varphi(t, x_{t_0})$ is dependent on the nature of Φ. Further, the flow can be described in the Lagrangian description as the propagation in the system Φ; its change can originate from the Reynolds transport theorem as $\psi(s, \Theta_{s_0})$ [32], which gives the flow's propagation at point s and initial condition of $\Theta = \Theta_{s_0}$. If $\psi(s, \Theta_{s_0})$ is near a constant in Φ, the medium of flow Φ is approximately homogeneous. In hydraulics, laminar flow is one such example. For a characteristic x to describe the field, its status is $\psi(s, x_{s_0})$.

Combining two description methods, the input of the flow is defined as the initial condition Θ_{t_0}, and output as the state of the flow Θ_{s_t}, where s_t is the point s in space at the point t in time.

In summary, the output of flow in time and space is influenced by the input (Θ_{t_0} or Θ_{s_0}) and intermedia (Φ). In the mechanical structure, the very simple but important example is the wave and vibration. The wave is the motion transmission in space from one point to another, and the vibration is a field varying in time from one moment to another. In Section 2.1, the flow in a vibrating structure is defined as the swarm of waves, as from the view of the Lagrangian description. In order to be closer to the concept of water flow in hydraulics, we will mainly continue using the Lagrangian description, but with some changes to meet our need to have a simple understanding of the vibration test data. Here are two assumptions:

1. Suppose for a system Φ, $\partial\Phi$ is the surface of this system. The system can store or release the same flow as the flow of input, and the flow stored or released by Φ is flow$_\Phi$.

$$\text{flow}_{\text{in}} = -\text{flow}_\Phi + \text{flow}_{\text{out}} \tag{2}$$

2. $h(t)$ and $h(t,\Phi)$ represent the input and output, respectively, and we record the input and output in the same period t, the effective input period $\left[t_{a_1}, t_{b_1}\right]$, and output period $\left[t_{a_2}, t_{b_2}\right]$:

$$h(t) = \int_{t_{a_1}}^{t_{b_1}} \left(\iint_{\partial\Phi} |\mathbf{j}_1| d(\partial\Phi) \right) dt; \, h(t, \Phi) = \int_{t_{a_2}}^{t_{b_2}} \left(\iint_{\partial\Phi} |\mathbf{j}_2| d(\partial\Phi) \right) dt \tag{3}$$

in which $\mathbf{j}_1$ and $\mathbf{j}_2$ are the flux at the differential surface $d(\partial\Phi)$ of the input and output, respectively, which is the function of time and system, $\mathbf{j} = g(t, \partial\Phi)$. For the same flow, the input lasts for (t_{b_1}, t_{a_1}) and the output lasts for (t_{b_2}, t_{a_2}). Then, to calculate the flow$_\Phi$, here is its function, $f(\Phi)$:

$$f(\Phi) = h(t, \Phi) - h(t) \tag{4}$$

Another comprehensive theoretical research work can be found on energy conservation law [33]. The change of the flow shows responses in the change of the system (intermedium) in turn. An infrastructure or any other open system [34] will have interaction with its environment in that the flow between the system and its environment is always changing but is often stable in statics by surveying data of times of measurement; the same relationships are seen among the interactions of different subsystems. However, in engineering applications, the process of flow propagation from the location of input to the location of output may often be impossible to investigate in practice, especially the sum of outputs in a structure with a complex shape of surface.

Then, the flow characteristics are suggested to approximately indicate the change of flow. For a specific system, if the flow can stably go through the system, these characteristics may not change as well. Furthermore, for the damage detection and identification, the FC-based method pays close attention to the damage spreading (e.g., the growth of cracks), structural integrity (e.g., structural disintegration process), signal propagation in the system (e.g., the change of frequency), migration of equilibrium between load effects, and resistance, etc. Kinds of structural parameters are also used to describe the change of flow in confined time or space [35], such as the changing of energy propagation or force distribution in a system. Sometimes in dynamics, flow can also be treated as the integration of waves. However, usually, in order to distinguish itself from other methods, the FC-based method mainly concerns the change of flow characteristics to evaluate the health of flow itself or indirectly diagnose the health condition of the system. According to the description of a general system, some basic concepts such as the development of interactions between a system and its environment, or among subsystems, frameworks, components or elements, boundaries, etc., will be considered [36,37]. Four classes of basic flow characteristics are proposed correspondingly to these four concepts of a general system:

Class 1: Spread of flow: The characteristics to describe the spread of flow in the system, such as acceleration, speed, travel or propagation time, displacement, etc.

Class 2: Channel for flow: The characteristics to describe the channel that the flow goes through. It is about the distribution of flow in the system and the route of flow in the system, such as its topology, hierarchy, fractal dimension, cross-sectional area, etc.

Class 3: Flow amount in the channel of flow: The characteristics to describe the amount of flow, such as mass, quantity, intensity, strength, etc.

Class 4: Expansion of flow in the flow system: The characteristics to describe the expansion of flow in a specific system or systems with or without clear boundaries, such as lifetime and propagation region (depth, width, or boundary).

Every flow has these four classes of characteristics, and according to some characteristics, flow can be classified conversely.

Of all criteria of differentiating kinds of flows, the intensity of flow is one simple standard. In a thought experiment, suppose there is an instrument with a certain accuracy which can be used to measure the response of the structure in any situation and there is a man who will observe the response of the structure with his naked eye. As the Figure 1 shows, when the flow's intensity increases, the structural response, which directly influences the observability and measurability, may be more easily observed and measured such that the efficiency of observation and measurement will increase, and the damage detection will also transition from a nondestructive experiment (which is similar to the nondestructive examination) to a destructive experiment (which is usually different from the destructive examination, such as the chemical analysis of a core sample).

When the intensity grows from small to large, the response of the structure can be divided into EPM (extremely poorly measurable), PM (poorly measurable), M (measurable), DO (destructive and observable), and SDO (severely destructive and observable). Moreover, flow is not only the cause of structural damage which interacts with the structure and influences the state of structure, but also is influenced by the structural damage. As a result, once the state of the structure changes, the flow may change immediately.

In the general understanding of Figure 1, for the DO and SDO, the response of the structure to the flow can be linear and even at an exponential level; in such cases, the efficiency of measurement and observation can be much higher and even near 1, such that people can directly notice the state change of the structure with minimal error in observation. While for the M and the PM, the response of structure is nonlinear, even index or logarithmic, which cannot be observed by the naked eye directly, and how to define the exact value or the value of boundary efficiency a, b depends on the specific measurement or different standard.

For the EPM, the efficiency often equals 0, and the general tools of measurement are useless. In fact, the increased accuracy, recognition, and sensitivity of the experimental equipment could further expand the range of the observable and measurable. Further, for the cases of the DO and the SDO, the change of the response can be directly measured. Moreover, of all kinds, for the case of the PM, the corresponding flow is much more important in that it can help to detect more changes of the structure. So, in the design of the experiment, there are two aspects to be specially considered.

On the one hand, for the reason of the interaction between some flows and structures, the structural damage should be clearly observed along with the intensity of flow increasing. On the other hand, some flows can be used to detect the change or the damage of the structure independently. In the following chapters, for the static loading experiment (of the DO and the SDO, which bring great damage to the structure), the response of the structure is displacement, and for the impact hammer experiment, or named as the vibration experiment or impact tests (of the PM, which does not bring new damage or only brings small damage to the structure), the response of the structure is acceleration.

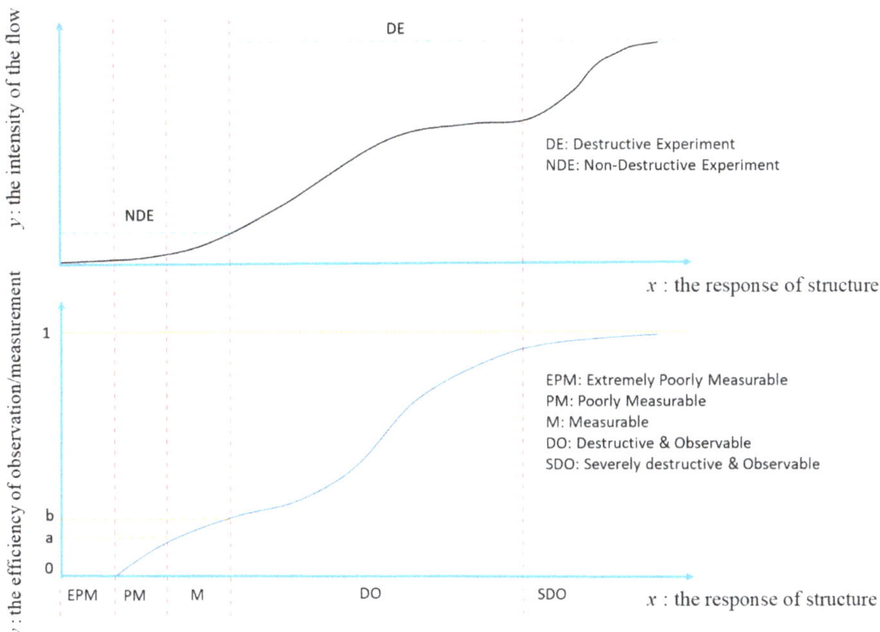

Figure 1. The relationship of a possible diagrammatic sketch between flow intensity and the structural response which directly influences the observability and measurability, or in other words, the efficiency.

3. Experiment and Analysis of Artificial Damage

3.1. Introduction to the Experimental Design and Process

Since the experiment in this research is to check the structural healthy condition (with artificial damage), the structural artificial damage should continue to increase while the residual capacity of resistance should continue to decrease. In general, static research, the stress–strain curve shows the response of the structure to the flow, in that the change of the displacement field illustrates the intrinsic variation of the structural resistance. Further, in the modal analysis for dynamics, with high probability, one lower-order natural frequency of the structure will approximately decrease within a limited magnitude [38], while the damping ratio of this frequency will usually approximately increase [39] if there are some continuous damages. The structural characteristics for comparison in these studies include displacement, frequency, damping ratio, and so on. The whole experiment contains two subexperiments: The static loading experiment and the impact hammer experiment. The following is a brief introduction to the experimental preparation, experimental progress, and experimental numerical representation.

First, the preparation for the experiment is discussed. The survey paid attention to the common and same data used in other research, such as the loading (kN) and displacement (mm) in the static loading experiment, and acceleration (m/s^2) in time series in the impact hammer experiment. Further, it was an indoor experiment so that the temperature and humidity would be in a controllable range that did not influence the process of the experiment.

Second, the experiment tested 3 nearly full-scale prestressed reinforced concrete open-section box girders (in short, box girder), named box girder 1, box girder 2, and box girder 3, by adding artificial damage. In this paper, only box girder 2 and box girder 3 (of the same design) were concerned especially for flow analysis. Both girders were in the same design, as shown by Figure 2 and Table 1: 8500 mm long, 2300 mm wide, and 1000 mm high. Further, there were 4 SDP-200 displacement meters, 8 SDP-100 displacement meters, and 10 ONOSOKKI-NP-2120 acceleration sensors set in different locations as shown in Figure 2, and the type of impact hammer used was the Brüel and Kjær type 8208.

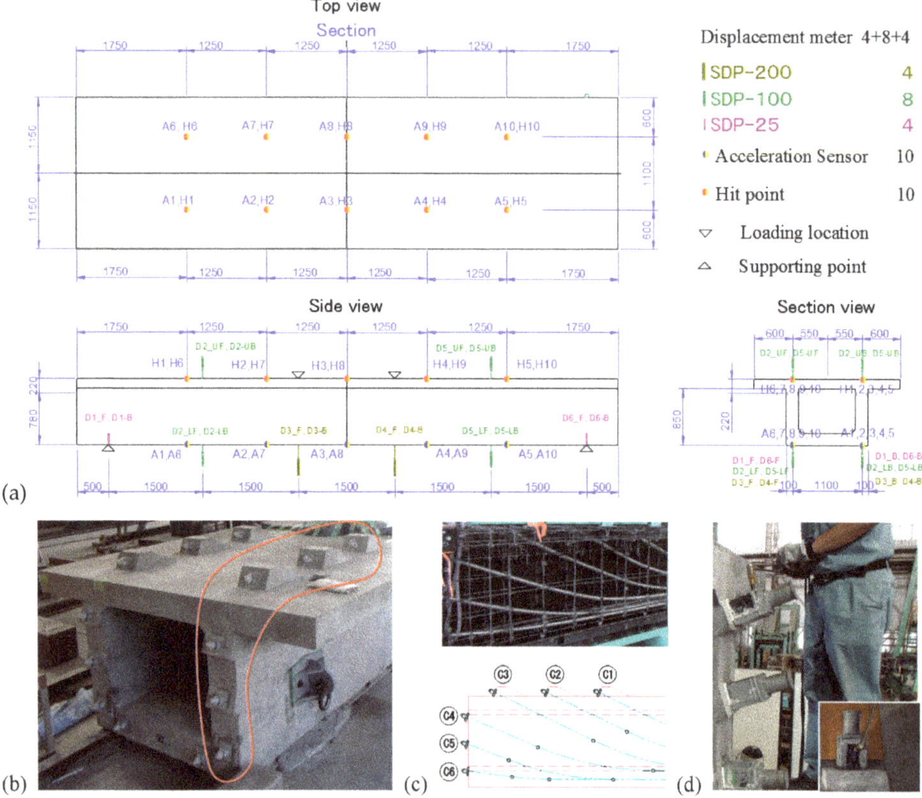

Figure 2. (**a**) The top view, side view, and section view of the experimental girder; also, the layout of displacement meters and acceleration sensors. (**b**) The real structure corresponding to (**a**) and the tendons marked by the red curve; (**c**) the reinforced bars and tendons, with C1, C2, C3, C4, C5, and C6 corresponding to the tendons marked by the red curve in (**b**). (**d**) The real operation of tendon cut.

Table 1. The description of the sensors.

Sensor	Type (Max Value)	Number	Purpose	Direction and Location	
Displacement meter	SDP-200 (200 mm)	4	For deflection	Vertical	Under the girder
	SDP-100 (100 mm)	8	For deflection	Vertical	Both under and up the girder
	SDP-25 (25mm)	4	Subsidence at the fulcrum	Vertical	Fulcrum of the girder
Acceleration sensor		10	Vibration at the fulcrum	Vertical	Under the girder

The concentrated force was loaded at the middle of the structure. The impact hammer experiment was conducted in 9 stages (Table 2, Figure 3). Meanwhile, in every stage, the impact hammer experiment was after every experiment of static loading or tendon cut. A stage here means a series of tests of static loading experiment (or one time of tendon cut), impact hammer experiment, or both of them, and there is a series of increasing loading from 0 to the greatest loading in this stage, which then decreases to 0 with a series of loading steps, or one whole process of tendon cut. A step means a step change in force (maintained for a period of time). Every step of loading or tendon cut takes less than 15 min or 1 min, respectively.

Table 2. Static loading and impact hammer experiment (stages 1–9) on box girder 2 and 3.

Object	Loading Stage	Pretest	Initial Load			Intermediate Load			Damage Load			
Box Girder 2	Stage	1	2	3	4	5	6	7	8	9	10	11
	Loading (kN)	-	816.1	-	-	840.1	-	-	838.4	973.9	1033.4	1427.3
Box Girder 3	Stage	1	2	3	4	5	6	7	8	9	10	11
	Loading (kN)	-	804.0	-	-	776.5	-	-	816.3	980.4	1051.00	1351.7

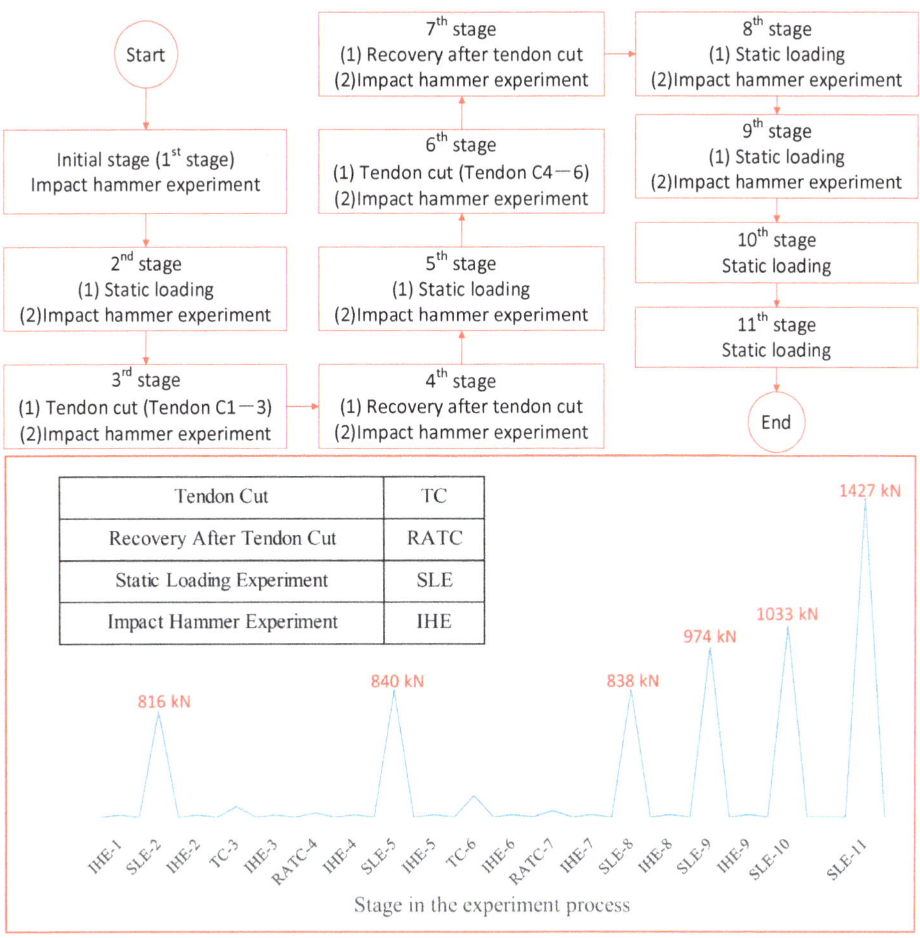

Figure 3. The experimental process corresponding to Table 1 for box girder 2.

The tendon cuts (in Figure 2) were conducted at stage 3 and stage 6, and the recovery after tendon cuts was at stage 4 and stage 7 for both box girder 2 and 3; for the tendon cuts of box girder 2, only 6 tendons were cut in total in the same side (the first time C1, C2, and C3; the second time C4, C5, and C6), while for the tendon cut of box girder 3, 12 tendons of both sides were cut; in the first time, they were C1, C2, and C3 in one side and C1, C2, and C3 in the other side; the second time they were C4, C5, and C6 in one side and C4, C5, and C6 in the other side.

Corresponding to the experiment, there are two kinds of damage: The damage caused by static loading and the damage caused by tendon cut. The crack development (Figure 4, the crack development in box girder 2) can be used to picture the damage development in the static loading. Since the damage happens inside the structure, we can only know that there is damage caused by the tendon cut but cannot observe it directly with the naked eye. However, we know that the damage is developing in the structure from stage 1 to 11 (or from steps 1 to 995 for box girder 2, and from steps 1 to 1150 for box girder 3).

Suppose the initial state of the structure is the healthy state (whose damage is 0), which means there is no damage, and the last state of the structure is a total failure (whose damage is 1), which means the system cannot satisfy its designed function. There is no need for us to know how much the damage is, and we just need to illustrate the increasing damage (from 0 to 1). If an indicator of damage can clearly indicate the damage development, it is qualified to be analyzed for the damage.

In this experiment, the acceleration sensors were located beneath the structure, while the same number of the hit points were designed further up the structure, directly above the corresponding sensors shown in Figure 5 (A1–A10).

Third, experimental numerical representation is discussed. There are 4 kinds of flow characteristics in data analysis.

(1) Intensity of distribution (IoD) in displacement flow (Figure 6): It means the displacement distribution in different locations for one step; in the analysis of the actual practical use, the reference step can be the step where the IoD corresponds to the load of 1/50–1/10 the design load.

Figure 7a shows the acceleration responses recorded by 10 acceleration sensors (corresponding to A1–A10). In the impact hammer experiment, there are another 3 kinds of characteristics for the acceleration flow shown in Figure 7b.

(2) Max-peak-time is from the first class of flow characteristics, i.e., spread of flow. In the acceleration time series, max-peak-time means the arrival time of "flood" (the max absolute value of amplitude in time series), starting from the time from the moment of the hit (contact between hammer and structure) to the moment of "flood".

(3) Max-peak belongs to the third class of flow characteristics, i.e., flow amount in the channel of flow. The max value is the max peak of the amplitude (the max absolute value of amplitude) of the time series. It means the intensity of "flood".

(4) Lifetime is classified in the fourth class of flow characteristics, i.e., expansion of flow in the flow system. The lifetime can be defined as the time from the moment of a hit to the moment of flow disappearance (the time from the moment when it is the end of the white noise to the moment when the flow vanishes and the new white noise begins again).

To investigate the whole system, we would take these data into a combined data space [40].

There, at every hit point, the impact hammer experiment would be repeated for many times. After many hits, a time distribution is found for those characteristics. Then, it is necessary to organize the data structure appropriately. In the following introduction, there are two kinds of matrices selected: The difference matrix for displacement flow and the combination matrix for acceleration flow.

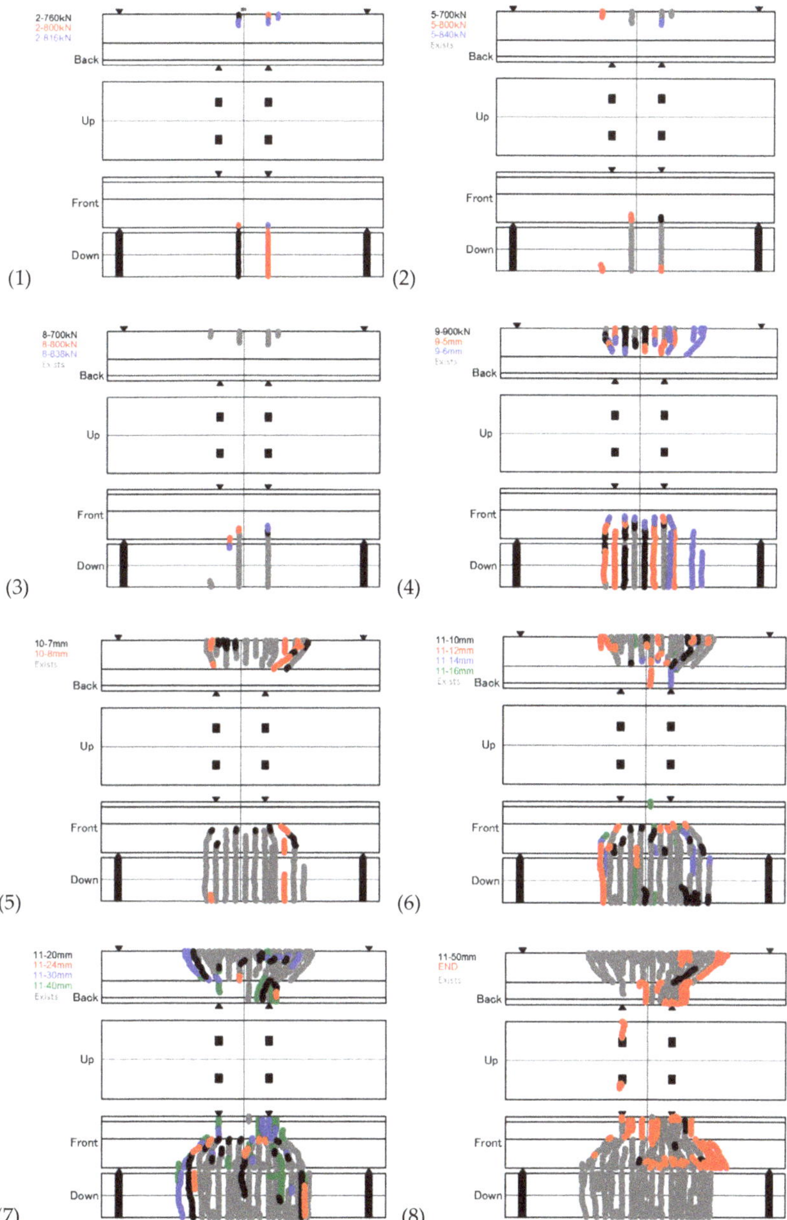

Figure 4. The subfigures show the evolution of the cracks, according to the real measurement of the cracks in the process of the experiment with box girder 2 in Table 1 at stage 2 (**1**), stage 5 (**2**), stage 8 (**3**), stage 9 (**4**), stage 10 (**5**), and stage 11 (**6–8**). Note for the control of loading, there are two methods: The first, in the notation x-a kN for stages 2, 5, 8, and 9, x denotes the ID of stage and a denotes the loading magnitude at this stage. For example, 2–760 kN in (1). The second, in the notation x-b mm for stage 11, x denotes the ID of stage and b mm denotes the average displacement by meters at this stage: D3_F, D3_B, D4_F, D4_B (D: Displacement, F: Front, B: Back). For example, 10–7 mm in (5).

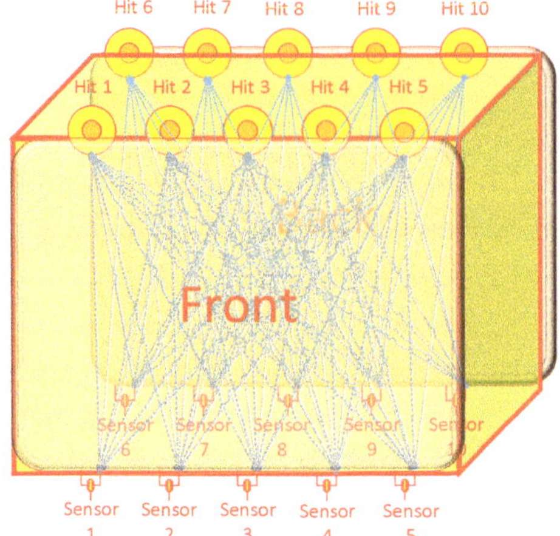

Figure 5. Diagrammatic sketch of the 3-dimensional layout of 10 hit points and 10 sensors, referring to Figure 2. The lines indicate the flow paths.

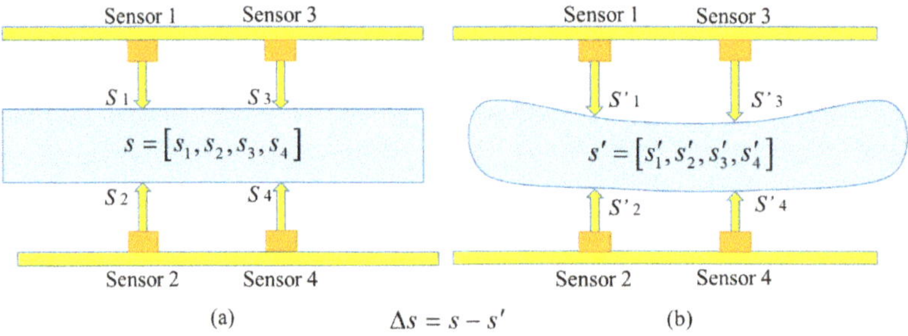

Figure 6. An example of the intensity of distribution (IoD) in displacement flow in (**a**) the initial reference stage, s, and (**b**) the test stage, s', and the change of the IoD is $\triangle s$.

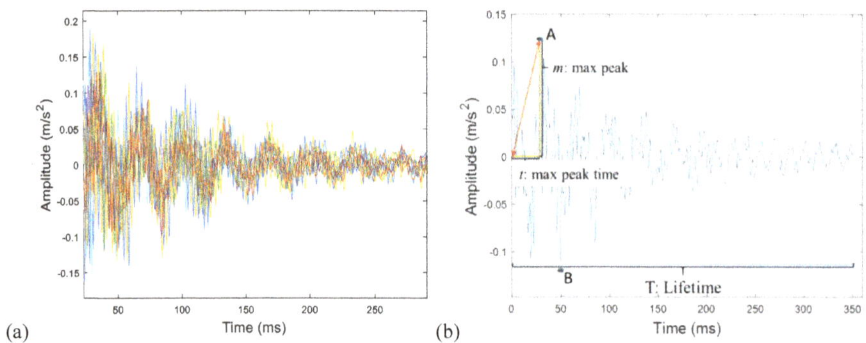

Figure 7. (**a**) An example of time series measured by the 10 sensors; (**b**) the diagrammatic sketch of max-peak, max-peak-time, and lifetime for one response time series, where amplitude A is greater than B.

On the one side, in the displacement flow, the static loading location is fixed at the right middle of the girder. There are 8 displacement meters of SDP-100, i.e., D2_UF, D2_UB, D2_LF, D2_LB, D5_UF, D5_UB, D5_LF, and D5_LB, located in different places. The record for meters is $s = [s_1, s_2, s_3, \ldots, s_8]$. Then, a difference matrix is used to show the difference of displacement between every two meters.

If the health condition of the structure is the same, the difference of displacement in the same situation (loading) may be the same, at least in a statistical sense. Then, the difference matrix is designed as:

$$
S = \begin{bmatrix}
s_{1,1} & s_{1,2} & s_{1,3} & \cdots & s_{1,j} & \cdots & s_{1,8} \\
s_{2,1} & s_{2,2} & s_{2,3} & \cdots & s_{2,j} & \cdots & s_{2,8} \\
s_{3,1} & s_{3,2} & s_{3,3} & \cdots & s_{3,j} & \cdots & s_{3,8} \\
\cdots & \cdots & \cdots & \cdots & \cdots & \cdots & \cdots \\
s_{i,1} & s_{i,2} & s_{i,3} & \cdots & s_{i,j} & \cdots & s_{i,8} \\
\cdots & \cdots & \cdots & \cdots & \cdots & \cdots & \cdots \\
s_{8,1} & s_{8,2} & s_{8,3} & \cdots & s_{8,j} & \cdots & s_{8,8}
\end{bmatrix}
$$

where $s_{i,j} = s_{j,i}$ means the absolute value of difference between displacements detected by the i-th and j-th meter ($s_{i,j} = |s_i - s_j|$); $i,j = 1,2,3, \ldots ,8$. When $s_{i,j} = 0$, $i = j$.

On the other side, to evaluate the change of acceleration flow in different health conditions of the structure, the reasonable combination matrices for the flow characteristics should circumspectly involve the information of the locations of hit points and locations of sensors to logically show the change of the flow at the surface of the structure. (1) For the same health condition of the structure with a specific location of input (hit point), the record of sensors in different locations should be statistically the same; (2) for the same health condition of the structure with different locations of input (hit point), the record of sensor in the same location should be statistically the same as well. Then, the characteristic's combination matrix contains both situations together, showing whether the structure is in the stabilization of the health condition or not. In our research, every expectation of max peak, max peak time, or lifetime (single value $x_{i,j}$) are used to build every combination matrix. The impact hammer would strike on the structure from hit point 1 to 10, and for every hit, there are 10 sensors ($j = 1,2,3, \ldots ,10$) recording the data, and after 10 hits ($I = 1,2,3, \ldots ,10$) there are two time matrices: The max peak time (t) or lifetime (T) in the form of combination matrices.

$$
t = \begin{bmatrix}
t_{1,1} & t_{1,2} & t_{1,3} & \cdots & t_{1,j} & \cdots & t_{1,10} \\
t_{2,1} & t_{2,2} & t_{2,3} & \cdots & t_{2,j} & \cdots & t_{2,10} \\
t_{3,1} & t_{3,2} & t_{3,3} & \cdots & t_{3,j} & \cdots & t_{3,10} \\
\cdots & \cdots & \cdots & \cdots & \cdots & \cdots & \cdots \\
t_{i,1} & t_{i,2} & t_{i,3} & \cdots & t_{i,j} & \cdots & t_{i,10} \\
\cdots & \cdots & \cdots & \cdots & \cdots & \cdots & \cdots \\
t_{10,1} & t_{10,2} & t_{10,3} & \cdots & t_{10,j} & \cdots & t_{10,10}
\end{bmatrix}
\& \ T = \begin{bmatrix}
T_{1,1} & T_{1,2} & T_{1,3} & \cdots & T_{1,j} & \cdots & T_{1,10} \\
T_{2,1} & T_{2,2} & T_{2,3} & \cdots & T_{2,j} & \cdots & T_{2,10} \\
T_{3,1} & T_{3,2} & T_{3,3} & \cdots & T_{3,j} & \cdots & T_{3,10} \\
\cdots & \cdots & \cdots & \cdots & \cdots & \cdots & \cdots \\
T_{i,1} & T_{i,2} & T_{i,3} & \cdots & T_{i,j} & \cdots & T_{i,10} \\
\cdots & \cdots & \cdots & \cdots & \cdots & \cdots & \cdots \\
T_{10,1} & T_{10,2} & T_{10,3} & \cdots & T_{10,j} & \cdots & T_{10,10}
\end{bmatrix}
$$

where $t_{i,j}$ and $T_{i,j}$ are the mean max-peak-time and lifetime of the i-th hit and j-th sensor, respectively. Meanwhile, the max-peak combination matrix is:

$$
m = \begin{bmatrix}
m_{1,1} & m_{1,2} & m_{1,3} & \cdots & m_{1,j} & \cdots & m_{1,10} \\
m_{2,1} & m_{2,2} & m_{2,3} & \cdots & m_{2,j} & \cdots & m_{2,10} \\
m_{3,1} & m_{3,2} & m_{3,3} & \cdots & m_{3,j} & \cdots & m_{3,10} \\
\cdots & \cdots & \cdots & \cdots & \cdots & \cdots & \cdots \\
m_{i,1} & m_{i,2} & m_{i,3} & \cdots & m_{i,j} & \cdots & m_{i,10} \\
\cdots & \cdots & \cdots & \cdots & \cdots & \cdots & \cdots \\
m_{10,1} & m_{10,2} & m_{10,3} & \cdots & m_{10,j} & \cdots & m_{10,10}
\end{bmatrix}
$$

where $m_{i,j}$ means the max peak of the i-th hit and j-th sensor.

Further, there are two kinds of data to indicate the change of the matrix in the acceleration matrix, value (such as the t, T, and m), and order. Each row of every matrix $[x_{i,1}, x_{i,1}, x_{i,3}, \ldots , x_{i,10}]$ means that in the i-th hit, all 10 sensors' detection values get the ID of the sensor with the value from the maximum to the minimum or from the minimum to the maximum (this case will be used in this paper), or according to the specific requirement of the data processing. Here is an example:

$$
n = \begin{bmatrix}
n_{1,1} & n_{1,2} & n_{1,3} & \cdots & n_{1,j} & \cdots & n_{1,10} \\
n_{2,1} & n_{2,2} & n_{2,3} & \cdots & n_{2,j} & \cdots & n_{2,10} \\
n_{3,1} & n_{3,2} & n_{3,3} & \cdots & n_{3,j} & \cdots & n_{3,10} \\
\cdots & \cdots & \cdots & \cdots & \cdots & \cdots & \cdots \\
n_{i,1} & n_{i,2} & n_{i,3} & \cdots & n_{i,j} & \cdots & n_{i,10} \\
\cdots & \cdots & \cdots & \cdots & \cdots & \cdots & \cdots \\
n_{10,1} & n_{10,2} & n_{10,3} & \cdots & n_{10,j} & \cdots & n_{10,10}
\end{bmatrix}
$$

where $n_{i,j}$ means the order of max peak time of the i-th hit and j-th sensor. The example above shows the combination matrix of the order of max peak time, which means the sensors detect the signal one by one from the minimum value to maximum value.

3.2. Experimental Data Analysis

There are matrices of characteristic values or orders, such as s, t, T, m, and n in the research, for a general matrix M:

$$
M = \begin{bmatrix}
x_{1,1} & x_{1,2} & x_{1,3} & \cdots & x_{1,j} & \cdots & x_{1,a} \\
x_{2,1} & x_{2,2} & x_{2,3} & \cdots & x_{2,j} & \cdots & x_{2,a} \\
x_{3,1} & x_{3,2} & x_{3,3} & \cdots & x_{3,j} & \cdots & x_{3,a} \\
\cdots & \cdots & \cdots & \cdots & \cdots & \cdots & \cdots \\
x_{i,1} & x_{i,2} & x_{i,3} & \cdots & x_{i,j} & \cdots & x_{i,a} \\
\cdots & \cdots & \cdots & \cdots & \cdots & \cdots & \cdots \\
x_{b,1} & x_{b,2} & x_{b,3} & \cdots & x_{b,j} & \cdots & x_{b,a}
\end{bmatrix}, \quad \begin{cases} i = 1,2,3,\ldots,a \\ j = 1,2,3,\ldots,b \end{cases}
$$

where $x_{i,j}$ means the value at location (i, j); in our case, $a = b$.

To evaluate the data obtained in the experiment, there are six variables chosen to describe every matrix.

Note: The initial stage or reference is often selected as the first stage. Any two matrices describing the initial state (reference) and the state considered (test) are named M_{initial} and $M_{\text{considered}}$, respectively.

1. Determinant of the status matrix (if it is a square matrix) Determinant can be treated as the scaling factor of the transformation from a status matrix. For a column (n), row (n) matrix, its determinant can be defined by the Leibniz formula or the Laplace formula. Here in this paper, we use the Leibniz formula [41]:

$$
\det(\mathbf{M}) = \sum_{\delta \in S_n} sgn(\delta) \prod_{i=1}^{n} a_{i,\delta_i} \tag{5}
$$

Here, all permutations δ are calculated as a summation of the set $\{1,2,3,\ldots,n\}$.

2. Norm of the status matrix; the matrix norm extends the notion from vector norm to the matrix. For a matrix, when it meets these conditions, it can have the matrix norm:

$$
\begin{cases}
|\mathbf{M}| \geq 0; \text{every matrix belongs to } (\mathbf{R}^{m \times n}) \text{ vector space} \\
|\mathbf{M}| = 0; \text{ if and only if } \mathbf{M} = 0 \\
|\alpha\mathbf{M}| = |\alpha||\mathbf{M}|; \text{every } \alpha \text{ is in } (\mathbf{R}^{1 \times 1}) \text{ vector space} \\
|\mathbf{M}_1 + \mathbf{M}_2| \leq |\mathbf{M}_1| + |\mathbf{M}_2|;
\end{cases} \tag{6}
$$

The calculation of status matrix norm is thus $|\mathbf{M}| = \sup\{|\mathbf{M}x|, x \in (\mathbf{R}^{n \times 1}), |x| = 1\}$. In this research, we use the 2-order norm (Euclidean norm): $|\mathbf{M}|_2 = \sqrt{\lambda_{\max}(\mathbf{M}^*\mathbf{M})}$; where $\mathbf{M}^*$ is the conjugate transpose of $\mathbf{M}$.

3. Max eigen or spectral radius of the status matrix, spectral radius: $SR(\mathbf{M}) = \max(|\lambda_i|)$.

4. 2D correlation coefficient (2D-CC) between every considered status matrix and initial status matrix. The 2D-CC is usually used to distinguish the status matrix change between the initial state (reference) and the state considered (test) in this research. The definition of this technology is as follows:

$$C = \frac{\sum \left(\mathbf{M}_{initial}^{T} - \overline{\mathbf{M}}_{initial}^{T} \right) \left(\mathbf{M}_{considered} - \overline{\mathbf{M}}_{considered} \right)}{\sqrt{\sum \left(\mathbf{M}_{initial}^{T} - \overline{\mathbf{M}}_{initial}^{T} \right)^{2} \sum \left(\mathbf{M}_{considered} - \overline{\mathbf{M}}_{considered} \right)^{2}}} \tag{7}$$

where $\overline{\mathbf{M}}_{initial} = \text{mean2}(\mathbf{M}_{initial})$ and $\overline{\mathbf{M}}_{considered} = \text{mean2}(\mathbf{M}_{considered})$; mean2 means the expectation of all data in this matrix (data at every row and column).

5. Distance between every considered status matrix and initial status matrix There are many methods to get the distance between two status matrices; here is an example:

$$D = \sqrt[4]{\left(\sum_{j=1} \mathbf{M}^{i,j}{}_{initial} - \sum_{i=1} \mathbf{M}^{i,j}{}_{considered} \right)^{2} \left(\sum_{i=1} \mathbf{M}^{i,j}{}_{initial} - \sum_{j=1} \mathbf{M}^{i,j}{}_{considered} \right)^{2}} \tag{8}$$

where $\mathbf{M}^{i,j}$ means the data in the matrix of row i and column j.

6. Procrustes analysis between every considered status matrix and initial status matrix

The Procrustes analysis [42] shows some change from the considered shape matrix $\mathbf{M}_{considered}$ to its initial shape matrix $\mathbf{M}_{initial}$.

In the 6 variables chosen to analyze the characteristic matrix changes, the variables from 1 to 3 are called as the intrinsic variables of the matrix, while variables from 4 to 6 are named the comparison variables (compared with the initial stage) of the matrix for convenience.

The data analysis processing of the experiment is shown in Figure 8.

1) Start of the data analysis.
2) Choose the flows to analyze (in this paper, we choose the displacement flow to represent the flow of high intensity and the acceleration flow to represent the flow of low intensity, respectively).
3) Choose the flow characteristics (max-peak-time from class 1, max-peak from class 3, and lifetime from class 4 are chosen for acceleration flow, and IoD from class 3 is chosen for displacement flow).
4) Choose the variables to represent the flow's characteristic matrices (6 kinds of variables will be used in the analysis, in which "determinant, norm, max eigen, 2D correlation coefficient, distance, Procrustes" are for acceleration flow and "2D correlation coefficient" is for displacement flow).
5) Data analysis.

> 5.1) Do the qualitative analysis for the acceleration flow. There are 9 stages (maximum amount = 9) in this data processing. Calculate the expectation of characteristics from the times recorded by different acceleration sensors. Construct the combination matrices. Calculate the variables to evaluate the characteristics. Conduct the qualitative analysis.
>
> 5.2) Do the quantitative analysis for the displacement flow. There are 11 stages (maximum amount = 11), or more than 1000 steps in this data processing. Collect the data recorded by different acceleration sensors at the same moment as arrays (loading ID, step). Construct the difference matrices. Calculate the variable (2D-CC) to evaluate the characteristics. Conduct the quantitative analysis.

6) Conduct correlation analysis and comparative analysis with modal analysis.

> 6.1) Calculate the correlation coefficient and conduct the correlation analysis for kinds of flow characteristics.

6.2) Compare the (acceleration) flow characteristics with the structural characteristics in modal analysis. These characteristics of both flow and structure all are using the same original data.

7) End of the data analysis.

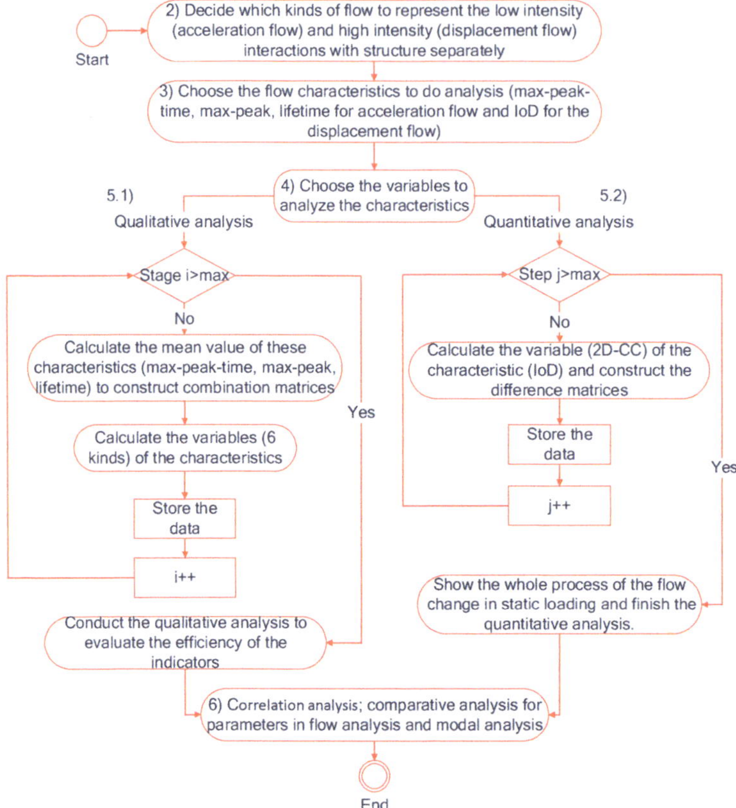

Figure 8. Flowchart of data analysis, in which "max" means the maximum stages for acceleration flow or maximum steps for displacement flow. 2D-CC: 2D correlation coefficient.

4. Results and Discussions

4.1. Calculation Method of Variables for Characteristics Matrices

In the experimental data processing, some variables used to describe the characteristic matrices may change from the former stage to the latter stage in the structural artificial lifetime:

$$\mathbf{M}_{\text{initial}} = \begin{bmatrix} x_{1,1} & x_{1,2} & \cdots & x_{1,j} & \cdots & x_{1,a} \\ x_{2,1} & x_{2,2} & \cdots & x_{2,j} & \cdots & x_{2,a} \\ \cdots & \cdots & \cdots & \cdots & \cdots & \cdots \\ x_{i,1} & x_{i,2} & \cdots & x_{i,j} & \cdots & x_{i,a} \\ \cdots & \cdots & \cdots & \cdots & \cdots & \cdots \\ x_{b,1} & x_{b,2} & \cdots & x_{b,j} & \cdots & x_{b,a} \end{bmatrix} \Rightarrow \mathbf{M}_{\text{considered}} = \begin{bmatrix} \overline{x}_{1,1} & \overline{x}_{1,2} & \cdots & \overline{x}_{1,j} & \cdots & \overline{x}_{1,a} \\ \overline{x}_{2,1} & \overline{x}_{2,2} & \cdots & \overline{x}_{2,j} & \cdots & \overline{x}_{2,a} \\ \cdots & \cdots & \cdots & \cdots & \cdots & \cdots \\ \overline{x}_{i,1} & \overline{x}_{i,2} & \cdots & \overline{x}_{i,j} & \cdots & \overline{x}_{i,a} \\ \cdots & \cdots & \cdots & \cdots & \cdots & \cdots \\ \overline{x}_{b,1} & \overline{x}_{b,2} & \cdots & \overline{x}_{b,j} & \cdots & \overline{x}_{b,a} \end{bmatrix}$$

Here is an example: A change of numerical value matrix: $\begin{bmatrix} 1 & 5 & 9 \\ 6 & 3 & 8 \\ 10 & 5 & 3 \end{bmatrix} \Rightarrow \begin{bmatrix} 1 & 5.4 & 9.5 \\ 7.5 & 4 & 6.3 \\ 10 & 4 & 2.5 \end{bmatrix}$

and its numerical order change: $\begin{bmatrix} 1 & 2 & 3 \\ 2 & 1 & 3 \\ 3 & 2 & 1 \end{bmatrix} \Rightarrow \begin{bmatrix} 1 & 2 & 3 \\ 3 & 1 & 2 \\ 3 & 2 & 1 \end{bmatrix}$. The results for both the order matrix

and value matrix of determinant, norm, max eigen, 2D-CC, distance, and Procrustes are shown in Tables 3 and 4.

Table 3. Evaluation of the characteristic matrix of order using kinds of variables (to 3 decimal places).

Stage	Determinant	Norm	Max Eigen	2D-CC	Distance	Procrustes
Initial Stage	12.000	6.059	6.000	1.000	0.000	0.000
Considered Stage	12.000	6.059	6.000	0.833	1.414	0.1875

Table 4. Evaluation of the characteristic matrix of value using kinds of variables (to 3 decimal places).

Stage	Determinant	Norm	Max Eigen	2D-CC	Distance	Procrustes
Initial Stage	279.000	16.946	16.690	1.000	0.000	0.000
Considered Stage	124.325	16.910	16.595	0.948	4.179	0.060

Just as the example above indicates, the value of each variable has changed from the initial stage to the second stage (named the considered stage) in both value and order according to the categories in Section 3.2. Even though the initial order matrix can be directly derived from the ideal state, it is still hard to guarantee the same effectiveness in practice. It is found in this example that the value matrix has a higher sensitivity than the order matrix, and the comparison variables have higher sensitivity than the intrinsic variables. However, since the sensitivity is too high in some cases, the data tend not to show the signs of change.

4.2. Quantitative Study of the Displacement Flow

In this section, 2D-CC is used to show the similarity [43] between the reference matrix (in the initial health stage) and some considered matrix (in the considered stage), describing the change of the IoD in displacement flow within a very clear range of $[-1,1]$.

Comparing Figures 9 and 10 with Figures 11 and 12, Figures 9a and 11a show the original data of the "displacement VS step" of box girder 2 and box girder 3, respectively; Figures 10 and 12 show the analysis results using 2D-CC for the difference matrices of IoD of displacements detected by meters (D2_UF, D2_UB, D2_LF, D2_LB, D5_UF, D5_UB, D5_LF, D5_LB) corresponding to the Figures 9a and 11a, respectively.

Comparing Figure 9a with Figures 10 and 11a with Figure 12, the variable of 2D-CC calculated according to Equation (7) is more sensitive than the displacement itself. In these figures, the reference matrix (initial) is set as some difference matrices at approximately 2–5% design load. From Figures 9a and 11a, the relationship between the loads and the structure response (displacement) is very clear in that in the experiment, when the loads on the structure increased, the displacement of the structure increased as well. However, for more detailed information in the whole experimental process, it is a little harder. Comparing Figure 9b with Figures 10 and 11b with Figure 12, it is known from the basic knowledge of structural mechanics that under a static load, if the basic shape is assumed to have not changed, the force applied to each part of the structure will be proportional to the force exerted by the loading device. Therefore, in Figures 9b and 11b, the elastic coefficient is calculated by $k = F/s$, where F is the force of static loading and s is displacement, and it is not necessary to calculate the forces at different positions. The main change in this study is the plasticity change, so the elastic coefficient is always changing in the static loading experiment. After taking the absolute value, the elastic coefficient of the structure itself undergoes a series of changes without loading after tendon cut, which is consistent with the results obtained by flow analysis. This process is a rebalancing process of the structure, which is the restabilization of the structure itself under gravity after the tendon cut. Moreover, even the asymmetric temperature can also influence the redistribution of structural internal force [44].

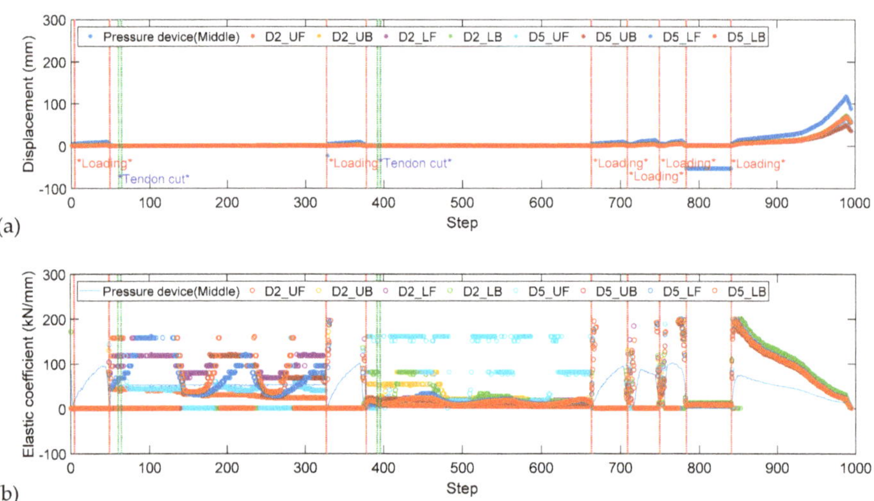

Figure 9. (**a**) The measured displacement and (**b**) elastic coefficient at displacement meters (loading device: D2_UF, D2_UB, D2_LF, D2_LB, D5_UF, D5_UB, D5_LF, and D5_LB) in every loading step of box girder 2. The green vertical line indicates tendon cut, and the red vertical line indicates the start to the end of static loadings.

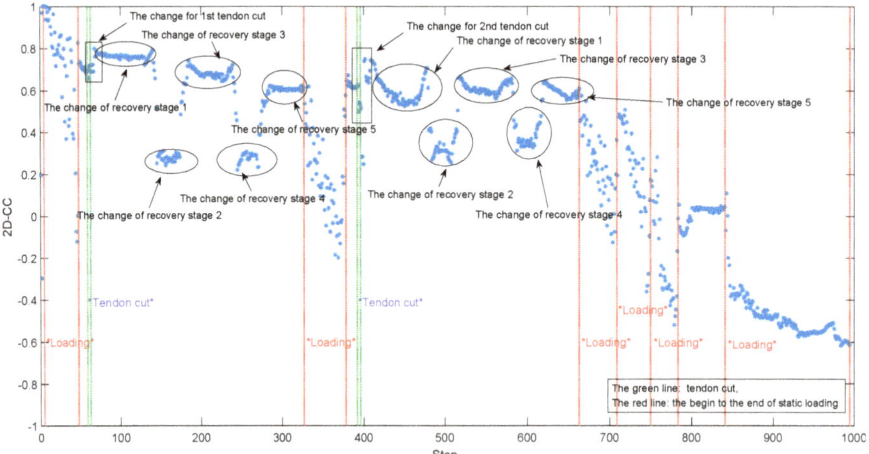

Figure 10. Max value of 2D-CC of the IoD matrix of box girder 2 between every considered step to every initial step at approximately 2–5% design load: 51 kN, 74 kN, 97 kN, and 118 kN (steps 4–7, respectively). The green vertical line indicates tendon cut, and the red vertical line indicates the start to the end of the static loading experiment.

In detail, in Figures 10 and 12, for stages 2, 5, and 8–11 of static loading, the difference matrix of displacement detected by different meters in different locations of the structure may slowly change, since the loading increases on the structure and the 2D-CC decreases; since the damages have occurred differently, the 2D-CC can still be maintained at a stable level after the static loading experiment. Once the loading disappears, there is a sudden but great change. For the 3rd and 6th stages of tendon cut, the 2D-CC changes suddenly, signifying the severe redistribution of internal force in the structure. Since the damage increases from the 3rd stage to the 6th stage, the reaction of the 6th stage will be more severe. From the 2D-CC of the intensity distribution, every time after the tendon cut, there is a short time in which the structure can migrate to a new balance (a new development of the artificial damage and a new resistance condition), and the flow in the system will change with high uncertainty. For the 4th and 7th stages of the recovery period (almost 3 days) after the tendon cut, from comparing Figure 10 with Figure 12, it is very clear that different kinds of tendon cut will have different results. In the result of box girder 2, from the tendency in Figure 10, just after the tendon cut, suddenly, the 2D-CC of the difference matrix increases, which means the structure may have an enhancement within a short time, and perhaps the structure releases some of its residual resistance at once. Then, the structure will be in a series of changes of alternating cycles, enhanced or weakened, and then achieve new balances. There are two shapes of "W" in both recovery periods of 3 days, and the 2D-CC experiences 5 recovery stages. For the reason that the difference matrix *S* shows the difference of displacement detected every two meters, it can be used in a variety of transferring equilibria of the displacement swarm (detected by different meters) in the structural artificial lifetime. Since in the asymmetric cut of C1, C2, and C3 in box girder 2, there are 3 equilibria transferring, there exists a "far from equilibrium state" according to dissipative structure theory, such as the change of recovery stage 2 and recovery stage 4 as seen in Figure 10. Meanwhile, in Figure 12, the box girder 3 is symmetrically cut, and from the original data, it may not be easy to find such kind of changes. About this phenomenon, from the perspective of flow analysis, it may contain the linear change near the real balance state, i.e., the so-called "near equilibrium state". After the tendon cut, the change of the system is possibly influenced by the action of environmental micro-perturbation, where the crack or the weak part in the structure slowly crept so that the structural damage occurred and developed in different locations, but due to the asymmetry of the tendon cut, the 2D-CC of the difference matrix can clearly indicate the damage effect for that the reference (initial) matrix is designed based on the initial symmetry structural state.

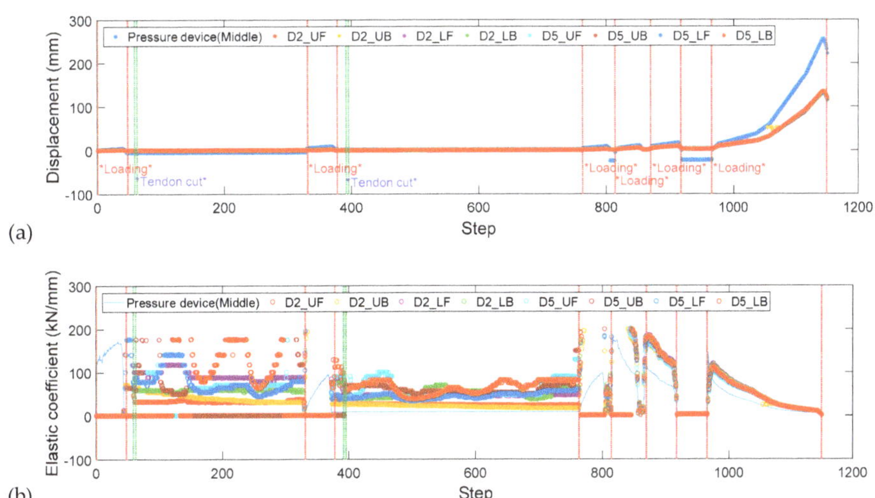

(a)

(b)

Figure 11. (**a**) The measured displacement and (**b**) elastic coefficient at displacement meters (loading device: D2_UF, D2_UB, D2_LF, D2_LB, D5_UF, D5_UB, D5_LF, and D5_LB) in every loading step of box girder 3. The green vertical line indicates tendon cut, and the red vertical line indicates the start to the end of static loadings.

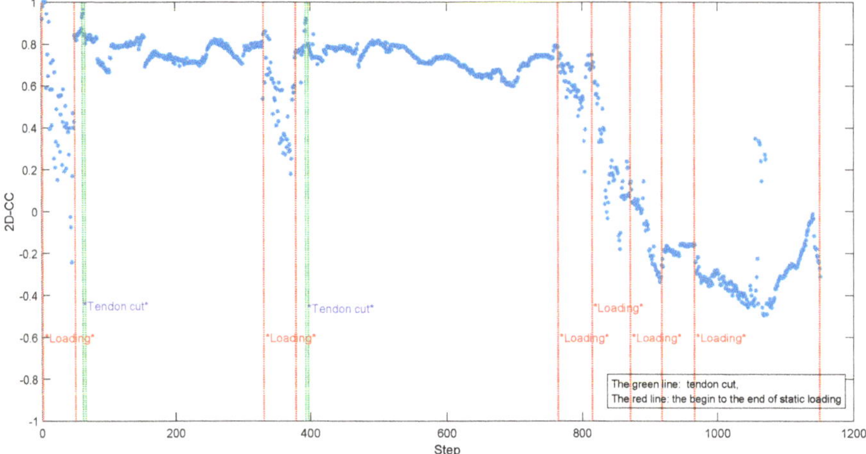

Figure 12. Max value of 2D-CC of the IoD matrix of box girder 3 between every considered step to every initial step at approximately 2–5% design load: 50 kN, 73 kN, 87 kN, and 117 kN (steps 3–6, respectively). The green vertical line indicates tendon cut, and the red vertical line indicates the start to the end of the static loading experiment.

In summary, since the test beams have the symmetrical size and commonly used type in the field, the damage detection based on 2D-CC of the characteristic of the displacement flow can investigate the continuous damage caused by the static loading and the tendon cut. Especially for the tendon cut, the special "W" shape is clearly seen.

4.3. Qualitative Study of the Acceleration Flow

The qualitative study involves an understanding of a phenomenon, situation, or event which comes from exploring the totality of the situation [45]. In this study, the impact hammer experiment was carried out manually. It is almost impossible to collect raw data continuously in all loading instants, so flow analysis can only be conducted discretely at nine major stages. The results of limited stages can only give a trend to approximate the damage changes in the structural artificial lifetime. So, in the flow analysis for the impact hammer experiment, the qualitative study is the mainstay, supplemented by the quantitative study to evaluate this trend. Further, inspiration is taken from the tendency of stock market research or other fields, where an average directional index (ADX) or directional movement index (DMI) is an indicator used in technical analysis as an objective value for the strength of a trend [46], wherein one kind of qualitative directional index (QDI) is proposed to identify the effectiveness of the methods. When comparing the reasonable development tendency and the analytic result, every change of forward and backward development will be defined by +1 and −1, respectively.

QDI was calculated as follows: First, in a sequence, from one point to the next point, if the derivative (or value of the post value minus the previous value) of the broken line is positive, the value of change is "+"; if the derivative of the broken line is negative, the value of change is "−"; if the derivative (k_i, $i = 1,2,3, \ldots ,N$) of the broken line is 0 or near 0 ($|k_i|/|k_{i+1}| \leq \varepsilon$, $i = 1$, ($|k_i|/|k_{i+1}| \leq \varepsilon$) $\cup$ ($|k_i|/|k_{i-1}| \leq \varepsilon$), $i = 2,3,4,\ldots,N-1$, and $|k_i|/|k_{i-1}| \leq \varepsilon$, $i = N$), where $|k_{i-1}|, |k_i|, |k_{i+1}|$ are three adjacent derivative values and ε is a very small number, or usually just according to the resolution of the naked eye on figures or curves), the value of change is 0. Second, according to the QDIs of all variables of characteristics (sum of data, whose absolute value should be not less than γ, and γ in this paper is 2/8) of the tendency, delete the result which cannot be clearly distinguished. Then, assign the weight: For the order, 1, and for the value, 3. Meanwhile, get the sum of these value, and if it is positive, it is +1, and if negative, −1. Last, get the accumulation of the value over the total number of stages. For example, for the approximate steps in 2D-CC of IoD matrices for the time in different stages of the impact hammer experiment (Figure 13), its QDI is 6/8.

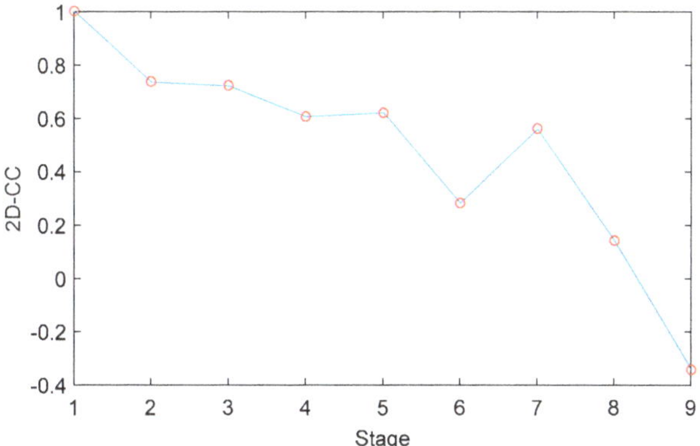

Figure 13. 2D-CC obtained in impact hammer experiments at step {X, 51, 67, 310, 380, 399, 650, 730, 770}, where X is any element from set {4, 5, 6, 7} in Figure 10 corresponding to stages 1–9 of the impact hammer experiment on box girder 2.

The indicators show the same tendency of damage development in the two girders using different kinds of variables. Take the 2D-CC, for instance. The characteristic matrices are introduced in Section 3.1, **M**, in which $x_{i,j}$ is the expectation of characteristic values of time series at one hit point.

All characteristics and variables to describe the characteristic matrices are analyzed by qualitative study, and the results are summarized in Tables 5 and 6. In Table 5b, since the determinant, distance, and Procrustes will increase, here, we just multiply it by -1 in the results of these variables. In Table 6b, since the determinant, distance, Procrustes will increase, here, we just multiply it by -1 in the results of these variables.

The final QDI of the acceleration flow described by these characteristics and variables are 3/8 (by variables) and 4/8 (by characteristics) in Table 7 (referring Table 5a and Figure 13), respectively. Surveying these tables, different characteristics have different sensitivities to reveal the tendency of damage development; from Table 6, the max-peak-time has the best results from three selected characteristics. Overall, there are many equilibrium states of flow (stages 1–2, stages 2–7, and stages 7–9) from Figure 13 and Tables 5 and 6 and at stage 2 and 7, there are large balance equilibria migrations, such that various characteristics will have larger differences than other stages.

Generally, but not strictly, [0, 0.3) means no correlation, [0.3, 0.5) is a weak correlation, [0.5, 0.8) is moderate correlation, and [0.8, 1.0] is a strong correlation. From Tables 8 and A2, the coefficients of the sequence correlation between every two characteristics of acceleration flow have many strong relationships. Since the correlation coefficient is related to cosine similarity [47], it also shows that the different intensity of flow will have different responses of structure in Section 2.2. In the experimental data analysis, different from the data (the IoD) obtained from the static loading experiment (displacement flow), the data (max-peak-time, max-peak, and lifetime) obtained from the impact hammer experiment (acceleration flow) shown in Figure 10 only had a weak correlation in the tendency of change. However, from observation of the curve in Figures 13 and 14, the QDI of displacement flow with 2D-CC is superior to any characteristic of acceleration flow in the impact hammer experiment with only 2D-CC in numerical value.

In summary, the damage indicators using flow analysis provide a new study of damage assessment. Although the results are not yet perfect, it is still possible to find that if the appropriate variables and characteristics are selected, satisfactory results can be obtained.

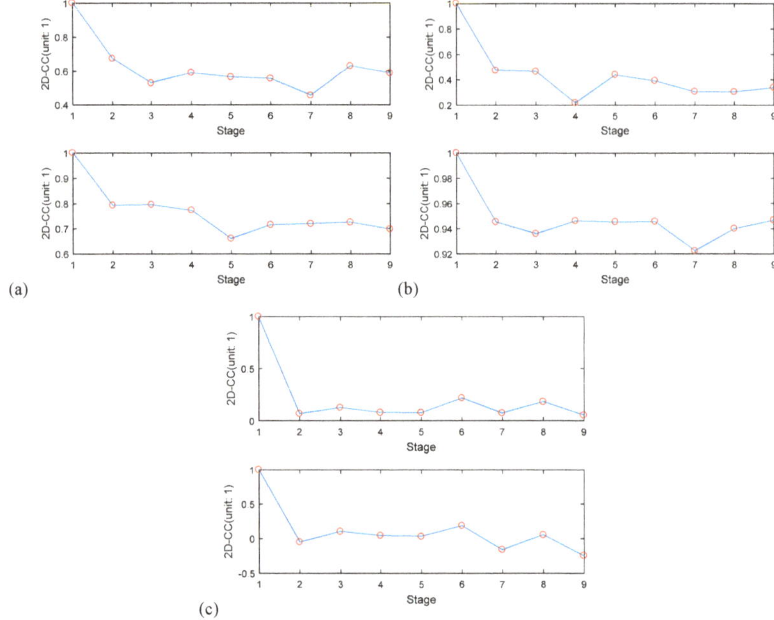

Figure 14. The mean value change of the 2D correlation coefficient between the test (considered) stage and reference (initial) stage matrix ((**a**), max-peak-time; (**b**), max-peak; (**c**), lifetime) in 9 stages, the top image of **a**, **b**, and **c** is the characteristic's order matrix, while the bottom is the characteristic's value matrix. In every figure, the y-axis has no unit, and the x-axis is the ID of the stage.

Table 5. Qualitative study of the flow analysis of acceleration flow from the viewpoint of the variables. (a) Qualitative study of the flow analysis of acceleration flow for 2D-CC for Figure 14. (b) Qualitative study for normalization of all "averages" for all variables.

(a)

Characteristic	Matrix Type	Stage								Sum Data
		1→2	2→3	3→4	4→5	5→6	6→7	7→8	8→9	QDI
Max-peak-time	Order	−1	−1	+1	−1	0	−1	+1	−1	3/8
	Value	−3	0	−3	−3	+3	−3	0	−3	4/8
Max-value	Order	−1	−1	−1	+1	−1	−1	−1	+1	4/8
	Value	−3	−3	+3	−3	0	−3	+3	+3	1/8
Lifetime	Order	−1	−1	−1	+1	+1	+1	+1	−1	0/8
	Value	−3	0	0	−3	−3	+3	+3	−3	2/8
Average	-	−1	−1	−1	−1	0	−1	+1	−1	-
Accumulation	-	−1	−2	−3	−4	−4	−5	−4	−5	5/8

(b)

Variable	Original Trend	Stage								Sum Data
		1→2	2→3	3→4	4→5	5→6	6→7	7→8	8→9	QDI
Determinant	Increase	−1	−1	−1	−1	0	+1	−1	−1	5/8
Norm	Decrease	−1	−1	−1	+1	−1	+1	0	0	2/8
Max eigen	Decrease	−1	−1	+1	+1	−1	−1	+1	+1	1/8
2D-CC	Decrease	−1	−1	−1	−1	0	−1	+1	−1	5/8
Distance	Increase	−1	−1	−1	−1	+1	+1	+1	−1	2/8
Procrustes	Increase	−1	−1	+1	−1	+1	+1	0	−1	1/8
Average	-	−1	−1	−1	−1	0	+1	+1	−1	
Accumulation	-	−1	−2	−3	−4	−4	−3	−2	-3	final QDI 3/8

Table 6. Qualitative study of the flow analysis of acceleration flow from the viewpoint of the characteristics. (a) Qualitative study of the flow analysis for the characteristic of max-peak-time. (b) Qualitative study for normalization of all "averages" for all characteristics.

(a)

Variable	Matrix Type and Original Trend	Stage								Sum Data
		1→2	2→3	3→4	4→5	5→6	6→7	7→8	8→9	QDI
Determinant	Order (Increase)	−1	−1	+1	−1	−1	+1	−1	+1	2/8
	Value (Increase)	−3	−3	−3	−3	+3	+3	−3	−3	4/8
Norm	Order (Decrease)	+1	−1	+1	−1	+1	−1	+1	−1	0/8
	Value (Decrease)	−3	−3	−3	+3	−3	+3	−3	+3	3/8
Max eigen	Order (Decrease)	0	0	0	0	0	0	0	0	0/8
	Value (Decrease)	−3	+3	−3	+3	−3	+3	−3	+3	0/8
2D-CC	Order (Decrease)	−1	−1	+1	−1	0	−1	+1	−1	3/8
	Value (Decrease)	−3	0	−3	−3	+3	−3	0	−3	4/8
Distance	Order (Increase)	−1	−1	+1	−1	−1	+1	−1	+1	2/8
	Value (Increase)	−3	−3	−3	−3	+3	+3	−3	−3	4/8
Procrustes	Order (Increase)	−1	+1	−1	+1	−1	+1	+1	−1	0/8
	Value (Increase)	−3	−3	0	+3	−3	0	0	0	2/8
Average	-	−1	−1	−1	−1	−1	+1	−1	−1	
Accumulation	-	−1	−2	−3	−4	−5	−4	−5	−6	QDI 6/8

(b)

Variable	Stage								Sum Data
	1→2	2→3	3→4	4→5	5→6	6→7	7→8	8→9	QDI
Max-peak-time	−1	−1	−1	−1	−1	+1	−1	−1	6/8
Max-peak	−1	−1	+1	−1	−1	−1	+1	−1	4/8
Lifetime	−1	−1	+1	−1	−1	+1	−1	−1	4/8
Average	−1	−1	+1	−1	−1	+1	−1	−1	
Accumulation	−1	−2	−1	−2	−3	−2	−3	−4	final QDI is 4/8

Table 7. Comparison between acceleration flow in Table 5a and displacement flow in Figure 13 in qualitative study (variable: 2D-CC).

Type of flow	Stage								Summary
	1→2	2→3	3→4	4→5	5→6	6→7	7→8	8→9	QDI
Acceleration Flow	−1	−1	−1	−1	0	−1	+1	−1	5/8
Displacement Flow	−1	−1	−1	0	−1	+1	−1	−1	5/8

Table 8. Correlation coefficients between every two characteristics in acceleration flow and displacement flow using the analysis data of the box girder 2 experiment by 2D-CC in Table A2.

Type of Flow	Characteristics	Displacement Flow	Acceleration Flow		
		IoD	Max-peak-time	Max-peak	Lifetime
Displacement Flow	IoD	1.0000	0.6257	0.3700	0.2325
	Max-peak-time	0.6257	1.0000	0.8175	0.8221
Acceleration Flow	Max-peak	0.3700	0.8175	1.0000	0.7663
	Lifetime	0.2325	0.8221	0.7663	1.0000

4.4. Comprehension between Flow Characteristics and Structural Dynamic Characteristics

The modal analysis and flow analysis can both be data analysis tools (Table A1), but even when analyzing the same data, they interpret structure and flow characteristics differently. Using the same data as the flow analysis, from the previous research, in all modes of modal analysis, the frequencies of lower modes (e.g., the 1st and 2nd bending modes in Figures 15 and 16 using stochastic subspace identification (SSI)) may be adequate to evaluate the structural damage degree. The QDIs of both modes are shown in Tables 9 and 10.

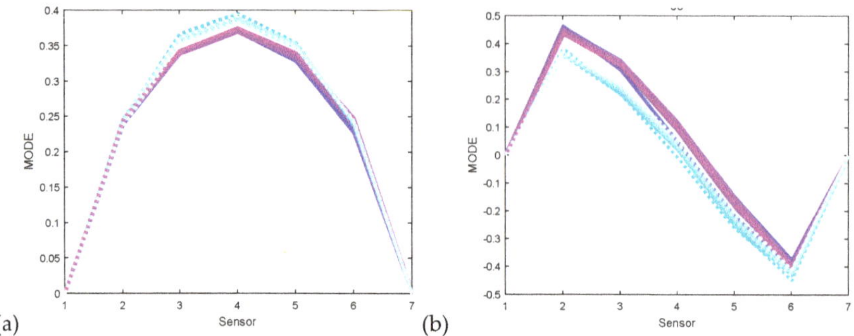

Figure 15. (**a**) The 1st bending mode, at around 29 Hz (29–31 Hz) and (**b**) the 2nd bending mode of box girder 2, at around 65 Hz (62–67 Hz). Different shades of color indicate changes in mode at different stages. Purple indicates sensors 1–5, and blue indicates sensors 6–10. Uniformity is indicated at the locations of sensors 2–6 in figures as well as at sensors 1 and 7 (if there is no sensor, assume the mode as 0).

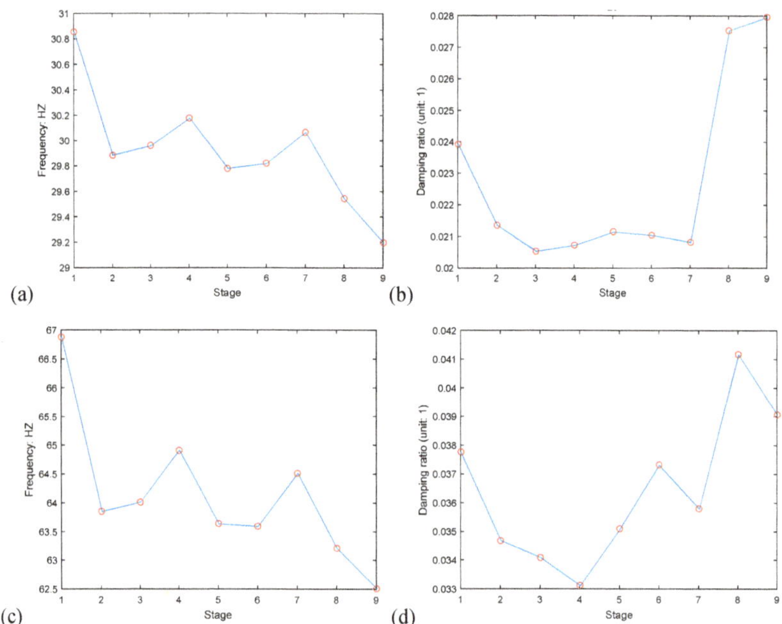

Figure 16. The mean value of (**a**) frequency and (**b**) damping ratio of the 1st bending mode and the mean value of (**c**) frequency and (**d**) damping ratio of the 2nd bending mode.

Table 9. Qualitative study of two selected modal parameters of the 1st bending mode.

Variable	Stage								QDI
	1→2	2→3	3→4	4→5	5→6	6→7	7→8	8→9	
Frequency	−1	+1	+1	−1	+1	+1	−1	−1	0/8
Damping Ratio	−1	−1	+1	+1	−1	−1	+1	+1	0/8

Table 10. Qualitative study of two selected modal parameters of the 2nd bending mode.

Variable	Stage								QDI
	1→2	2→3	3→4	4→5	5→6	6→7	7→8	8→9	
Frequency	−1	+1	+1	−1	−1	+1	−1	−1	2/8
Damping Ratio	−1	−1	−1	+1	+1	−1	+1	−1	−2/8

When the artificial damage gradually increases in the structure, the internal forces were redistributed, such that the cracks may be wider, deeper, and longer; the relationship between substructures may change; and the rate of energy dissipation may vary. The strength of associations among substructures will directly affect the occurrence and development of cracks. Then, the flow in the system may have different routes and rates of dissipation. There are some assumptions or presuppositions:

1. Once the crack has emerged, it will not disappear, and the depth and length will not shrink.
2. The redistribution of internal forces will change the strengths of associations among substructures as well as the width of cracks.
3. In statics, the energy dissipation is depending on the length of the route that the flow goes.
4. The natural frequency is determined by the structure itself, and the elastic modulus or coefficient of stiffness will directly influence the change of the natural frequency. Natural frequency in the experimental data processing using SSI is a holistic concept of the structure.
5. Global change is determined by the sum of the local changes.

In practice, the material characteristics of some systems are often nonlinear and nonstationary (for instance, the concrete box girder), and thus some structural characteristics may be ineffective. Further, the flow analysis of acceleration flow can be effective using the same original data of the modal analysis.

Comparing the results between modal analysis and flow analysis in Figures 14 and 16, and from the qualitative study in Tables 5, 6, 9 and 10, some observations can be made as follows: In the experiment, the overall size of the structure does not change visually, while the sizes of the existing cracks or new cracks, with their continuous development, will have undergone tremendous changes. From the results, the flow characteristics are more sensitive to the cracks' depth and length, while the structural characteristics through modal analysis are more sensitive to the redistribution of forces or the width of the cracks; both are specially related to the topology of the flow channel in class 2 of flow characteristics. That is to say, despite the topology of the channel being hard to obtain, its change will bring about the change of max-peak and max-peak-time in other classes of characteristics (1st and 3rd) and the structural characteristics. Among three kinds of flow characteristics, the max-peak and max-peak-time are relating to the change of "flood" of flow and the change of the flow's main part, respectively, while the lifetime is related to the whole process of the flow change in the structure. Put another way, the energy dissipation in the structure from the locations of input to the locations of output in the impact hammer experiment will follow different paths. When there is max flow (the max magnitude), it means there are "floods" in some locations.

As the natural frequency changes in Figure 16 shows, the coefficient of stiffness has changed for the reason of cracks change. To sum up, the natural frequency and the "max-peak and max-peak-time" have relatively strong coefficient correlations in Table 11. In Figure 16, in the process of every tendon cut and recovery after tendon cut (from stage 3 to stage 4 and from stage 6 to stage 7), the force may be redistributed again, the specific prestressing design leads to an increase in the elastic modulus of many positions, and the crack width becomes smaller, resulting in an increase of the low-order natural frequency and a decrease of the damping ratio. Through observation, the flow characteristics of max-peak and max-peak-time are not sensitive to the width of the cracks, thus they would not change abnormally. Concerning the damping ratio and the lifetime, they share the same idea that the intensity of the wave gradually decreases. In the assumption, the global change is determined by the sum of the local changes; one the one hand, the stiffness coefficient varies in different locations and differs from the global one, in that the different integrated methods (such as the modal parameters) will give different results, which will not correctly reflect the real damage of the structure; on the other hand, the flow analysis combined almost all original data space (data recorded at the surface of the structure) in different locations to investigate the change of public information and it can help get the result with higher precision. Furthermore, if we concern all kinds of structural characteristics, as well as many kinds of flow characteristics simultaneously, some better results may be obtained in the analytic hierarchy process.

Table 11. The correlation coefficient between every two characteristics in the acceleration flow of box girder 2 by 2D-CC and frequency and damping, referring to Table A2.

Method	Characteristics	Displacement Flow				Acceleration Flow		
		Fr_29	Fr_65	DR_29	DR_65	Max-peak-time	Max-peak	Lifetime
Modal	Fr_29	1.0000	0.9861	0.3227	0.8637	0.8255	0.6292	0.5490
Analysis	Fr_65	0.9861	1.0000	0.1858	0.7919	0.8564	0.6910	0.6575
	DR_29	0.3227	0.1858	1.0000	0.6102	−0.0929	−0.3238	−0.5776
	DR_65	0.7067	0.6379	0.7949	0.8552	0.3604	0.1273	−0.0547
Flow	Max-peak-time	0.8255	0.8564	−0.0929	0.6257	1.0000	0.8175	0.8221
Analysis	Max-peak	0.6292	0.6910	−0.3238	0.3700	0.8175	1.0000	0.7663
using 2D-CC	Lifetime	0.5490	0.6575	−0.5776	0.2325	0.8221	0.7663	1.0000

Note: Fr_29 and Fr_65 mean the frequency around 29 Hz and 65 Hz, respectively, while DR_29 and DR_65 are the damping ratio corresponding to the frequency around 29 Hz and 65 Hz, respectively.

In summary, through evaluating the order matrix and value matrix in flow analysis, both the intrinsic variables and comparison variables in different stages outperformed the modal characteristics (frequencies and damping ratios). By analyzing the correlation of two methods, it is obvious that both complement each other. If they are used in conjunction, some better structural damage assessment may be achieved, and then multi-criteria decision-making [48] can be conducted.

4.5. Measurement, Sampling, and Error

Generally, in order to know the behavior of flow in the structure, we measure the flow of import–export through the structure to indirectly assess the interaction (described as the dynamical field) between structure and environment. Moreover, this interaction can be formed between structure and flow, as well as between flow and environment. Further, in a global system, the measured data (often in time series) often contain information of structure, flow, and environment at the same time. For instances, for a time series, the natural frequency can be obtained from the modal analysis, the flow characteristics can be obtained from flow analysis, and some environmental characteristics can also be obtained from the error analysis, even when the whole environment is well controlled (e.g., in indoor experiments). Furthermore, every measurement might carry environmental update information. So, based on our assumption, inherent error analysis can also be partly treated as environmental analysis, although the experimental condition is strictly controlled in this indoor experiment.

In the data analysis, it is difficult to avoid some errors. To explain the reasons for these errors, besides the influence of the environment and random error of measurement, there are inherent errors in the data itself. Here, to explain the data accuracy, two possible reasons in this section are discussed.

Firstly, the arising phenomenon that causes the insufficient sampling rate of the experiment results is limited by the instrument [49].

Secondly, the assumed ideal situation and the situation in practice are sometimes totally different or the measurement error is so great that the information we want is hard to distinguish. In the construction and conservation period, this girder has experienced kinds of interactions with the environment, but also its own structural defects, and maybe there are numerous microcracks in the structure that cannot be observed with the naked eye, so the inner condition is not ideal.

Following is a pair of order matrices for acceleration flow with 7 input locations and 7 output locations in the initial stage: Real measurement versus the ideal situation (experiment in Figure 17):

$$
\begin{bmatrix}
2 & 3 & 1 & 4 & 5 & 6 & 7 \\
2 & 3 & 1 & 4 & 6 & 7 & 5 \\
3 & 4 & 2 & 1 & 7 & 5 & 6 \\
2 & 6 & 3 & 4 & 1 & 5 & 7 \\
5 & 6 & 4 & 3 & 2 & 1 & 7 \\
6 & 5 & 4 & 7 & 2 & 1 & 3 \\
4 & 1 & 2 & 7 & 5 & 6 & 3
\end{bmatrix}
\quad \text{VS.} \quad
\begin{bmatrix}
1 & 2 & 3 & 4 & 5 & 6 & 7 \\
2 & 1,3 & 3,1 & 4 & 5 & 6 & 7 \\
3 & 2,4 & 4,2 & 1,5 & 5,1 & 6 & 7 \\
4 & 3,5 & 5,3 & 2,6 & 6,2 & 1,7 & 7,1 \\
5 & 4,6 & 6,4 & 3,7 & 7,3 & 2 & 1 \\
6 & 5,7 & 7,5 & 4 & 3 & 2 & 1 \\
7 & 6 & 5 & 4 & 3 & 2 & 1
\end{bmatrix}
$$

For one input location, there may firstly be two possible sensors to detect the same prescriptive signal at the same time. For example (red font in the matrix above), for the input at hit 2, sensors 1 and 5 may detect this signal at the same time, but in the real measurement, either sensor 1 or 5 would detect the signal earlier rather than the two sensors detecting it at the same time.

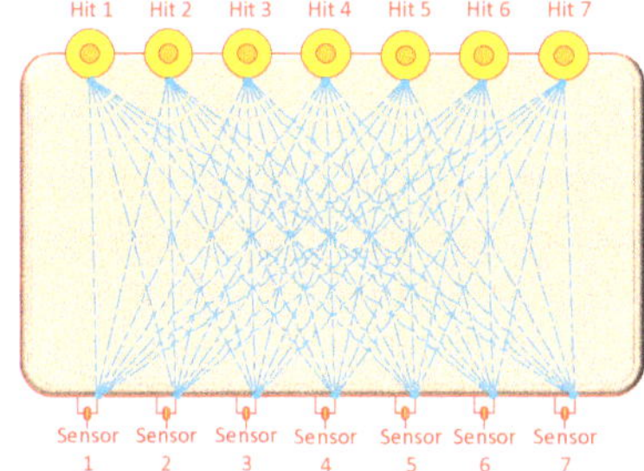

Figure 17. Two-dimensional layout of 7 hit points and 7 sensors on a beam.

Furthermore, in the quantitative study for the displacement flow, the unique data do not have statistical meaning. In the qualitative study for the acceleration flow, only investigating the expectation in statistics cannot provide the continuous and comprehensive change of the structural artificial lifetime. Further, in some cases, human error may also bring about some strange or wrong results. Fortunately, in this paper, the summarized results have a clear trajectory of change, which is enough to approximately show the overall change of the structure, and these indicators in flow analysis can be used to demonstrate the damage development.

4.6. The Application of the Flow Analysis in Practice

To apply the indicators of flow analysis to damage identification, there are several concerns, as follows.

First, the characteristics for decision-making should be carefully chosen. For different kinds of flow, how to select these characteristics properly depends on the structural types, materials, environment, and so on as well as the measurement practicability. Second, the experiment (detection or monitoring) should be well organized so that the data can carry enough information about the structure and flow. Third, if possible, multiple kinds of flow should be considered, which can aid in better analysis and comparison in practice. Last, besides flow analysis, other kinds of methods can also be conducted to help decision-making.

For the damage caused by static loading, if a structure is damaged, the structural equilibrium state may change. For example, in our case, about the quantitative study of the displacement flow, in stage 2, the structure firstly suffers a heavy loading and its inner structural rebalance occurs; after the loading, a permanent change appears in the curve "2D-CC versus step". Moreover, after static loading in stage 7, the 2D-CC is near or smaller than 0, so the system becomes totally different from its formal equilibrium and there is a high likelihood that the structure has experienced great damage. Further, in the discussion about the qualitative study of the acceleration flow, there are 3 equilibria. The first equilibrium is from stage 1 to stage 2. The second equilibrium is from stage 2 to stage 7. From stage 2, the structure has suffered a series of changes, for which almost all the indicators tend to be in disorder of complex curve-changing to a certain extent. The third equilibrium is from stage 7 to stage 9. After stage 7, the change of all variables describing reunification of all characteristics occurs, indicating that the structure may suffer a strong weakening. In practice, if there is an anomalistic change for all curves of indicators, it means there is an equilibrium shift of flow and there may be a great damage in the structure such that it asks the decision-maker to pay more attention to whether there is a need to maintain the structure.

In the tendon-cut damage case, the broken tendon is mostly asymmetrical in real structures. In our research, it is shown that an asymmetrical tendon cut will have a serial of rebalances in the structure and that the flow in the structure will have a large number of states far from equilibrium in the vicinity of the equilibrium state, and finally, a new balance will be formed. This kind of special change will be used to indicate the tendon breakage or not. The tendon here is cut at the end. More experiments should be done, as there may exist different kinds of modes when the tendons are broken in different ways or locations of prestressed structures.

Further, only concrete box girders were tested in this study. More tests should be conducted with other styles of structures and damage.

Fortunately, according to the dissipative structure theory, all structural changes follow the same law in system theory on equilibrium state migration. Meanwhile, if there is a big change in the structure, the efficient indicators can identify such changes as well. Since the flow exists in every system and the change of structure is just reasoned as the interaction between the structure and the flow, then the flow characteristics should have some changes corresponding to the changes in the structure.

The experiment method applying flow analysis can also be used in the common health monitoring of bridges or other structures, such that the curve of the flow characteristic may have a clearer and more continuous tendency. For the real bridge, for the displacement flow, the damage caused by tendon cut and the loads (vehicles or humans) can be detected by a swarm of meters. For the acceleration flow, it may use the data of vibration caused by wind, rain, or interaction between bridge and vehicles to show the change of flow in different characteristics. The arrangement of sensors can be decided according to the specific problem, and the data style can be varied in practice. Besides this research, to use the flow analysis with the data obtained by other methods, the data should be well organized into swarm data. Then, the reconstructed data matrix or some other form of data may be used in flow analysis.

5. Conclusions

In the global structural system, the interaction between structure and environment can be transformed into the interactions between structure and flow and between flow and environment; the interaction among substructures can be transformed into the interaction between flow and substructures. Flow can destroy the structure and can also be used to evaluate the structure. Often, but not strictly, a strong flow might damage structures, while a weak flow can be used to detect the existence and the level of damage. The flow analysis is mainly about the stability of flow characteristics. By evaluating the change of these characteristics, it is possible to assess the damage of the structure when the structural condition changes.

The flow analysis defines a system as a dissipative structure, utilizes the combined data space, provides a general description (usually matrices) of the system, and compares the similarity among different system development stages which can precisely identify the minor changes in the system. Kinds of flow characteristics can have relatively clear tendencies in the artificial structural lifetime, which indicates that the flow characteristics can be used to evaluate the structure indirectly with a higher QDI than some structural characteristics in the modal analysis with higher sensitivity in the damage diagnosis.

The main experiment objects are two nearly full-scale girders, investigations of which have been rarely conducted in past studies. In the experiment, two kinds of information flow were investigated in data analysis: Displacement flow and acceleration flow. The continuous aggravating damage was applied to the structure in different stages. In this process, the change of the flow characteristics can approximately indicate the tendency of the increasing damage, while some modal parameters (in modal analysis) failed to distinguish the change of structure, which shows the swarm behavior of displacement and the swarm behavior of waves, indicating the multistage redistribution of force and cracks in the structure.

Author Contributions: Conceptualization, X.-F.Y.; methodology, X.-F.Y.; validation, X.-F.Y., C.-W.K. and K.-C.C.; formal analysis, X.-F.Y., K.-C.C.; investigation, X.-F.Y.; resources, O.S.L.V., Y.O.; data curation, X.-F.Y.; writing—original draft preparation, X.-F.Y.; writing—review and editing, X.-F.Y., K.-C.C., C.-W.K., H.O., Y.O., O.S.L.V.; visualization, X.-F.Y.; supervision, C.-W.K., H.O.

Funding: This research received no external funding.

Acknowledgments: The first author would like to express his appreciation for the support from the China Scholarship Council in both life and research. This study was partly supported by a Japanese Society for Promotion of Science (JSPS) Grant-in-Aid for Scientific Research (B) under Project No. 16H04398. That financial support is gratefully acknowledged.

Conflicts of Interest: The authors declare no conflict of interest.

Appendix A

Table A1. Wave characteristics for damage diagnosis.

Wave Characteristics (Flow)		Damage Diagnosis	
Sound (Voice, Song)	**Wave (Flow)**	**Analyzing Methods**	**Data Processing**
Loudness	Amplitude/wavelength	Flow characteristics,	The data in flow analysis are
Acoustic pressure	Distance, sound pressure or intensity, damping	Structural characteristics, etc.	often in a matrix, evaluated by determinant, norm, max eigen,
Auditory impression	direction and speed of sound (velocity)	Frequency response, Time–history response,	2D correlation coefficient, distance, and Procrustes;
Tone	Frequency	etc.	for statistics: expectation,
Musical sound or noise	Harmonic wave	Extreme analysis, pattern recognition,	variance, mode, median; For other data forms, there are
Timbre	Wave pattern	analysis of	some other evaluation
Pitch interval	Wave interval	similarity/difference, etc.	methods.
Concerto, Symphony		Principal component analysis, association analysis, etc.	

Table A2. The characteristics of modal analysis and flow analysis.

Method	Stage	Stage 1	Stage 2	Stage 3	Stage 4	Stage 5	Stage 6	Stage 7	Stage 8	Stage 9
	Fr_29	30.8520	29.8801	29.9573	30.1747	29.7761	29.8173	30.0631	29.5403	29.1955
Modal	Fr_65	66.8703	63.8428	64.0005	64.8976	63.6303	63.5853	64.5042	63.2012	62.5008
Analysis	DR_29	−0.0284	−0.0241	−0.0229	−0.0240	−0.0238	−0.0242	−0.0240	−0.0311	−0.0314
	DR_65	−0.0493	−0.0486	−0.0502	−0.0474	−0.0496	−0.0535	−0.0492	−0.0584	−0.0577
Flow	IoD	1.0000	0.7368	0.7223	0.6059	0.6199	0.2811	0.5612	0.1408	−0.3431
Analysis	Max-peak	1.0000	0.7929	0.7951	0.7722	0.6594	0.7146	0.7184	0.7239	0.6980
using	Max-peak-time	1.0000	0.9453	0.9359	0.9461	0.9453	0.9456	0.9223	0.9399	0.9467
2D-CC	Lifetime	1.0000	0.1936	0.2128	0.2512	0.0047	-0.0276	0.2502	0.4485	0.3660

Note: Fr_29 and Fr_65 denote the frequency around 29 Hz and 65 Hz, respectively, while DR_29 and DR_65 are the damping ratios corresponding to frequency around 29 Hz and 65 Hz, respectively. Flow analysis (2D-CC) is applied on the difference matrix of IoD of displacement flow and the value matrix of max-peak-time, max-peak, and lifetime of acceleration flow.

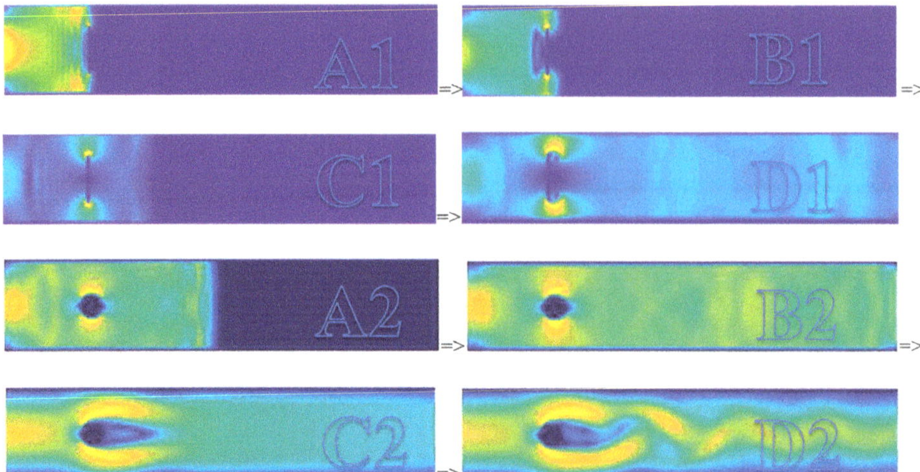

Figure A1. Simulation by the Lattice Boltzmann method (LBM) for the flow in the space in different stages (A, B, C, and D), in which A1, B1, C1, D1 simulate the line–crack barrier, while A2, B2, C2, and D2 simulate the round hole barrier. The flow is imported into the system with a limited boundary continuously.

Note for Figure A1: There are three basic methods to simulate the flow for different fineness: the graph method and element method (finite element method, FEM, and boundary element method, BEM, etc.), responding to the macroscale or LBM (Lattice Boltzmann method) responding to mesoscale (multiscales); and the field method, responding to the microscale. Based on the element method and LBM, the flow in systems can be shown up figuratively. Further, the graph method is often used in the search for the propagation path for energy flow in structures or systems that have clear boundaries, and the method using field theory to describe the flow more mathematically, which is an important aspect in fluid mechanics, wind mechanics, gravitation, and geomagnetics, etc.

References

1. Enright, M.P.; Frangopol, D.M. Survey and evaluation of damaged concrete bridges. *J. Bridge Eng.* **2000**, *5*, 31–38. [CrossRef]
2. Okeil, A.M.; Cai, C.S. Survey of Short and Medium-Span Bridge Damage Induced by Hurricane Katrina. *J. Bridge Eng.* **2008**, *13*, 377–387. [CrossRef]
3. Watanabe, E.; Furuta, H.; Yamaguchi, T.; Kano, M. On longevity and monitoring technologies of bridges: A survey study by the Japanese Society of Steel Construction. *Struct. Infrastruct. Eng.* **2014**, *10*, 471–491. [CrossRef]
4. Ph Papaelias, M.; Roberts, C.; Davis, C.L. A review on non-destructive evaluation of rails: State-of-the-art and future development. *Proc. Inst. Mech. Eng. Part F J. Rail Rapid Transit* **2008**, *222*, 367–384. [CrossRef]
5. Bayraktar, A.; Türker, T.; Tadla, J.; Kurşun, A.; Erdiş, A. Static and dynamic field load testing of the long span nissibi cable-stayed bridge. *Soil Dyn. Earthq. Eng.* **2017**, *94*, 136–157. [CrossRef]
6. Song, W.; Chunguang, X.U.; Pan, Q.; Song, J. Nondestructive testing and characterization of residual stress field using an ultrasonic method. *Chin. J. Mech. Eng.* **2016**, *29*, 365–371. [CrossRef]
7. Stepanova, L.; Roslyakov, P. Multi-parameter description of the crack-tip stress field: Analytic determination of coefficients of crack-tip stress expansions in the vicinity of the crack tips of two finite cracks in an infinite plane medium. *Int. J. Solids Struct.* **2016**, *100*, 11–28. [CrossRef]
8. Prigogine, I. Moderation et transformations irreversibles des systemes ouverts. *Bull. De La Cl. Des Sci. Acad. R. Belg* **1945**, *31*, 600–606.
9. Nicolis, G.; Prigogine, I. *Self-Organization in Nonequilibrium Systems*; Wiley: New York, NY, USA, 1977.

10. Prigogine, I.; René, L. Theory of dissipative structures. In *Synergetics*; Vieweg+ Teubner Verlag: Wiesbaden, Germany, 1977; Volume 1973, pp. 124–135.

11. Jeong, S.; Hou, R.; Lynch, J.P.; Sohn, H.; Law, K.H. An information modeling framework for bridge monitoring. *Adv. Eng. Softw.* **2017**, *114*, 11–31. [CrossRef]

12. Khajavi, R. A novel stiffness/flexibility-based method for euler–bernoulli/timoshenko beams with multiple discontinuities and singularities. *Appl. Math. Model.* **2016**, *40*, 7627–7655. [CrossRef]

13. Ralbovsky, M.; Flesch, S.D.R. Frequency changes in frequency-based damage identification. *Struct. Infrastruct. Eng.* **2010**, *6*, 611–619. [CrossRef]

14. Dammika, A.J. Experimental-Analytical Framework for Damping Change-Based Structural Health Monitoring of Bridges. Ph.D. Thesis, Saitama University, Saitama, Japan, 2014.

15. Pei, C.; Qi, S.; Tang, J. Structural damage identification with multi-objective DIRECT algorithm using natural frequencies and single mode shape. *Soc. Photo-Opt. Instrum. Eng.* **2017**, *10170*, 101702H.

16. Manuele, F.A. Hazard Analysis and Risk Assessment. In *On the Practice of Safety*, 3rd ed.; John Wiley & Sons, Inc.: Hoboken, NJ, USA, 2005; pp. 251–271.

17. Fadeyi, M.O. The role of building information modeling (BIM) in delivering the sustainable building value. *Int. J. Sustain. Built Environ.* **2017**, *6*, 711–722. [CrossRef]

18. Bajzecerová, V.; Kanócz, J. The Effect of Environment on Timber-concrete Composite Bridge Deck. *Procedia Eng.* **2016**, *156*, 32–39. [CrossRef]

19. Armbruster, D.; Martin, S.; Thatcher, A. Elastic and inelastic collisions of swarms. *Phys. D Nonlinear Phenom.* **2017**, *344*, 45–57. [CrossRef]

20. Sagasta, F.; Zitto, M.E.; Piotrkowski, R.; Benavent-Climent, A.; Suarez, E.; Gallego, A. Acoustic emission energy b-value for local damage evaluation in reinforced concrete structures subjected to seismic loadings. *Mech. Syst. Signal Process.* **2018**, *102*, 262–277. [CrossRef]

21. Mesnil, O.; Leckey, C.A.; Ruzzene, M. Instantaneous and local wavenumber estimations for damage quantification in composites. *Struct. Health Monit.* **2014**, *14*, 193–204. [CrossRef]

22. Wang, T.; Cheng, I.; Basu, A. Fluid vector flow and applications in brain tumor segmentation. *IEEE Trans. Biomed. Eng.* **2009**, *56*, 781–789. [CrossRef]

23. Metzger, R.J. The ergodic theory of Axiom A flows. *Invent. Math.* **2000**, *29*, 181–202.

24. Gottschalk, W.H. A survey of minimal sets. *Ann. Inst. Fourier* **1964**, *14*, 53–60. [CrossRef]

25. Glasner, S. *Proximal Flows//Proximal Flows*; Springer: Berlin/Heidelberg, Germany, 1976; pp. 17–29.

26. Iseli, A.; Wildrick, K. Iterated function system quasiarcs. *Conform. Geom. Dyn. Am. Math. Soc.* **2017**, *21*, 78–100. [CrossRef]

27. Sloan, E.D. Clathrate hydrate measurements: Microscopic, mesoscopic, and macroscopic. *J. Chem. Thermodyn.* **2003**, *35*, 41–53. [CrossRef]

28. Wiener, N. *Cybernetics or Control and Communication in the Animal and the Machine*; MIT Press: Cambridge, MA, USA, 1961.

29. Selleri, F. (Ed.) *Wave-Particle Duality*; Plenum Press: London, UK, 1992.

30. Shapiro, A.H. *The Dynamics and Thermodynamics of Compressible Fluid Flow*; John Wiley & Sons: Hoboken, NJ, USA, 1953.

31. Galbis, A.; Maestre, M. *Vector Analysis Versus Vector Calculus*; Springer: New York, NY, USA, 2012.

32. Reynolds, O. *Papers on Mechanical and Physical Subjects, Volume 3, The Sub-Mechanics of the Universe*; Cambridge University Press: Cambridge, UK, 1903.

33. Pascal, J.C.; Carniel, X.; Li, J.F. Characterisation of a dissipative assembly using structural intensity measurements and energy conservation equation. *Mech. Syst. Signal Process.* **2006**, *20*, 1300–1311. [CrossRef]

34. Von Bertalanffy, L. The theory of open systems in physics and biology. *Science* **1950**, *111*, 23–29. [CrossRef] [PubMed]

35. Haag, M.G.; Haag, L.C. *The Reconstructive Aspects of Class Characteristics and a Limited Universe // Shooting Incident Reconstruction*; Elsevier Inc.: Amsterdam, The Netherlands, 2011; Volume 2011, pp. 35–54.

36. Von Bertalanffy, L. *General System Theory*; George Braziller, Inc.: Manhattan, NY, USA, 1968; Volume 41973, p. 40.

37. Blanchard, B.S.; Fabrycky, W.J.; Fabrycky, W.J. *Systems Engineering and Analysis*; Prentice Hall: Englewood Cliffs, NJ, USA, 1990.

38. Hassiotis, S. Identification of Stiffness Reductions Using Natural Frequencies. *J. Eng. Mech.* **1995**, *121*, 1106–1113. [CrossRef]

39. Yamaguchi, H.; Matsumoto, Y.; Kawarai, K.; Dammika, A.J.; Shahzad, S.; Takanami, R. Damage detection based on modal damping change in bridges. In Proceedings of the ICSBE'12, Kandy, Sri Lanka, 14–16 December 2013.

40. Ye, X.F.; Ogai, H.; Kim, C.W. Discrepancy Analysis of Load–Displacement in the Combination Space for Concrete Box Girder Assessment. *Strength Mater.* **2018**, *50*, 695–701. [CrossRef]

41. Poole, D. *Linear Algebra: A Modern Introduction*; Brooks/Cole: Boston, MA, USA, 2011.

42. Goodall, C. Procrustes methods in the statistical analysis of shape. *J. R. Stat. Society. Ser. B* **1991**, *1991*, 285–339. [CrossRef]

43. Tao, L.; Wang, H.B. Detecting and locating human eyes in face images based on progressive thresholding// IEEE International Conference on Robotics and Biomimetics. *IEEE Xplore* **2008**, *2008*, 445–449.

44. Xu, Z.D.; Wu, Z. Simulation of the effect of temperature variation on damage detection in a long-span cable-stayed bridge. *Struct. Health Monit.* **2007**, *6*, 177–189. [CrossRef]

45. Ho RT, H.; Ng, S.M.; Ho DY, F. *The Sage Handbook of Qualitative Research// The SAGE Handbook of Qualitative Research*; Sage Publications: Newcastle upon Tyne, UK, 2005; Volume 2005, pp. 277–279.

46. Chong, O.; Sheng, O. Investigating Predictive Power of Stock Micro Blog Sentiment in Forecasting Future Stock Price Directional Movement. In Proceedings of the International Conference on Information Systems, Icis 2011, Shanghai, China, 4–7 December 2011.

47. Wang, D.; Lu, H.; Bo, C. Visual tracking via weighted local cosine similarity. *IEEE Trans. Cybern.* **2017**, *45*, 1838–1850. [CrossRef] [PubMed]

48. Wang, J.Q.; Wu, J.T.; Wang, J.; Zhang, H.Y.; Chen, X.H. Multi-criteria decision-making methods based on the hausdorff distance of hesitant fuzzy linguistic numbers. *Soft Comput.* **2016**, *20*, 1621–1633. [CrossRef]

49. Princen, J.; Bradley, A. Analysis/synthesis filter bank design based on time domain aliasing cancellation. *IEEE Trans. Acoust. Speechand Signal Process.* **1986**, *34*, 1153–1161. [CrossRef]

sustainability

Article

Time-Frequency Energy Distribution of Ground Motion and Its Effect on the Dynamic Response of Nonlinear Structures

Dongwang Tao [1,2], **Jiali Lin** [3] **and Zheng Lu** [3,*]

[1]   Institute of Engineering Mechanics, China Earthquake Administration, Harbin 150080, China; taodongwang@iem.ac.cn

[2]   Key Laboratory of Earthquake Engineering and Engineering Vibration, China Earthquake Administration, Harbin 150080, China

[3]   Research Institute of Structural Engineering and Disaster Reduction, Tongji University, Shanghai 200092, China; 1832418@tongji.edu.cn

*   Correspondence: luzheng111@tongji.edu.cn; Tel.: +(86)-21-65-986-186

Received: 28 December 2018; Accepted: 24 January 2019; Published: 29 January 2019

Abstract: The ground motion characteristics are essential for understanding the structural seismic response. In this paper, the time-frequency analytical method is used to analyze the time-frequency energy distribution of ground motion, and its effect on the dynamic response of nonlinear structure is studied and discussed. The time-frequency energy distribution of ground motion is obtained by the matching pursuit decomposition algorithm, which not only effectively reflects the energy distribution of different frequency components in time, but also reflects the main frequency components existing near the peak ground acceleration occurrence time. A series of artificial ground motions with the same peak ground acceleration, Fourier amplitude spectrum, and duration are generated and chosen as the earthquake input of the structural response. By analyzing the response of the elasto-perfectly-plastic structure excited by artificial seismic waves, it can be found that the time-frequency energy distribution has a great influence on the structural ductility. Especially if there are even multiple frequency components in the same ground motion phrase, the plastic deformation of the elasto-perfectly-plastic structure will continuously accumulate in a certain direction, resulting in a large nonlinear displacement. This paper reveals that the time-frequency energy distribution of a strong ground motion has a vital influence on the structural response.

Keywords: ground motion; matching pursuit decomposition; time-frequency energy distribution; ratcheting effect; nonlinear response

---

1. Introduction

1.1. Background

With the development of world economy, science, and technology, more and more structures have been constructed in the earthquake risk region, and the corresponding structural seismic design has also gained more attention. Structures undergo vibrations in response to earthquake excitations, which may lead to excessive lateral displacement or even collapse [1]. Consequently, innovative vibration control devices [2–7] have been developed to improve the resilience and sustainability of civil infrastructures under extreme loads.

The understanding of earthquake ground motion is crucial to structural seismic design [8]. Traditionally, a number of parameters, such as peak ground acceleration, peak ground velocity, peak ground displacement, spectral characteristics, duration, etc., are used to describe characteristics of earthquake ground motion. These parameters are not only used to describe the intensity of earthquake ground motion but also to estimate or judge the magnitude of the structural response [9]. In fact, the earthquake ground motion exhibits two types of non-stationarities, namely temporal and spectral non-stationarities. The temporal non-stationarity refers to the time variation of the intensity of the ground motion in the time domain and the spectral non-stationarity refers to the time variation of the energy distribution of the ground motion in the frequency domain. Both the temporal and spectral non-stationarities of the earthquake ground motion may have significant effect on structural response.

For the structural seismic response analysis, various studies have been taken to model and simulate the non-stationarity of earthquake ground motion. The studies concentrate on temporal non-stationarity and various temporal non-stationarity models [10–14] have been proposed to describe earthquake ground motion. However, only temporal non-stationarity of earthquake ground motions cannot fully explain the structural response under earthquake ground motion. Iyama and Kuwamura [15] constructed artificial earthquake ground motions with similar time history curves but found that the resultant structural responses and damages differed greatly. Beck and Papadimitriou [16] investigated nonlinear structural response during seismic waves with time-invariant frequency and time-varying frequency, and found that the time-varying frequency can cause larger response due to the inner resonance. Consequently, several methods considering spectral non-stationary in earthquake input were proposed, such as the Fourier transform [17,18], Hilbert transform [19], generalized harmonic wavelets [20], and wavelet transforms [21–23].

Nonlinear dynamic analysis is a required step in seismic response analysis of many structures, e.g., masonry buildings [24], high-rise buildings [25], cable-stayed bridge [26], and long-span bridge [27]. The focus on the causes of seismic response considering the non-stationarity of earthquake ground motion is helping to improve the engineering insight into earthquake-resistant structural design. With the development of ground motion modeling and elaborate structural analysis, the analysis of structural seismic response has been further developed. Cao and Friswell [28] used wavelet transform to establish local spectrum to reflect the time-frequency energy distribution of earthquake ground motion and found that the difference of energy concentration causes quite a large difference in nonlinear structural displacement. Yu et al. [29] investigated the influence of non-stationary characteristics of ground motion via performing the shaking table test under different ground motion inputs and the results show that the structural response will be increased when the earthquake energy focuses on a certain short period. Zhang et al. [30] used wavelet transform to analyze the impact of earthquake energy distribution in time-frequency domain on structural response and found energy concentration in the time domain can cause an adverse effect on inter-story isolation structures. Saman [31] used wavelet transform to represent the energy input of earthquake ground motions in time and frequency domains and found the amount of input energy and its corresponding frequency contents have a great influence on the level of damage. It indicates that the energy distribution of ground motion has a significant influence on the structural response.

Despite all these efforts, due to the complexity and randomness of the earthquake ground motion, the earthquake input effect concerning spectral non-stationarities has still not been fully investigated. Although it is generally recognized that the energy distribution of ground motion has a significant effect on the structural response, nevertheless, little targeted analysis has been done. In order to highlight the seismic effect of the energy distribution on the structural response, this paper implements a matching pursuit decomposition algorithm for time-frequency analysis. A series of artificial earthquake ground motions with the same peak ground acceleration, Fourier amplitude spectrum, and duration but with different time-frequency energy distribution are generated and chosen as an earthquake input. Hence, the difference in structural responses is mainly determined by the difference in time-frequency energy distribution. A study on the response of certain nonlinear structures excited by such artificial earthquake input, it can be found that the plastic deformation of the structure may accumulate in a certain direction, which is commonly known as the dynamic ratcheting phenomenon, under certain circumstances.

1.2. Scope

The contents of this paper are arranged as follows: In Section 2, a time-frequency analysis method for earthquake ground motion, matching pursuit decomposition algorithm, is implemented to analyze ground motion thus that the obtained time-frequency energy distribution has a high temporal and spectral resolution. Section 3 presents a brief introduction to the dynamic ratcheting phenomenon of the elasto-perfectly-plastic structure. In order to analyze the influence of time-frequency energy distribution of ground motion on the structural response, a series of simple signals and artificial earthquake ground motions are constructed as earthquake input of structure, and the structural responses are compared in Section 4. In Section 5, some conclusions are presented.

2. Matching Pursuit Decomposition Algorithm

For the purpose of constructing seismic waves with the same peak ground acceleration, spectral characteristics, and duration, but with different time-frequency energy distributions, a matching pursuit decomposition method is proposed. It is a common approach used in signal processing [32–39] for analyzing the non-stationarity. For the completeness of the article, this method is briefly introduced as follows.

Matching pursuit (MP) is time-frequency analysis method, as a new generation of spectral decomposition technology, has great application potential in seismic interpretation [40]. The original algorithm is based on the ultra-complete wavelet library composed of the Gabor function, and the time-frequency characteristics of the original signal are obtained by the sum of the Wigner-Ville distributions of the matching wavelets in the wavelet library.

Defined $D = g_{\gamma_n}(t)$ as an ultra-complete redundant time-frequency atom library, the module of each time-frequency atom is 1. The elements in the atomic library are orthogonal bases that are similar to the sine and cosine bases of the Fourier transform, except that these orthogonal bases have finite support intervals and frequency ranges, which allow the resolved signals to have different time and frequency resolutions. Here $g(t)$ is a normalized window function, given by Equation (1).

$$g_{\gamma_n}(t) = \frac{1}{\sqrt{s_n}} g\left(\frac{t - u_n}{s_n}\right) e^{i2\pi f_n t} \tag{1}$$

where $\gamma_n = (s_n, u_n, f_n)$ is the atomic control parameter group, where s_n is the scale, reflecting the time scale of the atom, the larger s_n the higher the frequency resolution, and on the contrary, the smaller s_n the higher the time resolution (lower frequency resolution); u_n is the phase, reflecting the time advance or lag of the atomic center deviating from the center of the parent atom; f_n is the atomic center frequency, reflecting the peak frequency of the atomic oscillation.

Assume that the input signal has a certain correlation with the atom in the dictionary library. This correlation is represented by the inner product of the signal and the atom in the atom library, that is, the larger the inner product, the greater the correlation between the signal and the atom in the dictionary library. Thus the atom can be used to approximate the signal $x(t)$:

$$x_0(t) = \langle x(t),\ g_{\gamma_n}(t)\rangle g_{\gamma_n}(t) \tag{2}$$

There is an error in this representation. The atom is subtracted from the original signal to obtain the residual:

$$Rx(t) = x(t) - x_0(t) \tag{3}$$

Then another atom is selected from the dictionary to represent the residual by calculating the correlation. Iteratively performing the above steps, the signal residual will become smaller and smaller. When the error is satisfied, the iteration is terminated, and a set of atoms are obtained. The set of atoms can be linearly combined to reconstruct the input signal.

Thus for any signal $x(t)$, a set of obtained atoms can be used to approximate the input signal [41]:

$$x(t) = \sum_{n=0}^{\infty} x_n(t) \tag{4}$$

$$x_n(t) = \langle R^n x(t), \, g_{\gamma_n}(t) \rangle g_{\gamma_n}(t) \tag{5}$$

$$R^n x(t) = \langle R^n x(t), \, g_{\gamma_n}(t) \rangle g_{\gamma_n}(t) + R^{n+1} x(t) \tag{6}$$

where $x_n(t)$ is the nth component of the decomposition; $R^n x(t)$ is the nth signal residual and $R^0 x(t) = x(t)$; $g_{\gamma_n}(t)$ is the nth optimal matching time-frequency atom, making the inner product of $R^n x(t)$ and $g_{\gamma_n}(t)$ is the largest.

Since the signal $x(t)$ to be decomposed is usually a real signal, each component obtained by the above decomposition algorithm must be converted into a real component. It can be realized by converting a complex time-frequency atom into a real time-frequency atom, shown in Equation (7).

$$g_{(\gamma_n,\phi_n)}(t) = \frac{C_{(\gamma_n,\phi_n)}}{2} \left(e^{i\phi_n} g_{\gamma_n}(t) + e^{-i\phi_n} g_{\gamma_n^-}(t) \right) \tag{7}$$

where $C_{(\gamma_n,\phi_n)}$ is the constant coefficient making the adjusted real time-frequency atom's modulus be 1; ϕ_n is the phase angle of the complex time-frequency atom; γ_n^- is the conjugate parameter group of γ_n, satisfying $\gamma_n^- = (s_n, u_n, -f_n)$.

Hence, the component signal obtained by the matching pursuit decomposition is obtained. (Equation (8)):

$$x(t) = \sum_{n=0}^{\infty} \left\langle R^n x(t), \, g_{(\gamma_n,\phi_n)}(t) \right\rangle g_{(\gamma_n,\phi_n)}(t) \tag{8}$$

Furthermore, the Wigner-Ville Transform is used to obtain the time-frequency energy distribution of signal.

Wigner-Ville Transform is defined as:

$$Wg(t, f) = \int_{-\infty}^{\infty} Hg\left(t + \frac{\tau}{2}\right) \overline{Hg}\left(t - \frac{\tau}{2}\right) e^{-if\tau} d\tau \tag{9}$$

where H stands for Hilbert Transform; $^-$ is the complex conjugate symbol; and the definition of Hilbert Transform is:

$$Hg(t) = \frac{1}{\pi} \int_{-\infty}^{\infty} \frac{g(u)}{t - u} du \tag{10}$$

Consequently, the time-frequency energy distribution of the input signal can be represented as:

$$Ex(t, f) = \frac{1}{2} \sum_{n=0}^{\infty} \left| \left\langle R^n x(t), g_{(\gamma_n,\phi_n)}(t) \right\rangle \right|^2 \left(Wg_{(\gamma_n,\phi_n)}(t, f) + Wg_{(\gamma_n^-,\phi_n)}(t, f) \right) \tag{11}$$

The above matching pursuit decomposition algorithm can be implemented by Matlab. Take an actual earthquake ground motion (Hachinohe) for example, the time-frequency energy distribution obtainment flow chart is shown in Figure 1. Due to the adaptability of the algorithm, the time-frequency energy distribution obtained by the matching pursuit decomposition algorithm has a relatively high time and frequency resolution.

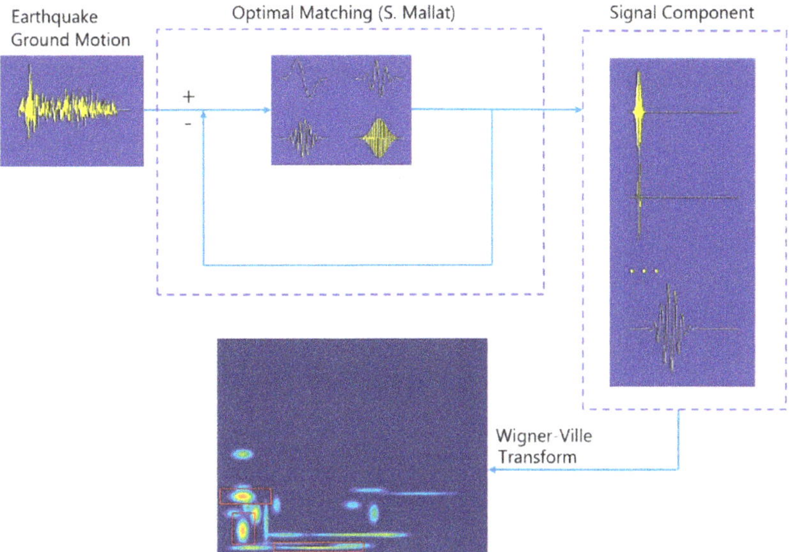

Figure 1. Flowchart of obtaining the time-frequency energy distribution.

3. Dynamic Ratcheting Phenomena

Dynamic ratcheting usually occurs in nonlinear systems. It means the structural plastic deformation may accumulate in a certain direction under certain excitations, which makes the structure produce large nonlinear displacement. The excessive displacement under seismic excitation may even cause progressive collapse, which should be avoided in practical projects. In order to facilitate the reader to understand the dynamic ratcheting phenomenon, this section briefly introduces the origination of dynamic ratcheting as follows.

The equation of motion for an elasto-perfectly-plastic single degree of freedom structure is given by Equation (12).

$$m\ddot{x} + c\dot{x} + R(x, \dot{x}) = -m\ddot{x}_g \tag{12}$$

where R is the nonlinear restoring force, satisfying:

$$\frac{dR}{dt} = \begin{cases} k\frac{dx}{dt}, & when \ |R| < |R_y| \\ 0, & else \end{cases} \tag{13}$$

where R_y is the structure yield force.

Structure ductility μ is:

$$\mu = \frac{x}{x_y} \tag{14}$$

where x_y is the structural yield displacement.

Assuming $r = \frac{R}{R_y}$, then

$$\frac{dr}{dt} = \begin{cases} \dot{\mu}, & when \ |r| < 1 \\ 0, & else \end{cases} \tag{15}$$

Substituting μ, r into Equation (12) obtains the equation of ductility, shown in Equation (16).

$$\ddot{\mu} + 2\xi\omega_0\dot{\mu} + \omega_0^2 r(\mu, \dot{\mu}) = -\frac{1}{x_y}\ddot{x}_g \tag{16}$$

where ζ is the damping ratio; ω_0 is the angular frequency before yield.

Take simple sinusoidal excitation as an example to explore the origination of the dynamic ratcheting phenomenon of an elasto-perfectly-plastic structure. Input $\ddot{x}_g$ is given by Equation (17).

$$\ddot{x}_g = \frac{1}{C}(sin(2\pi f_0 t) + sin(2\pi n f_0 t)) \tag{17}$$

where f_0 is the fundamental frequency of the sinusoidal excitation; n is the frequency ratio of two input excitation.

Supposing an elasto-perfectly-plastic structure model, whose mass is 1 kg, yield displacement is 0.01 m, structural frequency before yielding is 1 Hz, and the damping ratio is 0.05. Excite the elasto-perfectly-plastic structure by two sine waves. Since the fundamental frequency of the sine waves is 1 Hz, when the frequency ratio is rational, e.g., $1.5, 2, 2.5, 3, 3.5, 4, 5$, and 8 respectively, the ductility time-history is shown in Figure 2; when the frequency ratio is irrational, e.g., $\sqrt{2}, \sqrt{3}, \sqrt{5}, \sqrt{6}, \sqrt{7}, \sqrt{8}, \sqrt{10}$, and $\sqrt{11}$, respectively, the ductility time-history is shown in Figure 3. It can be seen from Figure 2 that when the frequency ratio is even, the dynamic ratcheting phenomenon is obvious, and the dynamic ratcheting effect is more significant with a smaller frequency ratio, and when the frequency ratio is 2 the structural ductility demand is far more than other cases. On the other hand, it is known from Figure 3 that when the frequency ratio is irrational, dynamic ratcheting does not occur.

In conclusion, the elasto-perfectly-plastic structure shows a very significant dynamic ratcheting phenomenon under sinusoidal excitation with a frequency ratio of 2; but this requires the multiple frequencies occurring within the same time period, if not at the same time, even if the frequency ratio is 2, there will be no dynamic ratcheting.

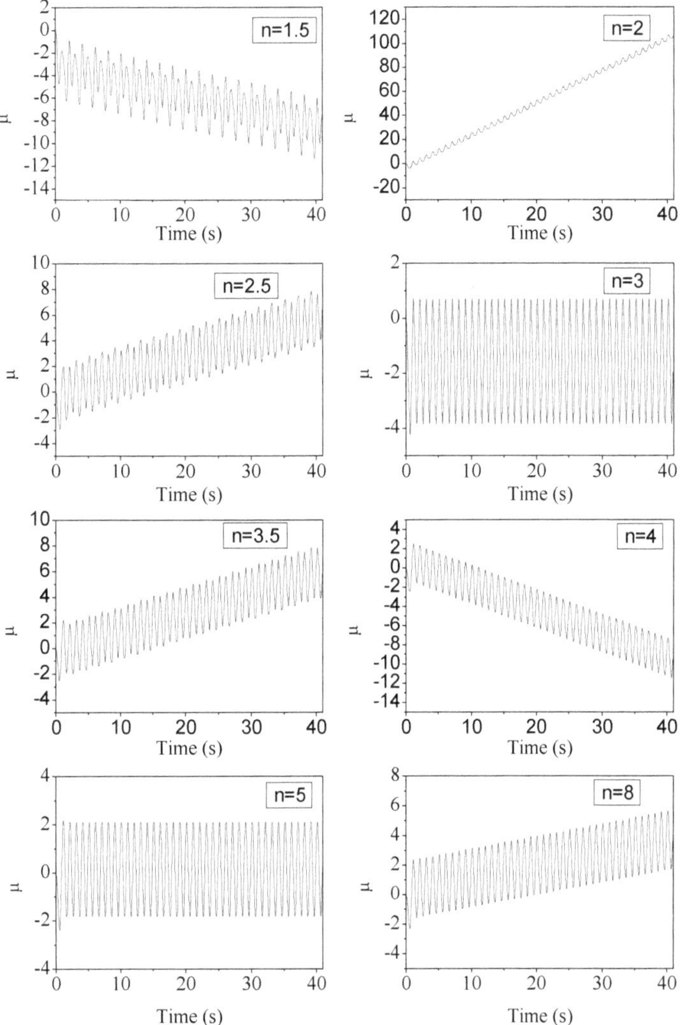

Figure 2. Ductility of Elasto-Perfectly-Plastic model under sine waves with rational frequency ratio (structural frequency is 1 Hz before yielding).

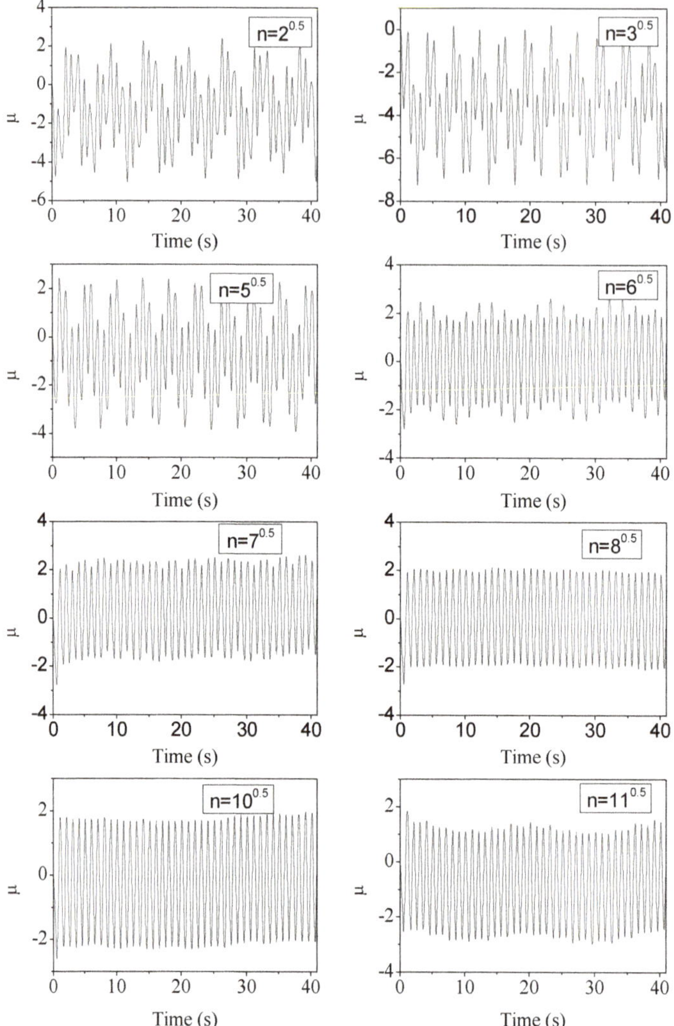

Figure 3. Ductility of Elasto-Perfectly-Plastic model under sine waves with irrational frequency ratio (structural frequency is 1 Hz before yielding).

4. Influence of Time-Frequency Energy Distribution of Ground motion on Structural Response

This section aims to analyze the influence of the time-frequency characteristics of ground motion on the structural response. It is well known that as structural members yielded, the dominant frequency of the nonlinear structure decreased [42,43], which may cause a resonance if the dominant frequency being equal to the time-varying frequency of the earthquake ground motion. Moreover, if the earthquake ground motion has multiple frequency components at the same time period, the dynamic ratcheting phenomenon will occur, resulting in a large nonlinear displacement of the structure [44–46]. For simplicity, only the elasto-perfectly-plastic structure is considered in this paper. The seismic nonlinear response of a practical project is much more complex.

4.1. Influence of Time-Frequency Energy Distribution of Simple Input on Structural Response

Construct two artificial signals with the same peak acceleration, duration, and power spectrum, but different time-frequency energy distributions. The power spectrum is considered here instead of the spectrum to prevent spectral leakage. The two signals have a sampling frequency of 100 Hz and the signal duration of 40.95 s. The peak ground acceleration (PGA) of the first signal is 1 m/s^2 at the time instant 9.99 s; the PGA of the second signal is also 1 ms^{-2}, while at the time instant 10.09 s. When using a 512-point Hanning window, a 256-point overlap, and the 4096-point Fast Fourier Transform, the smooth power spectrum density of the two signals is the same at 0–5 Hz. Their time history and smoothed power spectrum are shown in Figure 4.

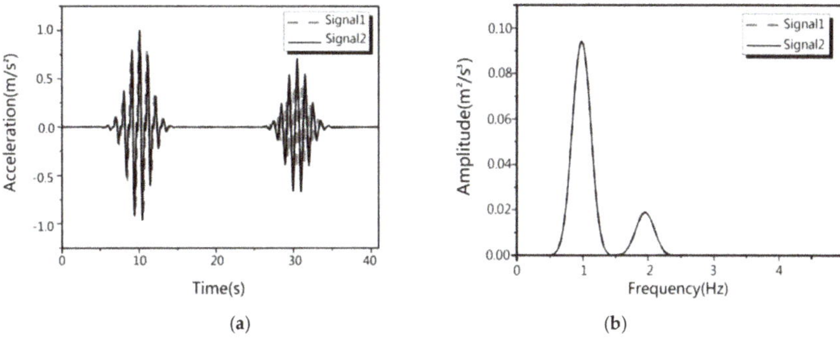

Figure 4. Two artificial signals with the same peak ground acceleration (PGA) and smoothed power spectrum density. (**a**) Time history. (**b**) Smoothed power spectrum density.

The time-frequency energy distribution of the above signals obtained by the matching pursuit decomposition is shown in Figure 5. It can be seen that although the smooth power spectra of the two signals are exactly the same, the respective frequency evolution processes are fairly different, so are the corresponding time-frequency energy distribution maps. The energy of the first signal in the 1 Hz band is concentrated within 5–15 s, and the energy in the 2 Hz band is concentrated in 25–35 s; while the energy of the second signal in the 1 Hz band is concentrated in both 5–15 s and 25–35 s, the energy in the 2 Hz band is concentrated within 5–15 s.

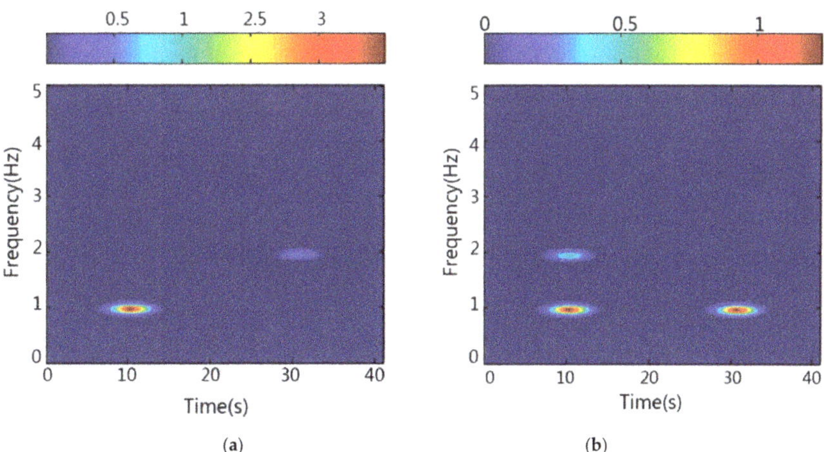

Figure 5. Time-frequency energy distribution of the two artificial signals. (**a**) First artificial signal. (**b**) Second artificial signal.

Analyze the structural response of the above elasto-perfectly-plastic structure under the two simple artificial signals. The ductility time history is shown in Figure 6, and the force-displacement relationship is shown in Figure 7, respectively.

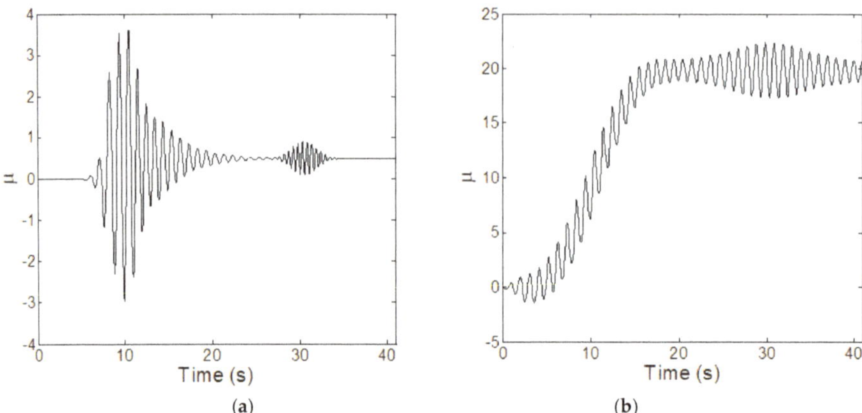

Figure 6. Ductility of the Elasto-Perfectly-Plastic model under the two artificial signals (structural frequency is 1 Hz before yielding). (**a**) First artificial signal. (**b**) Second artificial signal.

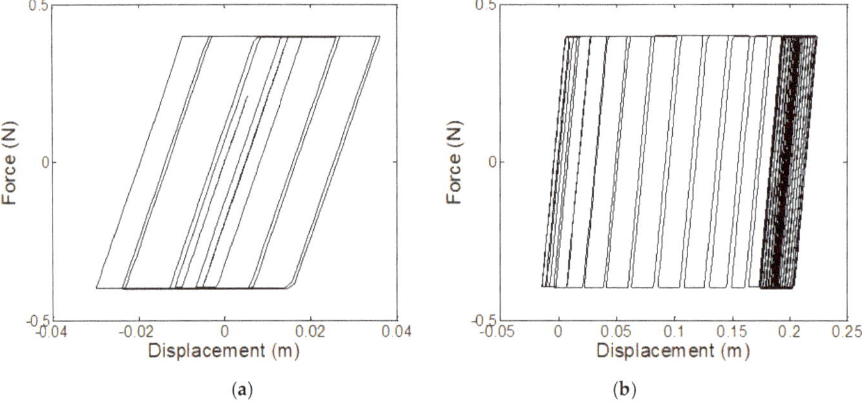

Figure 7. Force-displacement curve of the Elasto-Perfectly-Plastic model under the two artificial signals (structural frequency is 1 Hz before yielding). (**a**) First artificial signal. (**b**) Second artificial signal.

It can be seen from Figure 6 that the ductility of the elasto-perfectly-plastic structure under the second signal is much larger than that under the first one. The former has a peak ductility of more than 20 and a residual ductility of about 20; while the latter has a peak ductility of no more than 5 and a residual ductility of about 0.5. According to the time-frequency energy distribution (Figure 5), the second signal has both the energy of 1 Hz band and 2 Hz band concentrated in 5–15 s, and the frequency ratio is 2, resulting in dynamic ratcheting phenomenon; since there is only a 1 Hz band energy in 25–35 s, dynamic ratcheting phenomenon does not occur in the period of 25-35 s. In comparison, the first signal has only the 1 Hz band energy concentrated in 5–15 s, and only 2 Hz band energy in 25–35 s respectively. Without multiple frequencies occurring within the same time period, the structure does not show dynamic ratcheting under the first signal. This indicates that the occurrence of structural dynamic ratcheting is closely related to the time-frequency energy distribution input.

Meanwhile, the force-displacement curve in Figure 7a is more symmetrical, and the curve is more evenly distributed around zero displacement point; while the force-displacement curve in Figure 7b is concentrated on the right side of zero displacement point with a residual displacement of 0.2 m. This shows that the time-frequency energy distribution of the input signal has a great influence on the ductile response of the elasto-perfectly-plastic structure, and if there are frequency components with even frequency ratio at the same time, the structure may have a complex response such as a dynamic ratcheting effect.

When the elasto-perfectly-plastic structure frequency is 2 Hz before yielding and the yield displacement is 0.01 m, the structure ductility time history under the two artificial signals is shown in Figure 8, and the force-displacement relationship is shown in Figure 9. Results show that the structure also experienced dynamic ratcheting under the second signal, due to the fact that the second signal has both the energy of 1 Hz band and 2 Hz band in 5-15 s, and the frequency ratio is even, thereby the dynamic ratcheting phenomenon occurs.

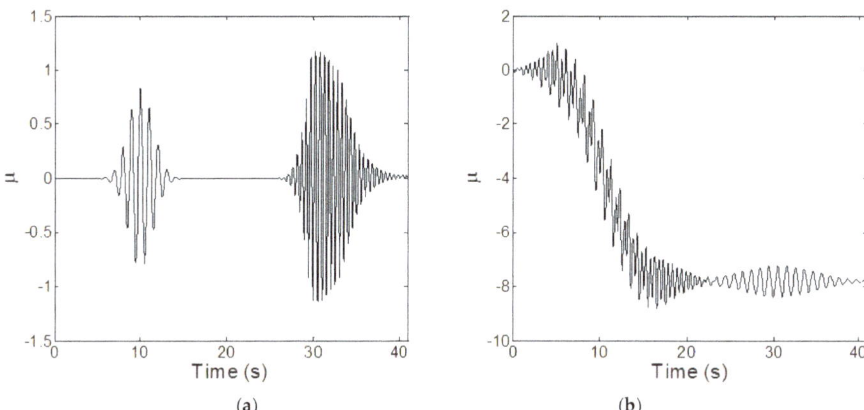

Figure 8. Ductility of Elasto-Perfectly-Plastic model under the two artificial signals (structural frequency is 2 Hz before yielding). (**a**) First artificial signal. (**b**) Second artificial signal.

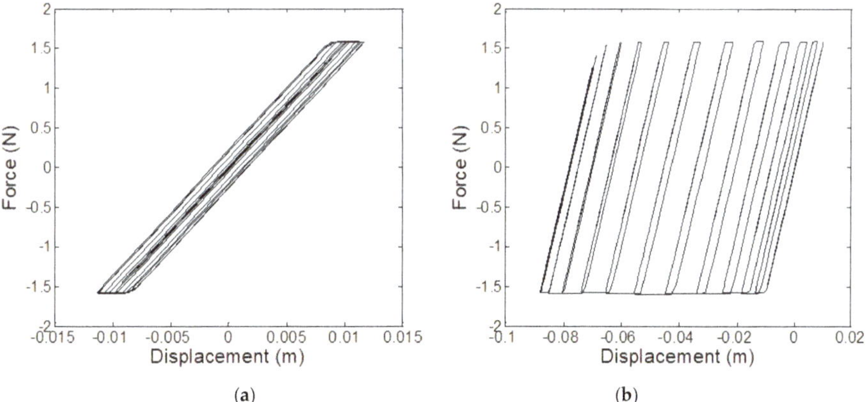

Figure 9. Force-displacement curve of Elasto-Perfectly-Plastic model under the two artificial signals (structural frequency is 2 Hz before yielding). (**a**) First artificial signal. (**b**) Second artificial signal.

4.2. Influence of Time-Frequency Energy Distribution of Ground Motion on Structural Response

Since it is difficult to find an actual earthquake ground motion with the same ground motion parameters as the above constructed two signals, this section constructs a series of artificial ground motions based on the actual earthquake ground motion records. Similarly, we construct the artificial ground motions with the same peak acceleration, spectral characteristics, and duration to fully analyze the seismic effect of the energy distribution of the ground motion.

The artificial earthquake ground motion generation flow chart is shown in Figure 10. The basic idea is that the Fourier amplitude spectra of the generated ground motions are from those of typical earthquake ground motions (for simplicity these ground motions are noted as GM-A) and the phase spectra are from those of ground motions with PGA larger than a certain value like 0.2 g (for simplicity these ground motions noted as GM-B). The generation procedures in details are as follows:

(1) Take Fourier transform of GM-A and get the amplitude and phase spectrum, meanwhile take time-frequency transform to get the center frequency with the largest time frequency atom near the time instant when PGA occurs;

(2) Substitute the phase spectrum of GM-A by that of GM-B and make an inverse Fourier transform to get the real sequence signals and take the time-frequency transform of the generated real sequence signals and get the time-frequency atom with the largest energy.

(3) Replace the time-frequency atom with largest energy by that of GM-A, and adjust the local PGA thus that the PGA of the new generated real time sequence is equal to that of GM-A;

(4) Take the Fourier transform of the new generated sequence signal to get the amplitude and phase spectrum and to check if the amplitude spectrum is same as that of GM-A in the specified interval. If it is true, the new generated sequence signal is a passable artificial earthquake ground motion. Otherwise modify the phase spectrum and go back to step (2).

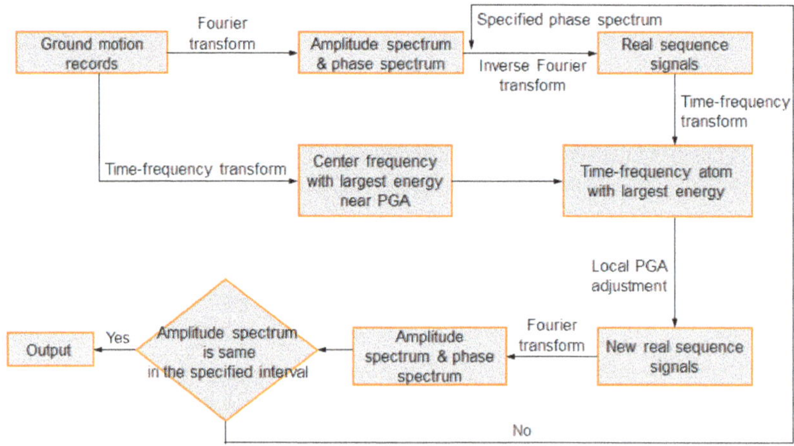

Figure 10. Flowchart of generating artificial ground motion.

In this study, the typical earthquake ground motion records referred to Chichi, El Centro, Northridge, Parkfield, and the Kobe earthquake ground motion records and their peak accelerations and peak frequencies are shown in Table 1. The phase spectra referred to the records in the horizontal direction with peak accelerations larger than 0.2 g in the Chichi, Imperial Valley, Northridge, Parkfield, and Kobe earthquakes, as shown in Table 2, for a total of 169 records.

Table 1. Typical ground motion records.

Site and Component, Ground Motion, Date	Peak Acceleration (g)	Peak Frequency (Hz)
TCU052-W, Chichi (20 September 1999)	0.348	0.452
I-ELC180, Imperial Valley (19 May 1940)	0.313	1.465
ORR090, Northridge (17 January 1994)	0.568	1.221
C02065, Parkfield (28 June 1966)	0.476	1.428
KJM000, Kobe (16 January 1995)	0.821	1.453

Table 2. Ground motion records generating the phase spectra.

Ground Motion	Number of Ground Motion Records
Chichi (20 September 1999)	71
Imperial Valley (19 May 1940)	2
Kobe (16 January 1995)	12
Northridge (17 January 1994)	77
Parkfield (28 June 1966)	7

Model an elasto-perfectly-plastic structure with a before-yielding frequency of 1 Hz, a yield displacement of 0.01 m, a damping ratio of 0.05, and a mass of 1 kg. Excite the structure by the generated artificial ground motions. If the ductility time history of the structure is provided with these characteristics: (1) The maximum ductility coefficient is greater than 3; (2) the plastic deformation of the structure continuously accumulates in a certain direction and the duration is greater than the three before-yielded period of the structure; the structure dynamic ratcheting phenomenon occurs. As a result, there are only seven artificial seismic waves generated according to above procedures, leading to the structural dynamic ratcheting phenomenon. The referred earthquake ground motion records are shown in Table 3.

Table 3. Ground motion records generating dynamic ratcheting.

Site and Component, Ground Motion, Date	Peak Acceleration (g)	Peak Frequency (Hz)
CHY006-E,Chichi (20 September 1999)	0.364	0.595
CHY028-N,Chichi (20 September 1999)	0.821	1.190
CHY029-N,Chichi (20 September 1999)	0.238	0.354
CHY029-W,Chichi (20 September 1999)	0.277	0.403
CHY101-W,Chichi (20 September 1999)	0.353	0.336
I-ELC180,Imperial Valley (19 May 1940)	0.313	1.465
KAK090,Kobe (16 January 1995)	0.345	0.586

Take CHY006-E and KAK090 for example to analyze the effect of the time-frequency energy distribution on the structural nonlinear response.

The time-frequency energy distribution of the CHY006-E ground motion is shown in Figure 11a. The ductility of the elasto-perfectly-plastic structure is shown in Figure 11b. As shown in Figure 11a, the center frequencies of the time-frequency atom #1, #2, #3, and #4, are 0.586 Hz, 1.563 Hz, 1.758 Hz, and 3.516 Hz, respectively. Not only that the atomic center frequency of the atom #2 is twice the center frequency of #1; similarly that of the atom #4 is twice of #3. It can be seen from Figure 11b that the dynamic ratcheting phenomenon occurs within 9–18 s, which is consistent with the time-frequency energy distribution in Figure 11a, because it exists time-frequency atoms whose center frequency ratio is twice during the period. Besides, no dynamic ratcheting occurred during other periods.

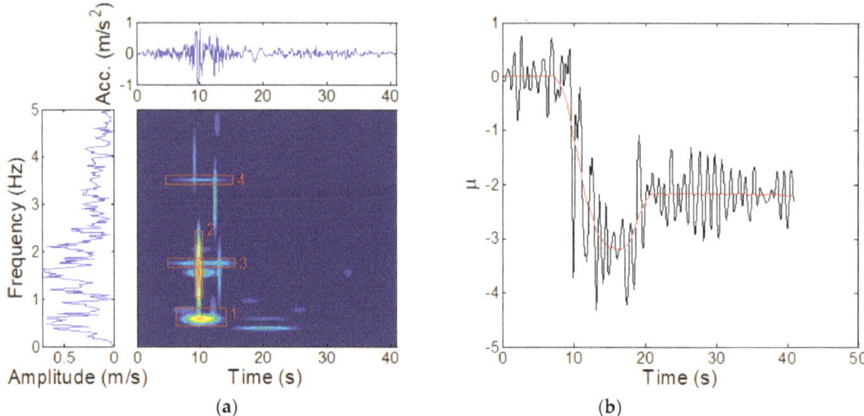

Figure 11. (**a**) Time-frequency energy distribution of CHY006-E ground motion. (**b**) Ductility of the Elasto-Perfectly-Plastic model under the CHY006-E ground motion.

The time-frequency energy distribution of the KAK090 ground motion is shown in Figure 12a. The ductility of the elasto-perfectly-plastic structure is shown in Figure 12b. As shown in Figure 12a, the center frequencies of the time-frequency atom #1, #2, #3, #4 are 0.488 Hz, 0.976 Hz, 1.824 Hz, and 2.734 Hz respectively. Moreover the atomic center frequency of the atom #2 is twice the center frequency of #1; and that of the atom #4 is 1.5 times of #3. It can be seen from Figure 12b that the dynamic ratcheting phenomenon occurs within 6–14 s, which is consistent with the time-frequency energy distribution in Figure 12a, because there is time-frequency atoms whose center frequency ratio is twice and 1.5 times during the period. Besides, no dynamic ratcheting occurred during other periods as well.

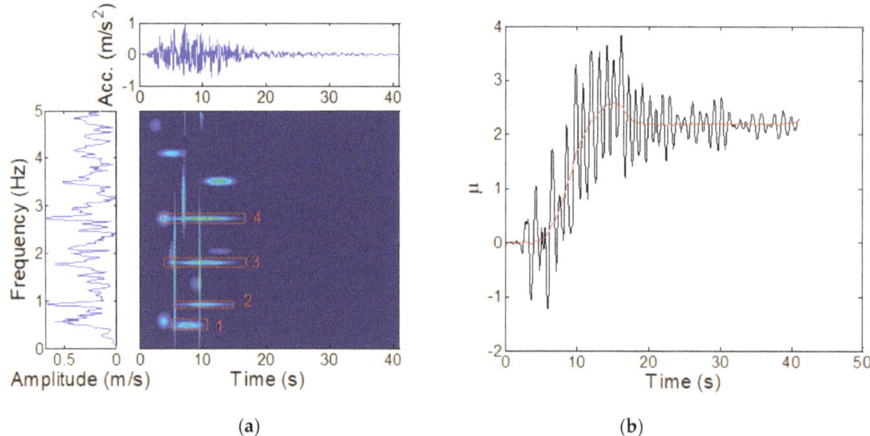

Figure 12. (**a**) Time-frequency energy distribution of KAK090 ground motion. (**b**) Ductility of Elasto-Perfectly-Plastic model under the KAK090 ground motion.

5. Conclusions

This paper proposes an earthquake ground motion generation method based on the matching pursuit decomposition algorithm and time-frequency energy distribution; constructs a series of simple signals and artificial earthquake ground motions with the same peak ground acceleration,

Fourier amplitude spectrum, and duration; explains the dynamic ratcheting phenomenon of the elasto-perfectly-plastic structure. The main conclusions are as follows:

(1) The proposed decomposition for the earthquake ground motion has both high time resolution and high frequency resolution. The obtained time-frequency energy distribution can effectively reflect the energy distribution of the frequency components at any time and also reflect the frequency components near the peak acceleration.

(2) If there are frequency components with an even frequency ratio in a certain time duration, the response of the elasto-perfectly-plastic structure experiences a dynamic ratchet phenomenon. If these frequency components appear at different time periods, the elasto-perfectly-plastic structure does not undergo dynamic ratcheting.

(3) The time-frequency energy distribution could be further considered as a characteristic variable for ground motion besides peak ground acceleration, frequency spectrum, and duration, which has a vital effect on the structural response, especially in the nonlinear range.

Author Contributions: Funding acquisition, Z.L.; investigation, D.T.; project administration, D.T. and Z.L.; software, D.T. and J.L.; writing – original draft, J.L.; writing – review and editing, Z.L.

Funding: This research was funded by Scientific Research Fund of Institute of Engineering Mechanics, China Earthquake Administration, grant number 2018D01, 2014B08, 2016A03. This research was also funded by China Scholarship Council, grant number 201704190039. The APC was funded by Scientific Research Fund of Institute of Engineering Mechanics, China Earthquake Administration, grant number 2018D01, 2014B08, 2016A03.

Acknowledgments: Financial support from the Scientific Research Fund of Institute of Engineering Mechanics, China Earthquake Administration (Grant No. 2018D01, 2014B08, 2016A03) is highly appreciated. The work is also supported by the China Scholarship Council (Grant No. 201704190039).

Conflicts of Interest: The authors declare no conflict of interest.

References

1. Lu, Z.; Chen, X.Y.; Lu, X.L. Shaking table test and numerical simulation of an RC frame-core tube structure for earthquake-induced collapse. *Earthq. Eng. Struct. Dyn.* **2016**, *45*, 1537–1556. [CrossRef]
2. Spencer, B.F.; Yao, J.T.P. Structural control: Past, present, and future. *J. Eng. Mech. (ASCE)* **1997**, *123*, 897–971.
3. Spencer, B.F., Jr.; Nagarajaiah, S. State of the Art of Structural Control. *J. Struct. Eng.* **2003**, *129*, 845–856. [CrossRef]
4. Housner, G.W.; Bergman, L.A.; Caughey, T.K.; Chassiakos, A.G.; Claus, R.O.; Masri, S.F.; Skelton, R.E.; Soong, T.T.; Li, S.; Tang, J. On vibration suppression and energy dissipation using tuned mass particle damper. *J. Vib. Acoust.* **2017**, *139*, 011008.
5. Lu, Z.; Chen, X.Y.; Li, X.W.; Li, P.Z. Optimization and application of multiple tuned mass dampers in the vibration control of pedestrian bridges. *Struct. Eng. Mech.* **2017**, *62*, 55–64. [CrossRef]
6. Lu, Z.; Chen, X.Y.; Zhou, Y. An equivalent method for optimization of particle tuned mass damper based on experimental parametric study. *J. Sound Vib.* **2018**, *419*, 571–584. [CrossRef]
7. Lu, Z.; Huang, B.; Zhou, Y. Theoretical study and experimental validation on the energy dissipation mechanism of particle dampers. *Struct. Control Health Monit.* **2018**, *25*, e2125. [CrossRef]
8. Hu, Y.X. *Earthquake Engineering*; Seismological Press: Beijing, China, 2006; pp. 380–382.
9. Bozorgnia, Y.; Bertero, V.V. *Earthquake Engineering: From Engineering Seismology to Performance-Based Engineering*; CRC Press: Boca Raton, FL, USA, 2004.
10. Cacciola, P.; Deodatis, G. A method for generating fully non-stationary and spectrum-compatible ground motion vector processes. *Soil Dyn. Earthq. Eng.* **2011**, *31*, 351–360. [CrossRef]
11. Sgobba, S.; Stafford, P.; Marano, G.; Guaragnella, C. An evolutionary stochastic ground-motion model defined by a seismological scenario and socal site conditions. *Soil Dyn. Earthq. Eng.* **2011**, *31*, 1465–1479. [CrossRef]
12. Zentner, I.; Poirion, F. Enrichment of seismic ground motion databases using Karhunen–Loève expansion. *Soil Dyn. Earthq. Eng.* **2012**, *41*, 1945–1957. [CrossRef]
13. Zhang, Y.; Zhao, F.; Yang, C. Generation of nonstationary artificial ground motion based on the Hilbert transform. *Bull. Seismol. Soc. Am.* **2012**, *102*, 2405–2419. [CrossRef]

14. Rezaeian, S.; Der Kiureghian, A. Simulation of orthogonal horizontal ground motion components for specified earthquake and site characteristics. *Soil Dyn. Earthq. Eng.* **2012**, *41*, 335–353. [CrossRef]
15. Iyama, J.; Kuwamura, H. Application of wavelets to analysis and simulation of earthquake motions. *Soil Dyn. Earthq. Eng.* **1999**, *28*, 255–272. [CrossRef]
16. Beck, J.L.; Papadimitriou, C. Moving resonance in nonlinear response to fully nonstationary stochastic ground motion. *Probab. Eng. Mech.* **1993**, *8*, 157–167. [CrossRef]
17. Stockwell, R.G.; Mansinha, L.; Lowe, R. Localization of the complex spectrum: The S transform. *IEEE Trans. Signal Process.* **1996**, *44*, 998–1001. [CrossRef]
18. Wu, Z.J.; Zhang, J.J.; Wang, Z.J.; Wu, X.X.; Wang, M.Y. Time-frequency analysis on amplification of seismic ground motion. *Rock Soil Mech.* **2017**, *138*, 685–695.
19. Zhang, Y.S.; Zhao, F.X. Validation of non-stationary ground motion simulation method based on Hilbert transform. *Acta Seismol. Sin.* **2014**, *36*, 686–697.
20. Wang, D.; Fan, Z.L.; Hao, S.W.; Zhao, D.H. An evolutionary power spectrum model of fully nonstationary seismic ground motion. *Soil Dyn. Earthq. Eng.* **2018**, *105*, 1–10. [CrossRef]
21. Naga, P.; Eatherton, M.R. Analyzing the effect of moving resonance on seismic response of structures using wavelet transforms. *Soil Dyn. Earthq. Eng.* **2013**, *43*, 759–768. [CrossRef]
22. Yazdani, A.; Takada, T. Wavelet-based generation of energy- and spectrum-compatible earthquake time histories. *Comput.-Aided Civ. Infrastruct. Eng.* **2009**, *24*, 623–630. [CrossRef]
23. Sun, D.B.; Ren, Q.W. Seismic damage analysis of concrete gravity dam based on wavelet transform. *Shock Vib.* **2016**, *2016*, 6841836. [CrossRef]
24. Altunisik, A.C.; Genc, A.F.; Gunaydin, M.; Okur, F.Y.; Karahasan, O.S. Dynamic response of a historical armory building using the finite element model validated by the ambient vibration test. *J. Vib. Control* **2018**, *24*, 5472–5484. [CrossRef]
25. Bai, Y.T.; Guan, S.Y.; Lin, X.C.; Mou, B. Seismic collapse analysis of high-rise reinforced concrete frames under long-period ground motions. *Struct. Des. Tall Spec. Build.* **2019**, *28*, e1566. [CrossRef]
26. Han, Q.; Wen, J.N.; Du, X.L.; Zhong, Z.L.; Hao, H. Nonlinear seismic response of a base isolated single pylon cable-stayed bridge. *Eng. Struct.* **2018**, *175*, 806–821. [CrossRef]
27. Li, X.Q.; Li, Z.X.; Crewe, A.J. Nonlinear seismic analysis of a high-pier, long-span, continuous RC frame bridge under spatially variable ground motions. *Soil Dyn. Earthq. Eng.* **2018**, *114*, 298–312. [CrossRef]
28. Cao, H.; Friswell, M.I. The effect of energy concentration of earthquake ground motions on the nonlinear response of RC structures. *Soil Dyn. Earthq. Eng.* **2009**, *29*, 292–299. [CrossRef]
29. Yu, R.F.; Fan, K.; Peng, L.Y.; Yuan, M.Q. Effect of non-stationary characteristics of ground motion on structural response. *China Civ. Eng. J.* **2010**, *43*, 13–20.
30. Zhang, S.R.; Du, Y.F.; Hui, Y.X. Seismic response analysis of isolation structures based on wavelet transformation. *China Earthq. Eng. J.* **2017**, *39*, 836–842.
31. Yaghmaei-Sabegh, S. Time-frequency analysis of the 2012 double earthquakes records in north-west of Iran. *Bull. Earthq. Eng.* **2014**, *12*, 585–606. [CrossRef]
32. Mallat, S.G.; Zhang, Z.F. Matching pursuits with time-frequency dictionaries. *IEEE Trans. Signal Proc.* **1993**, *41*, 3397–3415. [CrossRef]
33. Neff, R.; Zakhor, A. Very low bit-rate video coding based on matching pursuits. *IEEE Trans. Circuits Syst. Video Technol.* **1997**, *7*, 158–171. [CrossRef]
34. Gribonval, R. Fast matching pursuit with a multiscale dictionary of Gaussian chirps. *IEEE Trans. Signal Proc.* **2001**, *49*, 994–1001. [CrossRef]
35. Gribonval, R.; Bacry, E. Harmonic decomposition of audio signals with matching pursuit. *IEEE Trans. Signal Proc.* **2003**, *51*, 101–111. [CrossRef]
36. Tropp, J.A.; Gilbert, A.C. Signal recovery from random measurements via orthogonal matching pursuit. *IEEE Trans. Signal Proc.* **2007**, *53*, 4655–4666. [CrossRef]
37. Needell, D.; Vershynin, R. Signal recovery from incomplete and inaccurate measurements via regularized orthogonal matching pursuit. *IEEE J. Sel. Top. Signal Proc.* **2010**, *4*, 310–316. [CrossRef]
38. Yang, B.; Li, S.T. Pixel-level image fusion with simultaneous orthogonal matching pursuit. *Inf. Fusion* **2012**, *13*, 10–19. [CrossRef]

39. Liu, X.Y.; Zhao, Z.G. Research on compressed sensing reconstruction algorithm in complementary space. In Proceedings of the IEEE 17th International Conference on Communication Technology (ICCT), Chengdu, China, 27–30 October 2017.

40. Charles, I.P.; John, P.C. Comparison of frequency attributes from CWT and MPD spectral decompositions of a complex turbidite channel model. In Proceedings of the Expanded Abstracts of 78th SEG Annual International Meeting, Las Vegas, NV, USA, 9–14 November 2008; pp. 394–397.

41. Tao, D.W.; Li, H. Time-frequency energy distribution of wenchuan earthquake motions based on matching pursuit decomposition. In Proceedings of the 7th International Conference on Urban Earthquake Engineering (7CUEE) & 5th International Conference on Earthquake Engineering (5ICEE), Tokyo, Japan, 3–5 March 2010.

42. Montejo, L.A.; Kowalsky, M.J. Estimation of frequency-dependent strong motion duration via wavelets and its influence on nonlinear seismic response. *Comput.-Aided Civ. Infrastruct. Eng.* **2008**, *23*, 253–264. [CrossRef]

43. Chandramohan, R.; Baker, J.W.; Deierlein, G.G. Quantifying the influence of ground motion duration on structural collapse capacity using spectrally equivalent records. *Earthq. Spectra* **2016**, *43*, 927–950. [CrossRef]

44. Ahn, I.S.; Chen, S.S.; Dargush, G.F. Dynamic ratcheting in elastoplastic SDOF systems. *J. Eng. Mech.* **2006**, *132*, 411–421. [CrossRef]

45. Challamel, N.; Lanos, C.; Hammouda, A.; Redjel, B. Stability analysis of dynamic ratcheting in elastoplastic systems. *Phys. Rev. E* **2007**, *75*, 026204. [CrossRef]

46. Ahn, I.S.; Chen, S.S.; Dargush, G.F. An iterated maps approach for dynamic ratchetting in SDOF hysteretic damping systems. *J. Sound Vib.* **2009**, *323*, 896–909. [CrossRef]

 *sustainability*

Article

Shear Performance of Optimized-Section Precast Slab with Tapered Cross Section

Hyunjin Ju [1] , Sun-Jin Han [2], Hyo-Eun Joo [2], Hae-Chang Cho [2], Kang Su Kim [2,*] and Young-Hun Oh [3]

[1] Department of Civil Engineering and Natural Hazards, University of Natural Resources and Life Sciences, Vienna, Feistmantelstrasse 4, Vienna 1180, Austria; hyunjin.ju@boku.ac.at
[2] Department of Architectural Engineering, University of Seoul, 163 Siripdaero, Dongdaemun-gu, Seoul 130-743, Korea; sjhan1219@gmail.com (S.-J.H.); joo8766@uos.ac.kr (H.-E.J.); chang41@uos.ac.kr (H.-C.C.)
[3] Department of Medical Space Design & Management, Konyang University, 158 Gwanjeodong-ro, Seo-gu, Daejeon 35365, Korea; youngoh@konyang.ac.kr
* Correspondence: kangkim@uos.ac.kr; Tel.: +82-2-6490-5576

Received: 23 November 2018; Accepted: 22 December 2018; Published: 29 December 2018

Abstract: The optimized-section precast slab (OPS) is a half precast concrete (PC) slab that highlights structural aesthetics while reducing the quantity of materials by means of an efficient cross-sectional configuration considering the distribution of a bending moment. However, since a tapered cross section where the locations of the top and bottom flanges change is formed at the end of the member, stress concentration occurs near the tapered cross section because of the shear force and thus the surrounding region of the tapered cross section may become unintentionally vulnerable. Therefore, in this study, experimental and numerical research was carried out to examine the shear behaviour characteristics and performance of the OPS with a tapered cross section. Shear tests were conducted on a total of eight OPS specimens, with the inclination angle of the tapered cross section, the presence of topping concrete and the amount of shear reinforcement as the main test variables and a reasonable shear-design method for the OPS members was proposed by means of a detailed analysis based on design code and finite-element analysis.

Keywords: optimized section; precast slab; concrete; tapered cross section; shear performance

---

1. Introduction

The optimized-section precast slab (OPS) is a half precast concrete (PC) slab designed with a cross-sectional shape optimized by locating the top and bottom flanges differently at the centre and end of the member so that they can effectively resist the positive and negative moments [1]. As shown in Figure 1a, the lower section of the OPS has excellent structural aesthetics and thus can improve economic efficiency because there is no need for finishing work. As shown in Figure 1a–e, the concrete compression flange was placed at the upper part in the centre section of the span and the lower part in the end section of the spans so as to efficiently resist the compressive force generated due to the bending moment distribution caused by vertical loads. In addition, the structural performance was improved by introducing prestress into the member. Truss-type shear reinforcement was placed in each rib to improve the composite performance of the OPS and topping concrete as well as vertical shear resistance.

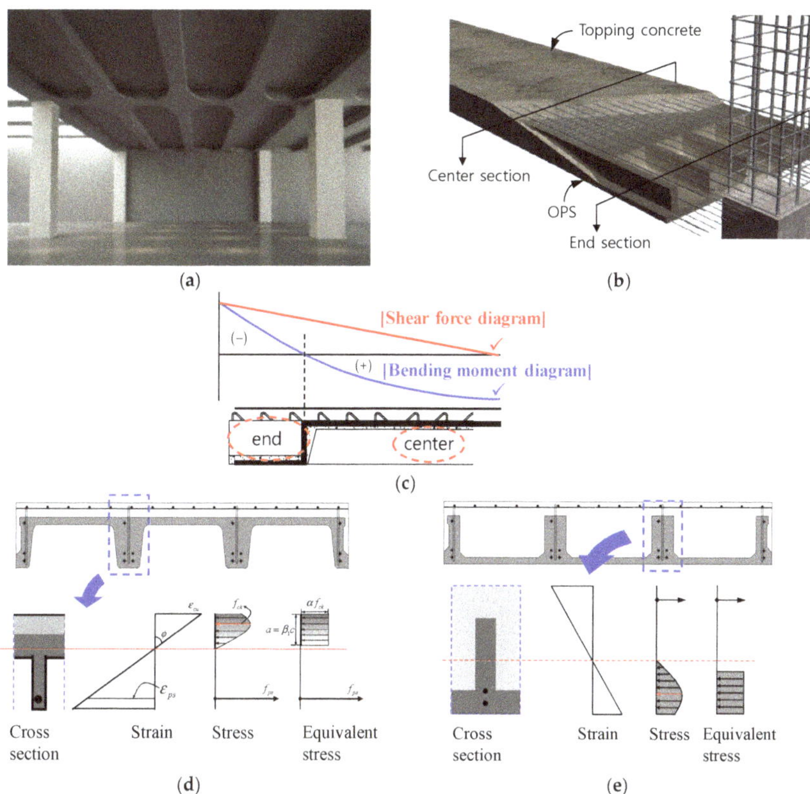

Figure 1. Optimized-section precast slab (OPS). (**a**) Bottom view of OPS system. (**b**) Description of OPS system. (**c**) Location of compression flange. (**d**) Centre section. (**e**) End section.

Nowadays, there is a high demand to reduce carbon emission and eco-friendly construction has become an important issue for sustainability of structures. Among structural members of a building, slabs are known to be heavily responsible for the carbon emission because they cause about 35% of the total carbon emission from structural members [2]. Considering that the PC method, used for producing the OPS members, can minimize waste of materials [3] and that their shapes are optimized by removing unnecessary parts of sections, the OPS members are productions of the sustainable and eco-friendly construction method. In construction markets, it is also very important to secure proper efficiency, constructability and short construction period. The optimized-section precast slabs (OPS) are produced in a precast concrete (PC) factory and assembled on sites. Thus, the OPS system does not require formworks or curing periods that are essential for any conventional reinforced concrete (RC) construction method, which is why it requires a very short construction period. The OPS system can also lead to save materials by adopting the efficient sectional shapes and to reduce manpower by utilizing automatic fabrication facilities [1,4]. On the other hand, it was confirmed that OPS system is more economical about 10% than the conventional RC slab system in a distribution centre and a basement garage construction because of its high load-carrying capacity [3].

According to previous researches [5–7], a large rotation at member ends causes many problems with serviceability such as cracks and large deflection, which depends on support types and joint details. In a fundamental point of view, a flexural member can be more efficiently designed by utilizing the continuous support conditions because the static moment can be distributed into positive and negative moments resulting in a larger load carrying capacity than a simply supported member.

For these reasons, many researches have been conducted on continuous PC members [5–10] and particularly in the authors' previous study [1], the flexural performance of continuous OPS members was studied in detail.

In the previous research [1], critical cracks occur in sections where the sectional area of the concrete is relatively small (i.e., near the tapered section). This is because the tapered cross section is formed where the positions of the top and bottom flanges change because of the characteristics of the OPS member. Since the crack angle is higher than 0.79 rad (45°), the truss bars placed in the ribs may not effectively resist the shear force. In addition, at the tapered section where the section changes drastically, local stress concentration or damage can occur with large shear forces [6], which is influenced by its shape and angle. Therefore, in this study, shear tests on the composite slab with topping concrete and the precast concrete (PC) slab unit were conducted to examine the shear performance of the OPS according to the shape and angle of the tapered cross-section. Numerical research using a finite-element analysis (FEA) and structural design code was also carried out. Furthermore, this study analysed the validity of the current design method in detail and proposed a reasonable shear-design method for OPS members.

2. Test Program

2.1. Details of Test Specimens

In this study, shear tests were planned considering the presence of topping concrete composite and the inclination angle of the tapered cross section as the main variables, as shown in Figure 2 and the amount of shear reinforcement was set as an additional variable in consideration of the spacing of the shear reinforcement for the OPS specimen with topping concrete. Figure 3 shows the section details of the US series specimens, which are PC unit members and CS series specimens, which are composite members with topping concrete. Table 1 summarizes the material properties and reinforcement details for each specimen. A total of four test specimens were fabricated and eight shear test results were obtained from tests conducted on the left and right side sections of each member. The angles of the tapered cross sections were planned to be 0.52 rad (30°), 1.05 rad (60°) and 1.57 rad (90°) for PC unit members and these specimens were named as US30, US60 and US90, respectively. Here, for the US60, tests on specimens with the same details were conducted twice; so the specimens were named US60a and US60b, respectively. For the composite members, the specimens in which the angles of the tapered cross sections were set at 0.52 rad (30°) and 1.57 rad (90°) were named as CS30 and CS90 and the specimens with more shear reinforcement than the CS30 and CS90 specimens were named as CSS30 and CSS90, respectively. As shown in Table 1, the concrete compressive strength of a PC slab unit was 46.9 MPa or 55.0 MPa and the compressive strength of topping concrete was 49.0 MPa. In addition, the yield strengths of D10, D16 and D19 rebars used in the test specimens were 505, 462 and 472 MPa, respectively.

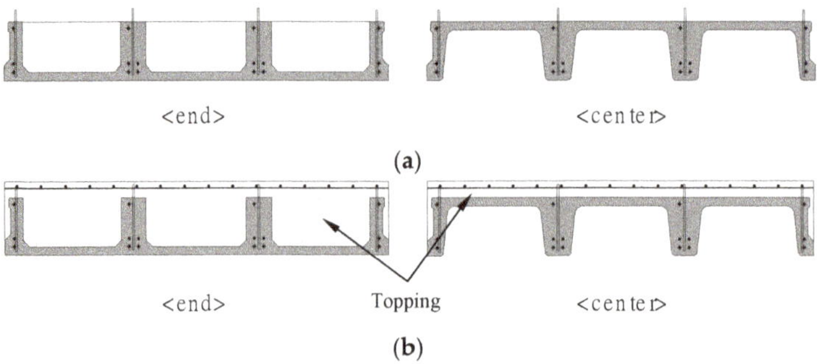

Figure 2. *Cont.*

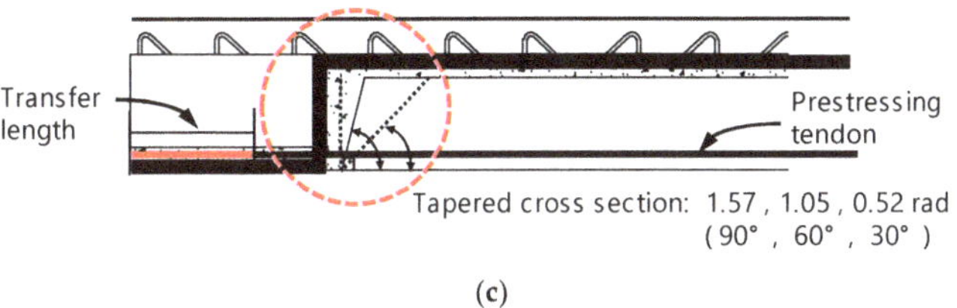

(c)

Figure 2. Summarized description of specimens. (**a**) Section of PC slab unit specimens (US series). (**b**) Section of composite slab specimens (CS series). (**c**) Tapered cross section according to inclination angle.

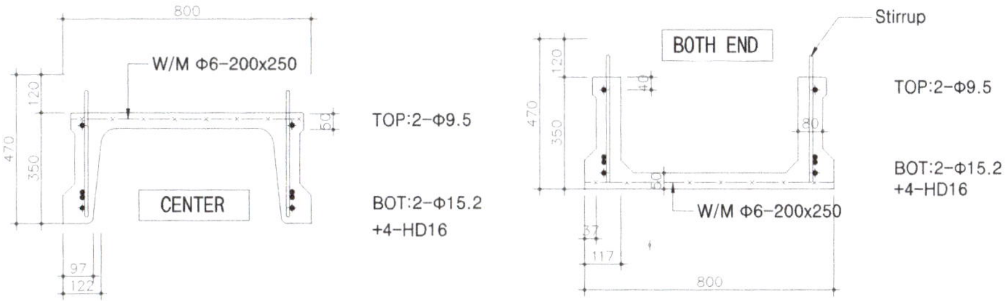

(a)

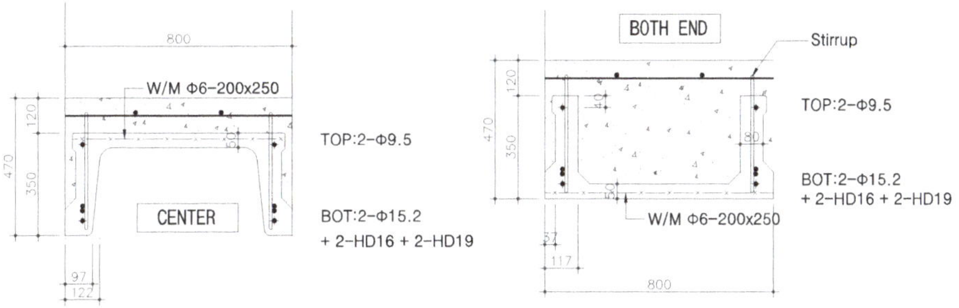

(b)

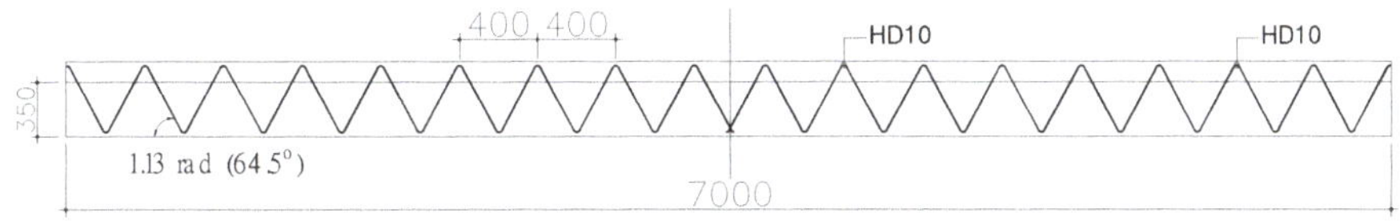

- CS30 and CS90 specimens

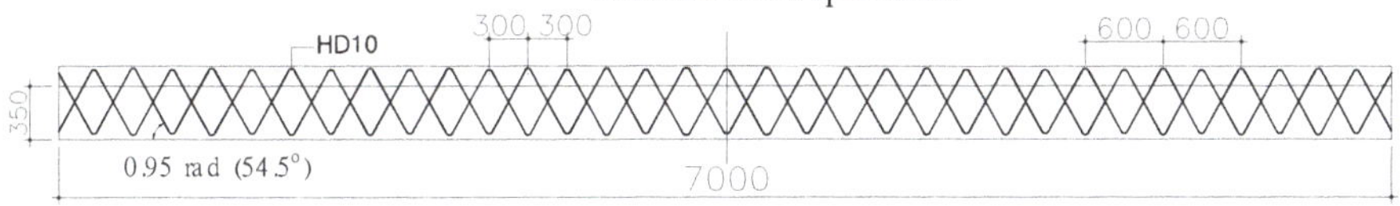

- CSS30 and CSS90 specimens

(c)

Figure 3. Section details of test specimens (unit: mm and degree). (**a**) US series specimens. (**b**) CS series specimens. (**c**) Shear reinforcement in CS series specimens.

Table 1. Material properties and reinforcement details of specimens (unit: mm and rad).

Specimen	Compressive Strength of Concrete, f_c' (MPa)		Compressive Reinforcement Sectional Area, mm^2)		Tensile Reinforcement (Sectional Area, mm^2)		Shear Reinforcement	Angle of Inclined Section (rad)
	PC	Topping	Rebar	Tendon	Rebar	Tendon		
US30	55.0	-	-	2-ϕ9.5 (110)	4-D16 (794)	2-ϕ15.2 (280)	D10@400	0.524 (30°)
US60a	46.9	-		2-ϕ9.5 (110)	4-D16 (794)	2-ϕ15.2 (280)	D10@400	1.047 (60°)
US60b	46.9	-		2-ϕ9.5 (110)	4-D16 (794)	2-ϕ15.2 (280)	D10@400	1.047 (60°)
US90	55.0	-	-	2-ϕ9.5 (110)	4-D16 (794)	2-ϕ15.2 (280)	D10@400	1.571 (90°)
CS30	55.0	49.0	2-D10 (143)	2-ϕ9.5 (110)	2-D16 +2-D19 (970)	2-ϕ15.2 (280)	D10@400	0.524 (30°)
CS90	55.0	49.0	2-D10 (143)	2-ϕ9.5 (110)	2-D16 +2-D19 (970)	2-ϕ15.2 (280)	D10@400	1.571 (90°)
CSS30	55.0	49.0	2-D10 (143)	2-ϕ9.5 (110)	2-D16 +2-D19 (970)	2-ϕ15.2 (280)	D10@600 +D10@600	0.524 (30°)
CSS90	55.0	49.0	2-D10 (143)	2-ϕ9.5 (110)	2-D16 +2-D19 (970)	2-ϕ15.2 (280)	D10@600 +D10@600	1.571 (90°)

As shown in Figure 3, the length and width of all specimens were 7 m and 800 mm, respectively. The height of the PC slab unit was 350 mm and that of the topping concrete 120 mm. The 7-wire strands with a tensile strength (f_{pu}) of 1860 MPa were used to produce the PC slab unit and two strands with diameters of 15.2 mm and 9.5 mm were placed at the top and bottom, respectively. The yield stress of strands (f_{py}) determined by using the 0.1% offset method was 1645 MPa and the initial prestress (f_{pi}) introduced in the strands was about $0.7f_{pu}$, which was estimated to be 137.3 kN and 63.8 kN for the 15.2 mm and 9.5 mm strands, respectively, in terms of prestressing forces. For the US specimens, four reinforcing bars with a diameter of 16 mm were additionally placed on the tension side to provide sufficient flexural capacity and induce shear failure. Meanwhile, in the design of the test specimens, the flexural and shear strengths were calculated based on the ACI318-14 code [11]. For the CS specimens with topping concrete, two deformed reinforcing bars with diameters of 16 mm and 19 mm were arranged on the tension side, as shown in Figure 3b. Figure 3c shows the details of the shear reinforcement placed in the CS series specimens. The shear reinforcement was fabricated in the form of a truss by bending D10 deformed bars and placed on each rib during the production of the PC slab unit. The spacing of shear reinforcement placed in the US and CS specimens was 400 mm. In the CSS specimens, two sets of truss bars with 600 mm spacing were placed crosswise, so that shear reinforcement was placed at intervals of 300 mm. The height of the truss bar was 420 mm and the net cover thickness was 25 mm.

2.2. Test and Measurement Setup

Figure 4 shows the loading and measurement setups for test specimens. As shown in Figure 4a, the specimens were simply supported at points installed 50 mm away and 3750 mm away from the end of the member. One-point loading was applied at a point 1200 mm away from the end of the member to induce shear failure near the end section where the tapered cross section is located. The load was applied by a 1000 kN actuator under displacement control and the loading rate was 1.5 mm/min. up to failure. Shear tests were carried out at one end and then at the other, undamaged opposite end. Figure 4b,c shows the inclination angles of the tapered cross sections for the US and CS series specimens, respectively. The US60 specimens with a 1.05 rad (60°) inclination angle have the same details for both ends, whereas the remaining specimens were designed to have inclination angles of

0.52 rad (30°) and 1.57 rad (90°) on the left and right ends, respectively. The deflections of the test specimens were measured using a LVDT installed under the section of the loading point.

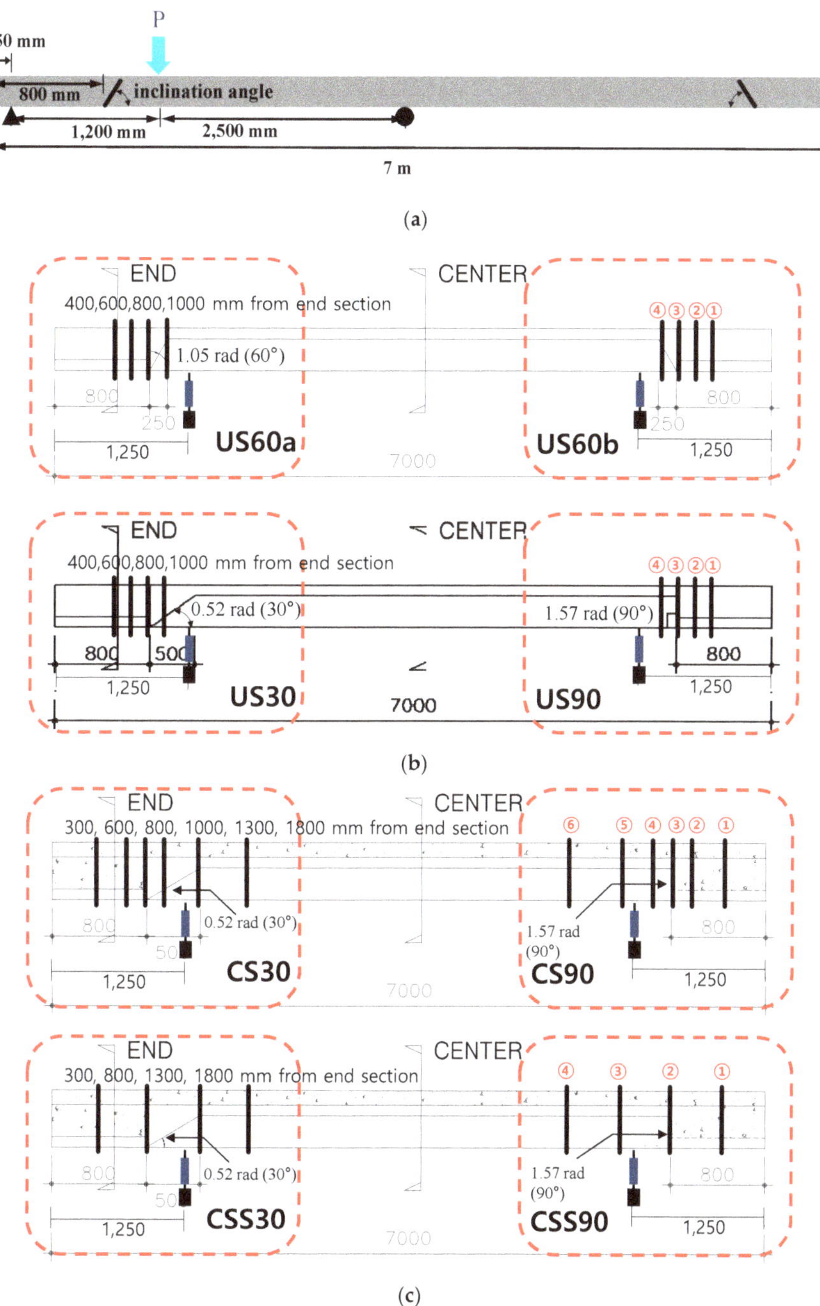

Figure 4. *Cont.*

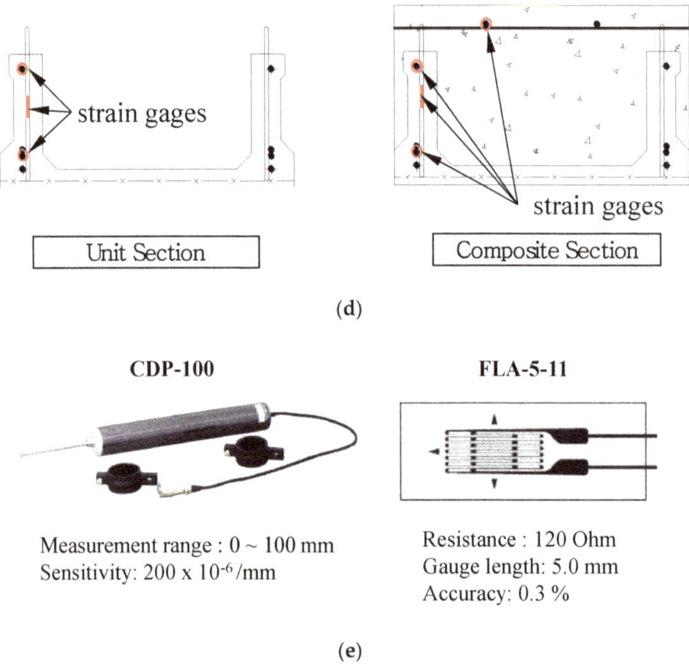

Figure 4. Test and measurement setup. (**a**) Loading setup. (**b**) PC slab unit specimens. (**c**) Composite slab specimens. (**d**) Location of strain gages in section. (**e**) Measurement instruments.

In Figure 4b,c, the thick vertical solid lines indicate the section positions where gauges were installed to measure the strains of longitudinal and shear reinforcements. Figure 4d shows the locations of gauges installed in sections. For the US specimens, strain gauges were installed in the tensile reinforcement, prestressing strands on the compression side and shear reinforcement in sections 400, 600, 800 and 1000 mm away from the end, as shown in the 'unit section' of Figure 4d. For the CS specimens, as shown in the 'composite section' of Figure 4d, strain gauges were installed in the tensile reinforcement, shear reinforcement, prestressing strands on the compression side and longitudinal reinforcement in the topping concrete at the sections indicated by the thick vertical solid lines in Figure 4c. The details of the instruments used in the test are summarized in Figure 4e.

3. Test Results

3.1. Shear Behaviour and Crack Pattern of Test Specimens

Figure 5 shows the load-deflection curves of the test specimens and Figure 6 shows the crack patterns of the specimens at the shear failure. Among the PC slab unit specimens (US series), the US90 specimen, which had the largest angle of the tapered cross section, showed the lowest maximum load of 169 kN and the remaining specimens exhibited similar maximum loads, ranging from 222 to 248 kN. In addition, the initial cracking loads of the US specimens ranged from 72 to 85 kN. The stiffness decreased after the occurrence of the initial cracks and the US90 specimen showed the greatest reduction in stiffness. The US60a and US60b specimens exhibited relatively ductile behaviour compared to the other US specimens. This is due to the significant contribution of flexural deformations as the ultimate failure modes shown in Figure 6c,d. On the other hand, the US90 specimen showed damage concentrated in the vicinity where the tapered cross section is formed and underwent premature failure as the top flange collapsed. This is because a relatively larger stress concentration occurred as the section was drastically changed due to the larger inclination angle of the tapered cross section.

The composite slab specimens (CS series) showed similar initial cracking loads, ranging from 142 to 156 kN and flexural-shear cracks occurred at a load of about 190 to 205 kN. Unlike the US specimens, the CS specimens did not show a significant reduction in stiffness even after the occurrence of cracks and underwent failure as the critical crack width gradually increased without an abrupt reduction in load even after reaching the maximum load. The CS30 and CS90 specimens exhibited the maximum loads of 394 kN and 404 kN, respectively and the maximum loads of the CSS30 and CSS90 specimens, which had much more shear reinforcement, were 440 kN and 441 kN, respectively. For the composite specimens, there was almost no difference in strength between the specimens according to the inclination angle of the tapered cross section. This is because as the topping concrete is placed, the change of sectional area in the vicinity of the tapered cross section is smaller than that of the US specimens. As shown in Figure 6e–h, no significant difference in crack patterns was observed between the CS specimens. However, as shown in Figure 5, the CS90 and CSS90 specimens with a 1.57 rad (90°) inclination angle were found to have smaller deformation capacities after the maximum load than those of the CS30 and CSS30 specimens with a 0.52 rad (30°) inclination angle.

PC slab unit specimen	US30	US60a	US60b	US90
Maximum load (kN)	225	248	222	169
Composite slab specimen	CS30	CS90	CSS30	CSS90
Maximum load (kN)	394	404	440	441

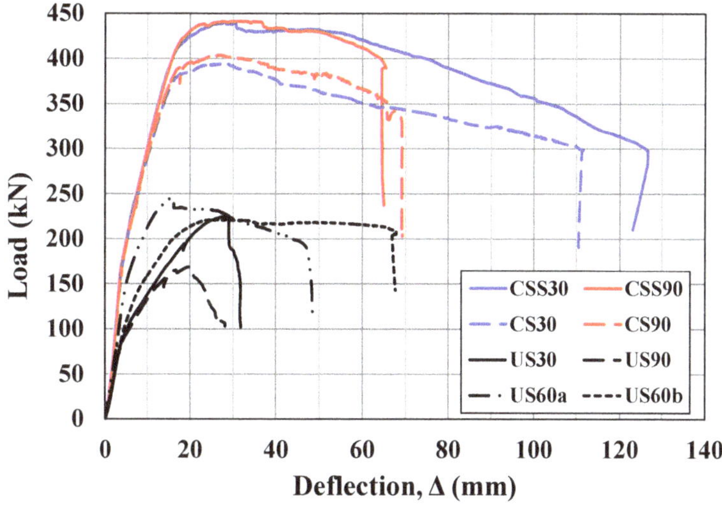

Figure 5. Load-deflection curves.

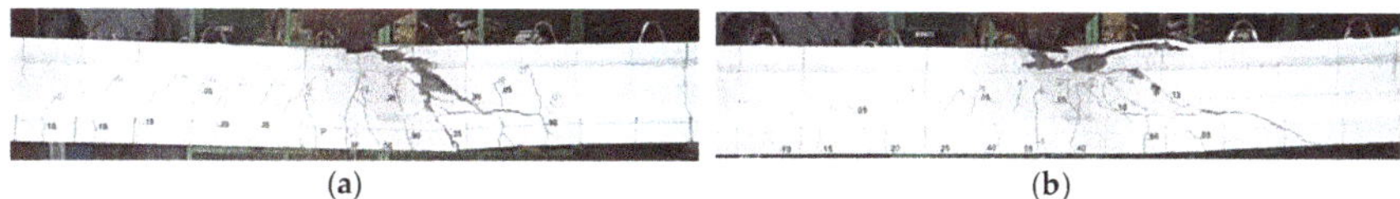

(a)   (b)

Figure 6. *Cont.*

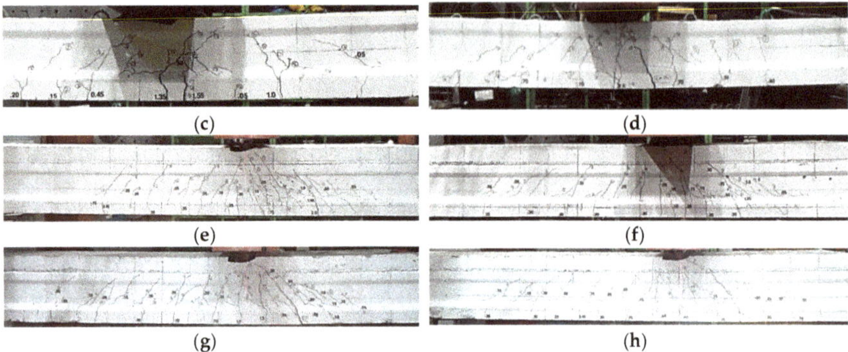

Figure 6. Crack patterns of specimens at failure. (**a**) US30 specimen. (**b**) US90 specimen. (**c**) US60a specimen. (**d**) US60b specimen. (**e**) CS30 specimen. (**f**) CSS90 specimen. (**g**) CSS30 specimen. (**h**) CSS90 specimen.

3.2. Measured Strains from Reinforcement

Figures 7 and 8 show strains measured in the US and CS specimens. Note that some gauges were damaged and failed to provide measurement values. In the graphs, "Compression," "Tension" and "Stirrup" represent the strain measured from gauges attached to the prestressing strands, longitudinal tensile reinforcement and shear reinforcement, respectively, while the gauge numbers signify the section locations shown in Figure 4c,d.

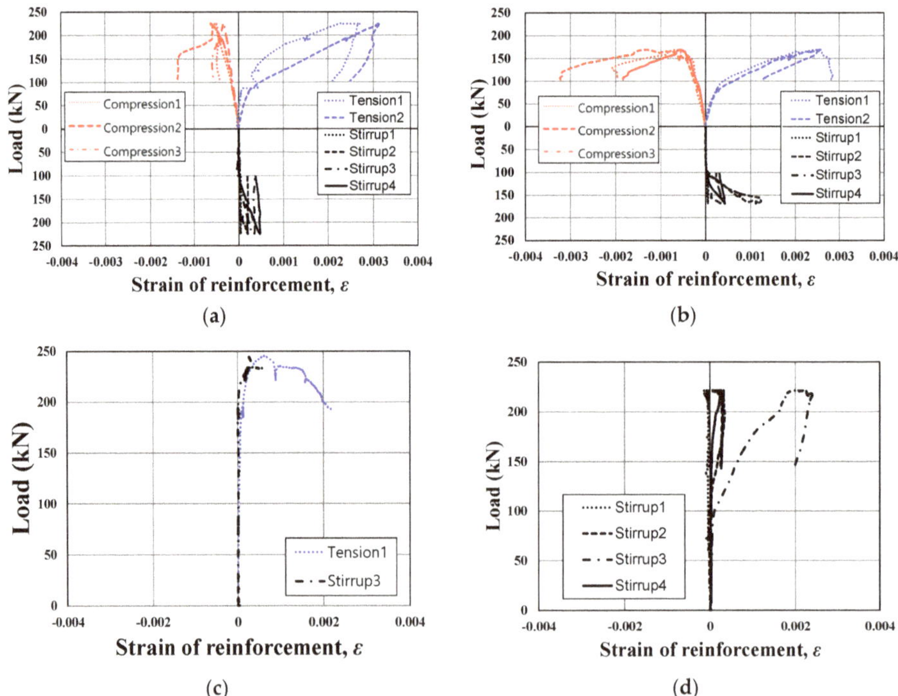

Figure 7. Strain behaviour of reinforcement in US specimens. (**a**) US30 specimen. (**b**) US90 specimen. (**c**) US60a specimen. (**d**) US60b specimen.

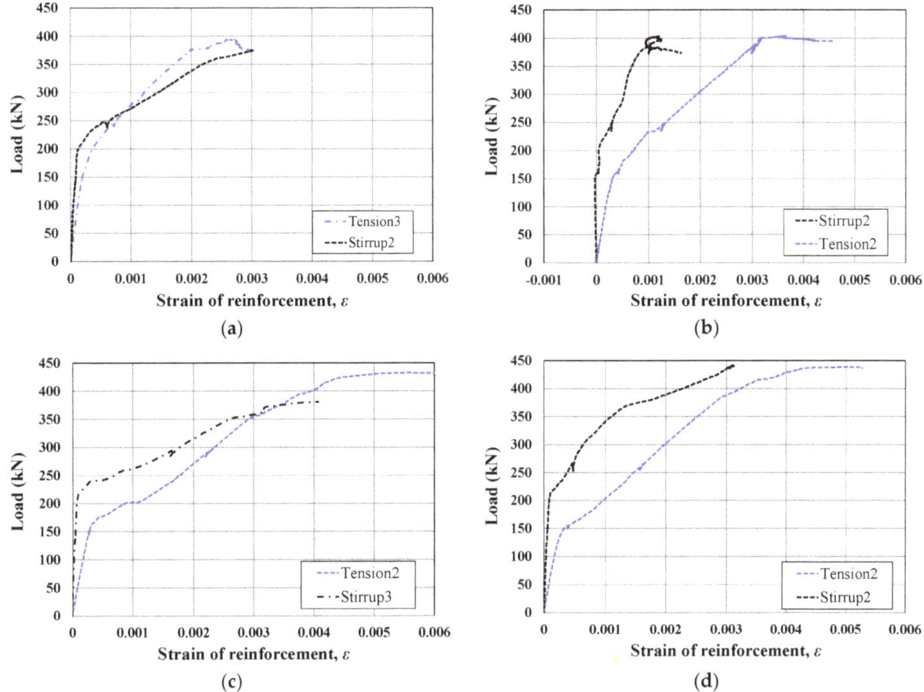

Figure 8. Strain behaviour of reinforcement in CS specimens. (**a**) CS30 specimen. (**b**) CS90 specimen. (**c**) CSS30 specimen. (**d**) CSS90 specimen.

As shown in Figure 7, the strain of shear reinforcement was considerably small in the US specimens, except for the US60b specimen and the strain of flexural tensile reinforcement increased greatly with the increasing loads. In particular, Figure 7b shows that the compressive strain of the upper strand was large enough to cause concrete crushing in the US90 specimen and the crushing failure of the concrete in the compression side occurred as shown in Figure 6b. In the case of the US60b specimen, only the strain of the shear reinforcement was measured due to the damage of the gauges and shear reinforcement located near the tapered cross section showed a larger strain than the yield strain.

Figure 8 shows the strains of the longitudinal tensile reinforcement and shear reinforcement, which are the largest strains among those measured from the CS specimens. Unlike in the US specimens, the strains measured from both the shear reinforcement and the flexural tensile reinforcement were greater than the yield strain in the CS specimens, except for a stirrup strain in the CS90 specimen. This suggests that the shear reinforcement effectively contributed to the shear resistance in the OPS with topping concrete.

4. Analysis for Design

4.1. Finite-Element Analysis Considering the Inclination Angle of the Tapered Cross Section

As observed in the test results, the shear performance of OPS is greatly influenced by the stress concentration that occurs near the tapered cross section and thus a detailed examination on the inclination angle of the tapered cross section is required [10]. In this regard, this study used a finite-element analysis (FEA) to investigate the effect of the inclination angle of the tapered cross section on the stress distribution in the OPS member. The finite-element analysis was performed using ABAQUS/CAE, a general-purpose analysis program [12] and it was aimed at comparing the stress distribution patterns according to the inclination angle of the tapered cross section rather than evaluating the performance of the members.

Figure 9 shows the finite-element analysis models for OPS. Only half of the member span was modelled using symmetric conditions [13]. As shown in Figure 9a, the concrete was modelled with solid elements and the prestressing strands and reinforcement were modelled with truss elements. The reinforcement materials were considered as being perfectly bonded with concrete. As shown in Figure 9b, Collins model and damaged plasticity model [12,14] were applied for the compressive and tensile behaviours of concrete and bilinear model and Ramberg-Osgood model [15–17] were applied to the reinforcement and prestressing strands, respectively. The material properties obtained from the experiments are summarized in Table 1. The proper mesh size was determined considering the mesh sensitivity and analysis efficiency and they were 50 mm and 140 mm for PC unit and topping parts, respectively, where the number of mesh elements was 25,729 and 1183, respectively. The concrete section was modelled by linear tetrahedral 3D solid elements (type C3D4) and the reinforcing bars were modelled by two-node 3D truss elements (type T3D2) [12,18]. The boundaries were modelled to simulate the simply supported conditions.

As shown in Figure 9c, the horizontal projection length of the slope (A) and the radius of the curve at the tapered cross section in the lower surface of the top flange (B) were set as the main analysis variables and the analysis for each variable was conducted on the cases shown in Table 2. The composite slab members (FEA2 series) were modelled with careful considerations on the characteristics of the contact surface between the PC slab unit and the topping concrete. In the production of the OPS, the surface is roughened on the upper part of the PC slab unit and the shear reinforcement to be placed in the member serves as a horizontal shear connector after the topping concrete composite. Therefore, the horizontal shear strength at the interface between the PC slab unit and the topping concrete can be estimated to be higher than 0.56MPa, which is the minimum horizontal shear strength presented in the ACI318-14 code [11]. In addition, no damage was observed at the interface between the PC slab unit and the topping concrete, even in the crack patterns shown in Figure 6. Therefore, in this study, 0.56 MPa and 0.6 were used for the cohesion (c) and the friction coefficient (μ), respectively, which are the interfacial contact characteristic between the PC slab unit and topping concrete [11].

Table 2. Variables for finite element analysis.

PC Slab Unit Case	A * (mm)	B * (mm)	Angle of Inclined Section (rad)	Composite Slab Case	A * (mm)	B * (mm)	Angle of Inclined Section (degree, °)
FEA1-1	200	200	0.983 (56.3°)	FEA2-1	200	200	0.983 (56.3°)
FEA1-2	250	200	0.876 (50.2°)	FEA2-2	250	200	0.876 (50.2°)
FEA1-3	400	200	0.644 (36.9°)	FEA2-3	400	200	0.644 (36.9°)
FEA1-4	250	150	0.876 (50.2°)	FEA2-4	250	150	0.876 (50.2°)
FEA1-5	400	150	0.644 (36.9°)	FEA2-5	400	150	0.644 (36.9°)

* refer to Figure 9c.

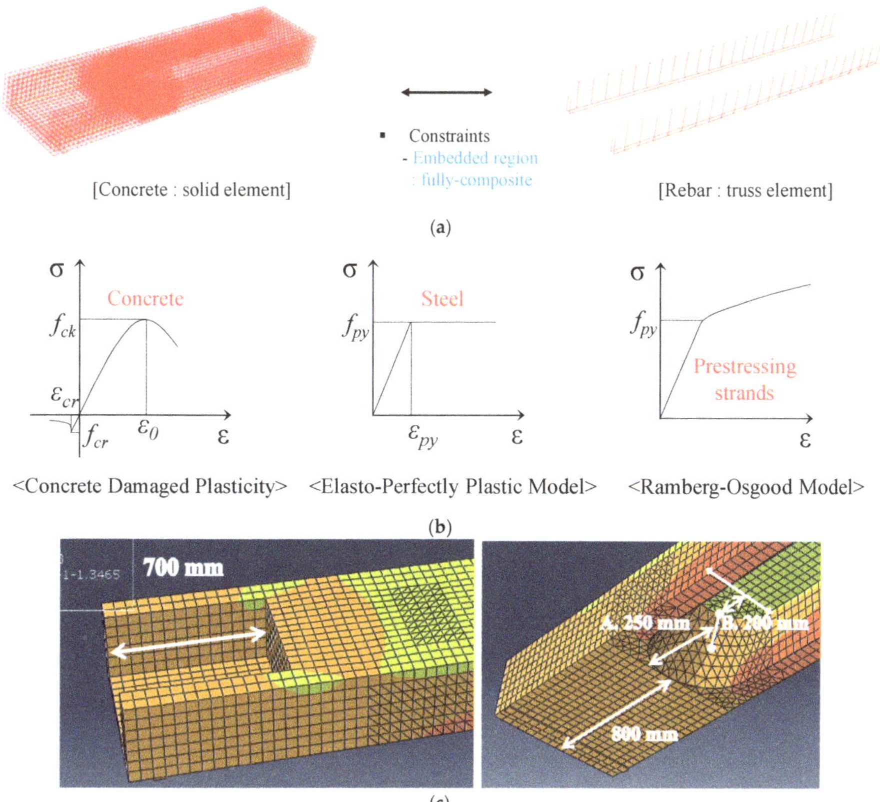

Figure 9. Finite element analysis model. (**a**) Interaction condition. (**b**) Material models. (**c**) Tapered cross section.

Figure 10a shows the analysis results on the FEA1 series, which are PC slab units without topping concrete. The cracked area, where the shear stress (v) is greater than the shear cracking strength (v_{cr}), was marked in a grey contour. In this case, when the shear cracking strength (v_{cr}), $0.33\sqrt{f_c'}$ is used [14] and 46.9MPa is applied as the concrete compressive strength (f_c') of the PC slab unit, v_{cr} is calculated to be 2.26 MPa. It can be confirmed from the analysis results that the stress is concentrated on the member end rib located at the support and on the web in the vicinity where the tapered cross section is formed. Figure 10b,c shows the analysis results on FEA 1-1 to 1-5 members according to variables A and B, respectively, where all analysis results are shown under the same deflection conditions. As shown in Figure 10b,c, the stress concentration tends to be mitigated as the inclination angle of the tapered cross section in the PC slab unit becomes smaller and the curvature of the curve at the tapered cross section in the lower surface of the top flange becomes smaller. As mentioned in Section 3.1, the US90 specimen with an inclination angle of 1.57 rad (90°) underwent premature failure along with damages due to the stress concentration.

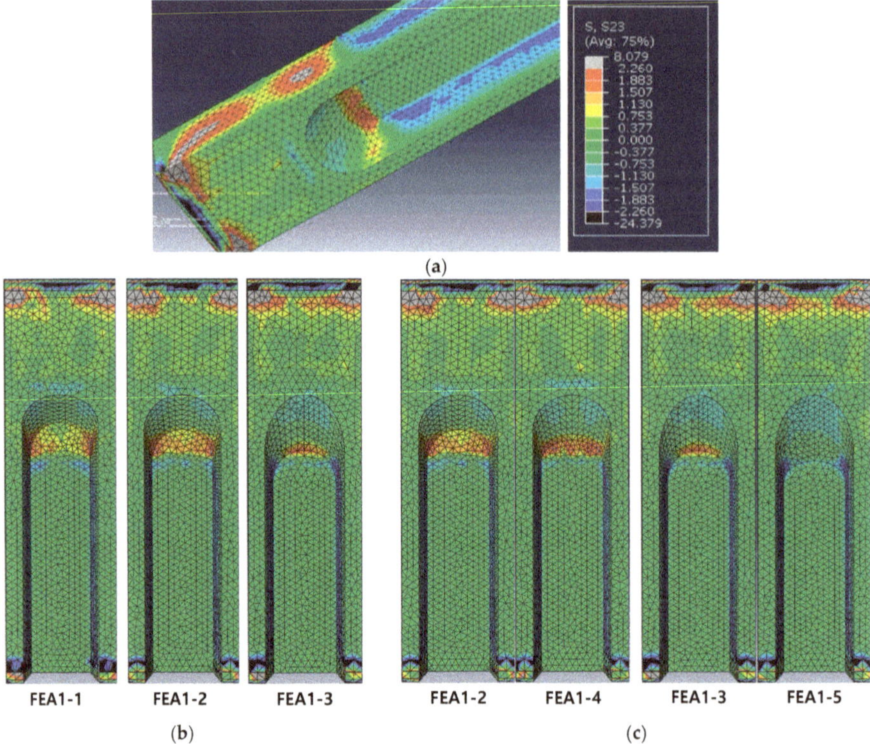

Figure 10. FEA results on PC slab units. (**a**) Shear stress distribution. (**b**) Shear stress distribution according to variable A. (**c**) Shear stress distribution according to variable B.

As shown in Figure 11, the finite-element analysis results on composite slabs (FEA2 series) also confirmed the mitigation of stress concentration with a smaller inclination angle of the tapered cross section and with a smaller curvature of the curve at the tapered cross section in the lower surface of the top flange, which were the same as in the FEA1 series. Therefore, the analysis and test results showed that, in order to alleviate the stress concentration occurring at the tapered cross section, it is desirable that the inclination angle of the tapered cross section should be less than 1.05 rad (60°) and the curvature of the curve at the tapered cross section in the lower surface of the top flange be as small as possible in the design step of OPS.

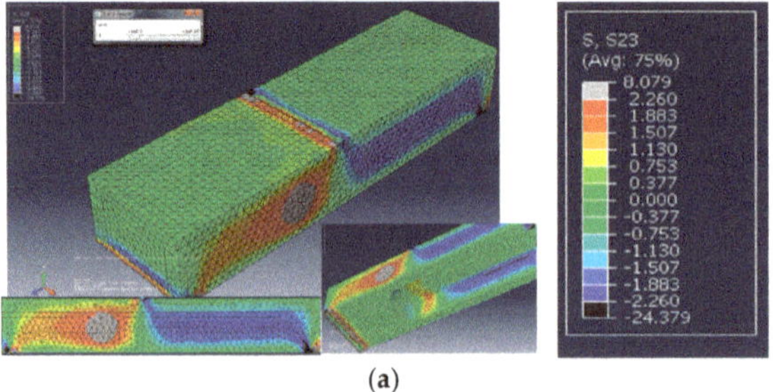

Figure 11. *Cont.*

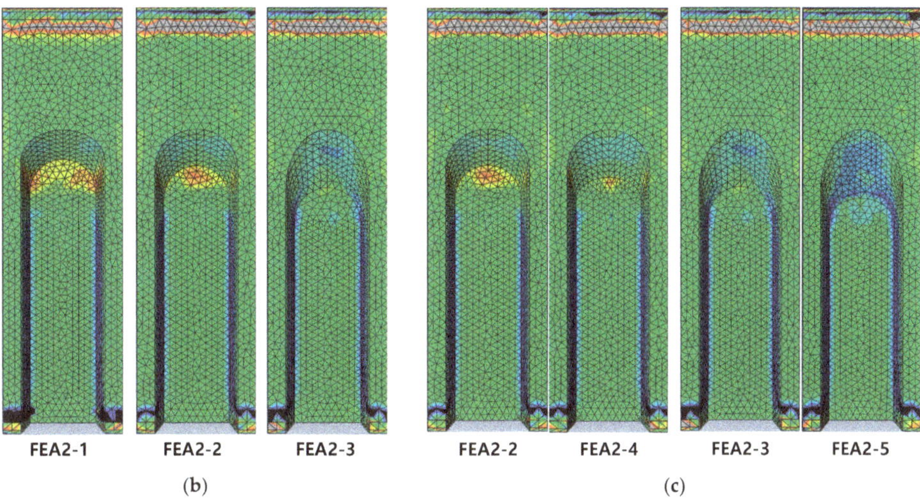

FEA2-1   FEA2-2   FEA2-3   FEA2-2   FEA2-4   FEA2-3   FEA2-5

(b)   **(c)**

Figure 11. FEA results on composite slabs. (**a**) Shear stress distribution. (**b**) Shear stress distribution according to variable A. (**c**) Shear stress distribution according to variable B.

4.2. Suggestion for Estimating Shear Strength of OPS

In this study, the shear strengths of the specimens were evaluated using the shear design equations in ACI318-14 [11], as shown in Table 3. The prestressing strands placed in the OPS specimen have a straight profile and thus the vertical component of effective prestress (V_p) in the web shear strength of concrete (V_{cw}) is zero. The actual inclined crack angles measured from the test specimens were used for shear crack angles (β). In addition, since the location of critical shear cracks occurred in the specimens was not within the transfer length region, the effective stress (f_{pe}) was used to calculate the compressive stress at the centroid of the cross section after prestress release (f_{pc}) and the PC slab unit and topping concrete were considered to be fully composite based on the experimental observations. In the calculation of the shear strengths of the composite slab specimens, the compressive strength of the topping concrete rather than that of the PC slab unit was applied conservatively, as in most practical application cases. In addition, the test results were analysed in detail using the modified compression-field theory (MCFT) [19–22], which is a shear analysis model with proven accuracy and the code equations.

Table 3. Code equations for estimating shear strength of OPS [11].

Type	Code Equation *
Web shear strength of concrete	$V_{cw} = \left(0.29\sqrt{f_c'} + 0.3f_{pc}\right)b_w d_p + V_p$
Shear contribution of transverse reinforcement	$V_s = \dfrac{A_v f_y d}{s}\left(\sin\alpha\cot\beta + \cos\alpha\right)$

* Note: A_v = sectional area of stirrup (mm²); b_w = web width (mm); d_p = depth of prestressing strands (mm); d = effective depth of reinforcement (mm); f_c' = compressive strength of concrete (MPa); f_{pc} = compressive stress at centroid of cross section after prestress release (MPa); f_y = yield strength of transverse reinforcement (MPa); s = stirrup spacing (mm); V_p = vertical component of effective prestress (N); α = angle of inclined stirrup (rad); β = angle of inclined shear crack (rad).

Table 4 and Figure 12 compare the shear strength test results (V_u) and analysis results (V_n). The ACI 318-14 code showed a tendency to overestimate the shear strengths of the US specimens, which are PC slab units. On the other hand, the MCFT provided a relatively approximate evaluation on the shear strengths of the US specimens, except for the US90 specimen, in which the inclination angle of the tapered cross section is 90°. Since the shear reinforcement placed in the OPS unit has a short embedment length and a stirrup hook at the top protrudes before the topping concrete is placed, as shown in Figure 3, the reinforcement may not develop up to its yield stress even if the shear cracks penetrate through the stirrups. This can be confirmed from the results of the shear reinforcement strain measurements shown in Figure 7. Therefore, the shear strengths of the US specimens represented as "Proposal" in Figure 12a were calculated by taking into account only the contribution of concrete (V_{cw}), excluding the contribution of shear reinforcement (V_s). This method provided a very accurate evaluation on the shear strengths of the PC slab unit specimens, except for the US90 specimen. It should again be noted that the US90 specimen underwent premature failure due to excessive stress concentration on the region of the tapered cross section.

In the case of the CS specimens with topping concrete, the MCFT provided conservative estimations on the shear strengths of the test specimens. On the other hand, the ACI 318-14 code showed a tendency to overestimate the test results to some extent but provided relatively accurate shear strengths when compared to the US specimens. As shown in Table 4, the critical shear crack angles observed in the CS specimens ranged from 0.93 rad (53.5°) to 1.00 rad (57.5°), which are considerably larger than the shear crack angles generated in ordinary reinforced concrete and prestressed concrete members [1,11]. The reason why the shear crack is formed at such a high angle is that the stress concentration phenomenon is greatly influenced by the inclination angle of the tapered cross section. In the case of having a steep angle of the tapered cross section as in the US90 specimen, therefore a premature failure can occur because of excessive stress concentration. Hence it is necessary to form the angle of the tapered cross section as low as possible and there is a need to assume that the crack angle (β) is larger than that of the typical slab member for safe shear design for OPS. Accordingly, $\beta = 1.05$ rad (60°) was applied in the calculation of shear strengths for "Proposal" shown in Figure 12b and this method was found to provide a very accurate evaluation on the shear strengths of the CS specimens. Therefore, if the shear strength is calculated by considering only the contribution of concrete (V_{cw}) in the construction phase and if the contribution of shear reinforcement (V_s) is estimated under the assumption that the crack angle (β) is 1.05 rad (60°) after the topping concrete is placed, then adequate safety can be ensured in the shear design of the OPS member.

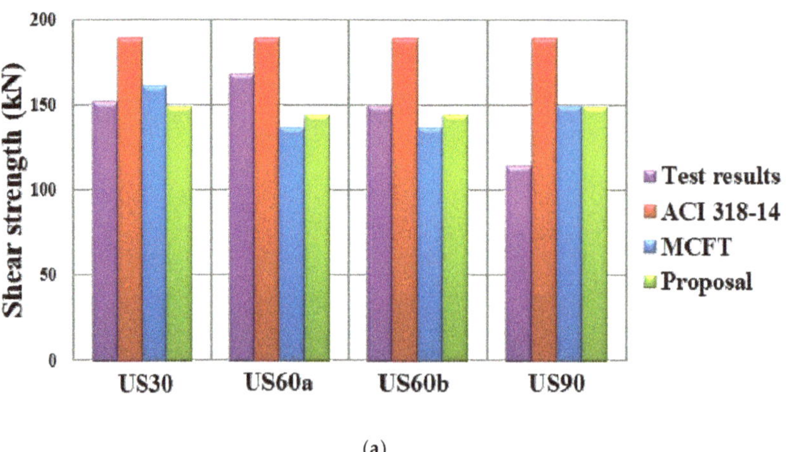

(a)

Figure 12. *Cont.*

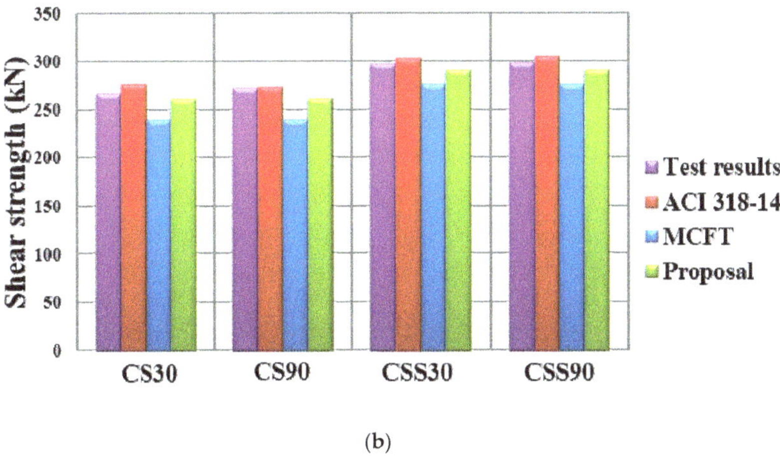

(b)

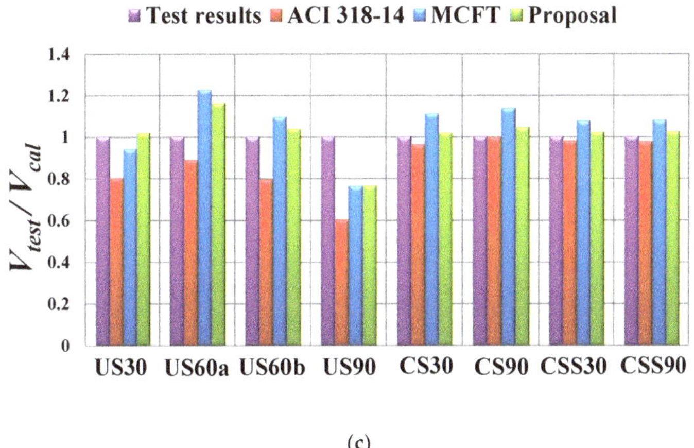

(c)

Figure 12. Shear strength evaluation results. (**a**) PC slab unit specimens. (**b**) Composite slab specimens. (c) Variations between analysis and test results.

Table 4. Comparison of test and analysis results.

Specimen	Critical Crack Angle (degree)	Compressive Strength of Concrete (MPa)	Spacing of Shear Reinforcement (mm)	Test Results, V_u (kN)	Analysis Results, V_n (kN)		
					ACI 318	MCFT	Proposal
US30	32.0	55.0	400	151.8	189.1	161.3	149.0
US60a	90.0	46.9	400	167.5	188.7	136.5	144.1
US60b	60.3	46.9	400	149.7	188.7	136.5	144.1
US90	36.9	55.0	400	113.9	189.1	149.3	149.0
CS30	54.5	49.0	400	266.2	276.2	239.5	260.8
CS90	56.3	49.0	400	272.7	273.2	239.5	260.8
CSS30	57.5	49.0	300	297.1	303.2	275.7	290.4
CSS90	56.3	49.0	300	298.1	305.4	275.7	290.4

5. Conclusions

In this study, an experimental investigation was carried out to examine the shear performance of the optimized-section precast slab (OPS). The effect of the shape of the tapered cross section on the stress concentration was analysed by means of a finite-element analysis. In addition, a shear-design method for the OPS member, which ensures adequate safety was proposed based on a detailed comparative analysis between ACI318-14 and the test results. The following conclusions were obtained from this study.

1. In the PC slab unit specimen where the inclination angle of the tapered cross section was 1.57 rad (90°), premature failure occurred because excessive stress was concentrated at the region of the tapered cross section. On the other hand, the inclination angle of the tapered cross section did not much influence the shear strength of the member in the composite slab specimen with topping concrete. This is because the composite slab specimens have a smaller change in sectional area near the tapered cross section due to the placement of topping concrete, as compared to the US specimens.

2. The finite-element analysis results showed that as the inclination angle of the tapered cross section and the curvature of the curve at the tapered cross section in the lower surface of the top flange became smaller, the stress concentration decreased. The shear tests also found that the PC slab unit specimens with an inclination angle of less than 1.05 rad (60°) had very similar shear strengths. Therefore, it is desirable that the tapered cross section be configured to have an inclination angle of less than 1.05 rad (60°) in the design of the OPS.

3. The shear reinforcement placed in the PC slab unit specimens did not yield even at the ultimate strength and thus the specimens showed lower strengths than those from the current code equations. Therefore, the shear contribution of shear reinforcement (V_s) should be excluded and only the contribution of concrete (V_{cw}) should be considered in the shear design of the PC slab unit members. This estimation method not only revealed that the shear strengths of the specimens are on the safe side but also provided a very accurate evaluation of them.

4. The shear crack angle observed in the composite slab specimens with topping concrete was about 0.96 rad (55°), which is steeper than those in typical reinforced concrete and prestressed concrete members. Therefore, it is expected that a proper safety can be ensured in the shear design of the composite OPS members when the crack angle (β) of 1.05 rad (60°) is used to calculate the contribution of shear reinforcement (V_s).

Author Contributions: Original draft manuscript, H.J.; Validation, S.-J.H. and H.-E.J.; Investigation, H.-C.C. and Y.-H.O.; Supervision and Review Writing, K.S.K.

Acknowledgments: This research was supported by Basic Science Research Program through the National Research Foundation of Korea (NRF) funded by the Ministry of Education (2016R1D1A3B03932214).

Conflicts of Interest: The authors declare no conflict of interest.

References

1. Ju, H.; Han, S.J.; Choi, I.S.; Choi, S.; Park, M.K.; Kim, K.S. Experimental Study on an Optimized-Section Precast Slab with Structural Aesthetics. *Appl. Sci.* **2018**, *8*, 1234. [CrossRef]
2. Korea Environmental Industry and Technology Institute. *Korea LCI Database Information Network*; Korea Environmental Industry and Technology Institute: Seoul, Korea, 2013.
3. Choi, I.S. Structural Performance of Precast Slab with Esthetics and Optimized Section for Positive and Negative Moment. Ph.D. Thesis, Inha University, Incheon, Korea, 2016.
4. Elliot, K.S.; Colin, K.J. *Multi-Storey Precast Concrete Framed Structures*, 2nd ed.; Wiley Blackwell: Hoboken, NJ, USA, 2013.
5. Mejia-McMaster, J.C.; Park, R. Test on special reinforcement for the end support of hollow-core slabs. *PCI J.* **1994**, *39*, 90–105. [CrossRef]
6. Gere, J.M.; Goodno, B.J. *Mechanics of Materials*, 8th ed.; Cenage Learning: Boston, MA, USA, 2012.
7. Rosenthal, I. Full scale test of continuous prestressed hollow-core slab. *PCI J.* **1978**, *23*, 74–81. [CrossRef]
8. Tan, K.H.; Zheng, L.X.; Paramasivam, P. Designing hollow-core slabs for continuity. *PCI J.* **1996**, *41*, 82–91. [CrossRef]

9.   Yang, L. Design of Prestressed Hollow-Core Slabs with Reference to Web-Shear Failure. *J. Struct. Eng.* **1994**, *120*, 2675–2696. [CrossRef]

10.   Lee, D.H.; Park, M.K.; Oh, J.Y.; Kim, K.S.; Im, J.H.; Seo, S.Y. Web-shear Capacity of Prestressed Hollow-Core Slab Unit with Consideration on the Minimum Shear Reinforcement Requirement. *Comput. Concr.* **2014**, *14*, 211–231. [CrossRef]

11.   ACI Committee 318. *Building Code Requirements for Reinforced Concrete and Commentary (ACI 318-14)*; American Concrete Institute: Farmington Hills, MI, USA, 2014.

12.   Hibbitt, H.; Karlsson, B.; Sorensen, P. *ABAQUS Analysis Users Manual*; Version 6.10; Dassault Systemes Simulia: Providence, RI, USA, 2011.

13.   Smith, I.M.; Griffiths, D.V. *Programming the Finite Element Analysis*, 4th ed.; John Wiley and Sons, Ltd.: Hoboken, NJ, USA, 2004.

14.   Collins, M.P.; Mitchell, D. *Prestressed Concrete Structures*; Prentice-Hall: Upper Saddle River, NJ, USA, 1991.

15.   MacGregor, J.G.; Wight, J.K. *Reinforced Concrete Mechanics and Design*, 4th ed.; Prentice-Hall: Upper Saddle River, NJ, USA, 2006.

16.   Mattock, A.H. Flexural strength of prestressed concrete sections by programmable calculator. *PCI J.* **1979**, *24*, 32–54. [CrossRef]

17.   Lee, D.H.; Oh, J.Y.; Kang, H.; Kim, K.S.; Kim, H.J.; Kim, H.Y. Structural Performance of Prestressed Composite Girders with Corrugated Steel Plate Webs. *J. Constr. Steel Res.* **2015**, *104*, 9–21. [CrossRef]

18.   Oh, J.Y.; Lee, D.H.; Cho, S.H.; Kang, H.; Cho, H.C.; Kim, K.S. Flexural Behavior of Prestressed Steel-Concrete Composite Members with Discontinuous Webs. *Adv. Mater. Sci. Eng.* **2015**, *2015*, 278293. [CrossRef]

19.   Vecchio, F.J.; Collins, M.P. Modified Compression-Field Theory for Reinforced Concrete Elements Subjected to Shear. *ACI J. Proc.* **1986**, *83*, 219–231. [CrossRef]

20.   Vecchio, F.J.; Collins, M.P. Predicting the Response of Reinforced Concrete Beams Subjected to Shear Using Modified Compression Field Theory. *ACI Struct. J.* **1988**, *85*, 258–268. [CrossRef]

21.   Bentz, E.C.; Vecchio, F.J.; Collins, M.P. Simplified Modified Compression Field Theory for Calculating Shear Strength of Reinforced Concrete Elements. *ACI Struct. J.* **2006**, *103*, 614–624. [CrossRef]

22.   Bentz, E.C. Sectional Analysis of Reinforced Concrete Members. Ph.D. Thesis, University of Toronto, Toronto, ON, Canada, 2000.

Article

A New Equation to Evaluate Liquefaction Triggering Using the Response Surface Method and Parametric Sensitivity Analysis

Nima Pirhadi, Xiaowei Tang, Qing Yang * and Fei Kang

State Key Laboratory of Coastal and Offshore Engineering, Dalian University of Technology, Dalian 116024, China; nima.pirhadi@yahoo.com (N.P.); tangxw@dlut.edu.cn (X.T.); kangfei@dlut.edu.cn (F.K.)
* Correspondence: qyang@dlut.edu.cn

Received: 3 December 2018; Accepted: 19 December 2018; Published: 26 December 2018

Abstract: Liquefaction is one of the most damaging functions of earthquakes in saturated sandy soil. Therefore, clearly advancing the assessment of this phenomenon is one of the key points for the geotechnical profession for sustainable development. This study presents a new equation to evaluate the potential of liquefaction (PL) in sandy soil. It accounts for two new earthquake parameters: standardized cumulative absolute velocity and closest distance from the site to the rupture surface (CAV_5 and r_{rup}) to the database. In the first step, an artificial neural network (ANN) model is developed. Additionally, a new response surface method (RSM) tool that shows the correlation between the input parameters and the target is applied to derive an equation. Then, the RSM equation and ANN model results are compared with those of the other available models to show their validity and capability. Finally, according the uncertainty in the considered parameters, sensitivity analysis is performed through Monte Carlo simulation (MCS) to show the effect of the parameters and their uncertainties on PL. The main advantage of this research is its consideration of the direct influence of the most important parameters, particularly earthquake characteristics, on liquefaction, thus making it possible to conduct parametric sensitivity analysis and show the direct impact of the parameters and their uncertainties on the PL. The results indicate that among the earthquake parameters, CAV_5 has the highest effect on PL. Also, the RSM and ANN models predict PL with considerable accuracy.

Keywords: liquefaction; response surface method; artificial neural network; Monte Carlo simulation

---

1. Introduction

Nowadays, it is crucially important for engineers and urban planners to take sustainability, livability, and social health into account when considering natural disasters. The sustainability framework includes principles of sustainability in conjunction with hazard mitigation. Further, evaluating hazards and revising plans are two of the main steps of the classic planning approach [1].

The subject of liquefaction started to garner a great deal of attention after two major earthquakes caused mortality and substantial damage in 1964. Massive landslides happened due to the liquefaction that occurred during the Alaska earthquake, which had a magnitude of 9.2 on the Richter scale, and demolished more than 200 bridges. During the Niigata earthquake, which had a magnitude of 7.5 on the Richter scale, excessive liquefaction caused major damage to more than 60,000 buildings and bridges, highways, and other urban city constructions. Liquefaction poses major threats to the sustainability of existing lifestyles.

Liquefaction occurs in saturated sandy soil due to the dynamic loads that occur during earthquakes. The rapid loads do not allow drainage, so pore water pressure increases, while at the same time effective stress goes down to be zero, or close to zero. Consequently, soil loses its shear strength, and liquefaction happens.

After several decades of intensive research on liquefaction, researchers have presented a variety of methods based on three approaches to evaluate its potential. These three approaches are (1) energy-based [2–12], (2) cyclic stress-based [13–20], and (3) strain-based [8–11,21–26]. The energy-based models estimate the energy that is dissipated into the soil by earthquake loads. In the stress-based method, first, the cyclic stress ratio (CSR) and cyclic shear strength (CRR) are calculated; then, the potential of liquefaction is evaluated by comparing them. The strain-based models suppose that the cyclic shear strain loads control the increase in the pore water pressure. Empirical and semi-empirical methods based on the stress-based approach are more popular among researchers [14,27,28]. Seed and Idriss [29] developed a semi-empirical equation for the earthquake load by calculating the uniform cyclic shear stress amplitude.

Moreover, to determine soil liquefaction resistance, two types of methods have been used to measure the cyclic resistance ratio (CRR) and then compare it with the cyclic stress ratio (CSR): laboratory test-based methods and in situ test-based methods.

Some studies, such as those of Robertson and Wride [13], Youd et al. [14], Juang et al. [27], and Rezania et al. [28], have conducted deterministic analysis to calculate the factor of safety (FS), but due to uncertainty in the liquefaction phenomena, some other researchers have performed probabilistic analysis [7,16,27,30–35]. Artificial neural network (ANN) is a strong technique in cases with non-linear correlations between parameters, such as the liquefaction phenomena. Hence, ANN has been used to investigate this issue [31,36–38]. However, the data division is conducted randomly in two groups for the testing and training phase, and there is no validating phase to prevent over-training the model. The validating phase is applied to minimize the over-fitting of the trained model. If the accuracy of the training dataset goes up, but the accuracy of the validation dataset decreases, then over-fitting occurs, and the training should be stopped. Goh [36] was the first to use a simple back-propagation ANN algorithm to find the relationship between the soil and seismic parameters, and the liquefaction potential. He examined the following eight parameters: SPT value ($N_{1,60}$), fines content (*FC*), mean grain size (*D50*), equivalent dynamic shear stress (τ_{av}/σ'_v), σ_v, σ'_v, *Mw*, and peak horizontal ground acceleration (*PGA*). His results showed that $N_{1,60}$ and *FC* are the most influential parameters, and he compared his ANN results with those of Seed et al. [39]. In another study, Baziar and Nilipour [38] used STATISCA neural networks to evaluate the potential of liquefaction. They confirmed that neural networks (NN) are a capable tool for analyzing liquefaction.

Applying fuzzy neural network models to predict liquefaction, Rahman and Wang [40] considered nine parameters: magnitude of the earthquake (*M*), vertical total overburden pressure, vertical effective overburden pressure, q_{c1N} value from CPT, acceleration ratio *PGA/g*, *CSR*, median grain diameter of the soil, critical depth of liquefaction, and water table depth. They combined fuzzy systems and NN to use both systems' advantages: the learning and optimization ability of NN, and the human-like analysis of the fuzzy system to consider the meaning of high, very high, or low. Their databases included 205 field liquefaction records from more than 20 major earthquakes between 1802–1990. On the other hand, Hanna et al. [31] used the generalized regression neural network (GRNN) to find the liquefaction potential of soil. They considered 12 soil and seismic parameters from the two earthquakes that occurred in Turkey and Taiwan.

All in all, ANN models that assess the potential of liquefaction are still lacking. A validating phase is missing to prevent the model from overtraining. Furthermore, data division has been performed randomly without attention to statistical aspects of the databases.

According to Kramer and Mitchell [41], the ground motion caused by earthquakes can be affected by an earthquake's fault type characteristics. Furthermore, the stress energy carried by earthquake waves decreases exponentially with epicentral distance (R_{epi}). Takahashi et al. [42] investigated the influence of the faulting mechanism on the motions caused by earthquakes, and defined, for example, reverse faulting in comparison with strike-slip and normal faulting shows around 30% higher spectra acceleration. Furthermore, near-fault zones experienced high peak horizontal ground acceleration (*PGA*) with high frequency and a short period, with a less active load on structures [43] and no liquefaction occurrence [44,45]. In an investigation of large-scale Japanese earthquakes, Orense [45] discovered some seismic records in Japan with a high frequency. Even with large acceleration, he observed no liquefaction. Therefore, he proposed the threshold for peak ground velocity (*PGV*) and *PGA* for the occurrence of liquefaction. Furthermore, Liyanapathirana and Poulos [44] analyzed some earthquakes from all around the world and compared them with characteristics of Australian earthquakes. They indicated that the low value of cumulative absolute velocity (*CAV* of Australian earthquakes, which is the time integration of the absolute values of the acceleration, is the key reason for their lack of liquefaction, even with high *PGA*.

It has been demonstrated that the earthquake intensity measure (*IM*) with the best combination of efficiency and sufficiency is standardized cumulative absolute velocity (CAV_5) [41,44]. However, no study has explored its influence and correlation with liquefaction in empirical or semi-empirical models. Furthermore, to the best of the present authors' knowledge, the studies that have been performed did not take into account the effectiveness of parameters and their uncertainties, according to the use of CRR and CSR, on the triggering of liquefaction.

In this study, the most essential parameters to evaluate the potential of liquefaction are selected according to background research studies [7,13,15,16,27,28,30,31,46–50]. Then, two new earthquake parameters added to the dataset to create a comprehensive database: CAV_5 and the closest distance from the site to the rupture surface (r_{rup}). So, strict consideration was given to the earthquake parameters in addition to the soil properties and geometry of the sites. The new earthquake parameters (CAV_5 and r_{rup}) were estimated through attenuation equations derived by Kramer and Mitchell [41] and Sadigh et al. [51], respectively. Consequently, this study considers the influence of earthquakes' causative fault type, the effect of earthquakes near the fault zone, and the frequency of earthquakes' load. In the first step, data are collected to develop an ANN model. To achieve this goal, the data are divided by considering their statistical aspects instead of random division. Furthermore, the data are divided into three groups for the training, testing, and validating phase to prevent the overtraining of the model.

Then, the design of experiment (DOE) is conducted to cover all of the ranges of parameters by points. The ANN model is applied to estimate the target of DOE sample points. Subsequently, the response surface method (RSM), which is a novel method regarding risk assessment and the hazard analysis of liquefaction, is used to establish a direct equation to evaluate the triggering of liquefaction instead of using CRR and CSR. This makes is easier for engineers and contractors to use. Finally, due to uncertainties in soil properties and earthquake parameters, on the basis of the ANN model that was developed in this study and through Monte Carlo simulation (MCS), sensitivity analysis is performed to show the effect of the parameters and their uncertainties on the potential of liquefaction in sandy soils. In Section 8, the proposed model to evaluate the potential for liquefaction triggering (PL) is described. The developed ANN and RSM models are presented in Section 9. The capability and accuracy of the models are confirmed by comparing their results with three extra well-known models. Finally, the sensitivity analysis results are illustrated in 10 graphs. Figure 1 illustrates the flowchart of the risk assessment performed in this study to evaluate PL and conducting sensitivity analysis.

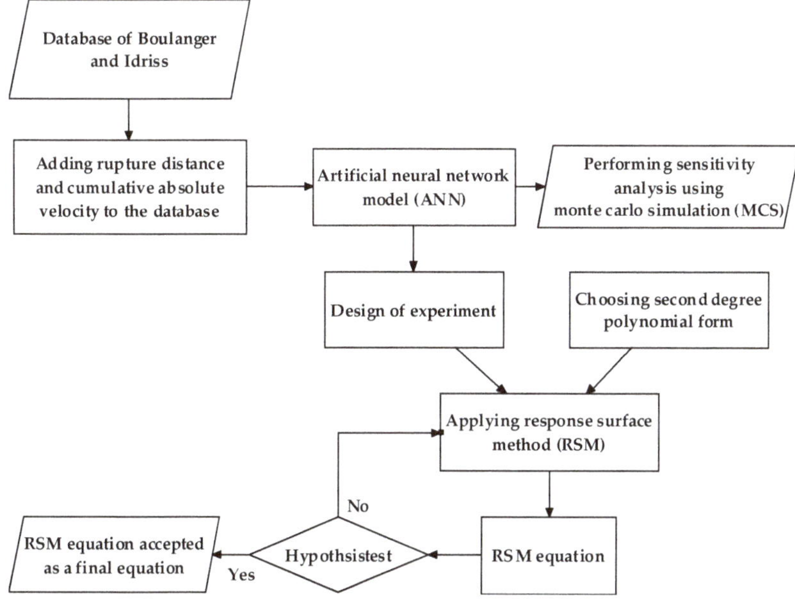

Figure 1. Flowchart of the model developed in this study for risk assessment according to the potential for liquefaction.

2. Analysis Methods

The liquefaction triggering analyses were carried out by comparing the strength of soil against the liquefaction or cyclic strength ratio (CRR) with the stress-loading effects or cyclic stress ratio (CSR). Sections 2.1 and 2.2 state the methods that have been defined by research studies.

2.1. Deterministic Methods

Through deterministic methods, the liquefaction potential is expressed as a factor of safety (FS), which is the ratio of CRR to CSR, without considering the uncertainty associated with the loading and resistance predictions. Robertson and Wride [13] divided the cone penetration test result range in two parts, and presented two equations for CRR:

$$CRR_{7.5} = 0.833\left[\frac{q_{c1N,CS}}{1000}\right] + 0.05 \text{ for } q_{c1N,cs} < 50 \tag{1a}$$

$$CRR_{7.5} = 93\left[\frac{q_{c1N,cs}}{1000}\right]^3 + 0.08 \text{ for } 50 \leq q_{c1N,cs} < 160 \tag{1b}$$

$$q_{c1N,cs} = K_1 q_{c1N} \tag{1c}$$

$$K_1 = 2.429(I_c)^4 - 16.943(I_c)^3 + 44.551(I_c)^2 - 51.497(I_c) + 22.802 \tag{1d}$$

$$q_{c1N} = \frac{q_c/100}{(\sigma_v'/100)^{0.5}} \tag{1e}$$

where I_c is the soil type index, σ_v' is the effective stress, and q_{c1N} is defined as above [13].

Youd et al. [14] presented the following relation for the measurement of *FS* in which K_σ is the overburden correction factor, K_α is the factor of static shear stress correction, and *MSF* is the magnitude scaling factor.

$$FS = (CRR_{7.5}/CSR)\cdot MSF\cdot K_\sigma\cdot K_\alpha \tag{2}$$

Juang et al. [27] generated artificial boundary points between a liquefied and a non-liquefied area onto a two-dimensional (2D) surface of CSR and CRR through a trained neural network. Then, they developed an empirical equation by regression analysis to present the limit state function as follows:

$$CRR = C_\sigma exp\left[-2.957 + 1.264\left(\frac{q_{c1N,cs}}{100}\right)^{1.25}\right] \tag{3a}$$

where:

$$C_\sigma = -0.016\left(\frac{\sigma'_v}{100}\right)^3 + 0.178\left(\frac{\sigma'_v}{100}\right)^2 - 0.063\left(\frac{\sigma'_v}{100}\right) + 0.903 \tag{3b}$$

Idriss and Boulanger [15] updated semi-empirical field-based procedures for evaluating the liquefaction potential of cohesionless soils during earthquakes. They re-examined the field data and updated the analytical framework, and presented standard penetration test results-based (SPT-based) and cone penetration test results-based (CPT-based) liquefaction correlations for cohesionless soils. Moreover, they developed a new equation:

$$CRR_{7.5} = exp\left\{\frac{(N_1)_{60cs}}{14.1} + \left(\frac{(N_1)_{60cs}}{126}\right)^2 - \left(\frac{(N)_{60cs}}{23.6}\right)^3 + \left(\frac{(N)_{60cs}}{25.4}\right)^4 - 2.8\right\} \tag{4}$$

By applying evolutionary polynomial regression (EPR), Rezania et al. [28] developed a model as a three-dimensional (3D) surface boundary to evaluate the potential for liquefaction:

$$CRR = exp\left[-6.994 + 7.9 * 10^{-6}q_{c1N}^{2.5} + 1.115 * Ln\left(\sigma'_v\right)\right] \tag{5}$$

The essential influence of the CPT test results on the potential for liquefaction has been illustrated in all of the defined models in this section.

2.2. Probabilistic Models

Due to uncertainties in soil properties, earthquake parameters, and models, $FS > 1$ does not always reveal non-liquefaction. Similarly, $FS \leq 1$ does not always correspond to the occurrence of liquefaction. In the records of Seed and Idriss [29] and Zhou and Zhang [52], there are 40 data points of 12 earthquakes that occurred in the time period of 1802–1976, among which 27 sites liquefied and 13 sites did not. Zhang et al. [46] considered the following five factors: magnitude of the earthquake (M), source distance (L), depth of the water table, depth of the sand deposit, and SPT blow count (N). Using the optimum-seeking method, the authors evaluated the liquefaction and influence of these five factors, and declared that the most effective parameters are SPT blow count (N) and the magnitude of the earthquake. Toprak et al. [30] analyzed a total of 79 data points of SPT by logistic regression. Fifty data points were from the 1989 Loma Prieta, California earthquake, and 29 data points were from the 1971 San Fernando, the 1979 Imperial, the 1987 Superstition Hills, and the 1994 Northridge earthquakes. They presented the following equation to estimate the potential for liquefaction (P_L):

$$\ln[P_L/1 - P_L] = 13.6203 - 0.2820(N_1)_{60,cs} + 5.3265\ln(CSR_{7.5}) \tag{6}$$

They also analyzed 48 data points from the United States (US) Geological Survey from the 1989 Loma Prieta earthquake by the logistic regression analysis of CPT, and presented the following equation:

$$\ln[P_L/1 - P_L] = 12.260 - 0.0567q_{C1N,cs} + 4.0817\ln(CSR_{7.5}) \tag{7}$$

where $CSR_{7.5}$ is the coefficient of the earthquake magnitude scale.

Chen et al. [7] presented two models for liquefaction by using seismic wave energy. They also used discriminate analysis with normal distribution and demonstrated the ability of this distribution to assess liquefaction potential. They presented the equations by using data collected from the Chi-Chi earthquake.

3. Attenuation Equations for Standardized Cumulative Absolute Velocity and Peak Ground Acceleration

Kramer and Mitchell [41] executed the regressive project to evaluate the influence of approximately 300 earthquake intensities (*IMs*) on the liquefaction. Their database contained more than 450 ground motions from 22 earthquakes. They stated that the *IM* with the best combination of efficiency and sufficiency is the cumulative absolute velocity with a threshold value of 5 cm/s^2, which is shown as CAV_5 (Equation (8)). Furthermore, they developed an attenuation equation to estimate CAV_5 as follows:

$$CAV_5 = \int_0^\infty \alpha \left| a_{(t)} \right| dt \quad \text{where} \quad \alpha = \begin{cases} 0 & \text{for } \left| a_{(t)} \right| < 5 \text{ cm/s}^2 \\ 1 & \text{for } \left| a_{(t)} \right| \geq 5 \text{ cm/s}^2 \end{cases} \tag{8}$$

$$\ln CAV_5 = C_1 + C_2(M - 6) + C_3 \ln(M/6) + C_4 \ln(\sqrt{r_{rup}^2 + h^2}) + f_1 F_N + f_2 F_R \tag{9}$$

The coefficients of the regression are C_1 = 3.495, C_2 = 2.764, C_3 = 8.539, C_4 = 1.008, f_1 = 0.464, f_2 = 0.165, and h = 6.155. The coefficients of the three fault types are given by:

Strike-slip ($F_N = F_R = 0$), normal faults ($F_N = 1$ and $F_R = 0$), and reverse or reverse-oblique ($F_N = 0$ and $F_R = 1$)

Sadigh et al. [51] used a strong motion database of California earthquakes for strike-slip and reverse causative fault with magnitudes of four to more than eight, and r_{rup} of up to 100 km. Based on this, they presented the following attenuation equation for rock and deep firm soil deposits:

$$\ln(\text{PGA}) = C_1 + C_2 M - C_3 \ln(r_{rup} + C_4 e^{C_5 M}) + C_7(8.5 - M)^{2.5} \tag{10}$$

where C_1 is an indicator of the causative fault type, the strike-slip value is −2.17, and the value of the reverse and thrust faults is −1.92.

C_2 = 1.0, C_3 = 1.70, C_6 = 0
C_4 = 2.1863, C_5 = 0.32 *for M* $\leq$ 6.5
C_4 = 0.3825, C_5 = 0.5882 *for M* > 6.5

4. Artificial Neural Network

In cases that exhibit non-linear relationships between input and output parameters, an ANN is often used in research. This technique simulates biological neuron processing in the human brain with simple artificial neuron connections. McCulloch et al. [53] were the first to present neural networks. After that, numerous researchers applied this technique. Multi-layer perceptron (MLP) [54], which is employed in the present study, is the most commonly used. The back-propagation network is the most popular among the MLP paradigms. A MLP contains three types of layers, which include: an input layer, a hidden layer, and an output layer or layers. However, it has been demonstrated that a single hidden layer is sufficient for non-linear analysis. There can be more than one hidden layer; however, theoretical works have shown that a single hidden layer is sufficient for MLP to approximate any complex non-linear function [55]. Therefore, an MLP with three layers and a single hidden layer is applied in the present study to predict the target.

There are three indicators to justify ANN: the root of the mean squared error (RMSE), the coefficient of determination (R^2), and the mean absolute error (MAE). However, R^2 is the most common, and it is considered herein as:

$$R^2 = 1 - \left(\frac{\sum_{i=1}^n (Y_{Estimated} - Y_{Measured})^2}{\sum_{i=1}^n (Y_{Measured})^2} \right) \tag{11}$$

5. Response Surface Methodology

The response surface method (RSM) consists of some statistical techniques, and supports Taguchi's philosophy [56] regarding mathematical processes to build a model and optimize a response. In the present study, the RSM is applied to establish a correlation between parameters. To achieve this goal, three main steps were followed:

1.   Choosing a reasonable design of experiment (DOE).
2.   Selecting an approximate mathematical form.
3.   Determining the significance of equation terms through hypothesis testing.

The first step is choosing a sampling point or experimental design. There are several kinds of DOE techniques, such as central composite design, minimum design, D-optimal designs, full factorial design, Taguchi's contribution design, star design, and Latin hypercube design. In this study, to examine a range of parameters and prevent a negative value in the DOE [57], the design of Box and Behnken [58] is applied. Furthermore, to find the sampling point space, an ANN is applied. According to the DOE and 10 input parameters, 170 sample points are produced by the RSM. To estimate the target (response) of these samples, which is called a coded value here, an ANN model that was trained with a case history database was applied. The coded values for any sample are as follows: The Max value is supposed to be (+1), the mean value is supposed to be (0), and the Min value is supposed to be (−1).

The second step of a RSM is choosing a suitable approximation form to present the relationship between the target (response) and inputs. This is the difference between the RSM and genetic programming.

The most common response surface form that researchers use is polynomial forms, which are given by:

$$R_{(X)} = a_0 + \sum_{i=1}^{n} b_i X_i, \text{ first-degree polynomial} \tag{12}$$

$$R_{(X)} = \sum_{i=1}^{n} b_i X_i + \sum_{i=1}^{n} C_i X_i{}^2, \text{ second-degree polynomial without cross terms} \tag{13}$$

$$R_{(X)} = a_0 + \sum_{i=1}^{n} b_i X_i + \sum_{i=1}^{n} C_i X_i{}^2 + \sum_{\substack{i=1 \\ j \neq i}}^{n} d_{ij} X_i X_j, \text{ second-degree polynomial with cross terms.} \tag{14}$$

To identify significant terms in the equation, a hypothesis test is conducted in the last step. A hypothesis test evaluates the terms relating to a population to determine which terms are the best supported by the sample data. By considering the p-value, some expressions are eliminated in order to achieve the most effective and meaningful terms to build the final equation.

It is common to use hypothesis testing based on p-values in statistics such as basic statistics, linear models, reliability, and multivariate analysis. The P-values have a range from 0 to 1, and show the probability of obtaining a test statistic that is at least as extreme as the calculated value when the null hypothesis is correct.

6. Monte Carlo Method

Researchers have used the Monte Carlo method to study both stochastic and deterministic systems [59–65]. It is a useful technic for both linear and non-linear systems. The only problem making it awkward to use is that it requires too large a number of simulations to present a reliable distribution of the response.

Monte Carlo methods can be divided into two general applications: first, the stochastic generation of samples; and second, solving problems that are too complicated to solve through analytical methods.

The latter is done by generating suitable random (or pseudo-random) numbers and calculating the ratio of the numbers that have the same properties to all of the numbers.

Jha and Suzuki [33] compared the first-order second-moment (FOSM) method, the point estimation method (PEM), and a Monte Carlo simulation (MCS) method, and also presented a combined method using both FOSM and PEM to find the probability of liquefaction. They showed that the safety factor of the potential for liquefaction that was obtained from the combined method was similar to the results from MCS.

7. Case History Database

This study uses an updated version of the liquefaction CPT database case histories from Boulanger and Idriss [19]. It contains a total of 253 cases in which $I_C < 2.6$. I_C is the soil behavior type index presented by Robertson and Wride [13] to classify soils based on fine content. The index ranges from 0.5 for sands to 1.0 for clays, and is given by:

$$I_c = \left[(3.47 - \log(Q))^2 + (1.22 + \log(F))^2 \right]^{0.5} \tag{15}$$

where:

$$Q = \left(\frac{q_c - \sigma_v}{P_a} \right) \left(\frac{P_a}{\sigma'_v} \right)^n, \ (Q) \text{ is normalized tip ratio} \tag{16}$$

$$F = \left(\frac{f_s}{q_c - \sigma_v} \right) \cdot 100\%, \ (F) \text{ is normalized sleeve friction ratio} \tag{17}$$

Of all cases, 180 contained surface evidence of liquefaction, 71 cases did not, and two included doubtful evidence. The common I_c border value of 2.6 [13] is applied to eliminate soils with a clay treatment; therefore, the 15 cases with $I_c \geq 2.6$ are not used in this study. Around 30% of all of the samples belonged to the 1989 Loma Prieta earthquake, including 76 case history samples. The map of liquefaction and its associated effects can be seen in Appendices A and B, in the southern part and the northern part, respectively. Next to that, Appendix C illustrates the locations of the ground failure and damage to the facilities on Treasure Island that were attributed to the 1989 Loma Prieta earthquake. These figures illustrate the ground failure sites, lateral spread, ground settlement, sand boil points, and cracks that were caused by liquefaction in this area.

The dataset contains two earthquake parameters: moment magnitudes (*Mw*) and *PGA*. Given that the purpose of this study is to consider the new earthquake parameters (*CAV*$_5$ and r_{rup}), Kramer et al.'s [41] equation is used. Due to the shortage of known parameters to apply this equation, the causative faulting mechanism of all of the earthquakes in the dataset is first characterized to utilize both equations. Then, Sadigh et al.'s [51] attenuation equation is applied to estimate the r_{rup}. Next to that, due to the magnitude range of the attenuation equations, the seven data samples from the Tohoku earthquake, with a moment magnitude of nine, are eliminated from the dataset. From all of the parameters in the source, only eight are selected according to studies reported in the literature [7,13,15,16,27,28,30,31,46–50]: momentum magnitude (*Mw*), *PGA(g)*, thickness of liquefiable soil (m) (measured from critical depth interval)(*T*), ground water table *GWT* (m), overburden stress (σ) (kPa), effective overburden stress (σ') (kPa), normalized con penetration test result tip (q_{c1N}), mean grain size (D_{50}), and fine content (*FC*) (%). The two extra parameters of *CAV* and r_{rup} are added to the dataset through Equations (9) and (10), Therefore, the study considers some of the essential aspects of the earthquakes, such as the near-fault earthquake zone and causative fault type of the earthquake. The summary of the dataset that is used in this study can be seen in Table 1.

Table 1. Summary of dataset used in this study. NZ: New Zealand, US: United States.

Earthquake	Total Data No.	Train Data No.	Test Data No.	Validation Data
Liquefied Sites				
1983 M = 6.9 Borah Peak, US	4	2	1	1
1999 M = 7.6 Chi-Chi, Taiwan	11	7	2	2
2011 M = 6.2 Christchurch, NZ	21	13	4	4
2010 M = 7.1 Darfield, NZ	18	12	3	3
1987 M = 6.6 Edgecumbe, NZ	12	8	2	2
1995 M = 6.9 Hyogoken-Nambu, Japan	16	10	3	3
1979 M = 6.5 Imperial Valley, US	2	2	0	0
1968M = 7.2 Inangahua, NZ	2	2	0	0
1999 M = 7.5 Kocael, Turkey	14	8	3	3
1989 M = 6.9 Loma Prieta, US	49	29	10	10
1983 M = 7.7 Nihonkai Chub, Japan	2	2	0	0
1964 M = 7.6 Niigata, Japan	2	2	0	0
1994 M = 6.7 Northridge, US	2	2	0	0
1987 M = 6.5 Superstition Hills, US	1	1	0	0
1976 M = 7.6 Tangshan, China	13	9	2	2
1980 M = 6.3 Victoria (Mexicali), US	4	2	1	1
1981 M = 5.9 West Morland, US	3	1	1	1
Total	176	112	32	32
Doubtful Sites				
1975 M = 7.0 Haicheng, China	1	1	0	0
1989 M = 6.9 Loma Prieta, US	1	1	0	0
Total No.	2	2	0	0
Non-Liquefied Sites				
2011 M = 6.2 Christchurch, NZ	4	4	0	0
2010 M = 7.1 Darfield, NZ	7	5	1	1
1987 M = 6.6 Edgecumbe, NZ	5	3	1	1
1975 M = 7.0 Haicheng, china	1	1	0	0
1995 M = 6.9 Hyogoken-Nambu, Japan	7	3	2	2
1979 M = 6.5 Imperial Valley, US	2	2	0	0
1999 M = 7.5 Kocaeli, Turkey	1	1	0	0
1989 M = 6.9 Loma Prieta, US	26	14	6	6
1983 M = 7.7 Nihonkai-Chubu, Japan	1	1	0	0
1964 M = 7.6 Niigata, Japan	1	1	0	0
1987 M = 6.5 Superstition Hills, US	5	3	1	1
1976 M = 7.6 Tangshan, China	6	4	1	1
1981 M = 5.9 West Morland, US	2	2	0	0
Total No.	68	44	12	12

8. Proposed Model and Equation to Evaluate Liquefaction Triggering

The dataset that was collected for this study is illustrated in Table 1. As can be seen, it includes 246 samples: 68 non-liquefied samples, 176 liquefied samples, and two doubtful samples [19]. The same portion of liquefied and non-liquefied samples is selected for division, instead of a randomness division. As can be seen from Table 1, around 18% of the liquefied samples, or 32 samples, are selected for the ANN testing phase, and an equal number is selected for the validating phase. A similar ratio of non-liquefied samples is chosen, too. Twelve samples from a total of 68 are chosen for testing, and an equal number is selected for the validating phase.

Furthermore, the parameters are divided into three parts based on similar statistical characteristics, such as the mean value and mean coefficient of variation (COV), in order to increase the accuracy and capability of the trained model. Single hidden layer neural network training was developed in conjunction with the back-propagation algorithm to train a model [36–38]. In addition, the ANN target to evaluate the potential for liquefaction according to the dataset is introduced as:

$$T = \begin{cases} 1 & for\ liquefied\ cases \\ 0.5 & for\ doubtful\ cases \\ 0 & for\ non-liquefied\ cases \end{cases} \tag{18}$$

The new equation is presented by applying the new tool for the response surface method (RSM). To the best of the authors' knowledge, no approach has yet used this tool to assess liquefaction in sandy soil. In the present study, the RSM is applied to develop correlations between the input parameters and the target. The ANN model is used to calculate the coded points, which are defined in Section 4. The new equation presented herein has two advantages over the processes and models that have been presented by other researchers:

1. Some aspects of earthquakes, such as the near-fault sites and the frequency content of earthquake motions, are considered by exploring and adding the two new parameters of CAV_5 and r_{rup}.
2. Since it uses direct parameters instead of just CRR and CSR, it is simpler to use in projects by engineers and city planners to obtain sustainable design.

Since the equation uses direct parameters, extra statistical and probabilistic aspects are considered during sensitivity analysis. In contrast, when analyzing liquefaction through CRR and CSR, it is only possible to consider model uncertainty, not parameter uncertainties.

In this study, the second-degree polynomial with cross terms (Equation (14)), which is the most frequently utilized form, is applied to derive a final equation. The alpha (α) level of 0.05 is commonly used by scientists and researchers, and is considered herein. This means that if the p-value of a test statistic is less than alpha, which is herein 0.05, then the null hypothesis is rejected. Here, the null hypothesis is that there is no meaningful correlation between the input value and the target, so the expressions with a P-value of more than 0.05 are eliminated. Furthermore, according to Equation (18), the following values are supposed to evaluate the results of an ANN and the RSM:

$$\begin{aligned} T < 0.4 &\quad Non-liquefied \\ T > 0.6 &\quad Liquefied \\ 0.4 \leq T \leq 0.6 &\quad Doubt\ (No\ acceptable\ result) \end{aligned} \tag{19}$$

9. Results

The developed ANN model is applied to measure the coded values of the DOE of the RSM, and is the basis of the sensitivity analysis using MCS. As illustrated in Table 2, 10 parameters are considered as input parameters, including earthquake parameters, geometry parameters, and soil properties. The target is introduced by Equation (19).

Table 2. Statistical aspects of variables.

	Statistical Parameters						
Input Variable	Mean	Std. Dev.	Min	Max	Mean / COV	Distribution Function	References
Mw	6.8	0.51	5.9	7.7	0.075	Normal distribution	[66]
PGA	0.465	0.06975	0.09	0.84	0.15	Normal distribution	[66]
r_{rup}	26.23	0	1	51.46	0	No uncertainty	*
CAV_5	29.78	2.978	1.2	58.36	0.1	Normal distribution	*
T	3.25	0	0.3	6.2	0	No uncertainty	*
GW	3.7	0	0.2	7.2	0	No uncertainty	*
σ_v	117	17.55	24	210	0.1	Normal distribution	[66]
σ'_v	83	12.45	19	147	0.15	Normal distribution	[66]
FC	42.5	8.5	0	85	0.3	Normal distribution	[67]
q_{c1N}	162.7	32.54	13.6	311.8	0.2	Normal distribution	[68]

r_{rup}, T and GW are supposed as a parameter and no variable. CAV is supposed variable with COV of 0.1 less than COV supposed for PGA [41] *.

Table 3 illustrates the certificate of the ANN model. In the second step, the target values of the DOE samples are predicted through the developed ANN model. In the third step, RSM analysis is performed to derive an equation. The second-degree polynomial with cross terms, which is the most complicated general equation and has the best capability and accuracy, and is illustrated in Equation (14), is utilized to present an equation that relates the 10 input parameters (as a coded value by DOE) and the response values (measured through ANN). Finally, MCS is applied to conduct a sensitivity analysis to demonstrate the influence of the parameters and their uncertainties on the probability of triggering liquefaction (PL) in sandy soil caused by earthquake motions.

Table 3. Correlation Coefficient of ANN model.

Total Data	Testing Data	Validating Data	Training Data
76.2	80.4	70	78

9.1. RSM Equation to Evaluate Liquefaction Triggering

In the first analysis using the RSM, the first generated equation has all of the expressions (terms) of the general equation. It contains 66 expressions, but using the hypothesis test, those with a *p*-value greater than 0.05 are eliminated. Then, the RSM analysis is performed repeatedly until the final equation is presented by 49 expressions, according to Table 4.

Table 4. Final RSM equation to evaluate the liquefaction potential.

Term	Constant	Mw	PGA	r_{rup}	CAV_5	T	GWT
Coefficient	0.12	0.074	0.611	0.098	1.41×10^{-5}	-0.13	0.167
Term	σ	σ'	FC	q_{c1N}	$PGA*PGA$	$\sigma'*\sigma'$	$FC*FC$
Coefficient	0.20	-0.159	0.181	-0.593	-0.364	0.09	0.214
Term	$q_{c1N}*q_{c1N}$	$M*PGA$	$Mw*r_{rup}$	$Mw*CAV_5$	$Mw*T$	$Mw*GWT$	$Mw*\sigma'$
Coefficient	0.181	-0.194	0.257	0.228	-0.107	-0.113	0.107
Term	$Mw*FC$	$Mw*q_{c1N}$	$PGA*r_{rup}$	$PGA*CAV_5$	$PGA*T$	$PGA*GWT$	$PGA*\sigma$
Coefficient	0.174	-0.267	0.313	0.347	0.254	-0.146	-0.144
Term	$PGA*\sigma'$	$PGA*FC$	$r_{rup}*CAV5$	$r_{rup}*T$	$r_{rup}*GWT$	$r_{rup}*\sigma$	$r_{rup}*FC$
Coefficient	0.126363	-0.18	-0.264	-0.164	0.199	-0.102	0.33
Term	$r_{rup}*q_{c1N}$	CAV_5*T	CAV_5*GWT	$CAV_5*\sigma$	CAV_5*FC	CAV_5*q_{c1N}	$T*\sigma$
Coefficient	0.37	-0.405	0.312	-0.111	0.126	0.111	0.14
Term	$T*\sigma'$	$T*FC$	$GWT*\sigma'$	$GWT*FC$	$\sigma*FC$	$\sigma*q_{c1N}$	$\sigma'*FC$
Coefficient	-0.149	0.106	0.259	0.253	-0.151	-0.178	0.203

The correlation coefficient (R) of the ANN model is illustrated in Table 3. Table 5 presents the certificate of the RSM equation that was developed in this study. It should be noted that R^2 (adjust) reveals the power of regression models that contain different numbers of predictors, and is always less than R^2. Closing their values indicates the high accuracy of the equation.

Table 5. Certificate of the RSM equation.

R^2	R^2 (Adjust)
70	65.55

To demonstrate the accuracy and capability of the RSM equation and ANN model, their results are compared to three well-known models that have been presented by researchers (Juang et al. [27]; Rezania et al. [28]; Robertson and Wride [13]); these models are discussed in Section 2.1. This is conducted using 44 testing datasets that were not used to train the ANN and develop the RSM equation. Table 6 illustrates these testing data samples. Table 7 compares the results of the present study with the other three models. Tables 8 and 9 present a summary of results of the ANN model and RSM equation, and the other three models.

For Equation (19), the results are summarized in Table 7. It is clear that in doubtful cases, the results of the models are not robust. Therefore, considering other models is recommended.

From Tables 8 and 9, it can be seen that among the 44 cases, the ANN model and RSM equation evaluate the triggering of liquefaction with only one wrong prediction, and with only one and two cases of doubt, respectively. In contrast, the models of Robertson and Wride [13], Juang et al. [27], and Rezania et al. [28] can assess the onset of liquefaction with five, four, and six wrong predictions. The present authors suggest that in the doubt cases predicted by the models that were proposed in this study, other models can be considered, too.

Table 6. Case history samples for testing capability and accuracy of presented models.

No.	Earthquake	Mw	a_{max}	r_{rup}	CAV_5	T	GWT	σ	σ'	Fc	q_{c1N}	Liq (1 Yes, 0 No)
1	Loma Prieta	6.93	0.12	43.77	3.26	1.6	6.4	131	123	9	49.6	0
2	Superstition Hills	6.54	0.18	17.93	3.62	0.6	2.1	42	39	18	97.4	0
3	Edgecumbe	6.6	0.26	11.35	3.66	0.6	4.4	83	80	5	149.7	0
4	Loma Prieta	6.93	0.28	17.74	7.72	1	3.4	68	63	1	191.3	0
5	Loma Prieta	6.93	0.28	17.74	7.72	0.6	3.7	83	73	1	107.7	0
6	Darfield	7	0.21	26.23	5.98	1.1	1	51	34	11	106.4	0
7	Tangshan	7.6	0.26	20.44	16.67	1.2	3.5	119	89	2	163.1	0
8	Loma Prieta	6.93	0.28	17.74	7.72	1	1.7	45	37	4	134.6	0
9	Loma Prieta	6.93	0.28	17.74	7.72	1	1.4	64	43	4	142.4	0
10	Loma Prieta	6.93	0.38	11.12	11.44	4	3.5	97	79	9	87.6	0
11	Hyogoken-Nambu	6.9	0.65	−1.33	18.81	1.3	2	68	51	0	165.1	0
12	Hyogoken-Nambu	6.9	0.7	1.00	19.00	2	2.5	93	68	0	163.5	0
13	Loma Prieta	6.93	0.13	40.72	3.50	1.4	5	160	124	13	35.1	1
14	Victoria	6.33	0.19	14.12	3.30	3.2	2.2	54	46	52	29	1
15	Christchurch	6.2	0.172	19.02	2.49	1.5	1.6	79	53	6	67	1
16	Loma Prieta	6.93	0.22	23.88	5.87	2.1	3.4	82	72	4	93.3	1
17	Christchurch	6.2	0.177	18.44	2.56	1.85	1.9	41	36	8	59.8	1
18	Loma Prieta	6.93	0.21	25.17	5.58	4.2	2.1	112	73	10	51.1	1
19	Darfield	7	0.231	23.52	6.63	1.15	1.4	47	35	5	70.9	1
20	Loma Prieta	6.93	0.28	17.74	7.72	1.5	3	116	84	3	61.6	1
21	Darfield	7	0.217	25.28	6.19	4.25	1.4	116	70	26	61.9	1
22	Loma Prieta	6.93	0.28	17.74	7.72	0.6	2.8	99	73	3	24.4	1
23	Loma Prieta	6.93	0.36	12.21	10.63	3	4.9	105	97	27	36.6	1
24	Loma Prieta	6.93	0.28	17.74	7.72	3.6	1.8	64	47	1	73.2	1
25	Westmoreland	5.9	0.26	5.37	3.47	4.4	1.2	90	54	30	70.9	1
26	Loma Prieta	6.93	0.28	17.74	7.72	0.6	2.4	204	120	30	42.1	1
27	Loma Prieta	6.93	0.28	17.74	7.72	0.6	1.8	92	60	4	114.7	1
28	Christchurch	6.2	0.339	7.53	5.15	2	2.3	81	60	7	34.4	1
29	Loma Prieta	6.93	0.28	17.74	7.72	0.6	1.4	155	87	30	46.5	1
30	Christchurch	6.2	0.346	7.25	5.27	3.2	2.4	162	101	11	60.4	1
31	Kocaeli	7.51	0.4	7.95	30.73	1	1.7	45	37	15	44.1	1
32	Kocaeli	7.51	0.37	9.81	26.65	1.2	1	33	25	16	21.8	1
33	Kocaeli	7.51	0.4	7.95	30.73	3	0.8	92	51	11	75.5	1
34	Chi-Chi	7.62	0.38	16.69	24.42	1	1	46	31	38	34	1
35	Hyogoken-Nambu	6.9	0.37	6.85	12.82	2	2	125	76	2	87.3	1
36	Hyogoken-Nambu	6.9	0.45	3.70	16.48	1.5	2.1	86	60	28	55.4	1
37	Hyogoken-Nambu	6.9	0.4	5.55	14.26	3	1.5	102	62	36	32.3	1
38	Edgecumbe	6.6	0.42	4.00	6.46	1.5	1.6	144	84	5	82.9	1
39	Edgecumbe	6.6	0.43	3.69	6.61	3	0.5	50	29	1	56.8	1
40	Borah Peak	6.88	0.5	2.12	11.09	1.6	0.8	44	29	20	96.7	1
41	Tangshan	7.6	0.64	1.00	56.23	1.1	3.7	109	85	5	70.9	1
42	Tangshan	7.6	0.61	1.00	57.89	1.2	0.9	36	24	9	64.6	1
43	Chi-Chi	7.62	0.25	30.79	13.77	3	3.5	98	79	61	23	1
44	Darfield	7	0.239	22.59	6.89	1.2	0.9	34	25	9	49.1	1

Table 7. Comparison between the results of this study with extra three models.

No.	Earthquake	Liq (1 Yes, 0 No)	CSR	CRR[1]	CRR[2]	CRR[3]	ANN	RSM
1	Loma Prieta	0	0.075	0.10	0.10	0.23	−0.26	0.23
2	Superstition Hills	0	0.123	0.32	0.27	0.11	0.07	0.20
3	Edgecumbe	0	0.165	0.48	0.41	1.06	0.12	−0.16
4	Loma Prieta	0	0.188	0.91	0.83	5.07	−0.07	−0.23
5	Loma Prieta	0	0.196	0.21	0.20	0.28	0.71	0.21
6	Darfield	0	0.201	0.24	0.22	0.12	0.47	0.42
7	Tangshan	0	0.215	0.47	0.53	2.00	−0.51	−0.33
8	Loma Prieta	0	0.216	0.34	0.29	0.27	0.19	0.32
9	Loma Prieta	0	0.26	0.39	0.34	0.41	0.26	0.36
10	Loma Prieta	0	0.283	0.16	0.16	0.21	1.04	0.39
11	Hyogoken-Nambu	0	0.545	0.59	0.51	1.17	−0.01	0.26
12	Hyogoken-Nambu	0	0.584	0.57	0.51	1.51	0.23	0.34
13	Loma Prieta	1	0.096	0.09	0.09	0.21	0.39	0.86
14	Victoria	1	0.139	0.15	0.13	0.07	0.62	0.69
15	Christchurch	1	0.157	0.12	0.10	0.10	0.69	0.67
16	Loma Prieta	1	0.157	0.16	0.16	0.21	0.68	0.19
17	Christchurch	1	0.16	0.10	0.09	0.06	0.81	0.64
18	Loma Prieta	1	0.193	0.10	0.09	0.13	0.61	0.65
19	Darfield	1	0.195	0.12	0.11	0.07	1.02	0.63
20	Loma Prieta	1	0.201	0.11	0.11	0.16	0.79	0.69
21	Darfield	1	0.218	0.19	0.19	0.13	0.86	0.60
22	Loma Prieta	1	0.232	0.07	0.06	0.11	0.95	0.95
23	Loma Prieta	1	0.237	0.12	0.13	0.16	0.98	0.62
24	Loma Prieta	1	0.24	0.12	0.11	0.10	0.98	0.71
25	Westmoreland	1	0.258	0.30	0.24	0.11	0.78	0.67
26	Loma Prieta	1	0.26	0.14	0.16	0.21	0.79	1.00
27	Loma Prieta	1	0.261	0.24	0.22	0.27	0.83	0.47
28	Christchurch	1	0.279	0.08	0.07	0.09	1.15	0.97
29	Loma Prieta	1	0.288	0.16	0.16	0.15	0.78	0.85
30	Christchurch	1	0.304	0.12	0.12	0.20	0.88	1.07
31	Kocaeli	1	0.311	0.10	0.09	0.06	0.89	0.87
32	Kocaeli	1	0.314	0.08	0.07	0.03	0.96	1.05
33	Kocaeli	1	0.338	0.14	0.13	0.11	1.03	0.74
34	Chi-Chi	1	0.36	0.13	0.13	0.04	0.98	0.74
35	Hyogoken-Nambu	1	0.36	0.15	0.14	0.20	0.90	0.93
36	Hyogoken-Nambu	1	0.396	0.18	0.17	0.11	0.95	0.75
37	Hyogoken-Nambu	1	0.396	0.13	0.13	0.10	0.86	1.03
38	Edgecumbe	1	0.413	0.14	0.14	0.21	0.94	1.09
39	Edgecumbe	1	0.48	0.10	0.09	0.05	1.30	1.61
40	Borah Peak	1	0.493	0.33	0.29	0.08	0.99	0.72
41	Tangshan	1	0.51	0.11	0.12	0.18	0.84	1.36
42	Tangshan	1	0.576	0.11	0.10	0.04	0.91	1.56
43	Chi-Chi	1	0.195	0.12	0.13	0.12	1.41	0.88
44	Darfield	1	0.212	0.10	0.08	0.04	1.14	1.83

CRR (Robertson and Wride) [13]. CRR (Juang et al.) [27]. CRR (Rezania et al.) [28].

Table 8. Summary of the results of the ANN model and the RSM equation.

Model	Total Cases	Right Prediction	Wrong Prediction	Doubt
ANN	44	42	1	1
RSM	44	41	1	2

Table 9. Summary of the results of the three other models.

Model	Total Cases	Right Prediction	Wrong Prediction
Robertson et al. [14]	44	39	5
Juang et al. [28]	44	40	4
Rezania et al. [29]	44	38	6

9.2. Sensitivity Analysis with the Monte Carlo Simulation Method

A sensitivity analysis determines how different values of a set of independent variables will certainly affect a specific dependent variable under a given set of presumptions. It should be mentioned that the analysis performed here is different from the classical sensitivity analysis, which usually leads to a so-called Tornado diagram or a sensitivity chart, such as that conducted for example by Hariri-Ardebil et al. [69].

In the context of sustainability, without considering uncertainties, decision making is made without a suitable understanding of the consequences. Therefore, sensitivity analysis is performed using the MCS method to analyze the influence of uncertainties on all of the soil properties, geometry conditions, and earthquake parameters on the PL. The main difficulty in using MCS is that it requires numerous samples. Due to the expensive procedures for testing and sampling and the scarcity, to overcome this shortage, in the first step, the ANN model is trained using existing samples from the prepared dataset. Then, sensitivity analysis is applied on the basis of the ANN model.

Table 2 illustrates variables with their mean values and their mean coefficient of variation (COV). Among all of the parameters, because they are directly measured values, r_{rup}, T, and GWT are supposed to be without uncertainty. In addition, a correlation matrix is built to apply MCS. Juang et al. [66] used standard statistical methods to find the correlation coefficient between variables. According to Juang et al. (2006), the correlation coefficients between σ_v' and σ'_v and between a_{max} and Mw are supposed to be 0.9. Moreover, the coefficients between $qc1N$ and σ_v and between $qc1N$ and σ_v' are supposed to be 0.2 and 0.3, respectively. Finally, according to Juang et al. [43], the correlation coefficients between σ_v and σ_v', q_{c1N} and σ_v', q_{c1N} and σ'_v, and a_{max} and Mw are assumed to be 0.95, 0.2, 0.3, and 0.9, respectively.

The previous studies by Lumb [70] and Tan et al. [71] demonstrated that in order to present soil properties as variables, the normal distribution function can be supplied considering that if the coefficient of variation is small in this function, then the error is negligible. Hence, all of the variables are assumed to possess a normal distribution function.

In the present study, according to Table 2, the 10 input parameters are defined as variables, and the probability of liquefaction (PL) is estimated as follows:

$$P_L = \frac{N_L}{N_T} \tag{20}$$

where N_L is the number of liquefied samples, and N_T is total number of the samples. Sensitivity analysis is presented by changing the mean or coefficient of variation of the target variable, whereas extra variables are confirmed in their mean and mean COV value. Figures 2–11 provide a visual representation of the sensitivity analysis. It can be seen in the range of the analysis that q_{c1N} provides an essential influence on the potential of liquefaction. It can be seen in Figure 11 that by increasing the q_{c1N} value from 136 to 311.8, which is the range of the collected dataset in this study, the probability of triggering liquefaction (PL) falls from 99.5% to near 0%. Moreover, by increasing the COV from 0.1 to 0.2 and then to 0.3, the probability of liquefaction triggering (PL) in the critical value of q_{c1N} =130 grows from 3% to 9% and then to 15.5%. This means that by increasing the COV by 20%, the PL illustrates a 12.5% growth.

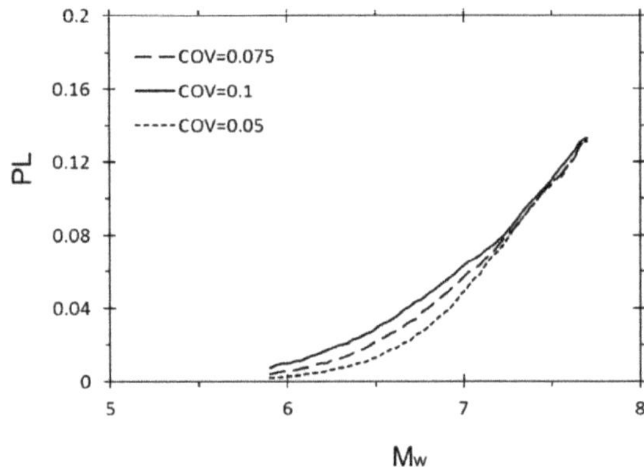

Figure 2. Moment magnitude vs. probability of triggering liquefaction.

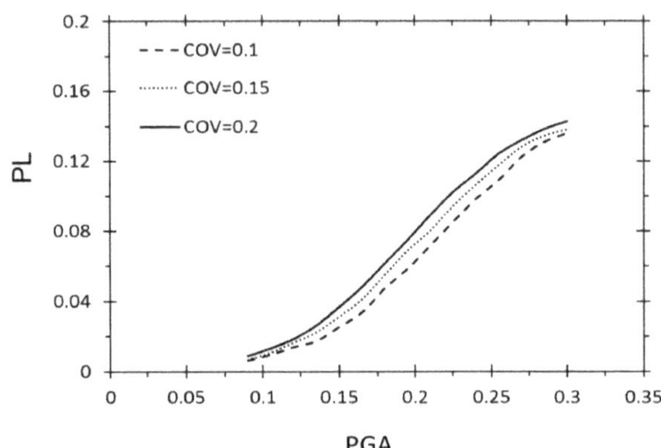

Figure 3. Peak horizontal ground accelerations vs. probability of triggering liquefaction.

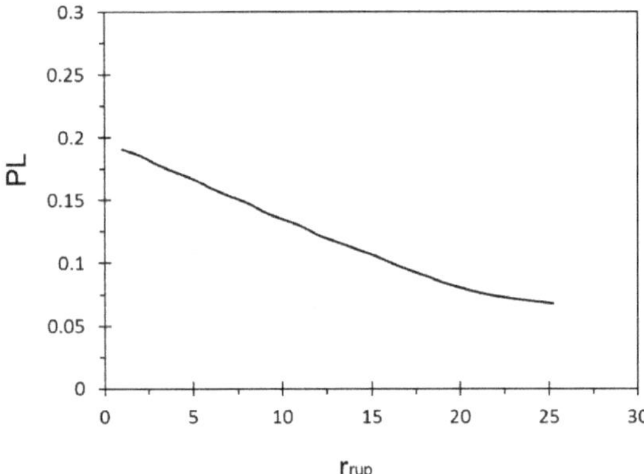

Figure 4. Closest rupture distance vs. probability of triggering liquefaction.

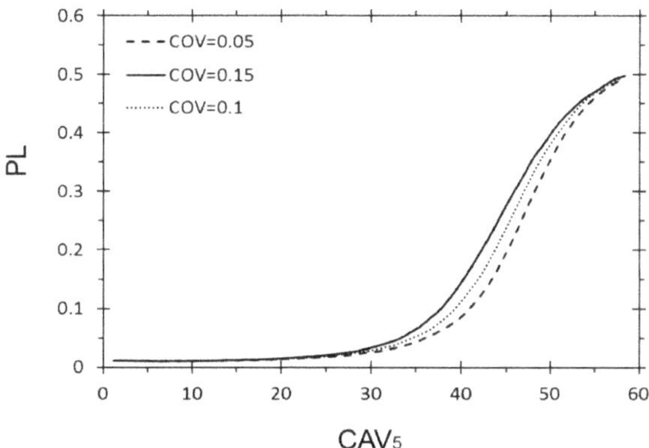

Figure 5. Standardized cumulative absolute velocity vs. probability of triggering liquefaction.

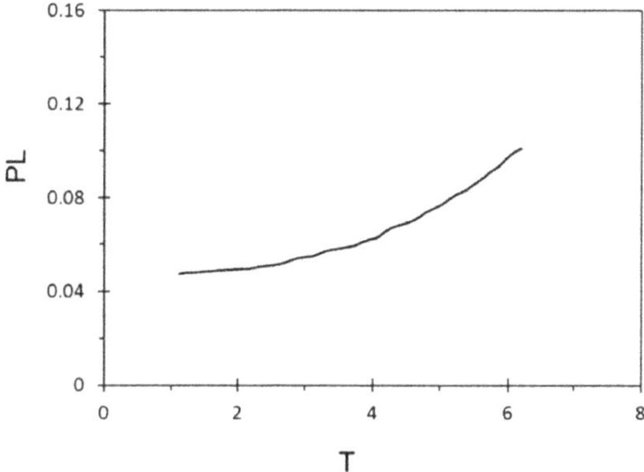

Figure 6. Thickness of liquefiable soil vs. probability of triggering liquefaction.

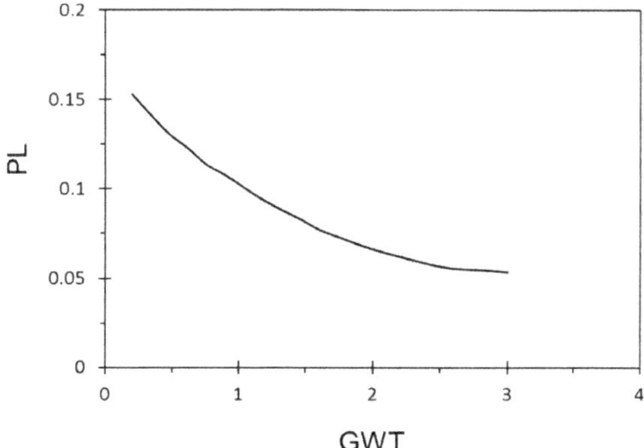

Figure 7. Ground water table (m) vs. probability of triggering liquefaction.

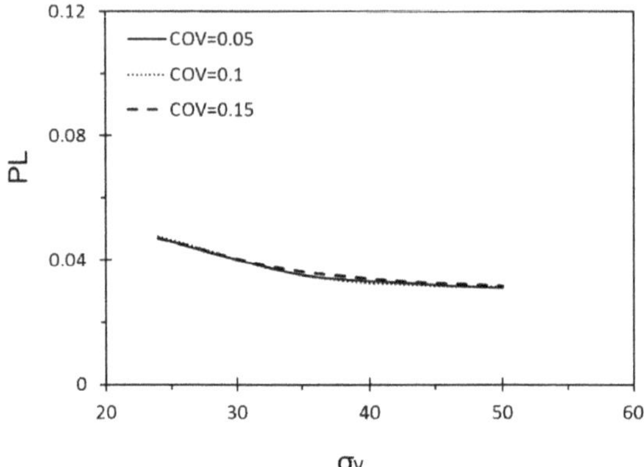

Figure 8. Overburden stress vs. probability of triggering liquefaction.

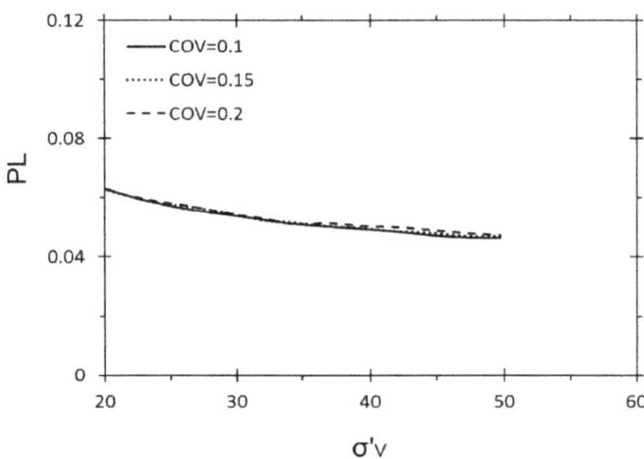

Figure 9. Effective overburden stress vs. probability of triggering liquefaction.

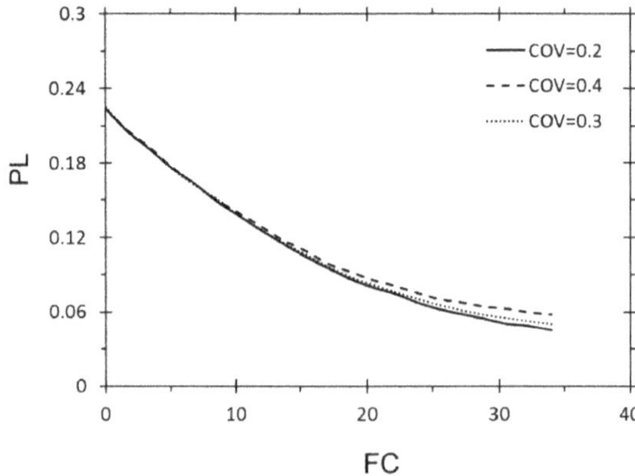

Figure 10. Fine content vs. probability of triggering liquefaction.

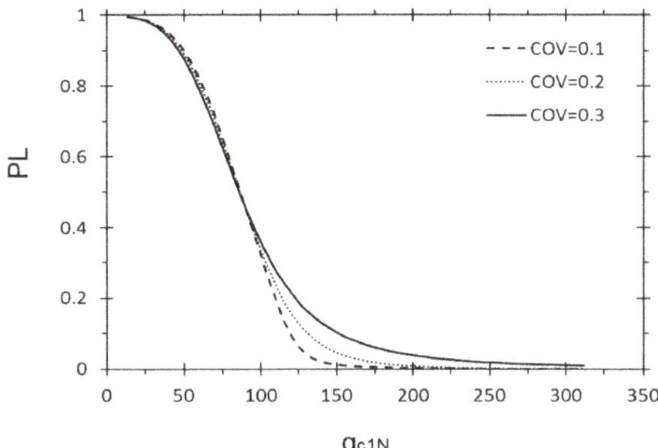

Figure 11. Normalized con penetration test result tip vs. probability of triggering liquefaction.

Figures 8 and 9 illustrate the negligible influence of uncertainties of σ_v and σ_v' on PL. Whereas, when CAV_5 is in the critical range from 40 m/s to 50 m/s, as shown in Figure 5, by increasing the COV by 10%, from 5% to 15%, the PL increases by around 8%. It is clear that this parameter provides a large effect on the PL by causing it to increase by 50% in the range of the presented CAV_5 in the dataset. Furthermore, when the *PGA* grows from 0.1 to 0.3, the PL shows 14% growth. By a 10% increase in COV, the results illustrate around a 1.5% growth in PL, as demonstrated in Figure 3. As *Mw* goes up from 5.9 to 7.7, the PL shows an increase from 0% to 14%. Furthermore, for a magnitude of around 6.8, the uncertainty affected in the maximum increases by around 1% by increasing any 2.5% in the COV (Figure 2).

10. Summary and Conclusions

The risk assessment models that are used for the mitigation of seismic hazards such as liquefaction constitute a fertile field for contributions by the geotechnical profession to sustainable development [72]. To achieve this goal, this paper presents an RMS-based method to evaluate the triggering of liquefaction by directly using influential parameters. These parameters include earthquake parameters (M, r_{rup}, CAV_5 and PGA), soil properties (normalized con penetration test result tip [q_{c1N}], FC), and geometry conditions (σ', σ, T, GWT). To develop this model, this study used a large database including various CPT database case histories of earthquakes. Strict attention was paid to earthquake aspects by adding the new parameters of CAV and r_{rup} to the database. Therefore, the study considered the influence of the causative fault type of the earthquakes, the near-fault zone effect, and the period of the earthquakes' load. Then, the derived equation and the ANN model were compared to three already existing models to show their capability and accuracy. The proposed RSM and ANN models obtain a reasonable performance in predicting the triggering of liquefaction, providing 41 and 42 correct predictions, respectively, out of 44 cases, and just one wrong prediction (extra cases as a doubt). Furthermore, we used soil properties, geometry conditions, and earthquake parameters to study their correlation with the probability of triggering liquefaction (PL) directly instead of using CSR and CRR. Thus, parametric sensitivity analysis explored to show their direct influence on PL. Based on our results, we conclude that:

1. RSM is a powerful tool and method to study the liquefaction phenomenon. This was shown from the comparison of its results with those of the other models.
2. Among all of the parameters, the normalized con penetration test result tip (q_{c1N}) is an essential factor, and has the highest effect on PL.
3. Among the earthquake parameters that were considered in this study, the standardized cumulative absolute velocity (CAV_5) provides the most significant influence on PL. In contrast,

the moment magnitudes (*Mw*), peak horizontal ground accelerations (*PGA*), and closest rupture distance (r_{rup}) illustrate lesser effects.

4.   Uncertainties of q_{c1N} and CAV_5 have a considerable influence on PL.

Author Contributions: Data curation, N.P.; Formal analysis, N.P.; Investigation, N.P. and X.T.; Methodology, N.P.; Project administration, Q.Y.; Software, F.K.; Supervision, X.T. and Q.Y.; Validation, N.P.; Visualization, N.P.; Writing—original draft, N.P.

Funding: This research was supported by the National Natural Sciences Foundation of China Granted No. 51639002 and National Key Research & Development Plan under Grant No. 2018YFC1505305. The authors are gratefully appreciated.

Conflicts of Interest: The authors declare no conflict of interest.

Appendix A

Locations of liquefaction and associated ground-failure effects related to the Loma Prieta earthquake, California of 17 October 1998—southern part.

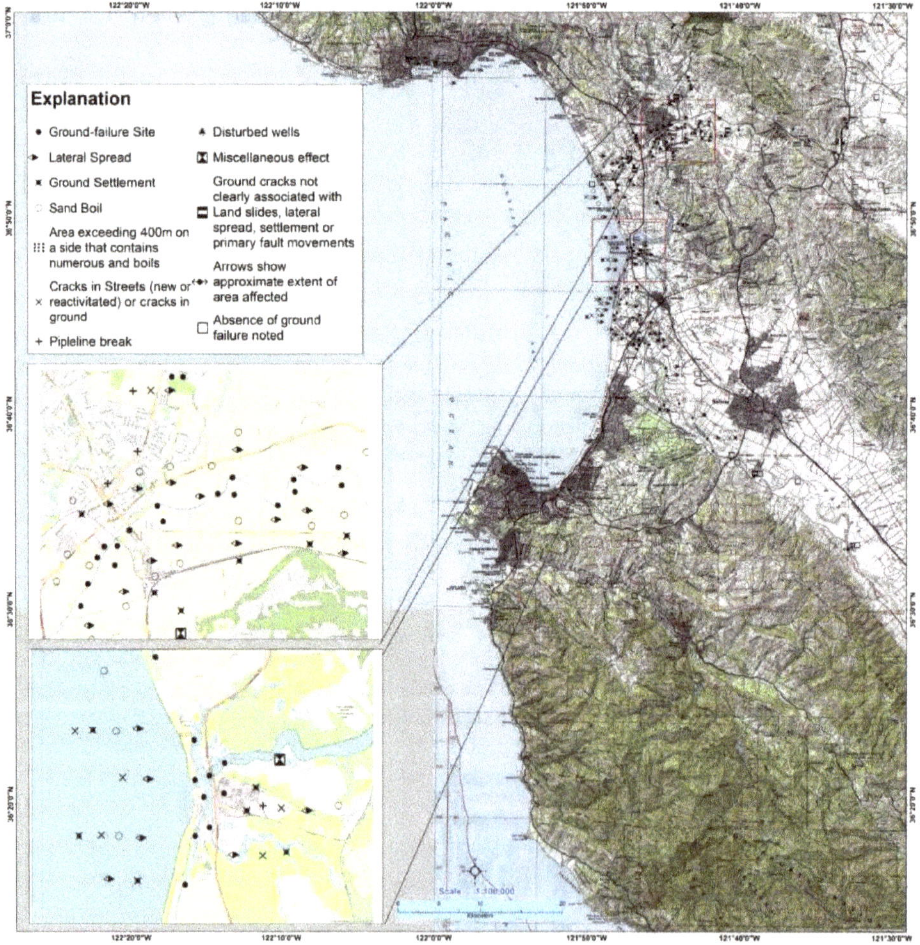

Appendix B

Locations of liquefaction and associated ground-failure effects related to the Loma Prieta earthquake, California of 17 October 1998—northern part.

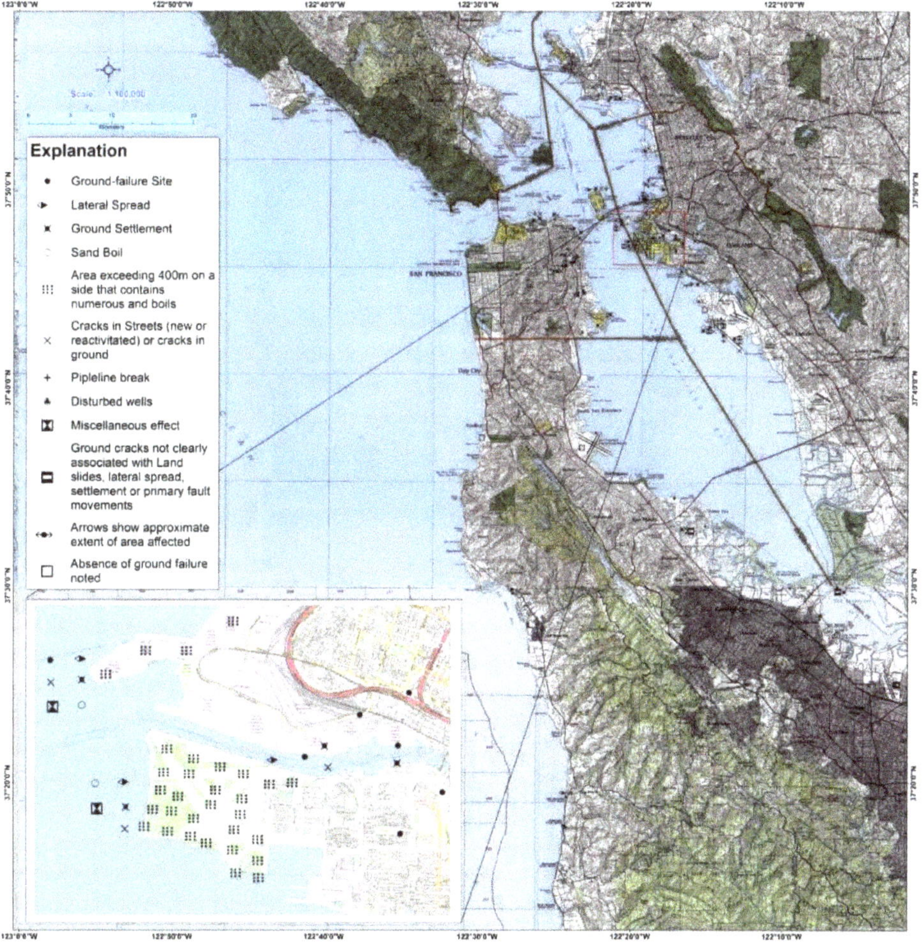

Appendix C

Locations of ground-failure and damage to facilities on Treasure Island attributed to the 1989 Loma Prieta earthquake.

References

1. FEMAF. *Planning for a Sustainable Future: The Link between Hazard Mitigation and Livability*; FEMAF: Los Angeles, CA, USA, 2006.
2. Liang, L. Development of an Energy Method for Evaluating the Liquefaction Potential of a Soil Deposit. Ph.D. Thesis, Department of Civil Engineering, Case Western Reserve University, Cleveland, OH, USA, 1995.
3. Liang, L.; Figueroa, J.L.; Saada, A.S. Liquefaction under random loading: Unit energy approach. *J. Geotech. Eng. ASCE* **1995**, *121*, 776–781. [CrossRef]
4. Kusky, P.J. Influence o f Loading Rate on the Unit Energy Required for Liquefaction. Master's Thesis, Department o f Civil Engineering, Case Western Reserve University, Cleveland, OH, USA, 1996.
5. Dief, H.M. *Evaluating the Liquefaction Potential of Soils by the Energy Method in the Centrifuge*; Reserve University: Cleveland, OH, USA, 2000.
6. Green, R.A. Energy-based evaluation and remediation of liquefiable soils. Ph.D. Thesis, Virginia Polytechnic Institute and State University, Blacksburg, VA, USA, 2001.
7. Chen, Y.R.; Hsieh, S.C.; Chen, J.W.; Shih, C.C. Energy-based probabilistic evaluation of soil liquefaction. *Soil Dyn. Earthq. Eng.* **2005**, *25*, 55–68. [CrossRef]
8. Baziar, M.H.; Jafarian, Y. Assessment of liquefaction triggering using strain energy concept and ANN model: Capacity Energy. *Soil Dyn. Earthq. Eng.* **2007**, *27*, 1056–1072. [CrossRef]
9. Baziar, M.H.; Jafarian, Y.; Shahnazari, H.; Movahed, V.; Tutunchian, M.A. Prediction of strain energy-based liquefaction resistance of sand–silt mixtures: An evolutionary approach. *Comput. Geosci.* **2011**, *37*, 1883–1893. [CrossRef]
10. Alavi, A.H.; Gandomi, A.H. Energy-based numerical models for assessment of soil liquefaction. *Geosci. Front.* **2012**, *3*, 541–555. [CrossRef]
11. Zhang, W.; Goh, A.T.C.; Zhang, Y.; Chen, Y.; Xiao, Y. Assessment of soil liquefaction based on capacity energy concept and multivariate adaptive regression splines. *Eng. Geol.* **2015**, *188*, 29–37. [CrossRef]
12. Kokusho, T. Liquefaction potential evaluations by energy-based method and stress-based method for various ground motions: Supplement. *Soil Dyn. Earthq. Eng.* **2017**, *95*, 40–47. [CrossRef]

13. Robertson, P.K.; Wride, C.E. Cyclic liquefaction and its evaluation based on SPT and CPT. In Proceedings of the NCEER Workshop on Evaluation of Liquefaction Resistance of Soils, Salt Lake City, UT, USA, 5–6 January 1996; The National Academies of Sciences, Engineering, and Medicine: Washington, DC, USA, 1997; pp. 41–88.

14. Youd, T.L. Liquefaction resistance of soils: Summary report from the 1996 NCEER and 1998 NCEER/NSF workshops on evaluation of liquefaction resistance of soils. *J. Geotech. Geoenviron. Eng.* **2001**, *127*, 297–313. [CrossRef]

15. Idriss, I.M.; Boulanger, R.W. Semi-empirical procedures for evaluating liquefaction potential during earthquakes. *Soil Dyn. Earthq. Eng.* **2006**, *26*, 115–130. [CrossRef]

16. Moss, R.E.; Seed, R.B.; Kayen, R.E.; Stewart, J.P.; der Kiureghian, A.; Cetin, K.O. CPT-Based Probabilistic and Deterministic Assessment of In Situ Seismic Soil Liquefaction Potential. *J. Geotech. Geoenviron. Eng.* **2006**, *132*, 1032–1051. [CrossRef]

17. Baxter, C.D.P.; Bradshaw, A.S.; Green, R.A.; Wang, J. Correlation between Cyclic Resistance and Shear-Wave Velocity for Providence Silts. *J. Geotech. Geoenviron. Eng.* **2008**, *134*, 37–46. [CrossRef]

18. Boulanger, R.; Idriss, I. *CPT and SPT Based Liquefaction Triggering Procedures*; Report UCD/CGM-10/02; Center for Geotechnical Modeling, Department of Civil and Environmental Engineering, University of California: Davis, CA, USA, 2010; 77p.

19. Boulanger, R.; Idriss, I. *CPT and SPT Based Liquefaction Triggering Procedures*; Report No. UCD/CGM-14/01; Center for Geotechnical Modeling, Department of Civil and Environmental Engineering, University of California: Davis, CA, USA, 2014.

20. Ghafghazi, M.; DeJong, J.; Wilson, D. Evaluation of Becker Penetration Test Interpretation Methods for Liquefaction Assessment in Gravelly Soils. *Can. Geotech. J.* **2017**, *54*, 1272–1283. [CrossRef]

21. Davis, R.O.; Berrill, J.B. Energy Dissipation and Seismic Liquefaction in Sands. *Earthq. Eng. Struct. Dyn.* **1982**, *10*, 50–68.

22. Law, K.T.; Cao, Y.L.; He, G.N. An energy approach for assessing seismic liquefaction potential. *Can. Geotech. J.* **1990**, *27*, 320–329. [CrossRef]

23. Figueroa, J.L.; Saada, A.S.; Liang, L.; Dahisaria, N.M. Evaluation of Soil Liquefaction by Energy Principles. *J. Geotech. Eng.* **1994**, *120*, 1554–1569. [CrossRef]

24. Kayen, R.E.; Mitchell, J.K. Assessment of Liquefaction Potential during Earthquakes by Arias Intensity. *J. Geotech. Geoenviron. Eng.* **1997**, *123*, 1162–1174. [CrossRef]

25. Okur, D.V.; Ansal, A. Stiffness degradation of natural fine grained soils during cyclic loading. *Soil Dyn. Earthq. Eng.* **2007**, *27*, 843–854. [CrossRef]

26. Cabalar, A.F.; Cevik, A.; Gokceoglu, C. Some applications of Adaptive Neuro-Fuzzy Inference System (ANFIS) in geotechnical engineering. *Comput. Geotech.* **2012**, *40*, 14–33. [CrossRef]

27. Juang, C.H.; Yuan, H.; Lee, De.; Lin, P.S. Simplified Cone Penetration Test-based Method for Evaluating Liquefaction Resistance of Soils. *J. Geotech. Geoenviron. Eng.* **2003**, *129*, 66–80. [CrossRef]

28. Rezania, M.; Faramarzi, A.; Javadi, A.A. An evolutionary based approach for assessment of earthquake-induced soil liquefaction and lateral displacement. *Eng. Appl. Artif. Intell.* **2011**, *24*, 142–153. [CrossRef]

29. Seed, H.B.; Idriss, I.M. Simplified procedure for evaluating soil liquefaction potential. *J. Geotech. Eng. Div. ASCE* **1971**, *97*, 1249–1273.

30. Toprak, S.; Holzer, T.L.; Bennett, M.J.; Tinsley, J.J. CPT- and SPT-based probabilistic assessment of liquefaction. In Proceedings of the 7th U.S.–Japan Workshop on Earthquake Resistant Design of Lifeline Facilities and Counter measures Against Liquefaction, Seattle, WA, USA, 15–17 August 1999; Multidisciplinary Center for Earthquake Engineering Research: Buffalo, NY, USA, 1999; pp. 69–86.

31. Hanna, A.M.; Ural, D.; Saygili, G. Neural network model for liquefaction potential in soil deposits using Turkey and Taiwan earthquake data. *Soil Dyn. Earthq. Eng.* **2007**, *27*, 521–540. [CrossRef]

32. Jha, S.K.; Suzuki, K. Reliability analysis of soil liquefaction based on standard penetration test. *Comput. Geotech.* **2009**, *36*, 589–596. [CrossRef]

33. Jha, S.K.; Suzuki, K. Liquefaction potential index considering parameter uncertainties. *Eng. Geol.* **2009**, *107*, 55–60. [CrossRef]

34. Chen, G.; Xu, L.; Kong, M.; Li, X. Calibration of a CRR model based on an expanded SPT-based database for assessing soil liquefaction potential. *Eng. Geol.* **2015**, *196*, 305–312. [CrossRef]

35. Hu, J.-L.; Tang, X.-W.; Qiu, J.-N. Assessment of seismic liquefaction potential based on Bayesian network constructed from domain knowledge and history data. *Soil Dyn. Earthq. Eng.* **2016**, *89*, 49–60. [CrossRef]
36. Goh, A.T.C. Seismic Liquefaction Potential Assessed by Neural Networks. *J. Geotech. Eng.* **1994**, *120*, 1467–1480. [CrossRef]
37. Wang, J.; Rahman, M.S. A neural network model for liquefaction-induced horizontal ground displacement. *Soil Dyn. Earthq. Eng.* **1999**, *18*, 555–568. [CrossRef]
38. Baziar, M.H.; Nilipour, N. Evaluation of liquefaction potential using neural-networks and CPT results. *Soil Dyn. Earthq. Eng.* **2003**, *23*, 631–636. [CrossRef]
39. Seed, H.B.; Tokimatsu, K.; Harder, L.F.; Chung, R.M. Influence of SPT Procedures in Soil Liquefaction Resistance Evaluations. *J. Geotech. Eng.* **1985**, *111*, 1425–1445. [CrossRef]
40. Rahman, M.S.; Wang, J. Fuzzy neural network models for liquefaction prediction. *Soil Dyn. Earthq. Eng.* **2002**, *22*, 685–694. [CrossRef]
41. Kramer, S.L.; Mitchell, R.A. Ground Motion Intensity Measures for Liquefaction Hazard Evaluation. *Earthq. Spectra* **2006**, *22*, 413–438. [CrossRef]
42. Takahashi, O.; Asano, A.; Okada, H.; Saiki, T.; Irikura, K.; Zhao, J.X.; Zhang, J.; Thio, H.K.; Somerville, P.G.; Fukushimaet, Y.; et al. Attenuation Models for Response Spectra Derived Japanese Strong Motion Records Accounting for Tectonic Source Types. In Proceedings of the 13th World Conference on Earthquake Engineering, Vancouver, BC, Canada, 1–6 August 2004.
43. Kramer, S.L. *Geotechnical Earthquake Engineering*; Hall, W.J., Ed.; Prentice Hall: Upper Saddle River, NJ, USA, 1996; 653p.
44. Liyanapathirana, D.S.; Poulos, H.G. Assessment of soil liquefaction incorporating earthquake characteristics. *Soil Dyn. Earthq. Eng.* **2004**, *24*, 867–875. [CrossRef]
45. Orense, R.P. Assessment of liquefaction potential based on peak ground motion parameters. *Soil Dyn. Earthq. Eng.* **2005**, *25*, 225–240. [CrossRef]
46. Zhang, L. Predicting seismic liquefaction potential of sands by optimum seeking method. *Soil Dyn. Earthq. Eng.* **1998**, *17*, 219–226. [CrossRef]
47. Idriss, I.M.; Boulanger, R.W. Semi-empirical procedures for evaluating liquefaction potential during earthquakes. In Proceedings of the 11th International Conference on Soil Dynamics and Earthquake Engineering, and 3rd International Conference on Earthquake Geotechnical Engineering, University of California Berkeley, Berkeley, CA, USA, 7–9 Januaty 2004.
48. Idriss, I.M. *Boulanger, Soil Liquefaction during Earthquakes. Monograph MNO-12*; Earthquake Engineering Research Institute: Oakland, CA, USA, 2008.
49. Kayadelen, C. Soil liquefaction modeling by Genetic Expression Programming and Neuro-Fuzzy. *Expert Syst. Appl.* **2011**, *38*, 4080–4087. [CrossRef]
50. Yazdi, J.S.; Moss, R.E.S. Nonparametric Liquefaction Triggering and Postliquefaction Deformations. *J. Geotech. Geoenviron. Eng.* **2017**, *143*. [CrossRef]
51. Sadigh, K.; Chang, C.-Y.; Egan, J.A.; Makdisi, F.; Youngs, R.R. Attenuation relationships for shallow crustal earthquakes based on California strong motion data. *Seismol. Res. Lett.* **1997**, *68*, 180–189. [CrossRef]
52. Zhou, S.G.; Zhang, S.M. *Liquefaction Investigation in Tangshan District*; Report to Ministry of Railway; China, 1979. (In Chinese)
53. McCulloch, W.S.; Pitts, W. A logical calculus of the ideas immanent in nervous activity. *Bull. Math. Biophys.* **1943**, *5*, 115–133. [CrossRef]
54. Haykin, S. *Neural Networks: A Comprehensive Foundation*, 2nd ed.; Prentice Hall: Upper Saddle River, NJ, USA, 1998.
55. Coulibaly, P.; Anctil, F.; Bobée, B. Daily reservoir inflow forecasting using artificial neural networks with stopped training approach. *J. Hydrol.* **2000**, *230*, 244–257. [CrossRef]
56. Montgomery, D.C. *Design and Analysis of Experiments*, 8th ed.; Wiley: Hoboken, NJ, USA, 2012.
57. Box, G.E.P.; Draper, N.R. *Empirical Model-Building and Response Surfaces*; Wiley: Hoboken, NJ, USA, 1987.
58. Box, G.E.P.; Behnken, D.W. Some New Three Level Designs for the Study of Quantitative Variables. *Technometrics* **1960**, *2*, 455–475. [CrossRef]
59. Chen, J.; Wang, J.; Baležentis, T.; Zagurskaitė, F.; Streimikiene, D.; Makutėnienė, D. Multicriteria Approach towards the Sustainable Selection of a Teahouse Location with Sensitivity Analysis. *Sustainability* **2018**, *10*, 2926. [CrossRef]

60. Cho, C.; Kang, S.; Kim, M.; Hong, Y.; Jeon, E. Uncertainty Analysis for the CH4 Emission Factor of Thermal Power Plant by Monte Carlo Simulation. *Sustainability* **2018**, *10*, 3448. [CrossRef]

61. Favi, C.; di Giuseppe, E.; D'orazio, M.; Rossi, M.; Germani, M. Building Retrofit Measures and Design: A Probabilistic Approach for LCA. *Sustainability* **2018**, *10*, 3655. [CrossRef]

62. Fregonara, E.; Ferrando, D.; Pattono, S. Economic–Environmental Sustainability in Building Projects: Introducing Risk and Uncertainty in LCCE and LCCA. *Sustainability* **2018**, *10*, 1901. [CrossRef]

63. Pan, X.; Hu, L.; Xin, Z.; Zhou, S.; Lin, Y.; Wu, Y. Risk Scenario Generation Based on Importance Measure Analysis. *Sustainability* **2018**, *10*, 3207. [CrossRef]

64. Wu, D.; Yang, Z.; Wang, N.; Li, C.; Yang, Y. An Integrated Multi-Criteria Decision Making Model and AHP Weighting Uncertainty Analysis for Sustainability Assessment of Coal-Fired Power Units. *Sustainability* **2018**, *10*, 1700. [CrossRef]

65. Yoo, J.-I.; Lee, E.-B.; Choi, J.-W. Balancing Project Financing and Mezzanine Project Financing with Option Value to Mitigate Sponsor's Risks for Overseas Investment Projects. *Sustainability* **2018**, *10*, 1498. [CrossRef]

66. Juang, C.H.; Rosowsky, D.V.; Tang, W.H. Reliability-Based Method for Assessing Liquefaction Potential of Soils. *J. Geotech. Geoenviron. Eng.* **1999**, *125*, 684–689. [CrossRef]

67. Gutierrez, M.; Duncan, J.M.; Woods, C.; Eddy, E. *Development of a Simplified Reliability-Based Method for Liquefaction Evaluation*; Final Technical Report, USGS Grant No. 02HQGR0058; Virginia Polytechnic Institute and State University: Blackburg, VA, USA, 2003.

68. Kulhawy, F.H.; Trautman, C.H. Estimation of insitu test uncertainty. In *Uncertainty in the Geologic Environment: From Theory to Practice*; GSP No. 58; ASCE: New York, NY, USA, 1996.

69. Hariri-Ardebili, M.A.; Saouma, V.E. Sensitivity and uncertainty quantification of the cohesive crack model. *Eng. Fract. Mech.* **2016**, *155*, 18–35. [CrossRef]

70. Lumb, P. The Variability of Natural Soils. *Can. Geotech. J.* **1966**, *3*, 74–97. [CrossRef]

71. Tan, C.P.; Donald, I.B.; Melchers, R.E. Probabilistic Slope Stability Analysis—State of Play. In Proceedings of the Conference on Probabilistic Methods in Geotechnical Engineering, Canberra, Australia, 10–12 February 1993.

72. Butlin, J. *Our common future.* By World commission on environment and development. (London, Oxford University Press, 1987, pp.383 £5.95.). *J. Int. Dev.* **1989**, *1*, 284–287. [CrossRef]

Article

A Multi-Objective Ground Motion Selection Approach Matching the Acceleration and Displacement Response Spectra

Yabin Chen [1], Longjun Xu [1,2,*], Xingji Zhu [1] and Hao Liu [1]

[1] Department of Civil Engineering, Harbin Institute of Technology at Weihai, Weihai 264209, China; 14B933032@hit.edu.cn (Y.C.); zhuxingji@hit.edu.cn (X.Z.); liuhao123369@163.com (H.L.)

[2] Cooperative Innovation Center of Engineering Construction and Safety in Shandong Blue Economic Zone, Qingdao 266033, China

* Correspondence: xulongjun80@163.com; Tel.: +86-0631-567-6076

Received: 11 October 2018; Accepted: 15 November 2018; Published: 7 December 2018

Abstract: For seismic resilience-based design (RBD), a selection of recorded time histories for dynamic structural analysis is usually required. In order to make individual structures and communities regain their target functions as promptly as possible, uncertainty of the structural response estimates is in great need of reduction. The ground motion (GM) selection based on a single target response spectrum, such as acceleration or displacement response spectrum, would bias structural response estimates leading significant uncertainty, even though response spectrum variance is taken into account. In addition, resilience of an individual structure is not governed by its own performance, but depends severely on the performance of other systems in the same community. Thus, evaluation of resilience of a community using records matching target spectrum at whole periods would be reasonable because the fundamental periods of systems in the community may be varied. This paper presents a GM selection approach based on a probabilistic framework to find an optimal set of records to match multiple target spectra, including acceleration and displacement response spectra. Two major steps are included in that framework. Generation of multiple sub-spectra from target displacement response spectrum for selecting sets of GMs was proposed as the first step. Likewise, the process as genetic algorithm (GA), evolvement of individuals previously generated, is the second step, rather than using crossover and mutation techniques. A novel technique improving the match between acceleration response spectra of samples and targets is proposed as the second evolvement step. It is proved computationally efficient for the proposed algorithm by comparing with two developed GM selection algorithms. Finally, the proposed algorithm is applied to select GM records according to seismic codes for analysis of four archetype reinforced concrete (RC) frames aiming to evaluate the influence of GM selection considering two design response spectra on structural responses. The implications of design response spectra especially the displacement response spectrum and GM selection algorithm are summarized.

Keywords: resilience-based design; dynamic structural analysis; GM selection; displacement response spectrum; structural response estimates; spectrum variance; probabilistic framework; reinforced concrete frames

---

1. Introduction

Dynamic response history analysis is commonly used in performance-based earthquake engineering (PBEE) to estimate the response of structure subjected to ground motion records representing potential earthquakes in future. The concept of PBEE indicates that spectrum match for a period range rather than entire spectrum match is sufficient to describe seismic behavior of an individual structure [1]. Nevertheless, records whose average response spectrum matches the target spectrum at a specified period range may produce significant uncertainty for estimating the response of structures in the same community, assuming that their natural periods are far away from that period range. In that case, resilience evaluation of community would be biased. In recent years, varieties of techniques have been developed to select recorded time histories for structural analysis [2–9]. One approach is to select recorded ground motions (GM) according to scenario earthquake [3], which apparently is not based on response spectral values. Another alternate approach is to select a group of most unfavorable real GMs that lead to the highest damage potential for given structures [4]. Likewise, it is not based on design response spectrum. The final alternate approach, more popularly used in seismic engineering, is to select "real" or artificial records whose response spectra match specified mean response spectrum [5–7,10].

For a given target spectrum, one approach is to select GM records that best match the target by minimizing the sum of squared difference between the response spectrum of scaled record and target [11]. The other alternate approach is to select a GM set scaled using a same factor, rather than one factor for each record, to match target response spectrum. It is more complicated for the latter one that some advanced technique is required to introduce, such as genetic algorithm (GA) [12] and evolutionary algorithms (EA) [13].

In most cases where the target response spectrum is the mean acceleration response spectrum, even though the acceleration response spectrum variance has been taken into consideration. In addition, it has been successfully solved in terms of greedy optimization algorithm despite both the mean and variance being taken into consideration [14]. The best set of GMs can be quickly selected using a greedy algorithm since the mean and variance of target response spectrum are synthesized into a single target function. Recently, a notable multi-objective optimization considering the acceleration mean and variance separately as target functions has been developed [15]. Nevertheless, it is computationally expensive for GA procedure when the size of the GM dataset is large. The solutions can hardly evolve towards the optimal direction in terms of crossover and mutation techniques as the size gets larger.

It is well known that the spectral demands of design acceleration response spectra at long period coordinates are usually much lower than those at short period ordinates. It would bias the estimates for long-period structures using records selected in terms of acceleration response spectra [16,17]. The displacement response spectrum, rather than the acceleration response spectrum, has been remarkably developed as an alternate target spectrum to select GM records in Italy recently [18]. The displacement response spectrum, instead of using standard pseudo-spectral criteria transferred from acceleration spectrum, is compatible with design acceleration response spectrum. However, the proposed GM selection method by selecting GM records merely in terms of displacement spectrum may limit its application where the displacement and acceleration response spectra are incompatible. It is because the acceleration response spectra of GM records may probably deviate from target response spectrum significantly. Furthermore, the displacement response spectra defined according to different seismic codes are significantly varied even though their corresponding acceleration spectra are similar. It will be illustrated in subsequent sections.

Additionally, the target spectrum either the acceleration spectrum or displacement spectrum would be preferred, assuming it is suitable up to long periods and constrained by results of full period probabilistic seismic hazard analyses (PSHA). In such a case, estimations of seismic resilience of a community including short-period and long-period systems would be more reliable. Spectrum match for entire period range is also in great need. To date, PSHA in terms of horizontal displacement response spectrum ordinates up to vibration period of 20 s was only found in Italy. The design spectra including displacement response spectrum can only be chosen as target spectra to identify the engineering demand parameters (EDP) under records selected in terms of the counterpart seismic intensity. Thus, it must increase the complexity of GM selection as both the means and variances of both response spectra are taken into account. It is a notable multi-objective optimization that the greedy algorithm is not easy to resolve since the number of GM records matching target response spectrum is predetermined by user who has no idea of actual number of GMs optimally matching. Even the GA for multi-objective optimization, its computational consumption would greatly increase if the sample size gets larger. Moreover, evolving speed would significantly drop down accordingly.

Irrespective of definition of the target response spectrum, GM selection algorithms that select GM records matching target acceleration or displacement response spectrum, individually, was presented previously. However, the GM selection approach that selects records matching both acceleration and displacement response spectra has rarely been illustrated. In this study, a computationally efficient approach based on probabilistic frame will be proposed. Firstly, the sample size will shrink promptly and several GM sets whose displacement response spectrum mean and variance matching the targets will be generated using a new method. Afterwards, a novel evolving technique was introduced to select GMs in each set matching target acceleration response spectrum mean and variance. Two scaling factors were derived accordingly to best match these two target spectra individually using the least square technique. The most compatible GM set, namely having the closest scaling factors for both spectra, was finally selected as the optimal solution.

After this introduction in Section 1, review of compatibility of design response spectra according to international codes and general flowchart of the proposed algorithm will be shown in Section 2. Then, the third section provides detailed information of a new GM selection algorithm and gives an example of selecting an optimal set of GMs matching given target spectra. Meanwhile, the efficiency of the proposed algorithm is proved by comparing with a developed GM selection tool. Application of the proposed GM selection approach for international seismic codes is carried out and compared with a popularly-used algorithm in the fourth section. Four reinforced concrete (RC) frames were modeled to compare the bias of structural response estimates under GMs selected from these two methods. Some important results are concluded in Section 5.

2. Compatibility of Response Spectra and GM Selection Framework

2.1. Review of Displacement Response Spectra from International Seismic Codes

As previously stated, the design displacement spectra of international seismic codes are significantly different, even though the corresponding design acceleration spectra of most seismic codes are similar, such as seismic codes, including Eurocode EC8; Italian NTC08; American Society of Civil Engineering (ASCE) 7-16 and New Zealand NZS 1170. Firstly, the corner periods for all these codes to define displacement response spectra by pseudo-spectral relationship mentioned previously are basically different. In addition, their spectral shapes are significantly different. The major feature of the latter two codes is that the displacement spectral shape can be represented as a bilinear shape in addition to the corner periods independent of site condition. The spectral displacement assumed to be constant beyond corner period, T_E, can be seemed as peak ground displacement (PGD). In contrast, the major feature of the first two codes is that a decreasing branch beyond a corner period has been introduced in spectral shape and afterwards the branch approaches a constant value. For EC8 code, the first corner period, T_D, depends on magnitude. While the corner period for NTC08 is dependent on site condition in addition to PSHA results. The displacement response spectra shapes of these codes have been illustrated in Figure 1 [18].

It is shown that the elastic acceleration spectra in Figure 1a were adjusted to be close to each other. For NZS2004 code, the hazard factor Z has been set as 0.3 g and the spectral shape was anchored in terms of rock site, namely C site type. Hence, the peak ground acceleration of NZS2004 code can be fixed to 0.3 g which also results in spectral accelerations $S_S = 1.125$ g and $S_1 = 0.4$ g for ASCE 7-16 code at periods of T_S and T_1, respectively. For the codes of EC8 and NTC08, the site type was defined as B and PGA was set as $a_g = 0.3$ g. These two codes have such similar expressions on acceleration spectra and definition of displacement spectra that their displacement spectra are much closer than those of the other two codes.

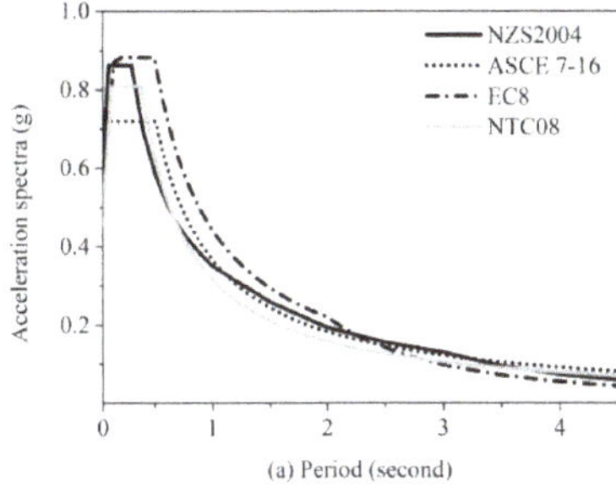
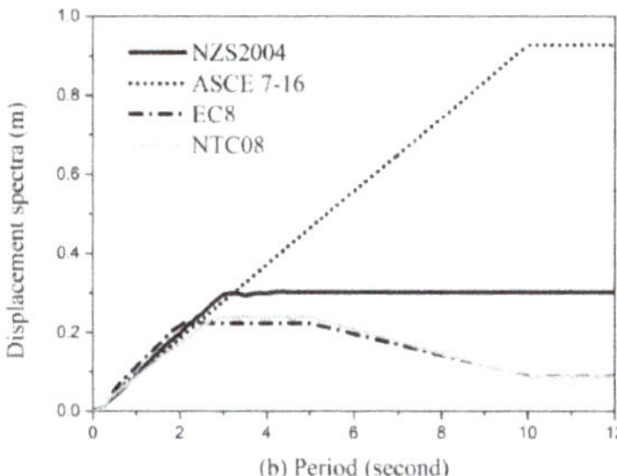

Figure 1. Elastic design acceleration (**a**) and displacement (**b**) spectra according to New Zealand NZS1170, United States, European and Italian seismic codes for 475-yr return period on ground category B.

The difference for both spectra at periods less than three seconds is not relevant since the spectra for four codes have marginal difference. Nevertheless, the ground motion selection results may be affected more seriously if the displacement response spectra are taken into account since the displacement spectra branches of codes NZS2004 and ASCE 7-16 at a long period range will be relevant to larger magnitude earthquake events than those from EC8 or NTC08 codes. Hence, it would be relevant to compare GM records selected according to acceleration response spectra in addition to their displacement response spectra.

2.2. Importance of Spectral Compatibility

The acceleration response spectrum other than the displacement response spectrum has conventionally been selected as target spectrum for GM selection. As we know, the acceleration response spectra of records are less sensitive to low frequency characteristic of GMs than the displacement response spectra. Thus, GM records selected merely based on target acceleration response spectrum may produce significant dispersion of displacement response spectra at long periods. On the contrary, GM selection only based on displacement response spectrum has been rarely studied [18]. Trivial but relevant differences between these two spectra are compared as the following.

Two pairs of GM records having the most similar spectral shape have been selected in terms of a similarity index as shown in Equation (1):

$$\delta_{ij} = \sqrt{\frac{1}{n-1}\sum_{k=2}^{n}\left[\frac{S_i(T_k) - S_j(T_k)}{S_j(T_k)}\right]^2},\tag{1}$$

where δ_{ij} represents the deviation of acceleration spectrum of *i*th record, i.e., S_i with respect to S_j of *j*th record. In addition, n is the number of periods within the considered period range. A comparison matrix was constructed with respect to δ_{ij}, from all records in the Next Generation Attenuation (NGA) database (https://ngawest2.berkeley.edu/). Finally, two pairs of records with the most similar acceleration and displacement response spectral shape were selected and shown in Figure 2. The damping ratio for the elastic response spectra equals 5% of all of the following.

For the first pair of GM records with NGA sequence numbers 144 (Dursunbey, Turkey earthquake in 1979) and 1738 (Northridge earthquake in 1994), the acceleration and displacement response spectra of these two records are shown in the first row of Figure 2. The acceleration response spectra are nearly the same at periods larger than 0.5 s. However, their displacement response spectra are significantly different at periods larger than 0.5 s. In addition, the displacement response spectra at periods less than 0.5 s are nearly the same on the contrary.

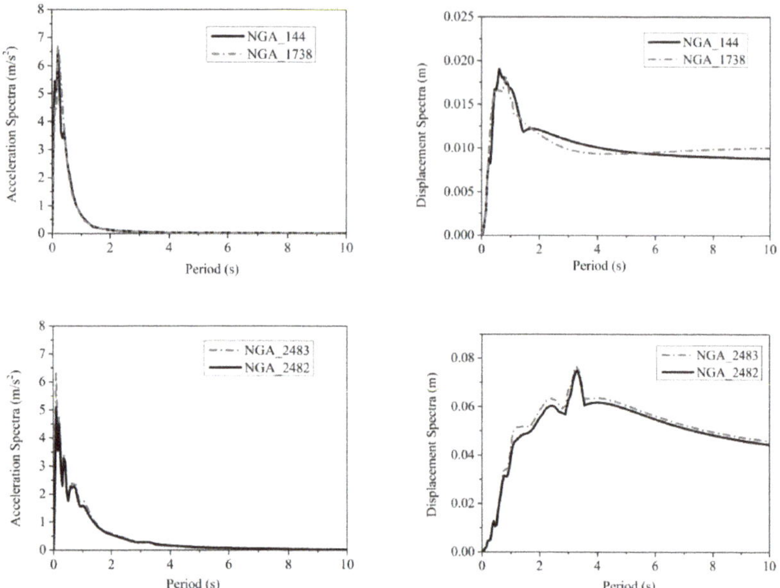

Figure 2. Acceleration and displacement response spectra of NGA_144, NGA_1738 records (top row plot), and NGA_2482, NGA_2483 records (bottom row plot).

In contrast, the other pair of records having the most similar displacement response spectrum is shown in the second row of Figure 2. Records of NGA_2482 (Chi-Chi Taiwan earthquake in 1999) and NGA_2483 (Chi-Chi Taiwan earthquake in 1999) derived in the same station have the most similar displacement response spectral shape in the NGA database. The displacement spectra of these two records at periods less than 0.8 s nearly overlap as shown in the right bottom panel of Figure 2. However, the corresponding acceleration spectra within this period range have a significant difference as shown in the left bottom panel of Figure 2. The largest gap of the spectral accelerations of these two records is more than 1 m/s^2. As such, the difference of displacement spectra at long periods cannot be clearly identified in corresponding acceleration response spectra.

As is clear from the above illustration, a trivial difference of spectral accelerations of records at long periods or the same spectral displacements at short periods probably produces a significant difference of spectral values of the counterparts. This spectral incompatibility of GM records may introduce significant dispersions of spectral values at this period range if one of target spectra is neglected.

Recently, the displacement response spectra at short periods have been proved compatible with corresponding acceleration spectra according to seismic hazard at a specified site in Italy. The target displacement response spectrum in Italy (TDSI), in accordance with NTC08 [19] at a short period range on one side, and, on the other side, with the framework of Project S5 (http://progettos5.stru.polimi.it/) at a long period range (attaining twenty second) has been well developed [18]. The TDSI can be expressed according to location, soil type, limit state and nominal life. Furthermore, the corresponding compatible spectra can be derived according to TDSI as shown in Figure 3.

The trend of acceleration spectrum beyond corner period, T_D, simply denoted as a function of period, is significantly different from that of displacement spectrum which is represented by two more corner periods, i.e., T_E and T_F in Figure 3a. It implies that the spectral compatibility at periods less than T_D may be more important for both spectra. Note that the factor, α, ranging from 0.85 to 1.4 in Italy, is so far less from two in Figure 3b that a significant drop will take place at period of T_D, whereas, it will be reduced as the corner period of T_D gets larger.

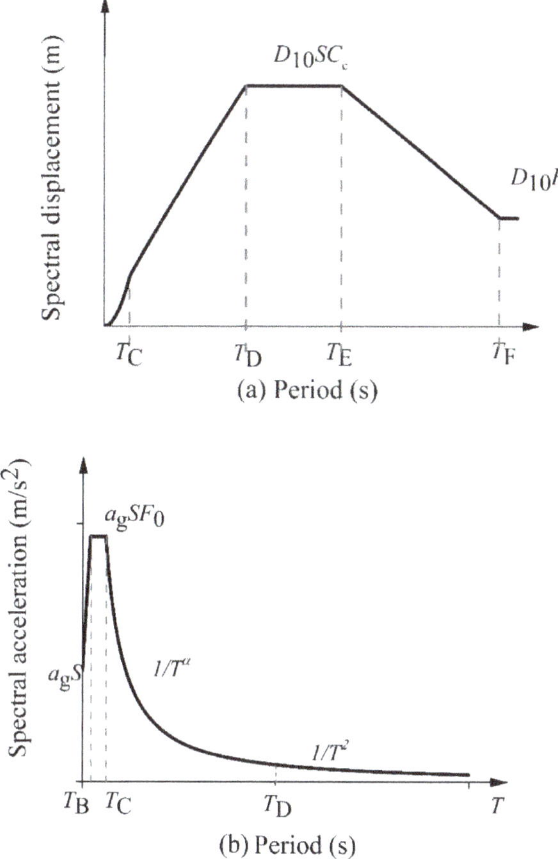

Figure 3. Compatible acceleration and displacement spectra derived from target displacement spectrum in Italy (TDSI).

A GM selection tool based on displacement response spectrum was successfully developed by Smerzini etc., [18]. The mean acceleration response spectrum of selected GMs matches the target acceleration response spectrum well because the target acceleration and displacement response spectra are compatible. It implies that the GM selection tool may be not suitable in other countries since the TDSI is only suitable in Italy. The TDSI additionally presented herein is majorly used for illustration of computational-efficiency of the GM selection algorithm newly proposed in this work.

2.3. Framework of the New GM Selection Algorithm

Traditional multi-objective optimization mainly includes four steps: i.e., (1) generating initial populations; (2) sorting the populations according to their non-dominated rankings; (3) evolving individuals in the place of the latter half using crossover and mutation techniques; and (4) finally resorting them together with the first half individuals. The solutions placed in the Pareto fronts can be traced to find the optimal solution user expects. Note that the first and third steps are carried out using random procedure, which, however, may increase computational consumption if the samples expand. However, the basic framework of multi-objective optimization is valuable and chosen for designing a new GM selection algorithm whose flowchart is shown in Figure 4.

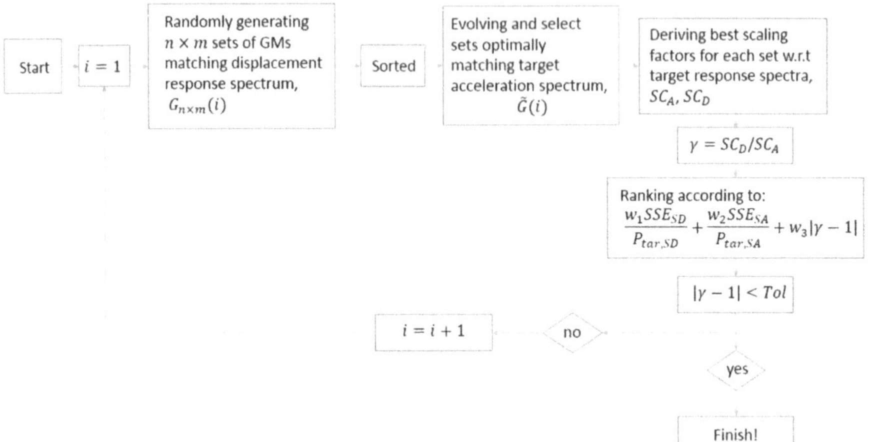

Figure 4. Flowchart of the ground motion selection algorithm.

The main steps for multi-objective optimization have been modified as shown in Figure 4 including:

(1) Initial populations are generated similarly with traditional method, but the generation method is changed using a Markov chain Monte Carlo (MCMC) sampling technique. Additionally, the total number of populations is divided into two parts, i.e., n and m, where variable n denotes the number of sub-target displacement response spectra and variable m denotes the sampling frequency for each target. Finally, a GM set $G_{n \times m}$ including $n \times m$ sub-sets of records is totally generated matching the target displacement response spectra.

(2) Non-dominated sorting technique is essential for this proposed algorithm since the displacement response spectrum mean and variance are taken as respective target functions. Note that the generation of initial populations is different from the traditional method. Populations generated from different sub-target response spectra may have different ranking levels.

(3) Thus, individuals ranking in the first level, rather than ranking at the latter half, are selected for the next evolving procedure since the displacement response spectrum is not the only target hereafter. The acceleration response spectrum mean and variance are also the targets so that potentially optimal sets of GMs are only interested. It is different from traditional multi-objective optimization that the evolved populations do not need to compare with the former original sets of GMs because the evolving procedure does not introduce samples outside the original set but only withdraws bad ones. The evolving procedure will be illustrated in the following section.

(4) GM sets after evolvement can probably improve the match of target acceleration and displacement response spectra. However, there is only one set of GMs finally selected as the optimal solution. Individuals evolved will be sorted according to their deviation from both target spectra and compatibility of the spectra which are synthesized as a function taking the weights of these variables into consideration. Details of this function will be stated in the following section. Note that the major target of the proposed GM selection algorithm is to derive a set of GMs whose response spectra are the most compatible. Therefore, computation will stop once the compatible tolerance is satisfied. Meanwhile, more iteration is needed if target tolerance cannot be satisfied. The solution can approach the optimal state to full extent if target tolerance is not specified.

2.4. Ground Motion Database and Scaling Approach

Ground motion database according to the SIMBAD database (Selected Input Motions for displacement-Based Assessment and Design) [18] including 220 records from Japan, 83 records from

Italy, 77 records from New Zealand, 44 records from United States, 18 records from Europe, 15 records from Turkey, seven records from Greece and three records from Iran were indispensable in this paper. Pre-selection of subsets having sufficient number of ground motions from general database according to site type or epicentral distance is impractical since the total number of earthquake events in SIMBAD is only 467 on one side, handful of records with specified characteristics on the other side. Hence, pre-selection of records from database of SIMBAD according to specified ground characteristics, such as site type and magnitude et al., was out of consideration in this paper.

Note that the candidate sets of ground motions were scaled with a same factor stated previously that the relative intensity of each ground motion record can be preserved. Furthermore, the largest scale factor was set equal to three, which ensures the scale factors for ground motions less five [20–23].

3. Ground Motion Selection Using the Passive Matching Distribution-Scaled Method (PMDS)

3.1. Generation of Initial Population

Recall that the purpose of generation of initial population is to reduce sample size and match the target displacement response spectrum. Additionally, the displacement response spectrum, rather than the acceleration response spectrum selected as a target, is because the displacement response spectrum at long periods may be harder to match for GMs compared with acceleration response spectrum at short periods. The scaling strategy, as stated previously, that retains relative energy distribution of ground motions enables utilization of a probabilistic sampling method. Furthermore, it implicates the target spectrum can be "moved" to match response spectrum mean of candidate bins of records inversely. That is, in fact, ground motion records do not need to be scaled but need only "passive" matched with a "moving" target spectrum. Hence, it is named as passive matching (PM) method hereafter.

The best solution can be found provided all the solutions with respect to each sub-spectrum were derived as shown in the second column of Figure 5. The target spectra including dispersion spectrum in the second column of Figure 5 show that the original target spectra were divided into n stripes by n scale factors. Furthermore, m sets of GM records were generated by a stochastic method corresponding to each stripe as shown in the third column of Figure 5. The initial populations were generated having totally $n \times m$ sets of GM records.

For each stripe, the traditional algorithm such as a greedy method, can easily select specified number of records to best match target spectra, which however is not proposed herein because the number of records, in fact, is unknown initially. In order to select ground motions optimally matching target spectra, a Monte Carlo sampling based on Markov chain (also known as the Metropolis algorithm) sampling technique has been used in this study to select records matching the target displacement spectra. In addition, the diversity of initial populations can be achieved using this random sampling technique.

Take TDSI, for example, a site locating at longitude 14.191, latitude 40.829 in Italy, the target displacement and tolerance spectra (blue and dotted blue lines individually) with respect to D site, Operability Limit State (SLO) and II function type structure were shown in Figure 6.

There were 78 GM records whose displacement response spectral ordinates at period of five second satisfying the target distribution function as follows:

$$p(x) = \frac{\beta^2}{\beta^2 + (x - \theta^2)} \, , \tag{2}$$

where β is the target deviation which is equal to 0.2 times the target mean, θ. In addition, x is the spectral values within a specified interval range. A similar proposal distribution function centered at θ, but dependent on previous state $p(x^{t-1})$, has been proposed according to the Metropolis algorithm. The proposed state at sample x^* will be accepted only if the target distribution, $p(x^*)$ is warranted with sufficiently large density, i.e., $p(x^*) > p(x^{t-1})$. However, on the contrary, the proposed sample may be rejected or accepted randomly. All of the samples will be traversed only once during each initial population selection process.

As shown in Figure 6a, the median (or mean) spectral value of selected records closely matches the target spectrum near the predefined period. Nevertheless, other median spectral points outside the specified period cannot match the target, which indicates MCMC sampling with a single period is not enough to select GM records matching spectral shape at long periods. The situation is improved when sampling at periods of ten seconds has been added to filter out records outside the tolerance as shown in Figure 6b. In order to cover all possible combinations of GM records, several sets of GM records will be generated for each sub-stripe to cover all possible combinations of GM records at the extreme. Totally, thirty sets of GM records will be generated when six random scaling factors and sampling five times for each target stripe are used.

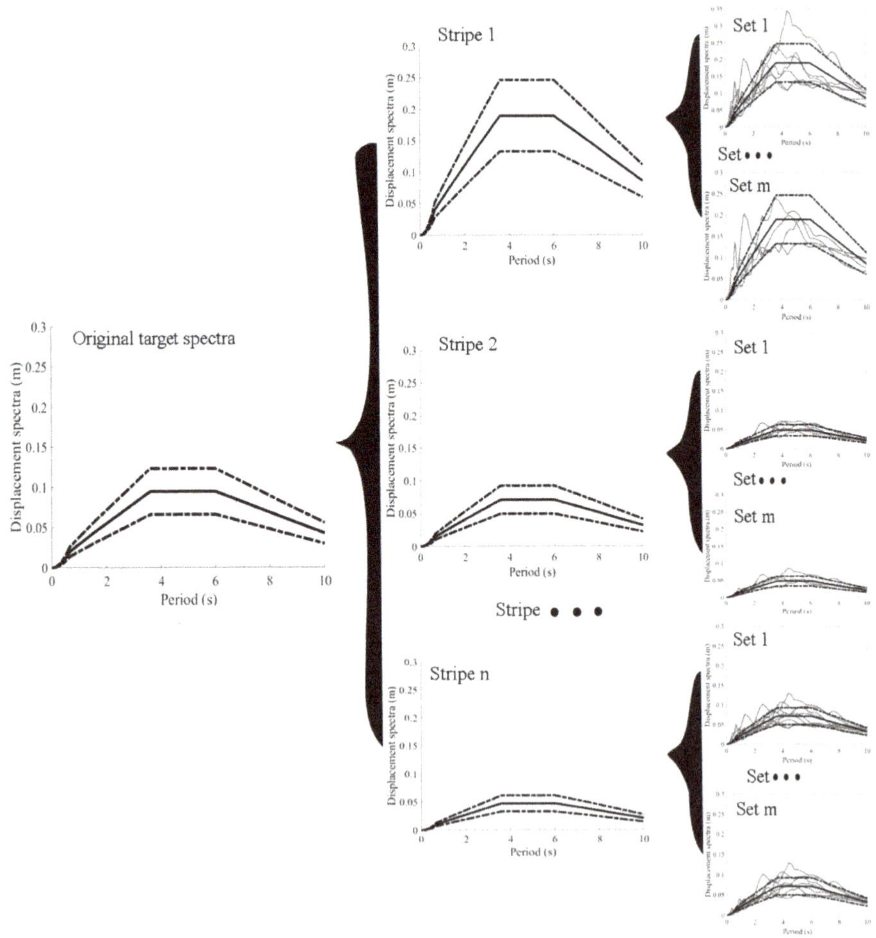

Figure 5. Illustration of pre-selection of GM records in terms of displacement response spectra.

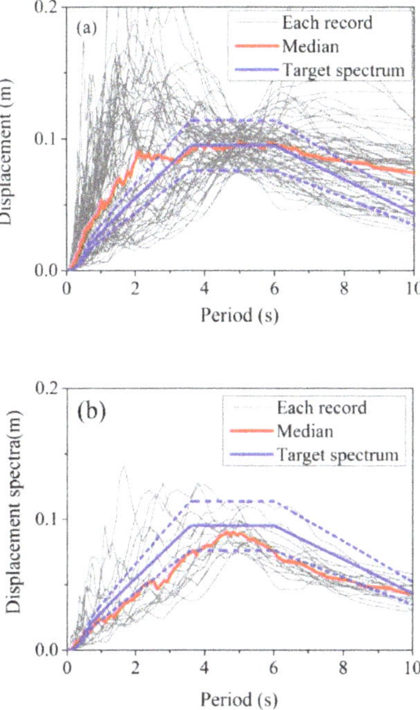

Figure 6. Pre-selection of GM records by the Markov chain Monte Carlo sampling technique using: (**a**) a reference period at five seconds and (**b**) two reference periods at five and ten seconds.

3.2. Non-Dominated Sorting of Initial Populations

Note that two target functions should be satisfied as shown in Figure 6. Namely, the dispersion spectrum written as $f_1 = \sqrt{\frac{1}{m} \sum_{j=1}^{m} \left(\sigma_j - \sigma_{t,j} \right)^2}$ and median displacement spectrum written as $f_2 = \sqrt{\frac{1}{m} \sum_{j=1}^{m} \left(\overline{SD}_j - \mu_{t,j} \right)^2}$ need to be matched. Herein, σ_j and $\overline{SD}_j$ are standard deviation and median spectral displacement of the scaled ground motion displacement spectra at *j*th point of period, respectively. Parameters of $\sigma_{t,j}$ and $\mu_{t,j}$ are target dispersion and median spectral displacement at the *j*th point of period.

In order to save computation time, the non-dominated sorting technique has been utilized to select candidate sets of GM records ranking in the first level [24]. Figure 7 is a candidate solution including eight records that match the target median and dispersion spectra. It shows that MCMC sampling and non-dominated sorting techniques demonstrate remarkable performance in selecting initial candidate solutions matching the target displacement spectra.

According to the displacement response spectral shape, two points of periods for MCMC sampling are proposed hereafter to select GM records matching the spectral shape. In addition, a relatively small amount of population (five herein) is necessary for MCMC sampling techniques. With the help of non-dominated sorting, more satisfactory candidate sets of ground motions can be well selected. However, whether the acceleration response spectrum mean of selected records at short periods significantly deviate from the target acceleration response spectra or not is still unknown.

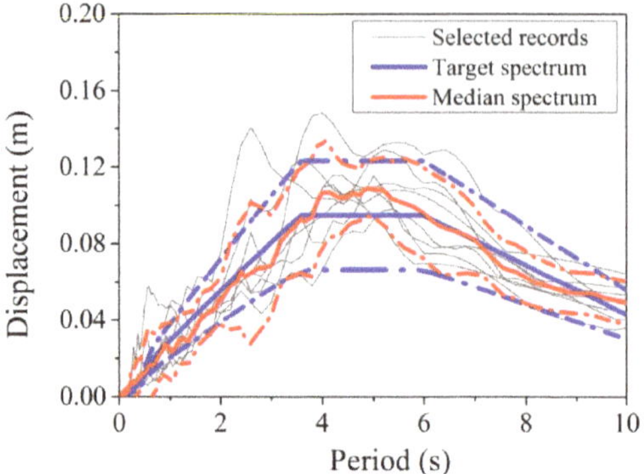

Figure 7. A solution selected by non-dominated sorting.

3.3. Evolving Based on the Distribution-Scaled Method

As previously stated, initial populations were from different target response spectra corresponding to different seismic intensities. Thus, traditional evolving techniques, such as crossover and mutation, are no longer usable. Assuming that the response spectra of GM records are in the place of sub-spectra derived from the initial target acceleration response spectrum as shown in Figure 8, these records would survive after evolvement in high probability. In order to further filtrate GM records significantly deviating from the acceleration response spectra at short periods, a distribution-scaled method was used.

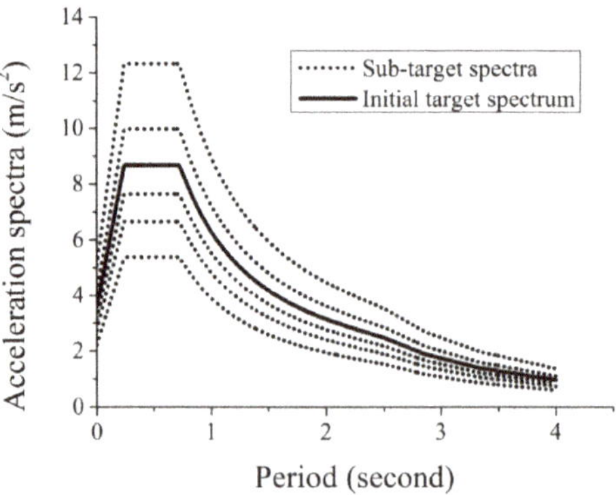

Figure 8. The acceleration sub-spectra derived according to the seismic distribution using the inverse of the normal distribution function.

The sub-spectra according to the intensity at the probability of P_i are defined as:

$$x_i = \widetilde{\theta} e^{\widetilde{\beta} \Phi^{-1}(P_i)}, \tag{3}$$

where $\Phi^{-1}(*)$ is the inverse standardized normal distribution function and P_i is the ith midpoint cumulative probability. Parameters of $\widetilde{\theta}$ and $\widetilde{\beta}$ denote the median and standard deviation of target spectral acceleration at a specified period. It is assumed that the acceleration spectra of ground motions were log-normally distributed. Then, parameter $\widetilde{\beta}$ was set as 0.2 times the target median spectral acceleration. The maximum allowed stripes in Figure 8 were set equal to the size of selected records representing the percentile probabilities, P_i, in corresponding individual. It can be inferred that the number of records finally selected would be smaller than the size of stripes initially defined.

Note that there was only one record selected for each stripe; meanwhile, the corresponding stripe should be deleted afterwards. In addition, ground motions whose response spectra are far away from the target stripes will be cleared. For simplicity, those records having peak response values at periods from 0.2 to two seconds three times more than or one third less than those of the target stripe at corresponding periods were removed additionally. None of the ground motions in each individual should be obsolete if they cannot get the deviation from target acceleration spectrum worse. On the contrary, records that amplify statistical deviation should be deleted. Herein, the scale factor both for displacement and acceleration spectra matching was derived by minimizing the summation Q:

$$Q = \sum_{i=1}^{m} \left(a \times \overline{S}_i - S_{\text{target},i} \right)^2,$$ (4)

where $\overline{S}_i$ and $S_{\text{target},i}$ represent the median spectrum and target spectrum at the ith period point, respectively. Furthermore, a is the scale factor required to be solved. Equation (4) achieves the minimum value when the first derivative of Q versus a equals zero.

3.4. Selecting the Optimal Solution

There are several sets of GM records ranking in the first level after non-dominated sorting to evolve. However, there is only one set of GMs can be left for each loop as shown in Figure 4 to determine whether the iteration is finished or not. In order to improve the solution, there is a trade-off between the target acceleration, displacement response spectra and their compatibility to rank the individuals as shown in Equation (5):

$$R(j) = \frac{w_1 SSE_{SD}(j)}{P_{tar,SD}} + \frac{w_2 SSE_{SA}(j)}{P_{tar,SA}} + w_3 |\gamma(j) - 1|,$$ (5)

where $SSE_{SD}(j)$ is the sum of squared errors (SSE) of displacement response spectrum mean and variance with respect to the target and, $P_{tar,SD}$ is the maximum target spectral displacement. Parameters of w denote the weighting values the target functions have. For simplicity, these three weighting factors are equally assumed, i.e., $w_1 = w_2 = w_3 = 1/3$. The variables of $SSE_{SA}(j)$ and $P_{tar,SA}$ have similar definition as $SSE_{SD}(j)$ and $P_{tar,SD}$. Parameter of $\gamma(j)$ denotes the ratio between the scaling factor for the displacement response spectra and that for the acceleration response spectra of jth set of GMs. Finally, the set of GM records having a smallest deviation between scaling factors for the acceleration and displacement response spectra will be only selected.

For example, a candidate set including twenty-four records was derived from the previous section and shown in Figure 9a where the median and dispersion of response spectra can well match the target displacement spectra. Nevertheless, the acceleration spectra of these accelerograms as shown in Figure 9b are significantly deviated from the target, especially those at short periods. Thus, twenty-four sub-target acceleration spectra derived according to Equation (3) were used afterwards to wide out records mismatching. The median value, $\widetilde{\theta}$, was derived to be equal to the target acceleration spectrum at 0.5 s, and the standard deviation, $\widetilde{\beta}$, was set equal to 0.2θ. There were twenty-four sub-target spectra produced for the acceleration spectra of records to match. After a few iterations, the solution successfully evolved as shown in Figure 9d. Apparently, the displacement spectra of selected records were also improved to match both the median and dispersion spectra as shown in Figure 9c.

It was stated that the mean displacement response spectrum of selected GM records may deviate from the target displacement spectrum after evolvement, and the scaling factor for displacement spectral match may also be different from that for acceleration spectral match. The iteration will stop if target compatible tolerance is satisfied. Furthermore, solutions can also evolve towards the perfect match if computation cost is out of consideration.

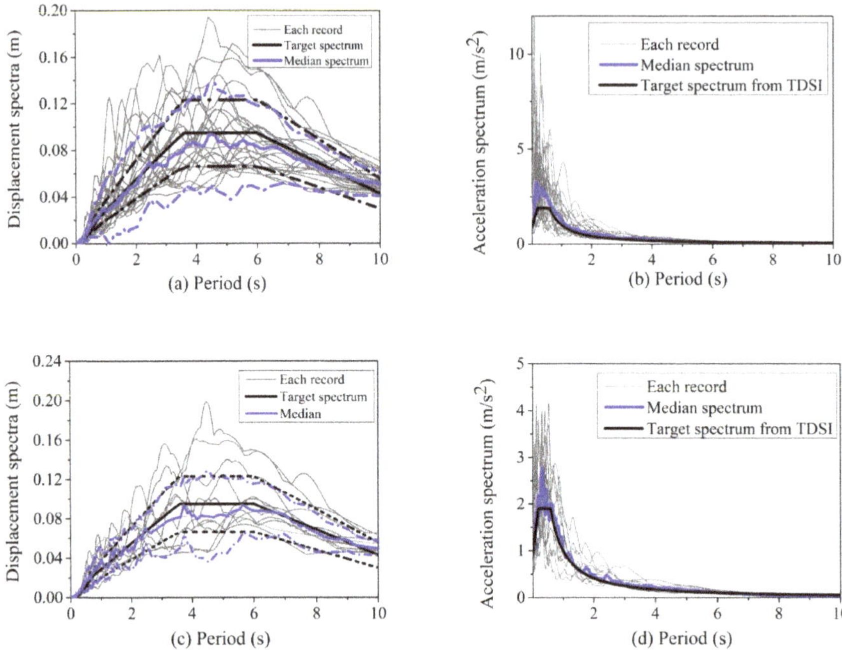

Figure 9. The displacement (**a**) and acceleration (**b**) response spectra derived before solution evolving; and the displacement (**c**) and acceleration (**d**) response spectra derived after the solution evolving using the distribution-scaled method.

3.5. Example of PMDS Method on GM Selection According to TDSI

As shown in Figure 4, iteration will not terminate if the target function or tolerance cannot be satisfied. An index coordinating the acceleration and displacement spectral matching was proposed in this study because the spectra of selected records can match the target spectra well separately. Nevertheless, the selected records may not match both target spectra simultaneously using the same factor. Therefore, a compatible index (simplified as CI) was defined as the ratio between scale factors, a derived from Equation (4), for displacement and acceleration response spectra, respectively. In order to prove the efficiency of PMDS method on GM selection, a developed GM selection tool, REXEL-DISP v 1.2 (Manufacturer, City, US State abbrev. if applicable, Country) [18], was used to selected GM records according to the same target spectra, i.e., TDSI.

Two sets of optimal GM records were found and shown in Figure 10 according to the target spectra the same with Figure 9. For the tool of REXEL-DISP v 1.2, the acceleration and displacement response spectra of records (Set 1) are shown in Figure 10a,b, respectively. In contrast, the acceleration and displacement response spectra of records (Set 2) selected according to the proposed PMDS method are shown in Figure 10c,d, respectively. The waveform IDs of these two sets of records were derived according to ground motion database compiled in REXEL v 3.5 [25] and shown in Table 1.

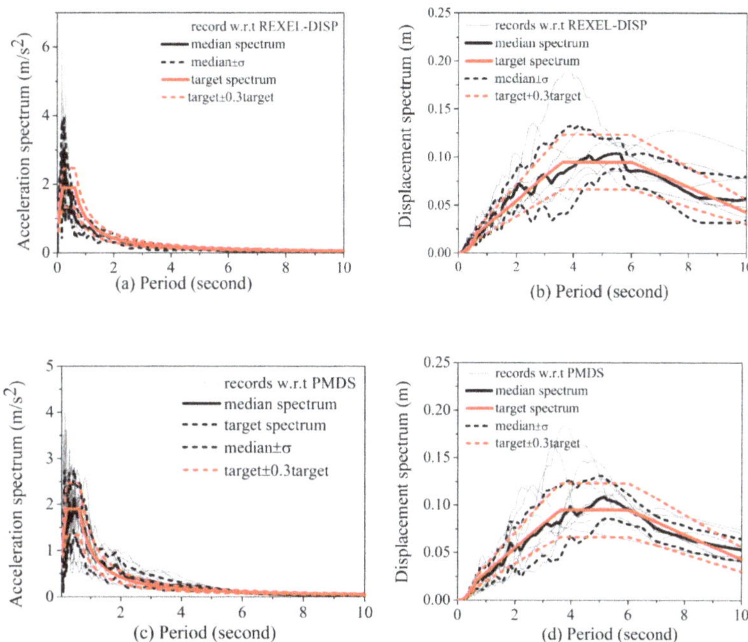

Figure 10. Acceleration and displacement response spectra of records derived using REXEL-DISP v1.2 ((**a**) and (**b**)) and the newly proposed method in this work ((**c**) and (**d**)) with respect to target displacement spectrum in Italy.

Table 1. Information of ground motion records selected based on the target spectra derived from TDSI.

GM Set	Waveform ID	Station ID	Earthquake Name	Date	M_w	Fault Mechanism	Epicentral Distance (km)
Set 1	014(y component)	TKY011	Near Miyakejima Island	2000-7-30	6.4	strike-slip	21.59
	114(y component)	ST_108	South Iceland	2000-6-17	6.5	strike-slip	13.23
	118(y component)	ST_105	South Iceland	2000-6-21	6.4	strike-slip	21.37
	243(y component)	FKS014	EASTERN FUKUSHIMA PREF	2011-4-11	6.6	normal	23.07
	244(y component)	FKS015	EASTERN FUKUSHIMA PREF	2011-4-11	6.6	normal	27.56
	383(y component)	CACS	Christchurch	2011-6-13	6.0	reverse	19.34
	389(x component)	RHSC	Christchurch	2011-6-13	6.0	reverse	14.76
Set 2	124(x component)	ST_306	South Iceland	2000-6-21	6.4	strike-slip	20.18
	242(x component)	FKS011	EASTERN FUKUSHIMA PREF	2011-4-11	6.6	normal	26.24
	242(y component)	FKS011	EASTERN FUKUSHIMA PREF	2011-4-11	6.6	normal	26.24
	313(x component)	SAN0	EMILIA_Pianura_Padana	2012-5-29	6.0	reverse	4.73
	345(x component)	TPLC	Christchurch	2011-2-21	6.2	reverse	19.97
	383(x component)	CACS	Christchurch	2011-6-13	6.0	reverse	19.34
	388(x component)	PPHS	Christchurch	2011-6-13	6.0	reverse	13.44
	388(y component)	PPHS	Christchurch	2011-6-13	6.0	reverse	13.44
	391(y component)	SMTC	Christchurch	2011-6-13	6.0	reverse	14.86
	438(x component)	ST_36445	Parkfield	2004-9-28	6.0	strike-slip	15.23

For REXEL-DISP v 1.2, default settings where the lower and upper tolerances equal to 20% and the matching period starts from 0.5 to 8 s were chosen to select optimal GM records best matching the target displacement response spectra. An option of "Seven records" included in the set size module was also expected in addition to "I'm feeling lucky" option. The final solution is illustrated in Figure 10a,b where the median displacement response spectrum of records matches the target displacement response spectrum better compared with that for the target acceleration response spectrum. The acceleration response spectrum median and variance at a short period range are significantly deviated from the targets as shown in Figure 10a. The corresponding median and variance at long period range much better match the targets in contrast due to a good match of the displacement response spectra at long periods. However, records selected according to the REXEL-DISP v 1.2 are not very compatible generally.

As previously stated, the acceleration and displacement response spectra at short periods are not easy to compatible even though the displacement response spectra at corresponding periods are overlapped. It is more reasonable for GM selection if the target acceleration response spectrum is taken into account as shown in Figure 10c,d. Herein, ten records in Set 2 where the compatibility index equals 0.972 were finally selected and listed in Table 1. As shown in Figure 10d, the median displacement response spectrum in addition to the dispersion spectrum of records evolved by the PMDS method matches the target spectra a little better than those derived by toolbox of REXEL-DISP v 1.2 as shown in Figure 10b. The more important thing is that the median and dispersion acceleration response spectra of records derived by PMDS method can better match the targets at whole periods. The median acceleration response spectrum of records at long period ordinates is a little higher than the target spectrum at these ordinates. It is implicated that these records have more energy at the low frequency as the CI is lower than one. In spite of this, both target the acceleration and displacement response spectra are generally better matched using the PMDS method.

4. Application of PMDS and Its Influence on Structural Response

4.1. GM Selection Based on PMDS

GM records selected based on PMDS method, for the first advantage, can select suitable records matching specified acceleration and displacement response spectra; for the second advantage, it does not consider whether or not these spectra are compatible. Remember that the displacement response spectra in Figure 1a are significantly different. The GM selection results would probably be different using the PMDS procedure even though the acceleration response spectra are adjusted to a similar magnitude. In such a case, records selected according to these codes may have significantly different intensities which lead to bias for structural response estimates. Meanwhile, a famous GM selection algorithm proposed by Baker et al. [26] was also introduced to compare both the response spectral match and its influence on structural response estimates.

There was no suitable set of GM records matching the spectra with respect to ASCE 7-16 due to its irrationally high spectral displacements at long period coordinates. The target spectra with respect to other three seismic codes concerned were left for GM record selection. The corresponding response spectra are illustrated in Figure 11.

Three pairs of target spectra have been well matched as shown in Figure 11. The acceleration response spectra are better matched in Figure 11a,c,e compared to those of corresponding displacement response spectra in Figure 11b,d,f, respectively. Without consideration of the displacement response spectra as shown in the right panel of Figure 11, dispersions of acceleration response spectra in Figure 11a,e are comparable except of those near two seconds. Thereinto, dispersion of acceleration response spectra of selected records according to EC8 ranks the lowest. Nevertheless, dispersion of acceleration response spectra of selected records in Figure 11a,c,e has a negligible difference at the period larger than four seconds. Note that this incompatibility has been stated in the previous section

and can be magnified in the displacement response spectra, which are shown in the right panel of
Figure 11.

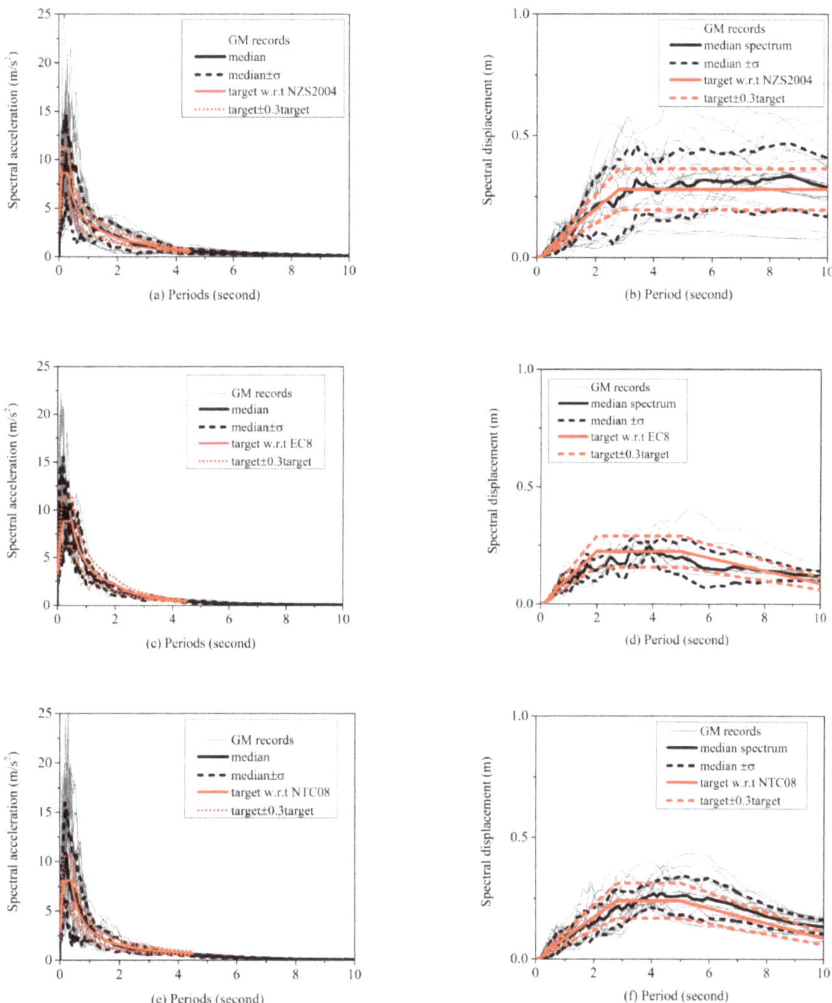

Figure 11. Acceleration and displacement response spectra of optimal sets of ground motion records
selected based on target spectra derived according to codes of NZS2004 (corresponding to (**a**) and (**b**)),
EC8 (corresponding to (**c**) and (**d**)) and NTC08 (corresponding to (**e**) and (**f**)).

The dispersion of displacement response spectra in Figure 11f is much less than that in Figure 11b
at most period coordinates especially at periods larger than one second. It may be attributed to the
fact that the number of GM records usable for the target displacement response spectrum is limited.
In addition, dispersion of displacement response spectra in Figure 11d at periods larger than four
seconds is much larger than that at periods lower than four seconds, whereas the corresponding
acceleration response spectra at periods larger than four seconds are significantly unified. It implies
that dispersion of structural response may be larger for structures designed with long periods (larger
than four second) than those with short periods. However, it cannot be totally determined by the

acceleration response spectra because the response spectra concerned above are within elasticity. Hence, further discussion of selected GMs on nonlinear structures is greatly needed.

4.2. GM Selection Based on Baker's Greedy Method

A greedy optimization proposed by Baker [26] has been popular to use in recent years. The target acceleration response spectrum mean and variance can be reduced maximally. However, the displacement response spectrum was out of consideration. In order to compare selection results with the PMDS method, GM sets including 22, 9 and 20 GM records (simplified as G22, G09 and G20) were derived in terms of Baker's greedy optimization method. The acceleration response spectrum corresponding with EC8 code for simplicity was chosen as the target spectrum. The corresponding numbers of records are consistent with those in Figure 11 of the previous section. Furthermore, the acceleration and displacement response spectra are shown in Figure 12.

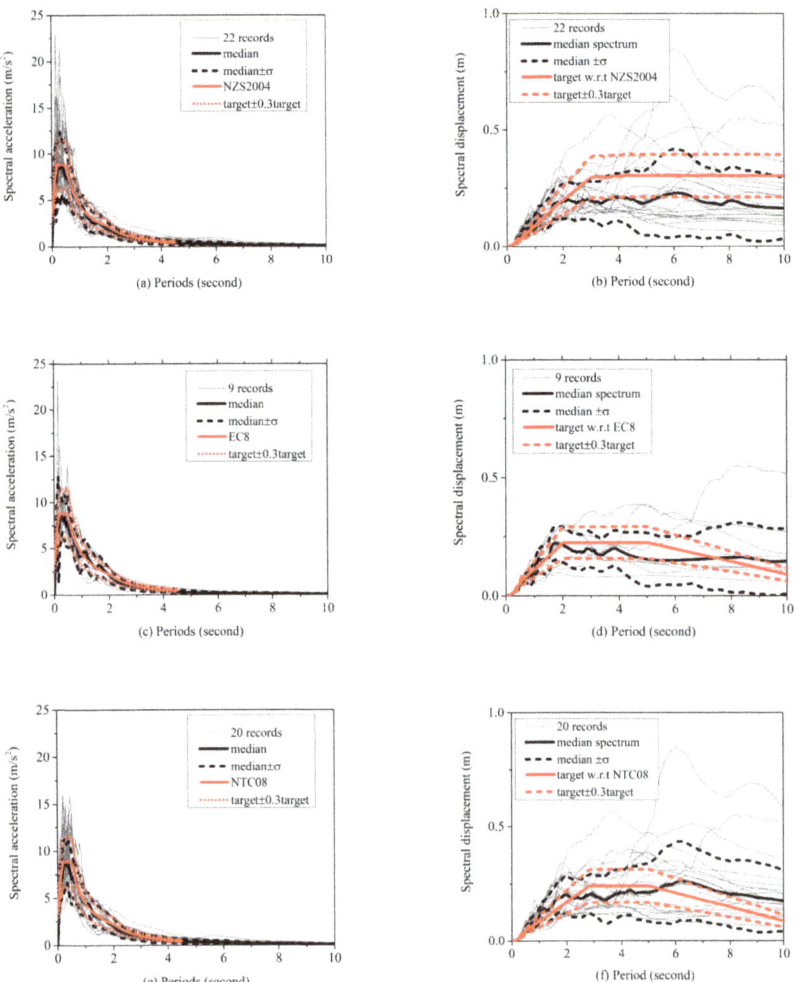

Figure 12. Acceleration and displacement response spectra of optimal sets of GM records selected in terms of Baker's greedy method for 22 records (corresponding to (**a**) and (**b**)), nine records (corresponding to (**c**) and (**d**)) and 20 records (corresponding to (**e**) and (**f**)).

Most records in G22, G09 and G20 overlap from the selection results. Eight records in G09 and sixteen records in G20 are included in G22, respectively. Target acceleration response spectrum means and variances in Figure 12a,c,e are well matched with samples. The G20 set derives the least variance with target variance as shown in Figure 11e. While G22 set in Figure 12a derives the worst variance among the other two sets even though it only has two more records, it implies that the target variance has no direct correlation with target number of records. Hence, G09 set in Figure 12c does not derive the least variance. Even so, the target acceleration response spectrum mean and variance are better matched than those in Figure 11a,c,e.

The displacement response spectra of selected records cannot match the target spectrum as shown in the right panel of Figure 12. The G22 set in Figure 12b derives the worst displacement spectrum variance relative with the target, and G20 set in Figure 12f is a little better off as some records excluded. The G09 set in Figure 12d derives the least variance, which, however, is larger than that of Figure 11d. The variances of displacement response spectra of these three groups of GMs at short period (two seconds herein) are generally smaller than those in Figure 11, especially the G22 and G20 sets. It is due to the fact greedy optimization method is emphasized on acceleration spectral matching.

Nevertheless, the acceleration response spectra variance of G22 set in Figure 12a is much larger than that in Figure 12e. It is inferred that solution of Baker's greedy algorithm is not "really" an optimal solution because the optimal number of GMs cannot be defined beforehand.

4.3. Influence of GM Selection Results on Structural Response

4.3.1. Numerical Model of Reinforcement Concrete Frames

In order to illustrate influence of ground motion selection in terms of different codes on structural response, four archetype RC frames, including 1-storey, 2-storey, 4-storey, and 8-storey, are referred to Haselton's models and were designed according to US codes, including ASCE 7-16, ACI 318-14 and IBC. [27–29]. All of the archetype frames were assumed to be located in California, coordinated as 33.855 N, -118.006 W on stiff soil site categorized as D per National Earthquake Hazards Reduction Program (NERHP). The corresponding mapped spectral accelerations for short periods and one-second periods are $S_{DS} = 1.0g$, $S_{D1} = 0.626g$ which were used to design structures.

A typical three-bay frame was considered in this study. The elevations and reinforcement details are illustrated in Figure 13. The height of first story is 15 ft and the other typical story height is 13 ft. Beam and column sizes were selected by joint shear requirements and beam-column dimensional compatibility. For all analysis, the P-Delta effect was accounted for, using leaning column members with gravity loads in addition to lateral resisting frame. The horizontal displacement of base leaning column is coupled with base floor of main RC frame. The design yield stress of reinforcements in beams and columns was both 60 ksi, and the compressive strengths of beams and columns were chosen in accordance with Haselton [30]. Expected dead and live loads were set to 175 psf and 12 psf, respectively [31]. The design parameters of archetype structures are illustrated in Table 2.

Table 2. Fundamental period and static pushover information for each archetype designs.

Structure	No. of Story	Bay Width (ft)	Period (s)	Design Base Shear (kips)	Peak Base Shear (kips)	Over-Strength
Str01	1	20	0.42	46.9	102	2.17
Str02	2	20	0.62	65	127.6	1.96
Str03	4	20	0.89	92.9	167.5	1.8
Str04	8	20	1.76	92	129.5	1.41

Figure 13. Design documentation for a one-story (top left), two-story (top right), four-story (bottom left) and eight-story (bottom right) space frame archetype with 20′ bay spacing.

Structural models were completed in Open Systems for Earthquake Engineering Simulation (OpenSees), [32] which can effectively simulate collapse of RC frame considering deterioration. The strength modification factor of fixed-base structures was chosen as eight, which corresponds with special moment frame. Space frames were considered in this study in which the ratio of tributary area for gravity and lateral loads was set to one.

Joint2D element, widely used for simulating RC frame sideway collapse was used in this study. A zero-length Ibarra lumped plastic hinge, which captured cycle and in-cycle deterioration hysteric modes located at the external nodes of Joint2D element well, was employed in these models. The tri-linear monotonic backbone curve and some hysteric rules can be used to model cyclic deterioration behavior of hinge from Federal Emergency Management Agency (FEMA) P695, [33] as shown in Figure 14b, while the model without consideration of degradation is compared in Figure 14a. Yield strength, a major parameter of backbone curve, was predicted by Panagiotakos and Fardis functions [34]. Other parameters of backbone curve were referenced with calibrated results from

FEMA P695. Additionally, a shear panel of Joint2D element, for simplicity, was modeled as elastic element whose cracked stiffness was only considered according to Sugano and Koreishi [35]. In this case, modifications in stiffness proportional damping have negligible effect on nonlinear response of RCF. Hence, Rayleigh damping expressed as linearly proportional to mass and initial stiffness was assigned only to elastic beams and columns [36]. Damping ratio, equal to 5% at the first mode period, was used and the factor applied to linear elements initial stiffness was amplified by 1.1 because the damping was only applied to linear elements.

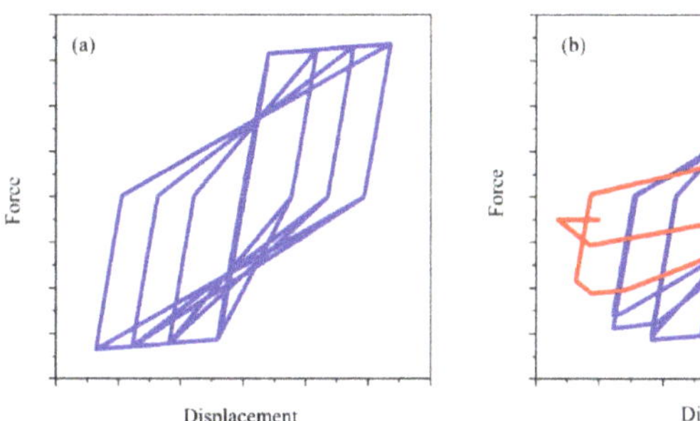

Figure 14. Force-displacement relation of beam-column joint (**a**) without considering deterioration; and (**b**) considering deterioration.

4.3.2. Impact of GM Selection on Structural Responses

Nonlinear time history analyses, up to 408 times, have been performed to study structural responses based on GM records selected. Six groups of ground motions as shown in Figures 11 and 12 were used to analyze four concrete steel frame structures, and the results are shown in Figure 15. Note that all the GM records were scaled the same as with Figures 11 and 12.

The Str01 derives comparable bias of structural response estimates as shown in the left panel of Figure 15a. Additionally, the G22 set derives larger median responses of Str01, Str02 and Str03 (as shown in the left panel of Figure 15b,c) than those of GMs relative with NZS2004, which may be due to the former set having relatively larger target spectral accelerations at fundamental periods of corresponding structures. However, the median structural response of Str04 under GMs selected by PMDS method as shown in the left panel of Figure 15d is larger than that under Baker G22 set. It may attribute to contribution of larger low frequency components as shown in Figure 11b. Generally, the bias of structural response estimate of GMs with respect to code NZS2004 is close to that of GMs derived by Baker's greedy method.

The structural response estimates are improved for GMs derived by PMDS method as shown in the middle panel of Figure 15 except Str01 model. The bias of structural response estimates decreases as the fundamental period of structure gets larger, such as Str02, Str03 and Str04. The structural response bias under GMs with respect to code EC8 derived by PMDS method is less than that derived by Baker's greedy method. The results are shown in the middle panel of Figure 15b–d. For Str03 model, the standard variance under GMs derived by PMDS method is nearly half of that derived by greedy method. The advantage of the former method with respect to Str04 model as shown in the middle panel of Figure 15d follows and the statistical results are illustrated in Tables 3 and 4.

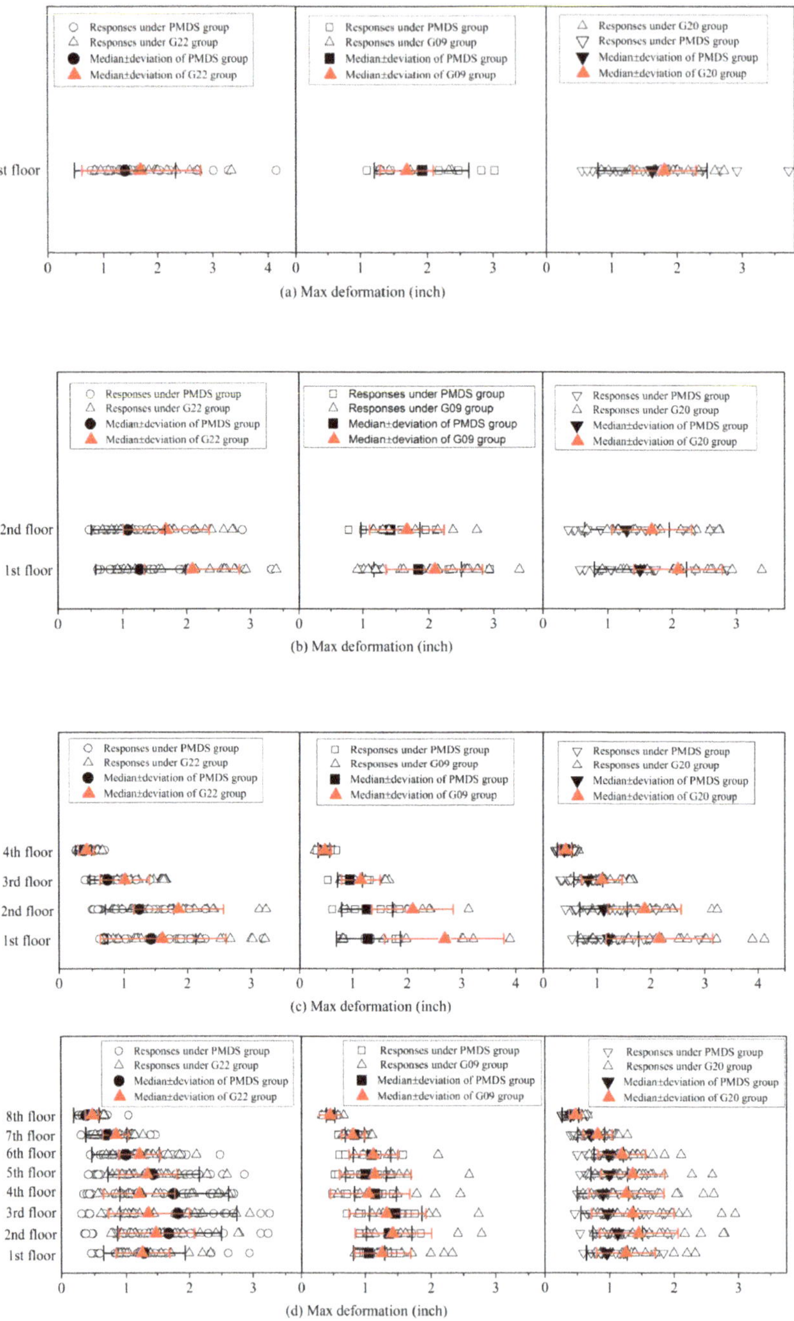

Figure 15. Maximum deformations of each floor of archetype Str01 (**a**), Str02 (**b**), Str03 (**c**) and Str04 (**d**) under sets of GM records with respect to NZS2004 and G22 in the first column, EC8 and G09 in the middle column, NTC08 and G20 in the third column, respectively.

Table 3. The statistical results of each archetype under excitations of three sets of records derived by the newly proposed ground motion selection method.

Structure	Story Number	NZS2004		EC8		NTC08	
		Median	Deviation	Median	Deviation	Median	Deviation
Str01	1	1.40	0.89	1.92	0.71	1.62	0.83
Str02	1	1.27	0.71	1.84	0.67	1.51	0.71
	2	1.09	0.58	1.42	0.45	1.30	0.65
Str03	1	1.44	0.69	1.28	0.59	1.20	0.57
	2	1.25	0.54	1.27	0.47	1.12	0.44
	3	0.74	0.27	0.95	0.23	0.83	0.26
	4	0.73	0.12	0.46	0.11	0.40	0.14
Str04	1	1.29	0.65	1.06	0.25	0.96	0.31
	2	1.68	0.82	1.36	0.35	1.13	0.39
	3	1.82	0.92	1.46	0.40	0.96	0.41
	4	1.76	0.86	1.16	0.32	0.89	0.40
	5	1.43	0.72	1.01	0.32	0.99	0.28
	6	1.00	0.53	1.10	0.29	0.99	0.23
	7	0.69	0.32	0.84	0.15	0.71	0.21
	8	0.38	0.19	0.46	0.07	0.40	0.13

Table 4. The statistical results of each archetype under excitations of three sets of records derived by greedy method.

Structure	Story Number	Baker G22		Baker G09		Baker G20	
		Median	Deviation	Median	Deviation	Median	Deviation
Str01	1	1.68	0.60	1.68	0.40	1.79	0.48
Str02	1	2.08	0.74	2.13	0.59	2.08	0.68
	2	1.67	0.67	1.66	0.56	1.67	0.62
Str03	1	1.61	0.99	2.67	1.09	2.15	0.99
	2	1.85	0.70	2.09	0.74	1.87	0.68
	3	1.00	0.39	1.14	0.36	1.08	0.37
	4	0.41	0.12	0.47	0.11	0.41	0.11
Str04	1	1.25	0.43	1.27	0.49	1.24	0.45
	2	1.47	0.60	1.41	0.58	1.44	0.61
	3	1.35	0.64	1.33	0.59	1.35	0.64
	4	1.20	0.58	1.06	0.61	1.25	0.58
	5	1.34	0.47	1.14	0.55	1.35	0.50
	6	1.21	0.33	1.12	0.37	1.18	0.37
	7	0.83	0.20	0.80	0.18	0.81	0.22
	8	0.46	0.10	0.44	0.11	0.47	0.09

The acceleration response spectrum is not the only target function for PMDS selection method that the acceleration spectrum variance of GMs selected would probably larger than that of GMs by the greedy method. The bias of structural response estimates, such as short-period structure Str01, would significantly depend on acceleration response spectrum variance. Therefore, the bias of Str01 under

GMs according to code NTC08 in the right panel of Figure 15a is larger than that of GMs derived by the greedy method. The biases of other models including Str02, Str03 and Str04, under GMs selected by PMDS method, are smaller than those under GMs derived by the latter method. The structural response means of all archetype RC models under G20 set of GMs are larger than those under GMs selected by PMDS method. Detailed results are illustrated in Tables 3 and 4.

The results as shown above show that the bias of long-period structural responses is tightly correlated with displacement spectra variance of selected records at long periods are expected. In addition, the GM group that has larger low frequency components would derive larger median structural response, especially long-period structures. The PMDS method generally performs better on structural response estimates than greedy method even though it has a little larger bias of structural response estimate for the Str01 model.

5. Conclusions

A computationally effective GM selection algorithm was proposed to select a set of recorded time histories whose acceleration and displacement response spectra match the targets. The algorithm initially probabilistically generates several sub-spectra from target displacement response spectrum, and then selects suites of GM records with the MCMC sampling technique individually match the sub-spectra. A distribution-scaling technique then further improves the match between the acceleration response spectra of records and targets. The evolved individuals were finally ranked in terms of a proposed normalized function to find the optimal one that deviates from the targets the least. It was shown that the proposed selection algorithm selects GM records whose acceleration and displacement response spectra generally better match the targets including means and variances by comparing with two mature GM selection algorithms. The GM records selected by the PMDS method are more compatible than those by the tool of REXEL-DISP v1.2 with respect to given TDSI. Furthermore, the proposed selection algorithm can select records without consideration of compatibility of target response spectra.

The proposed selection algorithm, then, was used to select GM records for structural seismic response estimates of four RC frames, aiming to assess influence of GM selection considering the target displacement response spectrum on structural response. It was seen that GM records selected merely according to acceleration response spectrum may overestimate seismic structural responses. The dispersion of displacement response spectra of records selected by Baker's algorithm is larger than those selected by proposed PMDS method. The increase on dispersion of displacement response spectra of records at long periods increases the dispersion of responses for Str02, Str03 and Str04 models except Str01 model. This is inevitable because the match of acceleration response spectrum mean and variance of records selected by Baker's algorithm at short periods is better than those by the PMDS algorithm. It can be improved if the match for the target acceleration response spectrum is better, which, however, needs more iteration. These results observed implicates that the application of GM selection including displacement response spectrum is important for many seismic performance evaluations since the design displacement response spectrum has a close correlation with long-period structures. It would be promising to quantify the influence of GM selection on RBD or PBEE estimation, which, however, has not been applied herein, and this new proposed algorithm will be facilitated in future study.

Author Contributions: Conceptualization and Methodology, Y.C.; Validation, X.Z. and H.L.; Resources, L.X. and X.Z.; Writing—Original Draft Preparation, Y.C.; Writing—Review and Editing, X.Z., L.X. and H.L.

Funding: This work was financially supported by the National Key R & D Program of China (No. 2017YFC1500602), the National Natural Science Foundation of China (No. 51678208) and the Natural Science Foundation of Shandong Province (No. ZR2018BEE044). We also appreciate the support of the Major Program of Mutual Foundation of Weihai City with the Harbin Institute of Technology (Weihai).

Conflicts of Interest: The authors declare no conflict of interest.

References

1. Cimellaro, G.P.; Renschler, C.; Bruneau, M. Introduction to resilience-based design (RBD). In *Computational Methods, Seismic Protection, Hybrid Testing and Resilience in Earthquake Engineering*; Springer: Cham, Switzerland, 2015; pp. 151–183.
2. Katsanos, E.I.; Sextos, A.G.; Manolis, G.D. Selection of earthquake ground motion records: A state-of-the-art review from a structural engineering perspective. *Soil Dyn. Earthq. Eng.* **2010**, *30*, 157–169. [CrossRef]
3. Jayaram, N.; Baker, J.W. Ground-motion selection for PEER Transportation Systems Research Program. In Proceedings of the 7th CUEE and 5th ICEE Joint Conference, Tokyo, Japan, 3–5 March 2010.
4. Zhai, C.H.; Xie, L.L. A new approach of selecting real input ground motions for seismic design: The most unfavourable real seismic design ground motions. *Earthq. Eng. Struct. Dyn.* **2007**, *36*, 1009–1027. [CrossRef]
5. Beyer, K.; Bommer, J.J. Selection and Scaling of Real Accelerograms for Bi-Directional Loading: A Review of Current Practice and Code Provisions. *J. Earthq. Eng.* **2007**, *11*, 13–45. [CrossRef]
6. Shantz, T.J. Selection and scaling of earthquake records for nonlinear dynamic analysis of first model dominated bridge structures. In Proceedings of the 8th U.S. National Conference on Earthquake Engineering, San Francisco, CA, USA, 18–22 April 2006.
7. Watson-Lamprey, J.; Abrahamson, N. Selection of ground motion time series and limits on scaling. *Soil Dyn. Earthq. Eng.* **2006**, *26*, 477–482. [CrossRef]
8. Dussom, K.B.; Hadj-Hamou, T.; Bakeer, R.M. Quake: An expert system for the selection of design earthquake accelerogram. *Comput. Struct.* **1991**, *40*, 161–167. [CrossRef]
9. Wang, G.; Youngs, R.; Power, M.; Li, Z. Design ground motion library: an interactive tool for selecting earthquake ground motions. *Earthq. Spectra* **2015**, *31*, 617–635. [CrossRef]
10. Al Atik, L.; Abrahamson, N. An improved method for nonstationary spectral matching. *Earthq. Spectra* **2010**, *26*, 601–617. [CrossRef]
11. Youngs, R.R.; Power, M.S.; Wang, G.; Makdisi, F.I.; Chin, C.C. Design ground motion library (DGML)—Tool for selecting time history records for specific engineering applications. In Proceedings of the SMIP07 Seminar on Utilization of Strong-Motion Data, Sacramento, CA, USA, 13 September 2007.
12. Naeim, F.; Alimoradi, A.; Pezeshk, S. Selection and scaling of ground motion time histories for structural design using genetic algorithms. *Earthq. Spectra* **2004**, *20*, 413–426. [CrossRef]
13. Georgioudakis, M.; Fragiadakis, M.; Papadrakakis, M. Multi-criteria selection and scaling of ground motion records using Evolutionary Algorithms. *Procedia Eng.* **2017**, *199*, 3528–3533. [CrossRef]
14. Jayaram, N.; Lin, T.; Baker, J.W. A computationally efficient ground-motion selection algorithm for matching a target response spectrum mean and variance. *Earthq. Spectra* **2011**, *27*, 797–815. [CrossRef]
15. Moschen, L.; Medina, R.A.; Adam, C. A Ground Motion Record Selection Approach Based on Multiobjective Optimization. *J. Earthq. Eng.* **2017**, *21*, 1–19. [CrossRef]
16. FEMA, Federal Emergency Management Agency. *Improvement of Nonlinear Static Seismic Analysis Procedures, FEMA 440*; Applied Technology Council (ATC-55 Project): Washington, DC, USA, 2005.
17. Powell, G.H. Displacement-based seismic design of structures. *Earthq. Spectra* **2008**, *24*, 555–557. [CrossRef]
18. Smerzini, C.; Galasso, C.; Iervolino, I.; Paolucci, R. Ground motion record selection based on broadband spectral compatibility. *Earthq. Spectra* **2014**, *30*, 1427–1448. [CrossRef]
19. Code-NTC08 I. Norme tecniche per le costruzioni in zone sismiche. *Ministerial Decree DM* **2008**, *14*, 9-04.02.
20. Kottke, A.; Rathje, E.M. A semi-automated procedure for selecting and scaling recorded earthquake motions for dynamic analysis. *Earthq. Spectra* **2008**, *24*, 911–932. [CrossRef]
21. Iervolino, I.; Cornell, C.A. Record selection for nonlinear seismic analysis of structures. *Earthq. Spectra* **2005**, *21*, 685–713. [CrossRef]
22. Heo, Y.A.; Kunnath, S.K. Abrahamson N. Amplitude-scaled versus spectrum-matched ground motions for seismic performance assessment. *J. Struct. Eng.* **2010**, *137*, 278–288. [CrossRef]
23. Seifried, A.E. Response Spectrum Compatibilization and Its Impact on Structural Response Assessment. Ph.D. Thesis, Stanford University, Stanford, CA, USA, 2013.
24. Deb, K.; Agrawal, S.; Pratap, A.; Meyarivan, T. A fast elitist non-dominated sorting genetic algorithm for multi-objective optimization: NSGA-II. In *Parallel Problem Solving from Nature PPSN VI*; Springer: Heidelberg, Germany, 2000; pp. 849–858.

25. Iervolino, I.; Galasso, C.; Cosenza, E. REXEL: Computer aided record selection for code-based seismic structural analysis. *Bull. Earthq. Eng.* **2010**, *8*, 339–362. [CrossRef]

26. Baker, J.W. The conditional mean spectrum: A tool for ground motion selection. *ASCE J. Struct. Eng.* **2011**, *137*, 322–331. [CrossRef]

27. American Society of Civil Engineers. *Minimum Design Loads for Buildings and Other Structures*; ASCE: Reston, VA, USA, 2010.

28. American Concrete Institute. *ACI 318–14: Building Code Requirements for Structural Concrete and Commentary*; American Concrete Institute: Farmington Hills, MI, USA, 2014.

29. International Code Council. International Building Code, USA. 2012. Available online: https://codes.iccsafe.org/content/IBC2012/toc (accessed on 18 November 2018).

30. Haselton, C.B. *Assessing Seismic Collapse Safety of Modern Reinforced Concrete Moment Frame Buildings*; Stanford University: Stanford, CA, USA, 2006.

31. Ellingwood, B. *Development of a Probability Based Load Criterion for American National Standard A58: Building Code Requirements for Minimum Design Loads in Buildings and Other Structures*; US Department of Commerce, National Bureau of Standards, 1980. Available online: https://nehrpsearch.nist.gov/static/files/NIST/PB80196512.pdf (accessed on 18 November 2018).

32. Mazzoni, S.; McKenna, F.; Scott, M.H.; Fenves, G.L. *OpenSees Command Language Manual*; Pacific Earthquake Engineering Research (PEER) Center: Berkeley, CA, USA, 2006.

33. Federal Emergency Management Agency. *Quantification of Building Seismic Performance Factors*; FEMA: Washington, DC, USA, 2009.

34. Panagiotakos, T.B.; Fardis, M.N. Deformations of reinforced concrete members at yielding and ultimate. *Struct. J.* **2001**, *98*, 135–148.

35. Sugano, S.; Koreishi, I. An empirical evaluation of inelastic behavior of structural elements in reinforced concrete frames subjected to lateral forces. *V WCEE* **1974**, *1*, 841–844.

36. Ibarra, L.F.; Medina, R.A.; Krawinkler, H. Hysteretic models that incorporate strength and stiffness deterioration. *Earthq. Eng. Struct. Dyn.* **2005**, *34*, 1489–1511. [CrossRef]

 *sustainability*

Article

Experimental Study on Mitigations of Seismic Settlement and Tilting of Structures by Adopting Improved Soil Slab and Soil Mixing Walls

Jianwei Zhang [1] and Yulong Chen [2,*]

[1] School of Port and Transportation Engineering, Zhejiang Ocean University, Zhoushan 316022, China; zhangjianwei@zjou.edu.cn
[2] State Key Laboratory of Hydroscience and Engineering, Tsinghua University, Beijing 100084, China
* Correspondence: chen_yl@tsinghua.edu.cn

Received: 11 October 2018; Accepted: 2 November 2018; Published: 6 November 2018

Abstract: Settlement of surface structures due to subsoil liquefaction is a big issue in geotechnical engineering. It has been happening during earthquakes in liquefaction-prone areas for many years. Mitigations have been proposed for this problem. The improved soil slabs and vertical mixing soil walls combined with lowering ground water levels (GWLs) were proposed in this study. Experiments were carried out by adopting a 1-g shaking table test. Two different soil densities with uniform and eccentric loads were included. Combined with lowering GWLs, three different soil slabs with a length of 40, 60 and 80 cm and two different mixing walls with soil and plastic were studied and compared. Results show that the horizontal soil slabs have good performance to reduce the settlement of structures. On the other hand, the vertical soil mixing walls did not reduce the settlement effectively, but its performance could be improved by lowering of GWLs.

Keywords: mitigation; shaking table test; liquefaction; settlement; ground improvement

---

1. Introduction

Liquefaction is a phenomenon which was first documented after the 1908 San Francisco earthquake [1]. More than 50 years later in 1964 two big earthquakes in Niigata, Japan and Alaska induced huge settlements to structures triggered by liquefaction, and the research to understand this phenomenon has become more important. Yoshimi and Tokimatsu [2] found and explained the basic mechanism behind the liquefaction through shaking table tests. It is shown that the liquefaction starts when the excess pore water pressure reaches the same value as the total stress in the same point. In that instant, the soil has zero bearing capacity, and it behaves like a fluid more than soil [3]. To stop or lower the damage caused by liquefaction effectively, mitigations should be investigated and then adopted.

Research has been focused on understanding the mechanism of structure settlement. Liu and Dobry [4] have carried out several centrifuge experiments to study the mechanism of settlement and effectiveness of sand densification. They noted that, by soil compaction under the structure, the settlement decreases while the earthquake motion is amplified. Dashti et al. [5,6] studied the mechanism of settlement by conducting shaking table tests. The ground and building settlement were categorized into volumetric and deviatoric groups and in each of them several governing mechanisms were listed. Effects of peak ground acceleration, relative density of ground, liquefiable layer thickness and foundation width were also examined. Mitigations such as installation of water barriers and rigid walls were also proposed and examined to reduce settlement of structures. Tsukamoto et al. [7] mainly focused on the mechanism of rigid circular foundations. Effects of neighboring foundation on settlement of buildings were also examined.

Previous studies show that the mechanism of liquefaction-induced settlement of structures was the main focus of researchers and proposing countermeasures for especially existing structures was not studied as extensively as the mechanism of the problem. Rasouli et al. [8] worked on effects of installation of sheet-pile walls against the problem by carrying out shaking table experiments. Rayamajhi et al. [9] examined the reinforcing mechanism of soil-cement columns. Olarte et al. [10] conducted an experimental study of the performance of 3-story structures with shallow foundations on a saturated soil profile including a thin liquefiable layer. Three different mitigation techniques were evaluated, including ground densification, enhanced drainage with prefabricated vertical drains and reinforcement with in-ground structural walls. Merits and drawbacks of the three mitigations were presented and discussed. Olarte et al. [11] also illustrated the potential tradeoffs associated with liquefaction mitigation involving ground densification with drainage control through a series of centrifuge experiments. They pointed out that enhancing drainage around the densified zone would notably reduce permanent foundation settlement and rotation during all motions but amplified accelerations.

Recently, numerical models have been studied to characterize the behavior of foundations on liquefiable ground. Naesgaard et al. [12] performed a numerical study of 2-D models representing strip foundations on liquefiable soils. The effects of foundation bearing pressure as well as the thickness and limiting strain of the liquefiable material in soil profiles consisting of a cohesive clay crust were investigated. Shahir et al. [13] numerically studied the correlations between the settlement of shallow, rigid box structures and their dimensions of the surrounding densified ground. Bray and Macedo [14] developed a 2-D, linear-elastic model to predict the influence of several parameters on settlement and characterized that influence in the model. Karimi et al. [15] presented results from a comprehensive numerical parametric study of soil foundation structure systems on layered liquefiable deposits. The numerical simulations involved fully-coupled, 3-D, nonlinear dynamic analyses of the soil-foundation-structure (SFS) system. It shows that the relative importance of parameters depend on ground motion intensity and soil relative density.

Generally, experimental and numerical methods have been used to investigate the mechanism behind the structure settlement and performance of different mitigation measures. In this study, to reduce the settlement and tilting of structures, the improved soil slab and deep soil mixing walls were proposed to install under the existing structures. The experiments were conducted using a 1-g shaking table to study the performance of the two mitigation measures. Different ground water levels (GWLs) were also considered in the experiments to address the effects of lowering GWL.

2. Method of Shaking Table Tests

Experiments were conducted on the shaking table with a square plate which can be accelerated in x and y directions. See in Figure 1, a rigid box with acrylic glass sides was fixed on top of the plate. Dimensions of the box are $2.65 \times 0.4 \times 0.6$ m (length $\times$ width $\times$ height). A wooden box filled with sand takes the role as a model structure in this set of experiments. The total weight of the box is always 20 kg. Together with the dimensions of 37.2×38.5 cm (length $\times$ width) an average pressure of $q = 1369.9$ N/m^2 resulted. In the case of eccentric load, the model structure has a weight of 7.5 kg in the negative direction of x axis and 12.5 kg on the opposite. That leads to a system with two different loads of $q_L = 1027.4$ N/m^2 and $q_R = 1712.4$ N/m^2.

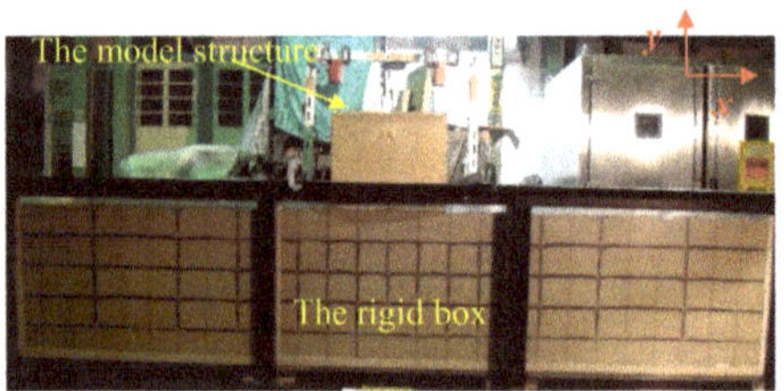

Figure 1. Shaking table with experimental set up.

The test set up is made up of five layers of sand of which the first bottom layer was compacted with a relative density D_r = 80%, and considered to be not liquefiable. The thickness of each layer is 10 cm. The overlying layers with a relative density D_r = 46% were prepared by water pluviation method. To achieve this relative density, water level was raised up to 10 cm above the bottom layer and then sand was poured uniformly into the water to make loose layers. To show the deformation of the sand visually, a mesh was formed with colored sand. For different GWLs, a drainage system at the bottom of the box was used. After finishing all five layers the model structure was placed on top of the sand in the middle of the box. In all experiments, silica sand No. 7 was prepared and used. The grain size distribution of silica sand No. 7 is shown in Figure 2. Its properties are shown in Table 1.

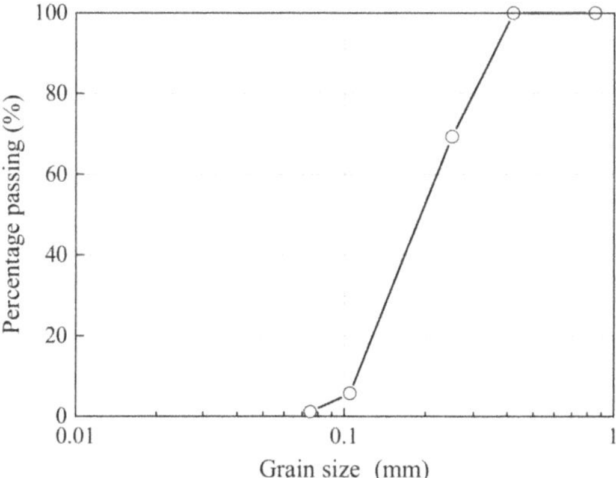

Figure 2. Grain size distribution of silica sand No. 7.

Table 1. Properties of silica sand No. 7.

Items	Value
Maximum void ratio, e_{max}	1.25
Minimum void ratio, e_{min}	0.749
Specific gravity, G_s (kg/m^3)	2650
Effective weight, γ' (D_r = 46%) (N/m^3)	7915
Effective weight, γ' (D_r = 80%) (N/m^3)	8649

For the data collection, a set of pore water pressure sensors, accelerometers, displacement transducers and a laser were used and mounted. Layout of the sensors is shown in Figures 1 and 3. The accelerometers and the pore water pressure sensors are fixed in the sand layers to record the behavior of the ground model. On every sand layer, a pore water pressure sensor was put in place and it is possible to measure the excess pore water pressure every 10 cm in height. The horizontal distance of the pore water pressure sensors is 20 cm, with a row of the sensors right underneath the model structure. Before putting the sensors in place, they were saturated in water to avoid false values for the influence of air bubbles. The accelerometers were also placed on every sand layer to measure the acceleration every 10 cm in height. The horizontal distance between the accelerometers is 30 cm, again with a set of sensors right underneath the model structure. With the results from the accelerometers it is possible to show the intensity of the shake in the soil and of the structure. The displacement transducer (LVDT) and laser sensors are fixed to the box to measure the movement of the structure. Apart from the settlement, it is possible to calculate the tilting of the structure.

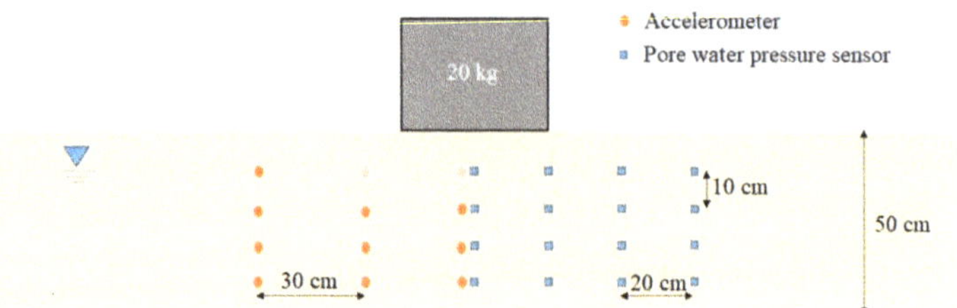

Figure 3. Layout of the sensors.

The input motion is assumed to have an amplitude of 300 Gal and a frequency of 10 Hz over 25 s, see in Figure 4. The duration is longer than most real earthquakes, and the maximum possible subsidence was be measured. The motion was applied in one direction, and the sensor measurement was always taken for 100 s.

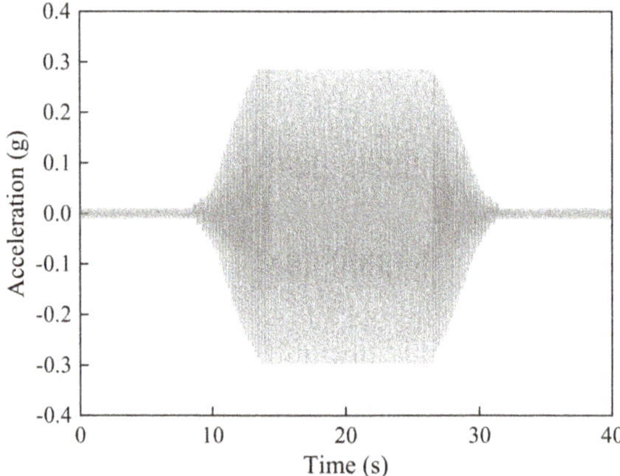

Figure 4. The input motion.

3. Results and Discussion

3.1. Performance of Improved Soil Slab against Liquefaction-Induced Settlement of Structures

The idea of the improved soil slab under the structure is to increase the cohesion of the soil and create an additional non-liquefiable layer. The length of the slab has a very big influence on its performance. A bigger area is able to distribute the induced load of the structure, and therefore decreases the actual load acting on the liquefiable soil layer. The thickness of the liquefiable soil layer would also be decreased and lead to less lateral flow of soil.

In this set of experiments, three different soil slabs were tested. The variable therefore was the length of the slabs. The height of 10 cm and the width of 38.5 cm were kept the same in all the three cases. The weight of the slabs should be the same as the sand layer and water should still be able to penetrate. Therefore, silica sand No. 7 with a relative density of $D_r = 46\%$ was mixed with 5% of two component epoxy glues and then put into the formwork in one layer. The masses of sand and epoxy glues for each geometry are listed in Table 2. The choice to use epoxy glue is to use the least possible amount of binder to create a rigid structure which would withstand the loads induced by the shaking during the experiment.

Table 2. Weight of soil slabs and epoxy glue.

Slab Length	Mass of Sand	Mass of Epoxy Glue	Total Mass
40 cm	20.19 kg	1.01 kg	21.2 kg
60 cm	30.28 kg	1.51 kg	31.8 kg
80 cm	40.37 kg	2.02 kg	42.4 kg

After preparing the sand layers, the improved soil slab was placed on top of the sand. The remaining sand was evenly distributed to the left and right side of the box. The slab in the water is sufficient saturated. The specifics of the soil slabs are given in Table 3. For the soil density, a different sieve with a bigger mesh size was used for the water pluviation method. Therefore, two different soil densities were achieved.

Table 3. Testing cases of improved soil slab.

Cases	Slab Length	Slab Thickness	Load	Soil Density
Case 1	60 cm	10 cm	Eccentric	52%
Case 3	40 cm	10 cm	Eccentric	46%
Case 4	80 cm	10 cm	Eccentric	46%
Case 8	60 cm	10 cm	Eccentric	46%
Case 11	50 cm	5 cm	Eccentric	46%
Case 12	none	−10 cm GWL	Eccentric	46%

After constructing the soil slab, the rigidity could be confirmed and its behavior was totally stiff. The slab works as a good mitigation because it reduced the liquefiable soil layer and the evenly distributed load. The subsidence of improved soil slabs in the four different cases is shown in Figure 5. As seen from the results, the subsidence reaches about 30 mm in the case of 80 cm slab and 63 mm in the case of 40 cm slab. The settlement rate too is lower in the cases of wider slabs. In the four cases, with the regular slabs of 10 cm thickness the settlement reaches the maximum shortly after the strong shaking.

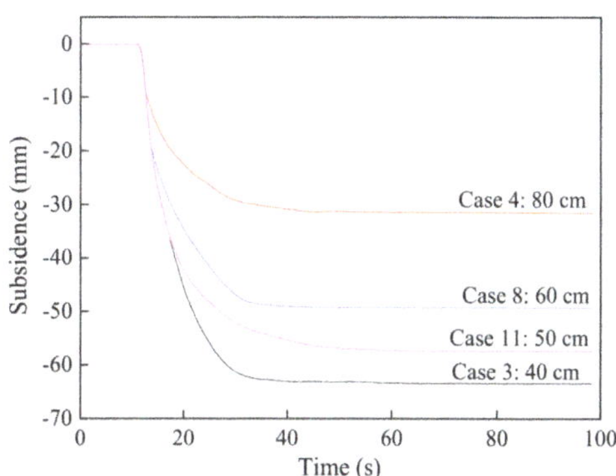

Figure 5. Subsidence of improved soil slabs.

Tilting of the structures with 10 cm thickness slabs are shown in Figure 6. With an eccentric load, the structure was always loaded with 12.5 kg in the positive direction of x axis and 7.5 kg in the other. As expected, the bigger slabs showed less tilting than the smaller one after the shaking. The main cause for the improvement is the smaller resulting moment under the bigger slab.

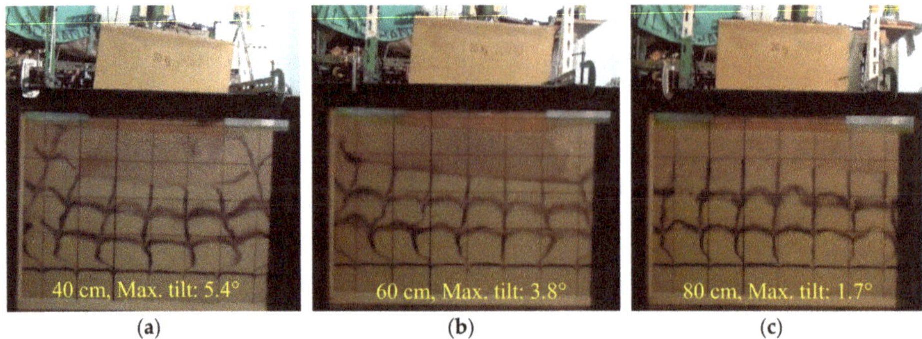

(**a**)   (**b**)   (**c**)

Figure 6. Tilting of the soil slabs with three different lengths, (**a**) 40 cm, (**b**) 60 cm, (**c**) 80 cm.

The horizontal displacement of soil under the 40 cm and 50 cm soil slab is shown in Figure 7. The soil on the right side, where the pressure under the slab is higher, has more displacement. The total displacement of soil in the case of 50 cm slab with a thickness of 5 cm is larger, because the displacement under a thinner slab is easier than the case of thicker slab which reduces the thickness of the liquefiable soil layer. The tilting over time in Figure 7 shows that after a maximum tilting in both cases, a recovery in the opposite direction starts and then tends to smooth for the time when the subsidence is deep enough that the uplift of sand soil mixture can support it. This point is reached at approximately 25 s. After that, the settlement under the lighter side still keeps going on, leading to a reduction of tilting towards the end of shaking.

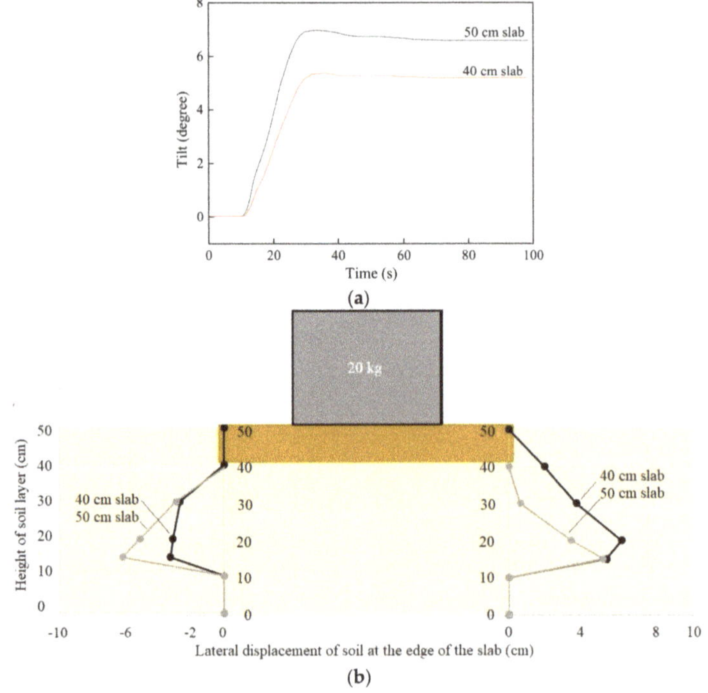

Figure 7. (**a**) Tilting over time of the structure and (**b**) horizontal displacement of soil for 40 cm and 50 cm soil slabs.

The lateral movement of the soil under the structure is triggered by the vertical subsidence. The sand during the liquefaction would be barely compacted or compressed. The settlement because of the compression is assumed to be around 3–5% of the liquefiable layer thickness [8]. The rest of the resulting subsidence is therefore caused by the lateral flow of soil away from the foundation. Lateral displacements of the soil at the edge of the slab with three different lengths are shown in Figure 8. The displacement of the 80 cm slab is the smallest compared to the other two slabs apart from the top and bottom parts. This effect is very advantageous to the resulting settlement under the wider structure for the same volume of moved soil and the bigger area of the 80 cm slab leads to less subsidence. The displacement on the left side of the slabs in the case of 40 cm and 60 cm is higher than the right side. The tilting of the structure increases the lateral movement in this direction. Most likely the eccentricity of the resulting load is responsible for this behavior. In the case of 80 cm slab, the tilting is reduced and the lateral displacement in this case is almost symmetric only with a difference in the top 20 cm soil layer.

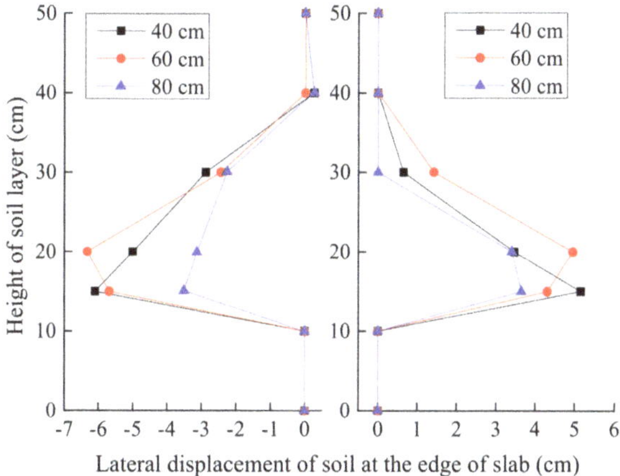

Figure 8. Lateral displacement of the soil at the edge of the slab.

The mitigation of lowering the GWL was also carried out and compared. Two different GWLs of −5 cm and −10 cm were considered. The load of the structure is 20 kg uniformly loaded. The case 12, also a reduction of the GWL of 10 cm was conducted with an eccentric load. The subsidence is shown in Figure 9. Based on the experimental results, it can be seen that the horizontal soil slab reduces the subsidence during an earthquake. Even the 40 cm soil slab, which is the same length as the structure foundation, shows a significant effect. The lowering of GWL also leads to a reduction of the liquefiable soil layer, but creates a non-liquefiable crust with a higher flexibility compared to the very stiff soil slabs. The 80 cm soil slab leads to a higher subsidence reduction than the −10 cm GWL case which creates a crust over a bigger area. The rigid soil slab seems to have a better pressure distribution than the still flexible and drained soil layer. Case 12 on the other hand shows that the lowering of GWL does not bring immediate results. The saturation of the top layer was still very high before the shaking. Therefore, the excess pore water which was pushed into the top layer because of the rise in pore water pressure in the lower layer led to immediate liquefaction. The higher settlement, even higher than the no mitigation case, can be related to the eccentric load, which led to a tilting of the structure. Higher pressure at the edges of the model structure followed by a larger subsidence shows strong post-shaking behavior.

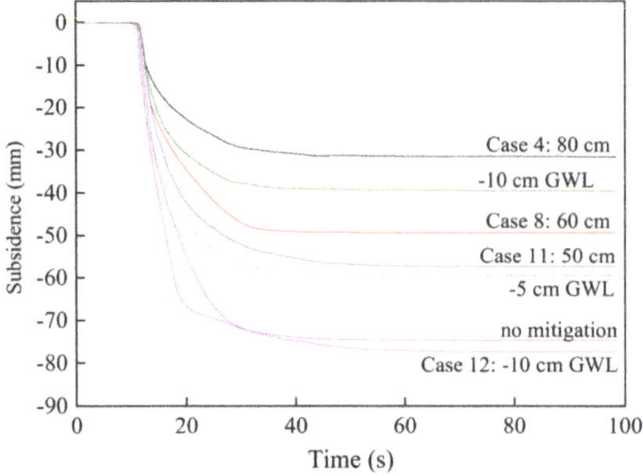

Figure 9. Subsidence of soil slab and lowering of ground water level (GWL).

The pore water pressure curves in Figure 10 of the sensor 20 cm under the surface level show that the pore water pressure has to rise to higher values. The difference under unloaded conditions is very small though and the measured difference correlates with the difference in total stress. Under the structure the difference of pore water pressure is approximately 0.5 kPa for the two sand densities. Further, the denser sand under the structure restricts the water flow. Together with the load of the structure, a pressure level higher than the unloaded conditions is built up. The lower settlement can also come from the bigger uplift created by the higher soil density of the sand. Therefore, the total possible settlement is reduced. The last factor is the consolidation and compaction of the sand during the shaking. The denser grain structure leads to less consolidation and therefore to a reduced consolidation settlement. It is noteworthy that with only two cases and a difference in density it is not possible to get significant results, because the difference can also just be in the range of uncertainty.

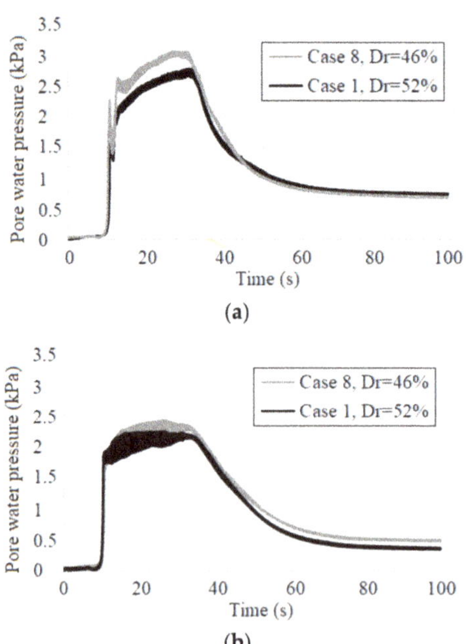

Figure 10. Pore water pressure for case 1 and 8, (**a**) under structure and (**b**) under unloaded area.

The settlement rate shows how fast the structure sinks into the soil. As shown in Figure 11, the settlement velocity reached its maximum in all cases almost at the same time, meaning a constant increase of settlement rate independent of the length of the slab. The maximum settlement velocity takes place before the input motion reaches its motion amplitude. The thinner slab with a length of 50 cm shows a bigger settlement velocity at the beginning compared to the slabs with 10 cm thickness. The inclination of the velocity curve has the same value as in the case of no mitigation. In general, the length of the slab defines the increase of settlement speed. The wider of the slab is the smaller of the settlement rate. The reason is that the distribution of the structure load over a wider area leads to smaller shear stresses in the soil and a smaller lateral flow of sand.

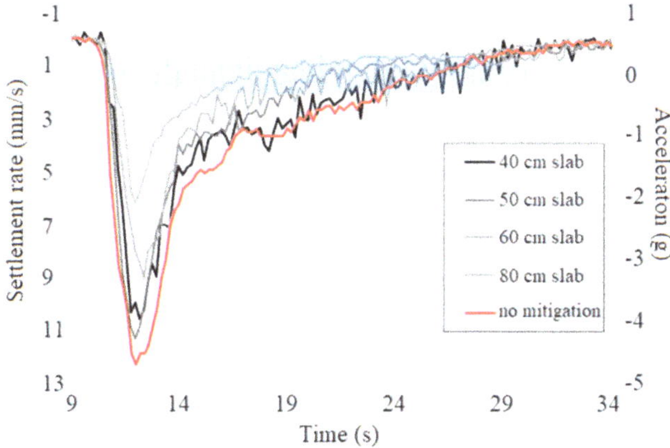

Figure 11. Settlement rate of improved soil slabs.

Comparison of the pore water pressure and the settlement rate is shown in Figure 12. The fast increase of settlement velocity is caused by the beginning of liquefaction and the fast rise of pore water pressure. During that time, the settlement mechanisms described by Dashti et al. [5] work and add up to the settlement velocity. After reaching the liquefied state, the main settlement mechanism is the lateral displacement of sand and then the settlement rate decreases steadily.

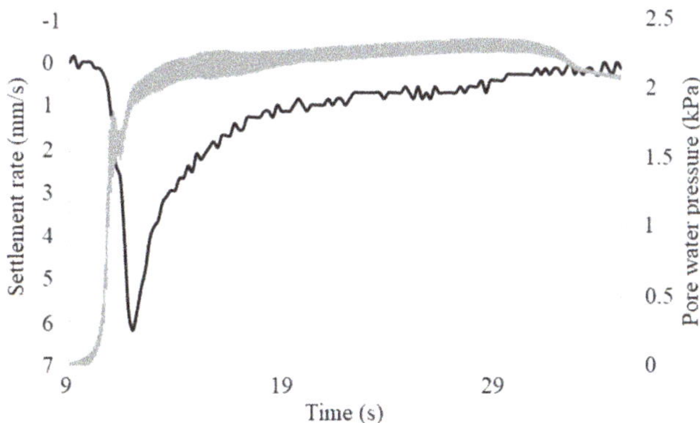

Figure 12. Settlement rate compared to the pore water pressure of 80 cm slab.

3.2. Performance of Vertical Deep Soil Mixing Walls against Liquefaction-Induced Settlement of Structures

Experimental set up for the vertical deep soil mixing walls is shown in Figure 13. The bottom was attached with duct tape to the box to create a fixed bottom and prevent the slab from moving during the compaction of the first non-liquefiable soil layer. The soil mixing walls are a few millimeters smaller than the width of the shaking box to allow the movement of the walls. Therefore, a small gap between soil walls and the acrylic glass of the box was unavoidable. Specifics of the testing cases for the vertical deep soil mixing walls are summarized in Table 4. In case 6, 7 and 11, the top layer was prepared to simulate lower GWL. In this section, to assess the effect of the deep soil mixing walls, results of the water barrier are also presented and compared. The latex sheets are used to simulate a water barrier. The latex sheet itself is a very soft material which could allow a lot of deformation of the liquefied sand, but still confine the water inside. Therefore a regular plastic bag with a lot of spare material was fixed to the box walls. This created a sector under the structure which could contain the water inside but allow movement of the soil.

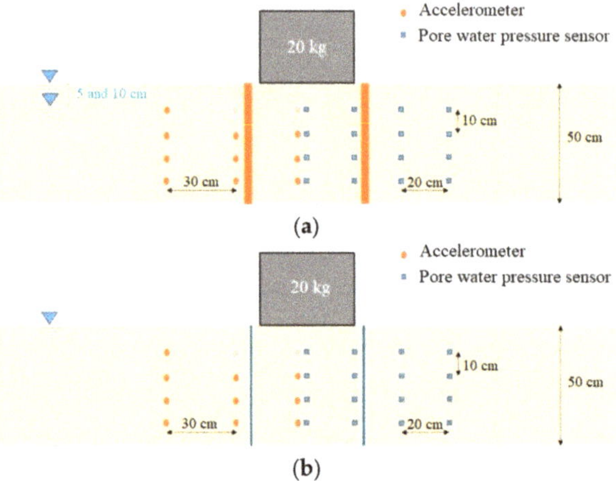

Figure 13. Experimental set up for vertical deep soil mixing walls, (**a**) plastics and (**b**) soil.

Table 4. Cases testing of vertical deep soil mixing walls.

Case	Wall Material	GWL	Load	Soil Density
Case 5	Soil	surface	uniform	46%
Case 6	Soil	−10 cm	uniform	46%
Case 7	Soil	−5 cm	uniform	46%
Case 10	plastics	surface	uniform	46%
Case 11	Soil	−5 cm	eccentric	46%

A comparison of the vertical deep soil mixing walls with the water barrier shows a very similar subsidence. Expected was a difference, because of the permeability of the soil walls. It can be assumed that in both cases the same effect led to the shape of settlement curve shown in Figure 14. Therefore the permeability of the soil wall seems to be non-existing, or the excess pore water pressure was decreased in different ways. The pore water pressure shown in Figure 15 leads to the assumption that the dissipation mechanism is similar, meaning that the settlement should show a very similar trend. In case of the water barrier, the pore water pressure dissipation starts a little later, leading to a later stop of settlement.

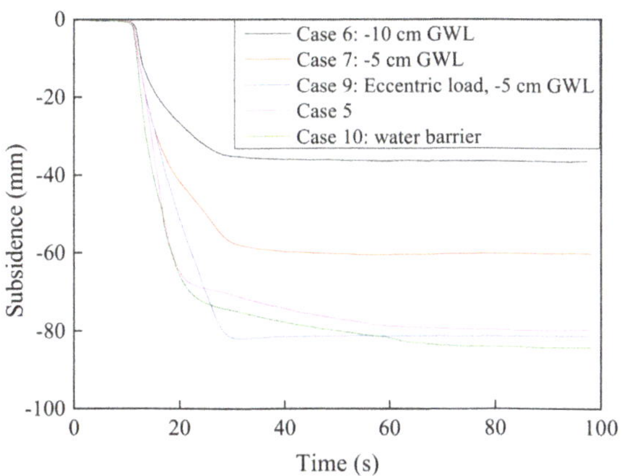

Figure 14. Subsidence of vertical deep soil mixing wall.

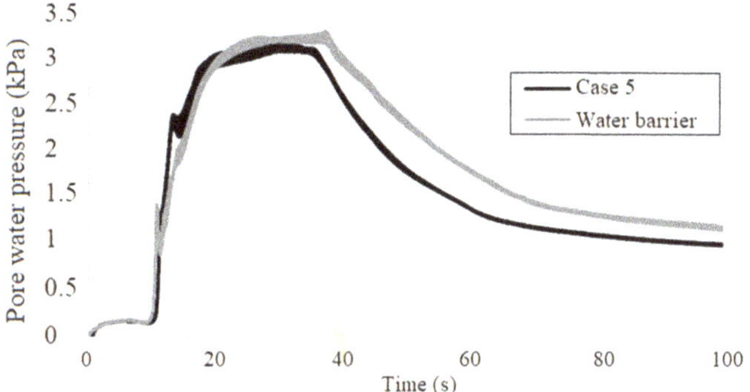

Figure 15. Pore water pressure of case 5 and water barrier.

Figure 16 shows the ultimate state seen by the colored sand. Both are very similar, which supports the assumption of similar settlement mechanisms. During liquefaction, between the structure and walls, water was pushed upwards and overtopped the walls. In both cases, the wall and the plastic bag too were pushed outwards. The type of movement is different though. In case of the plastic bag the shape of the bag was similar to the shape of the colored sand with large movement in the middle of the liquefiable soil layer. The walls rotated around the bottom and had the largest displacement at the top. This movement led to a movement of the untouched sand outside of the walls over the whole liquefiable layer height. The outwards movement of the walls created a bigger volume under the structure and decreased the density of the sand and the maximum possible uplift force was decreased, leading to higher subsidence, also after the shaking. After the biggest amplitude of the input motion the total settlement was still not reached, and the pore water pressure was still high enough to keep the sand in a liquefied state.

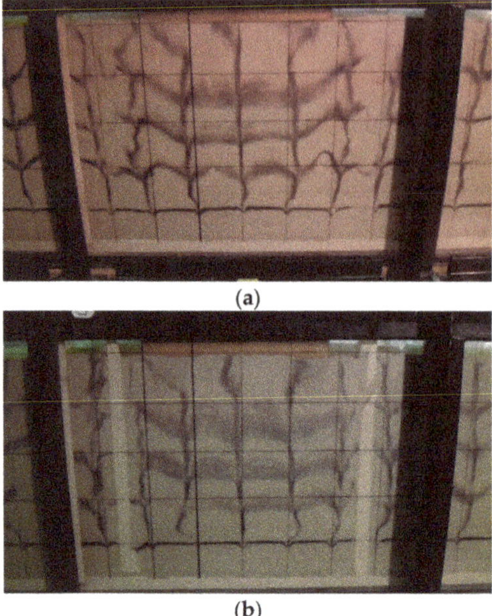

(a)

(b)

Figure 16. Ultimate state of (**a**) case 5 and (**b**) case 10.

Figure 17 shows the subsidence over time compared with the pore water pressure ratio of case 5 and 10. In the first stage a very rapid settlement happens until reaching around 70 mm. After that the settlement still increases for a long time after the end of shaking. The rapid settlement at the beginning can be correlated with the shaking, which supports the effect of settlement. After that the excess pore water pressure caused by the entrapped water and therefore the reduced effective stress still leads to a bearing capacity problem and to further settlement until the point of maximum possible settlement.

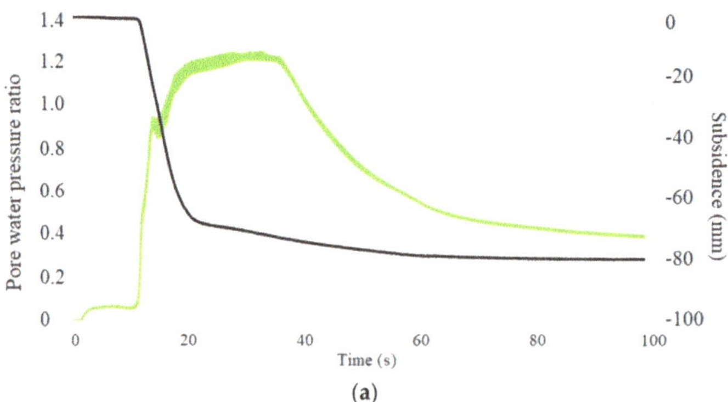

(a)

Figure 17. *Cont.*

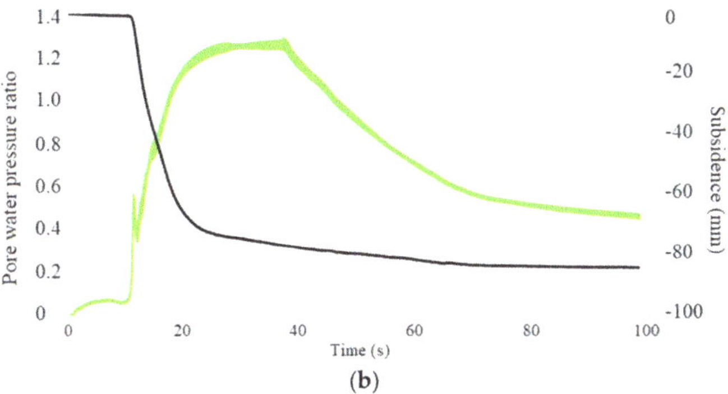

(b)

Figure 17. Subsidence and pore water pressure ratio for (**a**) case 5 and (**b**) case 10.

The dissipation of pore water pressure is also linked to the maximum possible settlement. The stop in settlement causes the pore water dissipation rate to change to smaller values because the total stresses in the soil do not change anymore. Additionally, the slower or the stopped settlement of the soil reduces the dissipation rate of the pore water pressure because the sensor does not change its position anymore and the hydraulic pressure caused by the ground water stays the same. The only measured difference at this point is the decreasing excess pore water pressure.

The trend of subsidence at the beginning differs in all the cases as shown in Figure 18. A big influence shows the lowering of the GWL. This is the same effect as already discussed in the previous section; the distribution of the structure load over a bigger area, and therefore a stronger influence of the uplift force. Results show that in case of the −5 cm GWL the eccentric and uniform load show a very similar behavior in the first stage of the shaking. After about 15 s, the eccentric load of the model structure starts to tilt and destroy the top layer and settlement keeps increasing. The water barrier case settles in the first stage faster than the vertical soil walls. The pore water pressure cannot be responsible for this behavior, because the development of pore water pressure in the case of soil walls is faster than the plastic bags. Therefore, the movement of the plastic bag would be the reason for the faster settlement velocity in the beginning. The vertical deep soil mixing wall in combination with lowering groundwater level shows very good performance. Case 6 with lowering the GWL −10 cm reached the smallest settlement of all the experiments.

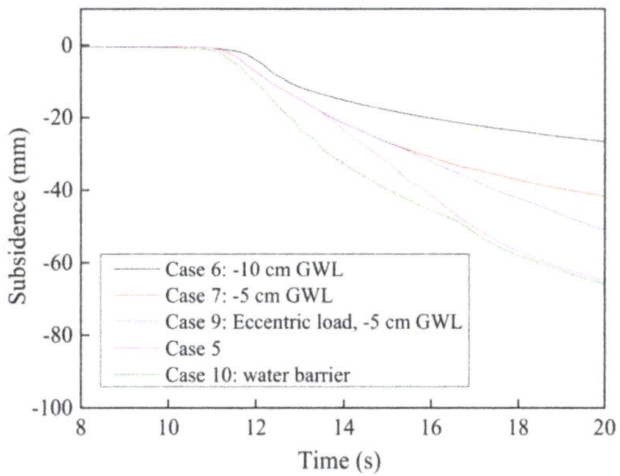

Figure 18. Subsidence of vertical soil mixing walls at the beginning.

In the previous section, the focus was mainly on the soil behavior between the walls. The idea of building a wall structure under a foundation is to decouple the overall liquefaction and produce a confined space, where with the same input motion less subsidence occurs. The area outside of the walls was untouched and experienced in the end just the regular consolidation, volumetric and sedimentation settlement. Only in case 5, without additional mitigation and the water barrier, the moving of the wall influenced the surrounding soil and induced significant lateral displacement of the top layer.

The pore water pressure inside the soil walls develops differently from outside of the walls (see Figure 19). In every case the pore water pressure directly under the structure rises slower than in other places. Outside of the walls the pore water pressure can rise easily to a certain value. The small rise during the shaking can be correlated to the volumetric, sedimentation and consolidation settlement and is most linear. In case 5, the curves are at different pressures. A pressure gradient is developing which is inclined compared to the soil level. This pressure gradient later influences the dissipation of pore water pressure.

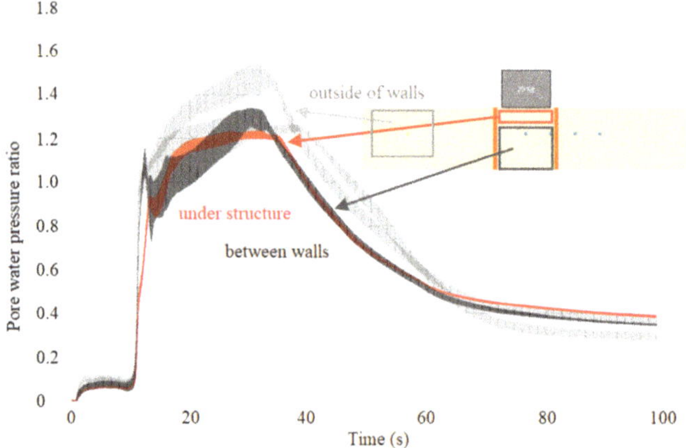

Figure 19. Pore water pressure of case 5, vertical deep soil mixing wall without additional mitigation.

The velocity diagram in Figure 20 of the settlement shows a very surprising behavior for the case 5 of no additional mitigation to the vertical deep soil mixing walls. In the case of lowering GWL, the peak velocity is reached after a short time, and decreases uniformly to zero. In the case of the water barrier, the velocity of subsidence keeps increasing and reaches the highest value, even higher than the case without mitigation. The settlement velocity decreases slower than in other cases. The soil flow is not restricted by the vertical walls and can expand, which leads to higher lateral displacement and in the end to vertical subsidence. In the cases of lowered GWL which fixed the top of the soil walls, the decrease in settlement velocity is much higher.

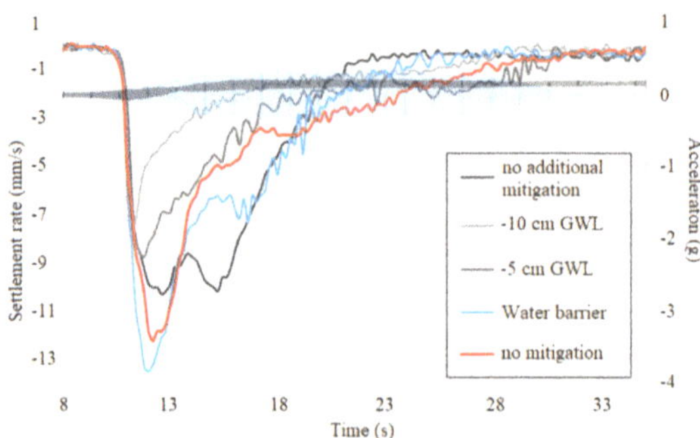

Figure 20. Settlement rate of vertical deep soil mixing walls and water barrier.

An unexpected event is the increase of settlement speed in case of no additional mitigation to the vertical deep soil mixing walls. After reaching a first peak at the same time as the no mitigation and the water barrier case, the settlement velocity decreases, and at the time the shaking reaches the maximum amplitude of 300 gal, the settlement velocity increases again.

The rate develops in the same way for both cases in the beginning of shaking (see in Figure 21). After 14 s, the settlement velocity is not decreasing at the same rate anymore. In the case of the eccentric load, the decreasing of the velocity stays at the same rate, which can also be seen in the subsidence curve. Responsible is the tilting of the structure, which increases the pressure on the soil in a smaller area than the case of a uniform loaded model structure. The uplifting soil cannot support the structure, the drained layer breaks and saturates. Therefore the settlement rate is higher over the whole time of shaking and does not decrease until the end of the shaking. Then the shear stress reaches almost zero, and the settlement rate decreases very fast to zero. Additionally, the walls tilted in the case of eccentric load. As explained before, because of the destruction of the drained top layer, movement of the soil wall became possible, which leads to higher lateral soil flow, same as in the case of no additional mitigation.

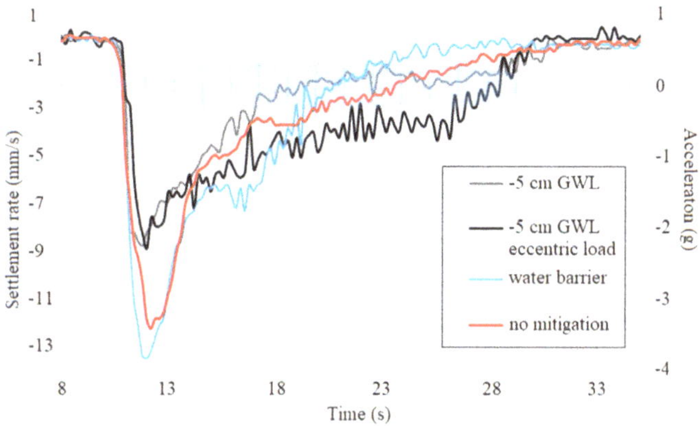

Figure 21. Settlement rate for eccentric and uniform load in the cases of vertical sheet pile walls with −5 cm GWL.

4. Conclusions

To mitigate liquefaction-induced settlement and tilting of structures, the improvement soil slabs and soil mixing walls combined with lowering GWLs were proposed and investigated. Model tests were carried out based on a 1-g shaking table. Results of this study would be helpful for practical implementation of soil slabs and mixing walls against liquefaction-induced settlement and tilting of structures in future. Important findings of the experiments can be drawn and listed as follows.

1. The horizontal improved soil slabs show a very good performance. The ultimate settlement, tilting and settlement rate was reduced obviously compared to the case of no mitigation. With a length of 80 cm, the subsidence is 30 mm which is the smallest among the investigated cases. When the length is 40 cm, the subsidence increases to nearly two times of the case of 80 cm. The reduction of subsidence was strongly influenced by the length of the slab. The thickness on the other hand was not the main deceiving factor.

2. The vertical deep soil mixing walls showed a disappointing behavior. The settlement was not reduced at all compared to no mitigation. In the case of soil and plastic walls, the subsidence of the model structure are large and reach to 85 mm and 80 mm, separately. However, lowering of the GWL in combination with the vertical soil walls performed well in reducing the subsidence. Combined with −10 cm GWL, the subsidence is reduced to 35 mm. From the results, it can be seen that the lowering of GWL is the main reason behind the reduction of subsidence.

3. The tilting of the structure can be easily reduced by building a horizontal soil slab under the structure. As expected, the length of the slab governs the resistance against tilting. The wider the slab is, the less of the tilting. The thickness of the slab too has an influence.

4. All the cases showed that the settlement rate depends on the length of the substructure. The distribution of the structure load led to different rates of settlement. In each case during the quick rise of pore water pressure the highest settlement rate is measured. In the first stage, the subsidence is mainly governed by the lateral soil flow and loss of bearing capacity of the soil. A slow settlement rate can be advantageous because the duration of an earthquake and the intensity is usually shorter than the applied motion in these experiments. Generally, if liquefaction occurs, settlement of the structure also occurs. With a smaller settlement rate, the maximum subsidence is reduced.

Author Contributions: J.Z. wrote the manuscript and Y.C. performed the experiments.

Funding: Support for this research was provided partly by the National Natural Science Foundation of China under Award no. 51809237.

Acknowledgments: The authors thank the two anonymous reviewers, who have given valuable recommendations to improve the paper.

Conflicts of Interest: The authors declare no conflict of interest.

References

1. Youd, T.L.; Hoose, S. *Historic Ground Failures in Northern California Triggered by Earthquakes*; U.S. Government Publishing Office: Washington, DC, USA, 1978.
2. Yoshimi, Y.; Tokimatsu, K. Settlement of Buildings on Saturated Sand during Earthquakes. *J. Jpn. Soc. Soil Mech. Found. Eng.* **1977**, *17*, 23–38. [CrossRef]
3. Terzaghi, B.K. *Theoretical Soil Mechanics*; Chapman & Hall: London, UK, 1959.
4. Liu, L.; Dobry, R. Seismic response of shallow foundation on liquefiable sand. *J. Geotech. Geoenviron. Eng.* **1997**, *6*, 557–567. [CrossRef]
5. Dashti, S.; Bray, J.D.; Pestana, J.M.; Riemer, M.; Dan, W. Mechanisms of Seismically Induced Settlement of Buildings with Shallow Foundations on Liquefiable Soil. *J. Geotech. Geoenviron. Eng.* **2009**, *137*, 151–164. [CrossRef]
6. Dashti, S.; Bray, J.D.; Pestana, J.M.; Riemer, M.; Dan, W. Centrifuge Testing to Evaluate and Mitigate Liquefaction-Induced Building Settlement Mechanisms. *J. Geotech. Geoenviron. Eng.* **2010**, *136*, 918–929. [CrossRef]
7. Tsukamoto, Y.; Ishihara, K.; Sawada, S.; Fujiwara, S. Settlement of Rigid Circular Foundations during Seismic Shaking in Shaking Table Tests. *Int. J. Geomech.* **2012**, *12*, 462–470. [CrossRef]
8. Rasouli, R.; Towhata, I.; Hayashida, T. 1-g Shaking Table Tests on Mitigation of Seismic Subsidence of Structures. In Proceedings of the International Conference on Physical Modeling in Geotechnics 2014, Perth, Australia, 14–17 January 2014.
9. Rayamajhi, D.; Tamura, S.; Khosravi, M.; Boulanger, R.W.; Wilson, D.W.; Ashford, S.A.; Olgun, C.G. Dynamic Centrifuge Tests to Evaluate Reinforcing Mechanisms of Soil-Cement Columns in Liquefiable Sand. Available online: https://ascelibrary.org/doi/10.1061/%28ASCE%29GT.1943-5606.0001298 (accessed on 5 November 2018).
10. Olarte, J.; Paramasivam, B.; Dashti, S.; Liel, A.; Zannin, J. Centrifuge modeling of mitigation-soil-foundation-structure interaction on liquefiable ground. *Soil Dyn. Earthq. Eng.* **2017**, *97*, 304–323. [CrossRef]
11. Olarte, J.C.; Dashti, S.; Liel, A.B.; Paramasivam, B. Effects of drainage control on densification as a liquefaction mitigation technique. *Soil Dyn. Earthq. Eng.* **2018**, *110*, 212–231. [CrossRef]
12. Naesgaard, E.; Byrne, P.M.; Huizen, G.V. Behaviour of light structures founded on soil 'crust' over liquefied ground. In *Geotechnical Earthquake Engineering and Soil Dynamics III*; American Society of Civil Engineers: Reston, VA, USA, 1998.
13. Shahir, H.; Pak, A.; Ayoubi, P. A performance-based approach for design of ground densification to mitigate liquefaction. *Soil Dyn. Earthq. Eng.* **2016**, *90*, 381–394. [CrossRef]
14. Bray, J.D.; Escudero, J.M. Ishihara Lecture: Simplified Procedure for Estimating Liquefaction-Induced Building Settlements. In Proceedings of the International Conference on Soil Mechanics and Geotechnical Engineering 2017, Seoul, Korea, 17–22 September 2017.

15. Karimi, Z.; Dashti, S.; Bullock, Z.; Porter, K.; Liel, A. Key predictors of structure settlement on liquefiable ground: a numerical parametric study. *Soil Dyn. Earthq. Eng.* **2018**, *113*, 286–308. [CrossRef]

Article

Evaluation of Progressive Collapse Resistance of Steel Moment Frames Designed with Different Connection Details Using Energy-Based Approximate Analysis

Sang-Yun Lee [1], **Sam-Young Noh [1,*]** and **Dongkeun Lee [2]**

[1] Department of Architectural Engineering, Hanyang University, 55 Hanyangdaehak-ro, Sangrok-Gu, Ansan, Gyeonggi-Do 15588, Korea; yongsha@hanyang.ac.kr

[2] Department of Civil Engineering, National Chiao Tung University, 1001 University Road, Hsinchu, Taiwan; ddklee@nctu.edu.tw

* Correspondence: noh@hanyang.ac.kr; Tel.: +82-31-400-5182

Received: 15 September 2018; Accepted: 18 October 2018; Published: 20 October 2018

Abstract: This study evaluates the progressive collapse resistance performance of steel moment frames, individually designed with different connection details. Welded unreinforced flange-bolted web (WUF-B) and reduced beam section (RBS) connections are selected and applied to ordinary moment frames designed as per the Korean Building Code (KBC) 2016. The 3-D steel frame systems are modeled using reduced models of 1-D and 2-D elements for beams, columns, connections, and composite slabs. Comparisons between the analyzed results of the reduced models and the experimental results are presented to verify the applicability of the models. Nonlinear static analyses of two prototype buildings with different connection details are conducted using the reduced models, and an energy-based approximate analysis is used to account for the dynamic effects associated with sudden column loss. The assessment on the structures was based on structural robustness and sensitivity methods using the alternative path method suggested in General Services Administration (GSA) 2003, in which column removal scenarios were performed and the bearing capacity of the initial structure with an undamaged column was calculated under gravity loads. According to the analytical results, the two prototype buildings satisfied the chord rotation criterion of GSA 2003. These results were expected since the composite slabs designed to withstand more than 3.3 times the required capacity had a significant effect on the stiffness of the entire structure. The RBS connections were found to be 14% less sensitive to progressive collapse compared to the WUF-B ones.

Keywords: progressive collapse; abnormal loads; sudden column removal; seismic connection detail; energy-based approximate analysis; structural robustness; structural sensitivity

---

1. Introduction

Building structures are designed to resist loading combinations specified in building codes. However, in many cases, structures fail due to progressive collapse. In other words, even though structures are properly designed to bear design loads, structural redundancy suddenly decreases because of unexpected abnormal loads, which consequently causes the entire structures to collapse. The following cases are representative examples of progressive collapse: the collapse of the Ronan Point Apartment in London caused by a gas explosion, the collapse of the Alfred P. Murrah Federal Building in the U.S. caused by a truck bomb attack, the collapse of the World Trade Center caused by the impact of large passenger jets in the U.S., and the collapse of the Sampung Department Store in Republic of Korea caused by the punching shear failure of a flat plate system. Based on these catastrophic accidents, the American Society of Civil Engineers (ASCE) 7, the GSA 2003, and the U.S.

Department of Defense (DoD) 2009 [1–3] have introduced analysis and design methods to prevent the progressive collapse of buildings. Additionally, the British Standard (BS) 5950 [4] in the U.K. provides guidelines to prevent disproportionate collapses by ensuring sufficient material strength of structural members and by reinforcing ties between connections in an effort to resist such extreme loads. Even though terror threats have been increasing around the world, many countries do not have proper guidelines to prevent progressive collapses. For example, in Republic of Korea, the guidelines by the Korea Ministry of Land, Infrastructure and Transport (KMLIT) [5] provide a design method that can prevent or minimize the damage caused by terrorist attacks for large-scale and high-rise buildings. However, this method focuses on the site and interior planning, security facilities, evacuation, and the planning for building facilities. At the moment, there is no effective guide that prevents the collapse of a structure due to local failure of a main structural member after a terrorist attack. Therefore, the safety evaluation of structures under abnormal loads such as impact and blast caused by terrorist attacks is a significant concern for residents and nations, and it is indispensable to study and propose reliable design methods to prevent the progressive collapse of damaged buildings.

Design methods used to prevent the progressive collapse of buildings include the tie force, alternative path (AP), and the specific local resistance (SLR) methods. The typical progressive collapse resistance design guidelines of the GSA 2003, the DoD 2009, and the BS 5950 [2–4], commonly recommend the AP method. It is a design method that increases the stiffness and strength of the member and structure under the column removal scenario so that the load initially supported by the removed column can be replaced by the adjacent member. Many researchers have applied the AP method to assess the progressive collapse resistance performance of buildings. Marjanishvili [6] compared the characteristics of each method presented in the guideline and emphasized that the results of nonlinear dynamic analysis with the dynamic effect of sudden column loss and nonlinear behavior (material nonlinearity, large deformation, etc.) of the beam and connection could be most reliable from the standpoint of the collapse mechanism under the column removal scenario. However, this method requires repeated analysis for each load step, and the time required for analysis is much greater than static analysis. In nonlinear static analysis, the adopted procedure is simple, and it takes a relatively short time for analysis compared to nonlinear dynamic analysis. However, the reliability of the analysis is degraded considering the dynamic amplification factor due to column removal. The disadvantages of these methods increase with the order of the element and the size of the structure. Sadek et al. [7] and Bao et al. [8] developed the 1-D element modeling method to assess the resistance performance against progressive collapse based on gravity load for steel and reinforced concrete structures. In these modeling approaches, the stiffness of the panel zone with the hinge and the rigid element were applied as the nonlinear uniaxial and shear springs, and the columns and beams with beam elements were connected to the rigid element by constraint. In addition, Main [9] proposed the modeling method using a composite slab with a 2-D shell element and verified the applicability to the 2×2 bay floor system structure previously used by Sadek et al. [10] and Alashker et al. [11]. These methods are reliable and reduce time and data consumption in progressive collapse analyses for the entire 3-D structure. As research for the efficient analysis and assessment of the AP method, Izzuddin et al. [12] proposed energy-based approximate analysis with the dynamic effect of sudden column loss using nonlinear static response. Using this analysis, Main and Liu [13] performed column removal scenarios for steel moment frames until the initial failure of the connections to compare the results from the dynamic and energy-based approximate analyses and to evaluate the collapse resistance of the structure. Similarly, Bao et al. [14] analyzed the collapse resistance performance for reinforced concrete structures and evaluated the structural robustness as the maximum residual capacity of the structural system that can resist progressive collapse induced by sudden column loss. Starossek and Haberland [15] analyzed the sensitivity of the structure to local failure on the basis of energy released and required, quantification of damage progression, and system stiffness. Based on the aforementioned studies, Noh et al. [16] evaluated sensitivity using structural robustness under the column removal scenario for seismically designed steel moment frames. Structural sensitivity is a relative evaluation index used for the reduction of load-bearing capacity depending on the column removal location and for the determination of the column, which is sensitive. It is also advantageous to determine which system is sensitive to progressive collapse in different structural systems, comparing various structures on the basis of seismic design. Thus, structural sensitivity can be categorized into structural systems depending on collapse resistance performance among their various structures on the basis of seismic design and can be directly helpful for structural design. However, there was no previous research reporting the assessment of the sensitivity of progressive collapse to different structural systems, while Noh et al. [16] indicted that an analysis of sensitivity is required for steel moment frames with different connection details or structural systems.

Based on these backgrounds, this study has been conducted to evaluate the resistance performance of a steel structure, constructed using the moment frames individually designed with WUF-B and RBS connections, in order to analyze the sensitivity for progressive collapse of different structural systems. Herein, the evaluation method of the structures was based on the structural sensitivity proposed by Noh et al. [16], which can be quantitatively and clearly evaluated. The steel moment frames were seismically designed as per KBC 2016 [17]. The entire structures were modeled using reduced models of 1-D and 2-D elements for beams, columns, connections, and composite slabs. After performing nonlinear static analyses on these structures, progressive collapse resistance performance was evaluated using structural robustness and sensitivity calculated from the energy-based approximate analysis with the dynamic effect of sudden column loss.

2. Evaluation Methods for Progressive Collapse Resistance Performance

2.1. Acceptance Criteria for Progressive Collapse Resistance Performance

The guidelines of GSA 2003 and DoD 2009 [2,3] provide acceptance criteria to evaluate the failure of key structural components in progressive collapse analysis. In linear analyses, the acceptance criteria are based on the demand-capacity ratio for each structural member or connection in the design. In contrast, the possibility of collapse in nonlinear analyses is evaluated by means of the chord rotation angle θ specified in the guidelines. Table 1 lists the acceptance criteria for the chord rotation angle of the steel moment frames specified in GSA 2003 [2]. In the present study, the AP method was applied to the steel moment frames individually designed with WUF-B and RBS connections and progressive collapse resistance performance was evaluated for allowable rotation limits $\theta = 0.025$ rad and $\theta = 0.035$ rad.

Table 1. Acceptance criteria for nonlinear analyses in accordance with GSA 2003 [2].

Component	Rotation [% rad]
Steel Frames	3.5
Steel Frame Connections: Fully Restrained	
Welded Beam Flange or Coverplated (all types)	2.5
Reduced Beam Section	3.5

2.2. Alternative Path Method

The AP method of GSA 2003 [2] applied in this study is divided into static and dynamic analyses; and loading conditions are different, depending on the chosen analysis method. In static analysis, the amplified load combination $2.0(1.0DL + 0.25LL)$ is applied to the tributary area of the column removal span considering the dynamic effect of sudden column loss, and the load combination $1.0DL + 0.25LL$ without the dynamic effect is applied to all other areas. In dynamic analysis, load combination $1.0DL + 0.25LL$ is applied to the entire floor area, but dynamic effect attributed to the sudden column loss is considered, using the internal reaction forces of the lost column to the supported structure (Figure 1). The column removal locations in the guideline are the exterior columns located at the corners, and center and interior columns on the ground level of buildings. Since the most vulnerable column removal locations that could lead to progressive collapse are different, depending on the plan and structure type of the building, it is noted that the removal scenarios for each of the column locations should be performed to accurately assess collapse resistance performance.

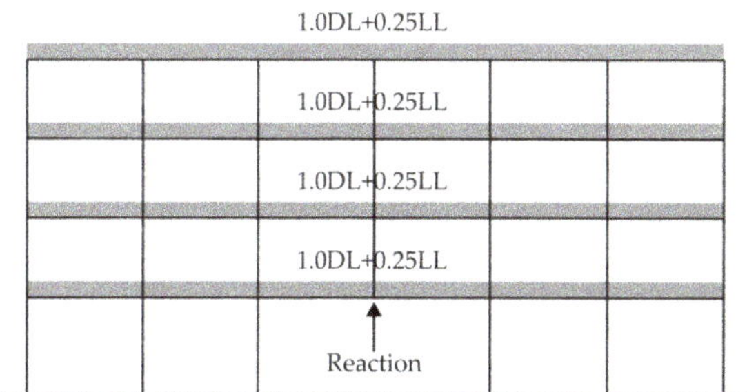

Figure 1. Dynamic analysis loading in accordance with GSA 2003 [2].

2.3. Energy-Based Approximate Analysis

Dynamic analysis representing the most realistic simulations and responses of structural behavior is complex and time-consuming. Considering these disadvantages, Izzuddin et al. [12] proposed an energy-based approximate analysis that can efficiently calculate the load-displacement response with the dynamic effect of sudden column loss using nonlinear static analysis. This analysis is based on the assumption that the predominant deformation responds in a single mode by the gravity load applied to the structure, and this system can be considered as a single-degree-of-freedom. At dynamic displacement δ after sudden column loss, the external work $W_{Dyn}(\delta)$ performed by the dynamic loads P_{Dyn} can be expressed as Equation (1):

$$W_{Dyn}(\delta) = \alpha P_{Dyn}\delta = \alpha\left\{\frac{1}{2}m[\delta'(t)]^2 + \frac{1}{2}k[\delta(t)]^2\right\} \tag{1}$$

where α is a constant that depends on the shape of the deformation mode, and m and k are the mass and stiffness of the structure, respectively. Since the kinetic energy in the state of reaching the peak dynamic displacement δ_p is zero, the external work $W_{Dyn}(\delta_p)$ is equal to the internal energy $U(\delta_p)$ absorbed by the structure, as seen in Equation (2):

$$W_{Dyn}(\delta_p) = \alpha P_{Dyn}\delta_p = \frac{1}{2}\alpha k\delta_p^2 = U(\delta_p) \tag{2}$$

Assuming the same deformation mode response under the static loads, the external work $W_{St}(\delta_p)$ at displacement δ_p can be expressed as Equation (3) and the load-displacement curve $P_{St}(\delta)$ in Figure 2 can be obtained from the nonlinear static analysis:

$$W_{St}(\delta_p) = \alpha \int_0^{\delta_p} P_{St}(\delta)\,d\delta = \alpha \int_0^{\delta_p} k\delta\,d\delta = \frac{1}{2}\alpha k\delta_p^2 = U(\delta_p) \tag{3}$$

Equating the work performed under static and dynamic loading at peak dynamic displacement δ_p allows the constant α to be eliminated, yielding the following equation:

$$P_{Dyn}\delta_p = \int_0^{\delta_p} P_{St}(\delta)\,d\delta \tag{4}$$

The left-hand side of Equation (4) represents the hatched area in Figure 2 and the right-hand side represents the shaded area. The dynamic load P_{Dyn} is expressed as the following equation by dividing the peak dynamic displacement δ_p on both sides. Consequently, the function $P = P_{Dyn}(\delta)$ can be obtained by using Equation (5) with varying δ_p:

$$P_{Dyn} = \frac{1}{\delta_p} \int_0^{\delta_p} P_{St}(\delta) \, d\delta \tag{5}$$

The function $P = P_{St}(\delta)$ in Figure 2 continues to increase beyond displacement δ_u corresponding to the ultimate resistance due to the residual bearing capacity of the structural system. However, the uncertainties of analysis in the post-ultimate response significantly increase due to accelerations and increasing dynamic effects produced under force-controlled loading. Considering these reasons and conservatism, the ultimate capacity $P_{Dyn,u}$ under sudden column loss, as shown in Equation (6), is obtained at displacement δ_u corresponding to the ultimate static load:

$$P_{Dyn,u} = P_{Dyn}(\delta_u) \tag{6}$$

The progressive collapse resistance performance of the structures in this study was evaluated by the previously described energy-based approximate analysis.

2.4. Structural Robustness of Progressive Collapse

The structural robustness of progressive collapse is a measure of the extent to which a structure can resist collapse due to abnormal loads. To use the structural robustness as a structural performance measure, it is necessary to satisfy usefulness and validity. Starossek & Haberland [15] presented the general requirements for structural robustness as follows:

- Generality for applicability to any type of structure
- Expressiveness of all aspects of robustness with clear distinction between robust and non-robust
- Objectivity by independence from user decision and reproducibility
- Simplicity to gain objectivity, generality, as well as acceptance with users
- Ability to calculate the properties and behaviors of the structure with sufficient accuracy and without excessive effort

From this viewpoint, structural robustness should be uniquely defined so that it can be quantitatively measured and clearly evaluated with general validity. Bao et al. [14] proposed an index r to assess the robustness of a structure. This can be expressed as the ratio of the maximum resistance capacity C to the required capacity D under sudden column loss, as shown in Equation (7), where D can be defined according to an applicable code or standard. The acceptance criteria for robustness to prevent disproportionate collapse require that $r \geq 1$, which indicates that as index r increases, the structure becomes more robust:

$$r = \frac{C}{D} \tag{7}$$

The structural sensitivity defined by Noh et al. [16], as shown in Equation (8), is a relative evaluation index for the reduction of the load-bearing capacity depending on column removal location, or an index for assessing the extent of safety of progressive collapse depending on column removal scenarios in different structural systems, in terms of structural materials, connection details, etc. The structural sensitivity index S can be evaluated as the ratio of the residual capacity $(rD - D)$ under gravity loading of the structure with a missing column to the value of $(r_0D - D)$ of the undamaged structure, which can be expressed using the structural robustness index r. A value of $0 \leq S \leq 1$ is obtained if the structure has the required capacity after sudden column removal, and the sensitivity index S of the structure that does not satisfy the acceptance criteria for robustness has a negative value.

$$S = \frac{(rD - D)}{(r_oD - D)} = \frac{r - 1}{r_o - 1} \tag{8}$$

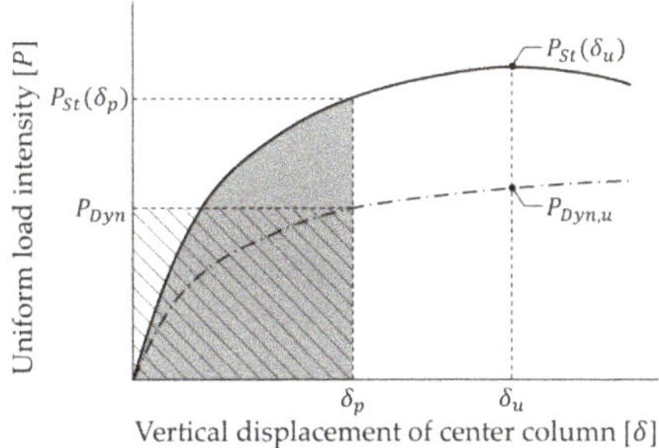

Figure 2. Concept of load-displacement responses in the energy-based analysis (Noh et al. [16]).

3. Analysis Modeling Approach

3.1. Reduced Modeling

Modeling of the entire structure is required to confirm resistance performance against progressive collapse for the seismically designed steel moment frame. There are detailed and reduced models for the modeling approach of composite slabs and seismic connections of beam-columns, which are components that determine the behavior of the structure [18]. The detailed model using the 3-D solid element can easily visually confirm the geometry, deformation, and stress distribution of the structure, but time and data consumption for analysis is highly demanded due to the large number of elements. In contrast, the reduced model using 1-D and 2-D elements, such as springs, beams, and shells, is easier to model and requires less analysis time than the detailed one. However, for reliable simplification of the 1-D and 2-D modeling, the geometric and structural behaviors of real 3-D members and connection details should be considered using additional material parameters. For instance, there are varying depths of concrete and steel deck in composite slab, panel zone behavior in beam-column connection, and width of the radius-cut section in RBS connection. This section introduces the modeling approach for the composite slab, WUF-B, and RBS connections. Additionally, comparisons between the analysis results of the reduced models and the experimental results were conducted to verify the applicability of the models. Numerical analyses considering both geometrical and material nonlinearities were performed using the finite element software ABAQUS, while the loading condition involved the application of a concentrated static load to the prescribed location under displacement control until failure occurred.

3.2. Reduced Modeling for Composite Slab

This study considered the modeling and analysis of the seismically designed moment frames including the composite slab. The modeling approach of the composite slabs proposed by Main [9] was adopted and the concrete slab on the steel deck was represented in the reduced model using alternating strips of shell elements referred to as "strong" and "weak" strips, as illustrated in Figure 3. The weak strips include only the concrete above the top of the steel deck, while the strong strips include the full depth of concrete and the steel deck. The thickness of the steel deck used in the strong strips is calculated by considering the average width of the bottom segment of each rib. This thickness t_d can be expressed as:

$$t_d = t \cdot \frac{\omega_B}{\omega_M} \tag{9}$$

where t is the actual deck thickness, ω_M is the average rib width, and ω_B is the bottom segment width of each rib. Conversely, no contribution from the steel deck is included in the weak strips.

The connection of the composite slabs and the floor beams is modeled using the rigid links and the beam elements: the former represents a half of the depth of beams (or girders), while the latter represents a shear stud.

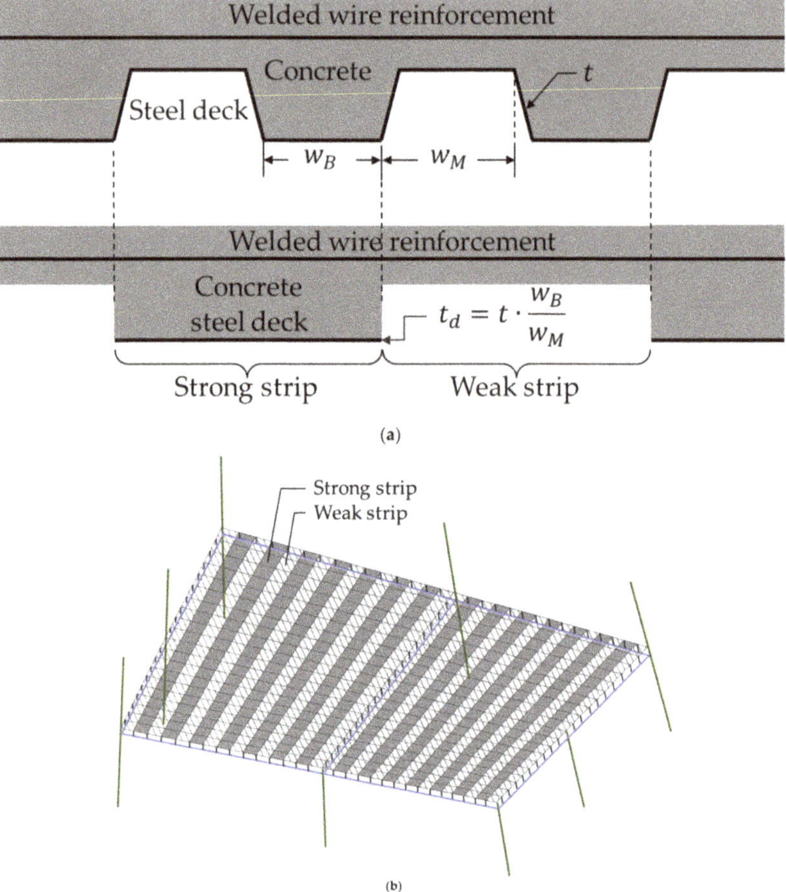

Figure 3. Reduced model of composite slab (Main [9]). (**a**) scaled thickness of steel deck; (**b**) reduced model of 2 × 2 bay composite slab system.

3.3. Verification of Reduced Modeling for Composite Slab

To validate the reduced model of the composite slab, the test specimen (see Figure 4) conducted by Kim et al. [19] was modeled based on finite element model (FEM) and the experimental data, and simulated results were compared. The composite slab was modelled using four-node quadrilateral shell elements with reduced integration (S4R) and the wire mesh was considered using a rebar option from the ABAQUS library. C21 concrete with a compressive strength of 21 MPa, SD400 reinforcing bars with a yield strength of 400 MPa, and a QL600 steel deck with a minimum specified yield strength of 205 MPa were used for material properties. The concrete material model used in the analysis was concrete damage plasticity in ABAQUS, which could provide an effective method for modeling the concrete behavior in tension and compression. This study employed the stress-strain relationship proposed by Hognestad [20] in compression and Mondal and Prakash [21] in tension, as shown in Figure 5a. An elastic perfectly-plastic material was used for the steel deck and rebar with an identical behavior in tension and compression (see Figure 5b). The roller and pin supports for the composite slab were placed at a distance of 100 mm away from the span ends. The two-point displacement load was applied, and the vertical deflection at the mid-span was measured. Figure 6 shows a comparison of the vertical load-displacement curves obtained from the FEM and the tests. The analytical and experimental results were in good agreement with each other for the ultimate load, initial stiffness, and elicited behavior, but the descending branch was not developed in the large-deformation behavior of the analysis model after reaching the ultimate load. To improve this result, the concrete tension stiffening model, which decreases linearly to reach zero stress at a strain of 0.0035 in the softening behavior (Figure 6b), was applied to this analysis. As a result, the curve computed from the reduced model is generally in good agreement with the experimental one. The initial tensile cracks of the concrete occurred at the lower surface of the loading location under $P = 11.04$ kN (Figure 7a), followed by the yielding of the steel deck of the strong strip when the vertical load reached $P = 37.05$ kN (Figure 7b).

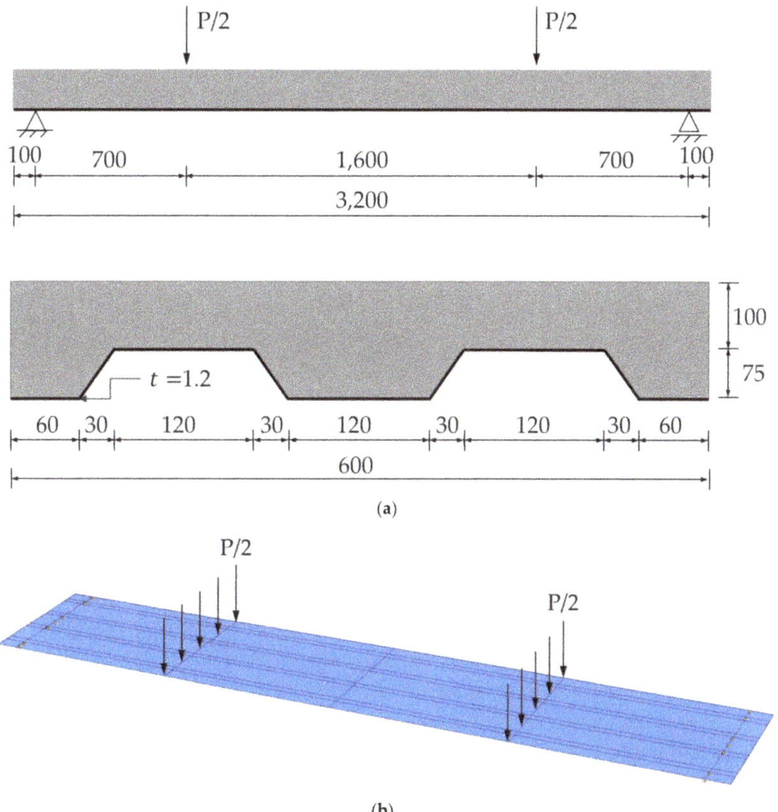

Figure 4. Composite slab test under two-point concentrated loading (Kim et al. [18]). (**a**) schematic of test setup; (**b**) ABAQUS-model of composite slab.

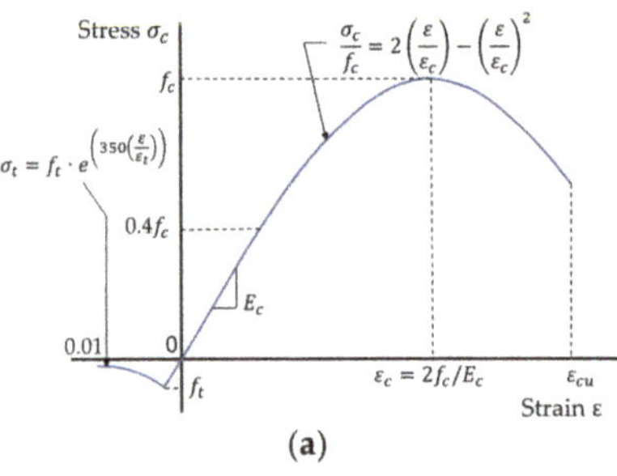

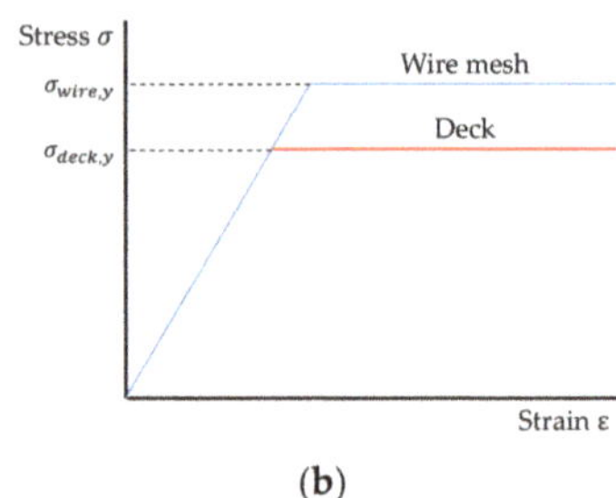

Figure 5. Stress-strain relationship for concrete, steel deck, and wire mesh. (**a**) concrete; (**b**) steel deck and wire mesh.

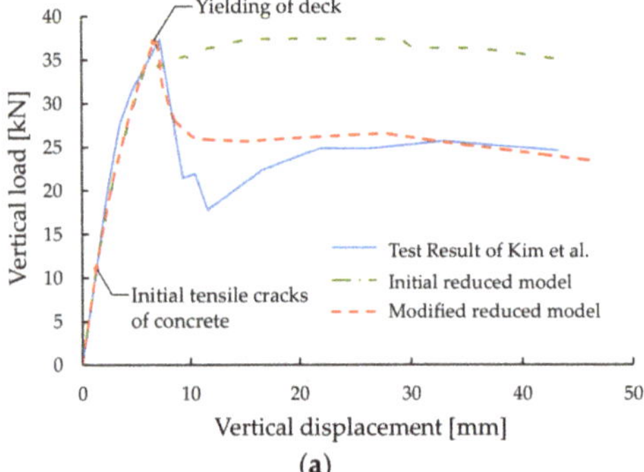

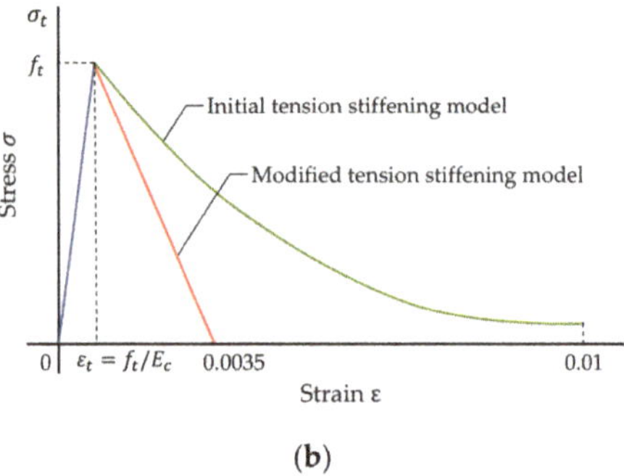

Figure 6. Comparison of results from the test and analysis of the composite slab. (**a**) vertical load-displacement curves; (**b**) applied tension stiffening models.

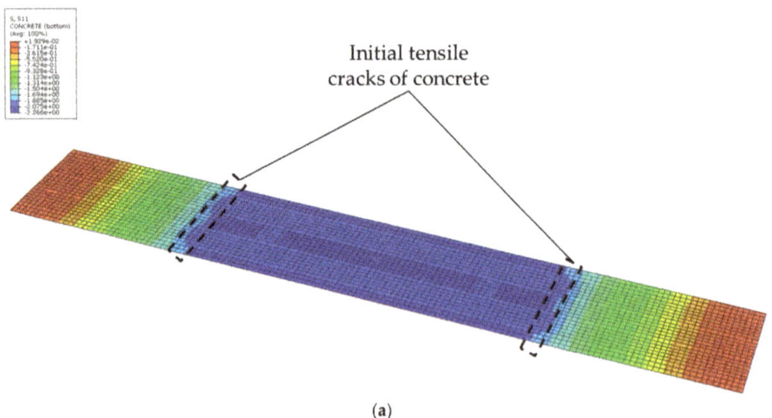

Figure 7. *Cont.*

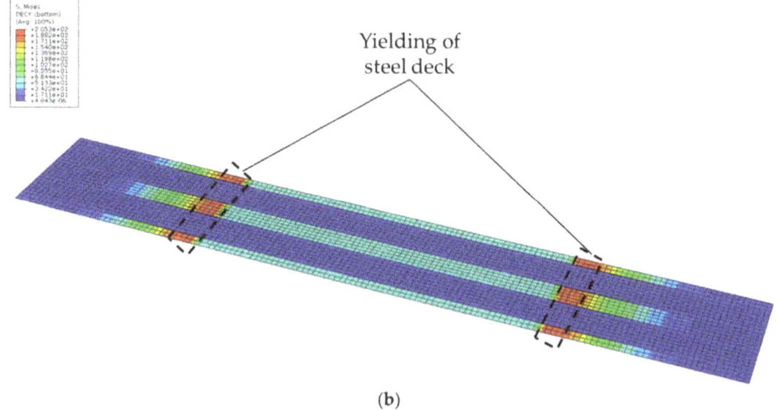

(b)

Figure 7. Stress contour results from the analysis of the composite slab. (**a**) axial stress of concrete under P = 11.04 kN; (**b**) von Mises stress of steel deck under P = 37.05 kN.

3.4. Reduced Modeling for WUF-B and RBS Connection

In this study, the prequalified connection details of WUF-B and RBS specified in FEMA 350 [22] and AISC 341 [23] for use in ordinary moment frames were selected to compare the structural sensitivity of different structural systems for the steel structure. In FEMA 355F [24], there is also the reduced model proposed by Krawinkler [25] (Krawinkler model) as a representative analytical model to simulate the non-linear behavior of the beam-column connections, as shown in Figure 8. This model developed to resist lateral forces due to earthquake loads is most commonly used in seismic-related analytical research studies. Furthermore, it is also extensively used in analytical studies for progressive collapse based on gravity loads. It holds the full dimension of the panel zone with rigid links and controls the deformation of the panel zone using two rotational springs and hinges. The yielding and plastic properties of the panel zone are calculated as follows:

$$k_\theta = \frac{M_y}{\theta_y} \text{ where}: \ M_y = 0.55F_y d_c t_{pz} d_b, \ \theta_y = \frac{F_y}{\sqrt{3}G} \tag{10}$$

$$M_p = 0.55F_y d_c t_{pz} d_b \left[1 + \frac{3b_{cf} t_{cf}^2}{d_b d_c t_{pz}} \right], \ \theta_p = 4\theta \tag{11}$$

where d_c is the column depth, d_b is the beam depth, t_{cf} is the column flange thickness, b_{cf} is the column flange width, and t_{pz} is the panel zone thickness.

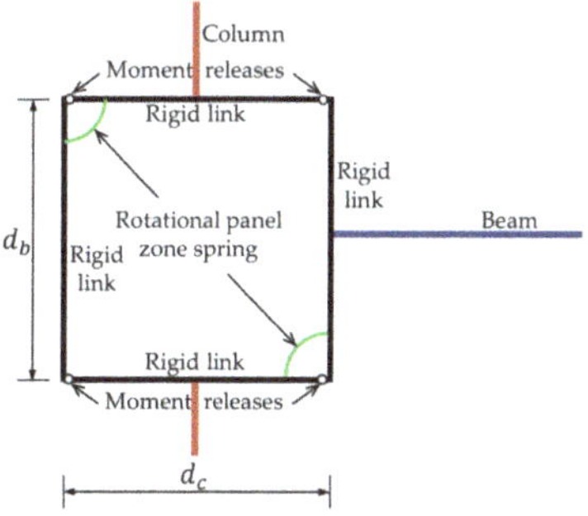

Figure 8. Reduced model proposed by Krawinkler [25].

211

As depicted in Figure 9, Sadek et al. [7] (Sadek model) developed the various reduced models with WUF-B and RBS connections suitable for progressive collapse analyses based on gravity loads. Compared with the Krawinkler model, this WUF-B reduced model (Figure 9a) further considers the upper and lower flanges and shear tap components whereby the panel zone behavior is defined as a uniaxial spring element determined by the section properties as follows:

$$k_{pz} = \frac{G\left(d_c - t_{cf}\right)t_{pz}}{\left(d_b - t_{bf}\right)\cos^2\theta} \quad \text{where}: \ \cos^2\theta = \frac{\left(d_c - t_{cf}\right)^2}{\left(d_c - t_{cf}\right)^2 + \left(d_b - t_{bf}\right)^2} \tag{12}$$

$$f_{pz} = \frac{0.6F_y d_c t_{pz}}{\cos\theta}\left[1 + \frac{3b_{cf}t_{cf}^2}{d_b d_c t_{pz}}\right] \tag{13}$$

where G is the shear modulus of steel and t_{bf} is the beam flange thickness.

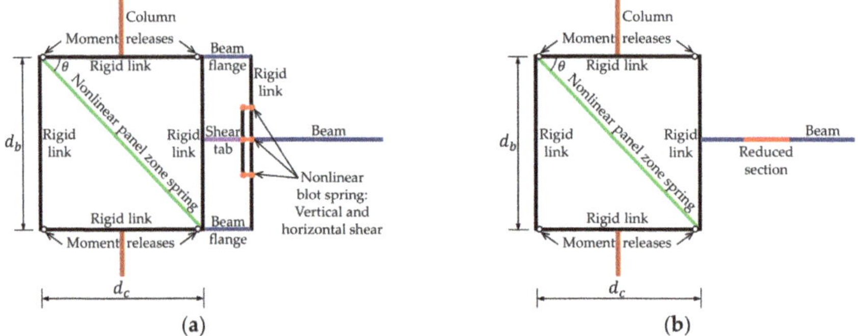

Figure 9. Reduced models developed by Sadek et al. [7]. (**a**) WUF-B model; (**b**) RBS model.

Sadek et al. [7] used the bolt spring connection based on the test results of the shear tab and beam web. However, in this study, the biaxial spring model was employed with the vertical and horizontal axes proposed by Main and Sadek [26]. Zero-length spring elements were used to model the shear behavior of the bolts, and each bolt spring was defined as an element with properties based on the axial load-deformation relationship. Figure 10 shows a typical shear load-deformation relationship and the parameters for bolt springs, which exhibit a steeper drop in resistance after the ultimate load is reached in both tension and compression. The yield and ultimate capacities of each spring, i.e., $t_y(= c_y)$ and $t_u(= c_u)$, are calculated using equations listed in the AISC 360 [27], and the deformation at the ultimate load δ_u is obtained as $\delta_u = \theta_{max} \cdot y_{max}$, where θ_{max} is the plastic rotation capacity and y_{max} is the distance from the center of the bolt group to the most distant bolt. Failure deformation is applied as $1.15\delta_u$ referring to Main and Sadek [26] in order to consider the rapid drop after the ultimate load.

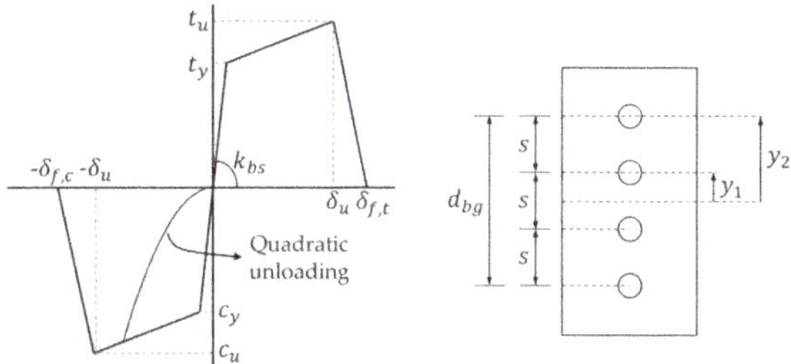

Figure 10. Shear load-deformation relationship and parameters for bolt spring (Main and Sadek [26]).

The initial stiffness of a spring k_{bs} can be estimated using Equation (14) in the FEMA 355D [28] based on the linear regression of rotational stiffness data from seismic testing.

$$k_{bs} = \frac{\kappa}{\Sigma_i\, y_i^2} \text{ where : } \kappa = 124550\left(d_{bg} - 142\right) \tag{14}$$

where κ is the initial rotational stiffness of a shear tab, $d_{bg} = s(N-1)$ is the depth of the bolt group, s is the vertical spacing between bolts, N is the number of bolts, and y_i is the vertical distance of the ith bolt row from the center of the bolt group.

The RBS model is the same as the modeling approach used for the panel zone of WUF-B and the rigid links connect the end of the beam elements along the girder centerline, as shown in Figure 9b. In this study, the equivalent effective width model proposed by Lee [29] in Figure 11 was employed to consider the radius-cut section that led to the plastic hinge of the beam. This model was verified using 3-D finite element analyses for RBS specimens subjected to cyclic loads. It can be applied to the Krawinkler model because it considers only the radius-cut section of the beam. Additionally, it is assumed that the initial elongation equals the elongation of the section replaced with the equivalent constant width of the radius-cut section, and the equivalent effective width b_{eq} is derived as Equation (15).

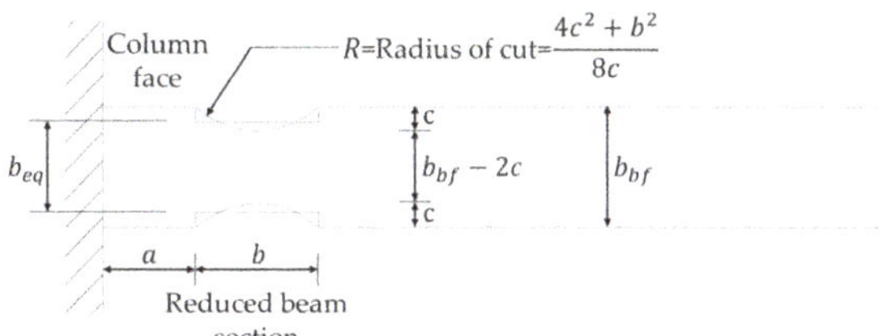

Figure 11. Geometries and parameters of radius-cut RBS (Lee [29]).

$$b_{eq} = \frac{X1}{X2} \text{ where : } X1 = b\left[\frac{L_b}{2} - \left(a + \frac{b}{2}\right)\right], \ X2 = \frac{(L_b - 2a - b)}{\sqrt{\frac{b_{bf}-2c}{R}}} \times \tan^{-1}\left(\frac{b}{2R\sqrt{\frac{b_{bf}-2c}{R}}}\right) \tag{15}$$

where a is the horizontal distance from the face of column flange to the start of the RBS cut, b is the length of an RBS cut, c is the depth of the RBS cut at the center of the reduced beam section, R is the radius of the RBS cut, L_b is the length of the beam, and b_{bf} is the beam flange width.

3.5. Verification of Reduced Modeling for WUF-B and RBS Connection

To verify the reduced models for WUF-B and RBS connections, the test specimen in Figure 12 was modeled using ABAQUS, as shown in Figure 13. These physical tests were carried out by Sadek et al. [7] to investigate the response of steel beam-column assemblies with moment resisting connections under the monotonic loading conditions expected in the column removal scenarios. The sizes of the specimens were two spans with a length of 6.10 m and columns with a height of 3.66 m, and two diagonal braces for each column were rigidly constrained to the top of the column and the strong floor of the test facility. Diagonal braces were installed to simulate the lateral restraint effect provided by the continuous upper-floor framing system of the multistory building. Displacement transducers and load cells were used to measure the displacement and vertical load applied by the hydraulic actuator with a capacity of 2700 kN and a stroke length of 500 mm at the center column of the test specimens. The strain gauges were attached to calculate the axial forces developed at the midspan of the beams, columns, and diagonal braces during the test. In the analysis model based on the test specimens, the shear tab, flange, beam, and column consisted of a two-node linear beam element (B31), and rigid links of the panel zones were assigned to the discrete rigid element (RB3D2). The CONNECTOR option in ABAQUS were used to model the panel zone spring, bolt spring, and diagonal braces installed for the lateral support of the upper column. The ASTM A992 structural steel with a yield strength of 344.8 MPa was used in all beams, columns, and doubler plates in the panel zones. The ASTM A36 steel with a yield strength of 248.2 MPa was used for the shear tabs and continuity plates at beam-column connections. The ASTM A490 high strength bolts were used for the bolted moment connections. Figure 14 shows the nonlinear panel zone spring properties for each test specimen calculated on the basis of the reduced modeling approaches described in Section 3.4. Using Equation (15), the equivalent effective width b_{eq} of the RBS specimen was calculated to be equal to 147 mm and the finite element analysis was performed under displacement-controlled loading of the unsupported center column.

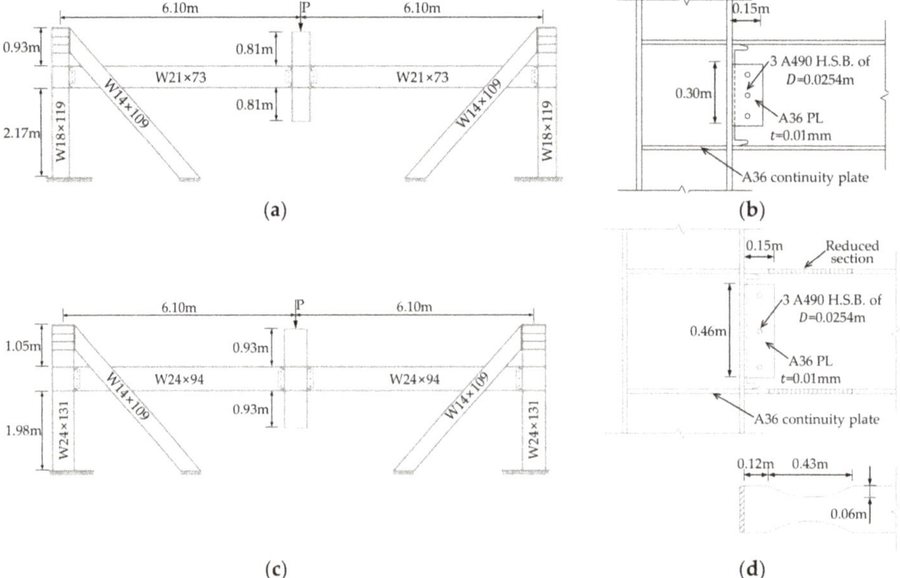

Figure 12. Resistance performance test of steel moment connections proposed and conducted by Sadek et al. [7]. (**a**) test setup for WUF-B connection; (**b**) WUF-B connection details; (**c**) test setup for RBS connection; (**d**) RBS connection details.

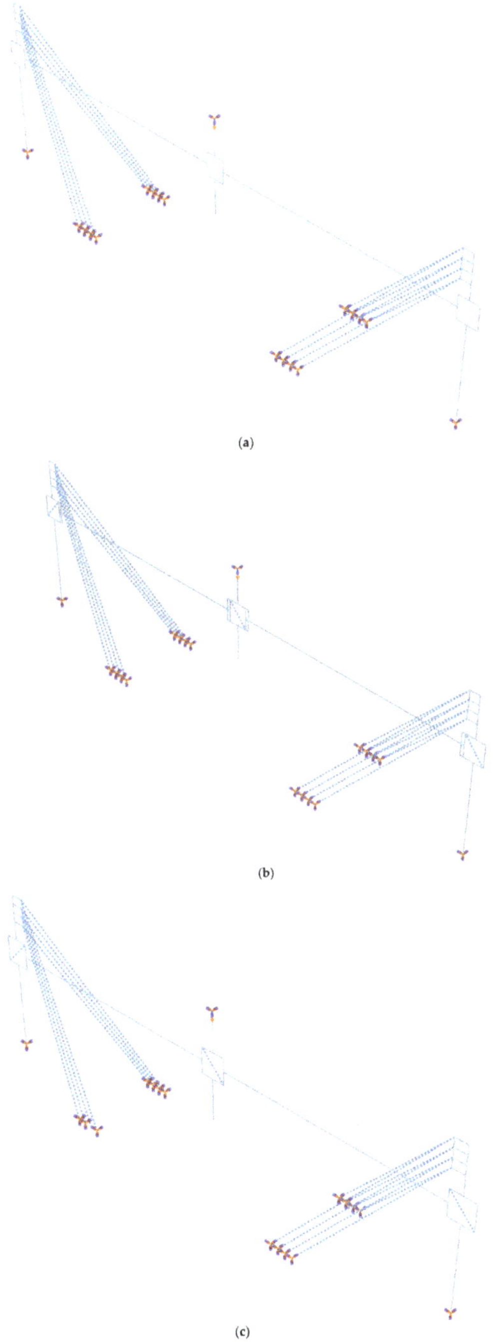

(a)

(b)

(c)

Figure 13. ABAQUS modeling for tests with steel beam-column assemblies. (**a**) WUF-B specimen using Krawinkler model; (**b**) WUF-B specimen using Sadek model; (**c**) RBS specimen using Sadek model.

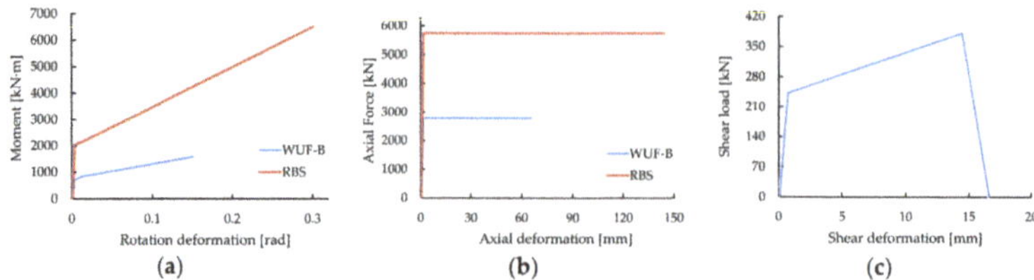

Figure 14. Nonlinear panel zone spring and bolt spring properties. (**a**) panel zone spring behavior for Krawinkler model; (**b**) panel zone spring behavior for Sadek model; (**c**) bolt spring behavior for Sadek model.

Figure 15 represents a comparison between the numerical results from the FEM and experimental results from the WUF-B and RBS specimens. Compared to the experimental results, the Krawinkler model curves show that the yielding loads increased by 45% and 31% in the load-displacement curves, respectively, and that the development of catenary action in the beams is quite different from the experimental and Sadek model results. In contrast, the Sadek model curves indicate good agreement with the experimental load-displacement and catenary action curves. Figures 16 and 17 show the failure mode of test specimens and the deformation of the WUF-B and RBS test specimens under failure load, respectively. In the experiment, the dominant deformation of the WUF-B test specimen occurred at the flanges near the weld access hole of the beam, and the failure was characterized by the fracture of the bottom flange near the center column. In the case of the RBS test specimen, dominant deformation occurred at the reduced section of the beam and panel zone, and the failure was characterized by the fracture of the bottom flange in the reduced section of the connection near the center column. The observed behaviors of tests described previously are fairly well described by Sadek models, but are completely different to those by the Krawinkler models. This indicates that the Krawinkler model is not appropriate for the progressive collapse analysis based on gravity loads. Therefore, this study employs the Sadek model to accurately assess collapse resistance performance.

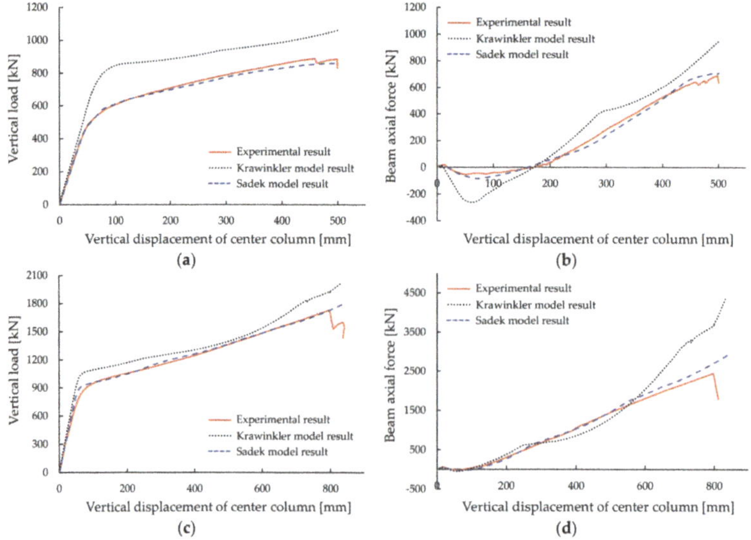

Figure 15. Results from test and analysis of WUF-B and RBS specimens. (**a**) vertical load-displacement curves for WUF-B specimen; (**b**) beam axial force-displacement curves for WUF-B specimen; (**c**) vertical load-displacement curves for RBS specimen; (**d**) beam axial force-displacement curves for RBS specimen.

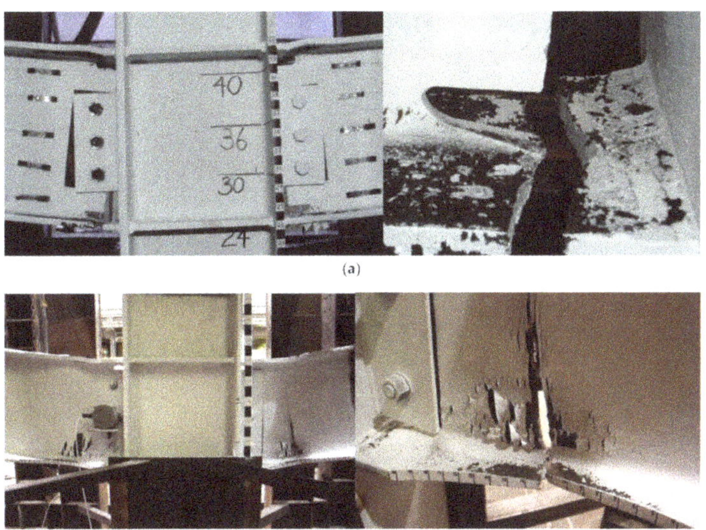

(a)

(b)

Figure 16. Failure mode of test specimens conducted by Sadek et al. [7]. (**a**) WUF-B specimen; (**b**) RBS specimen.

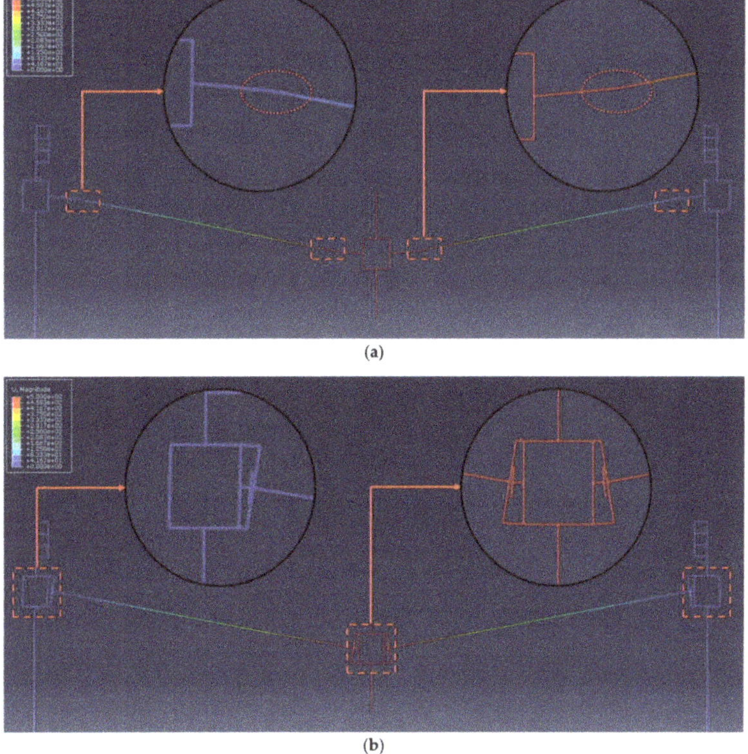

(a)

(b)

Figure 17. *Cont.*

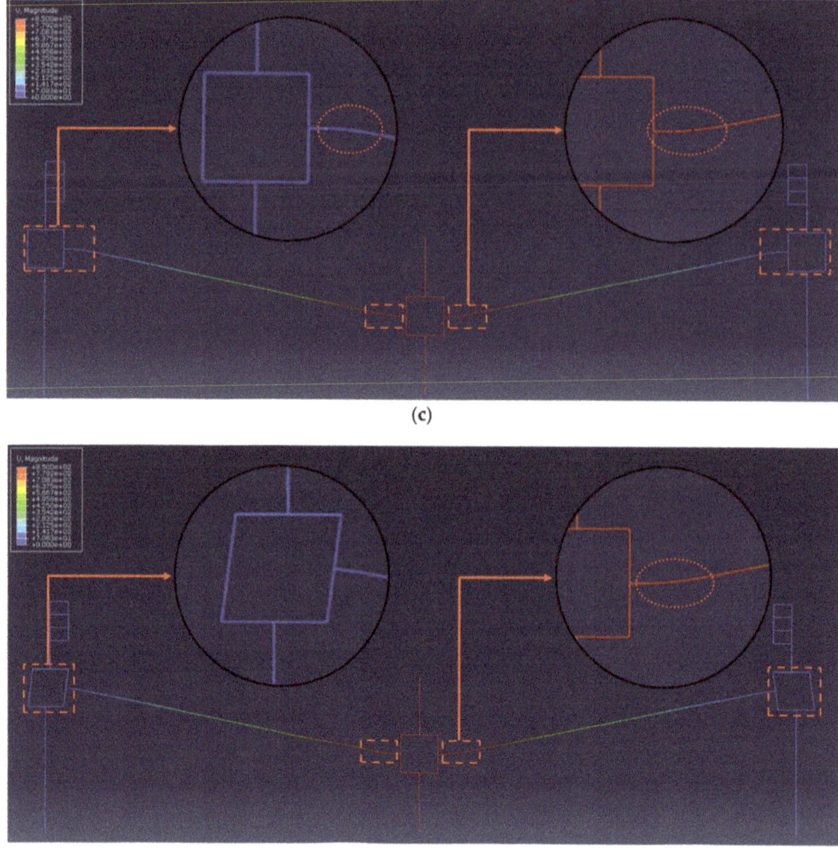

Figure 17. Deformations of WUF-B and RBS test specimens under failure load. (**a**) Krawinkler model for WUF-B specimen; (**b**) Sadek model for WUF-B specimen; (**c**) Krawinkler model for RBS specimen; (**d**) Sadek model for RBS specimen.

4. Evaluation of Progressive Collapse Resistance Performance of Selected Structure

4.1. Description of Prototype Building Design

To evaluate the structural sensitivity proposed by Noh et al. [16] for 3-D steel moment frames with different seismic connections, a prototype building was designed. Considering the guidelines by the KMLIT [5], the steel structure was assumed to be used as office space with a total floor area that exceeded 20,000 m^2, as shown in Table 2. As a result, the building had seven stories and plan dimensions of 40 × 72 m with 5 × 8 bay, as shown in Figure 18. Using the design loads based on the KBC 2016 [17], the structural members were designed as per KSSDC 2016 [30]. The building was assumed to be in Seismic Design Category D, and the lateral loads were resisted by seismically designed ordinary moment frames (OMFs) located at the exterior of the building. All interior frames were designed to support gravity loads only. In the design of the prototype building, SS400 structural steel with a yield strength of 235 MPa was used in all the beams and girders, whereas SM490 steel with a yield strength of 325 MPa was used in all the columns. Table 3 shows the member sizes of the moment resisting frames, while the gravity loads considered in the design are listed in Table 4.

Table 2. Prototype building information.

Region	Seoul, Korea
Sum of floor area	20,160 m^2
Scale of structure	7 stories
Seismic design category	D
Type of frame	Steel ordinary moment frames
Seismic connection type	WUF-B or RBS

Table 3. Member sizes of moment resisting frames.

Member	Floor and Section		Member	Floor and Section	
Girder G1	1-3F	H594×302×14/23	Girder G2	1-3F	H912×302×18/34
	4-6F	H594×302×14/23		4-6F	H700×300×13/24
	7F	H244×175×7/11		7F	H400×200×8/13
Girder G3	1-3F	H596×199×10/15	Girder G4	1-3F	H582×300×12/17
	4-6F	H596×199×10/15		4-6F	H582×300×12/17
	7F	H496×199×9/14		7F	H496×199×9/14
External column	1-3F	H428×407×20/35	Internal column	1-3F	H428×407×20/35
	4-6F	H406×403×16/24		4-6F	H300×305×15/15
	7F	H300×300×10/15		7F	H200×200×8/12

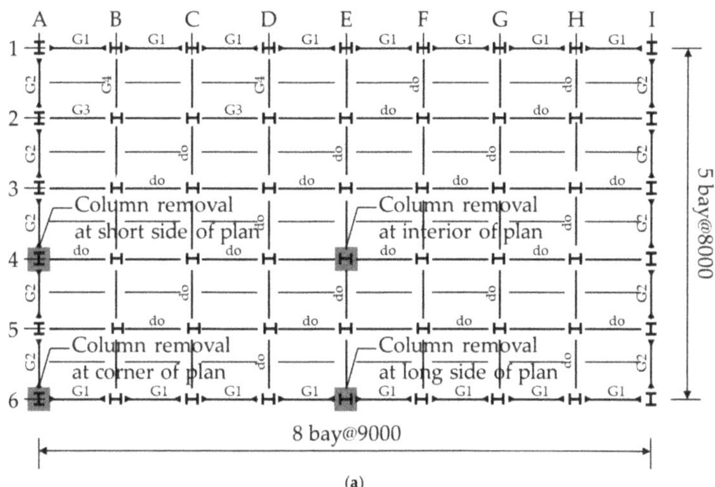

(a)

Figure 18. *Cont.*

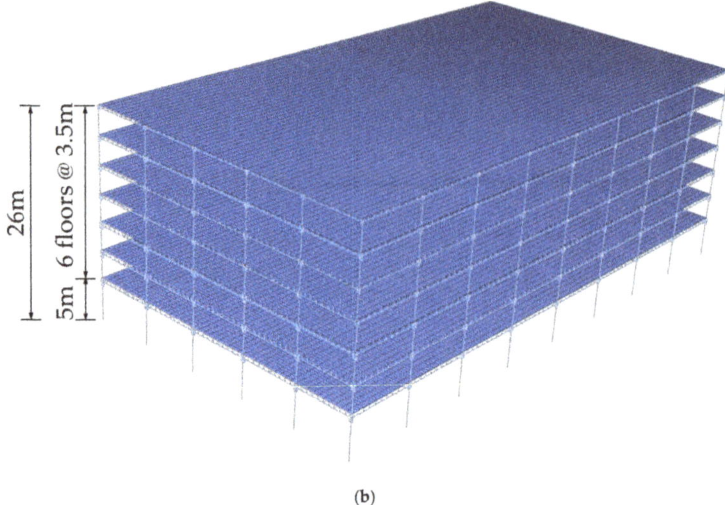

(b)

Figure 18. Overview of structure model (Noh et al. [16]). (**a**) plan of prototype structure; (**b**) ABAQUS-model of analyzed structure.

Table 4. Applied gravity load.

Type	Dead Load	Live Load
Floor [kN/m^2]	4.0	2.5
Roof [kN/m^2]	2.0	1.0

4.2. Design of Composite Slab and Shear Stud Anchor

The composite slab of the prototype building was designed using the material properties of the test specimen in Section 3.3. Considering bearing capacity and serviceability, the thickness of the topping concrete and steel deck was determined to be 100 mm and 1.2 mm, respectively. The designed strength of the composite slab was more than 3.3 times the required strength to satisfy the acceptance criteria for natural frequency considering noise and vibration in the serviceability check. The shear strength of the stud anchor connecting the beam flange to the steel deck was calculated using Equation (16) in the KBC 2016 [17].

$$(D_{max})_{push} = 0.50 A_{sh} \sqrt{f_{ck} E_c}, \ E_c = 5050 \sqrt{f_{ck}} \tag{16}$$

Shear stud anchors with a diameter of 19 mm were used to develop a full composite action between the steel beams and the concrete slab. Considering the nominal shear strength of the shear stud and the requirements in the standard, the arrangement of the stud anchor was equally spaced every 900 mm in the beam at the long side of the plan and every 800 mm in the beam at the short side of the plan, as shown in Figure 19. The physical test for the progressive collapse of the composite slab was conducted by Astaneh-Asl et al. [31], who reported that the stud anchor hardly deformed at large displacements under the monotonic loading conditions. Additionally, Kim et al. [32] used the spot welding option in ABAQUS to model stud anchors in an analytical study of the progressive collapse resistance of the 2 × 2 bay subassembly frames. Based on these studies, the present study employed the spot welding option to model stud anchors.

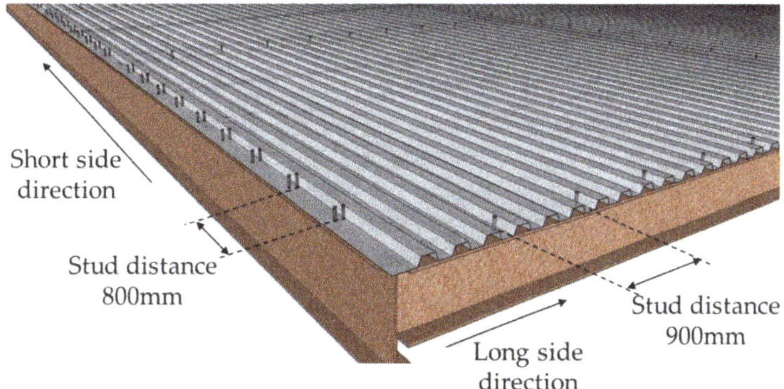

Figure 19. Arrangement of shear stud anchors.

4.3. Design and Modeling of RBS Connections

The WUF-B connection is prequalified by FEMA 350 and AISC 341 [22,23] for use only in ordinary moment frames and provides a fully rigid interconnection between the beam and column. In contrast, the RBS connection is applicable to ordinary, intermediate, and special moment frames because of its excellent ductility. AISC 358 [33] provides the requirements and dimensions ($a \cong (0.50 \sim 0.75)b_f$, $b \cong (0.65 \sim 0.85)d_b$, and $c \leq 0.25b_f$) for the radius-cut section of the RBS connection to induce plastic hinges in the beam and to limit the moment and inelastic deformation developed at the face of the column. The RBS connections of the moment frames in the prototype building were designed as per AISC 358 [33] and were modeled using the reduced modeling approaches explained in Section 3.4. Equivalent effective widths b_{eq} and dimensions of the RBS connections for moment resisting frames are listed in Table 5. Figure 20 depicts an example of the RBS connection for G1 at 1-3F.

Table 5. Equivalent effective widths and dimensions of RBS connections for moment resisting frames.

Member	Floor	a [mm]	b [mm]	c [mm]	b_{eq} [mm]
	1-3F	155	390	70	197
Girder G1	4-6F	155	390	70	197
	7F	90	159	43	108
	1-3F	155	600	75	191
Girder G2	4-6F	180	460	75	188
	7F	120	280	50	125

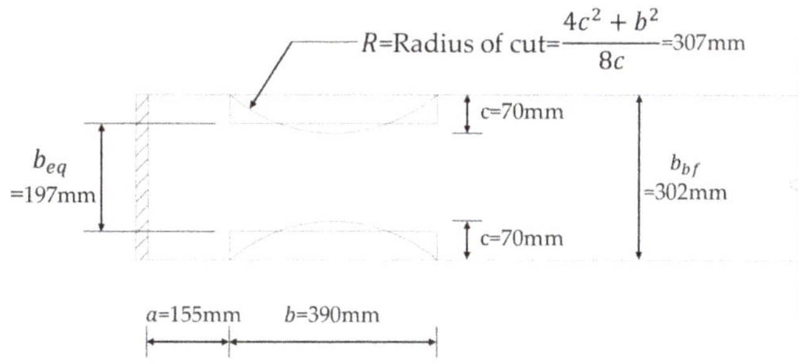

Figure 20. Example of RBS connection for G1 at 1-3F.

4.4. Energy-Based Approximate Analysis Based on Nonlinear Static Analysis

Uniformly distributed loading of the gravity load combination $\lambda(1.0DL + 0.25LL)$ was applied to the entire floor slab of the prototype buildings, individually designed with WUF-B and RBS connections. With the use of a force-controlled pushdown approach, the load factor λ was gradually increased until failure occurred. The column removal scenarios were performed for each column location specified in the GSA 2003 [2], as shown in Figure 18a. Figure 21 shows the load factor-vertical displacement curves at the removed column location. Failure of the structure was determined by the divergence in numerical analysis due to instability of the structure after successive yielding of the beam-column connections or the beam members adjacent to the removed column at all the levels. The yielding of the 1-D element was evaluated using the von Mises stress in the finite element program used in this study. Herein, the von Mises stress was obtained from the maximum bending stress and axial stress calculated through corresponding section forces (M, N).

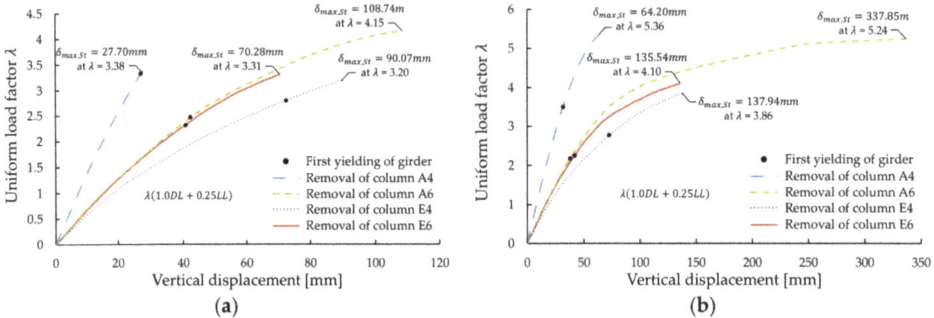

Figure 21. Load factor-vertical displacement from nonlinear static analyses for column removal scenario. (**a**) elicited response for the building with WUF-B connections; (**b**) corresponding response for the building with RBS connections.

In the column removal scenario, the maximum and minimum failure load factors of the building with WUF-B connections were λ_{A6} = 4.15 and λ_{E4} = 3.20 for the case where columns A6 and E4 were removed, respectively, and the corresponding displacements were δ_{A6} = 108.74 mm and δ_{E4} = 90.07 mm, as shown in Figure 21a. In all the scenarios, the flange yielding was observed for all the WUF-B connections located at the exterior of the roof floor level (Figure 22b). It was confirmed that the yielding of the flange preferentially occurred since the design redundancy of the beam members in the roof floor level was smaller than that of floors 1 through 6, as listed in Table 6. When columns A4, A6, and E6 were removed, plastic hinges increased due to the yielding of the WUF-B connections from the lower to the upper floor level until the failure load factor was reached. Correspondingly, the analysis was terminated due to the instability of the whole structure (Figure 22a). In the case of the removal of the column E4, the analysis was terminated since the structure was unstable because of the successive yielding of the beam members adjacent to the removed column at all the levels.

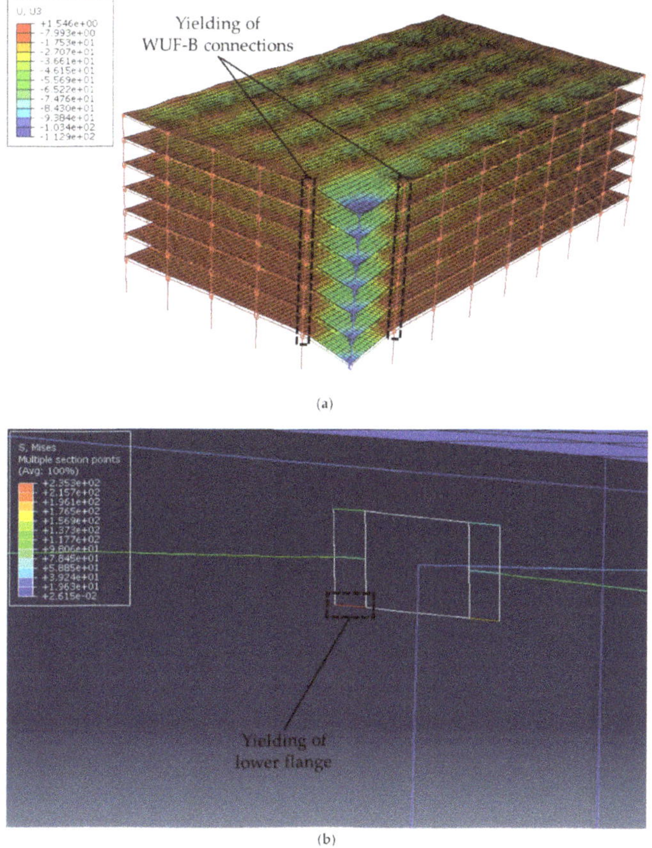

Figure 22. Elicited results from the building with WUF-B connections in accordance to the removal scenario of column A6. (**a**) deformation of structure for $\lambda = 4.15$; (**b**) von Mises stress for first yielding of the WUF-B connection at the roof floor.

Table 6. Design redundancy of girder members.

Floor	Redundancy
Floors 1-3	1.283
Floors 4-6	1.425
Roof Floor	1.131

The load-displacement curves and failure modes for the column removal scenario of the building with the RBS connections in Figures 21b and 23a were almost similar to the results of the WUF-B connections, and the yielding of the moment resisting frames located at the exterior of the roof floor level was observed. However, the maximum failure load factor was verified in the case of the removal of the column A4, while the yielding of the moment frames occurred within the reduced section of the RBS connections (Figure 23b). In the column removal scenario, the maximum and minimum failure load factors of the building with the RBS connection after the removal of columns A4 and E4 were $\lambda_{A4} = 5.36$ and $\lambda_{E4} = 3.86$, respectively, and the corresponding displacements were $\delta_{A4} = 64.20$ mm and $\delta_{E4} = 137.94$ mm. These results show that the load factor and vertical displacement increased by 20~58% and 53~211%, respectively, compared with the results of the WUF-B connections.

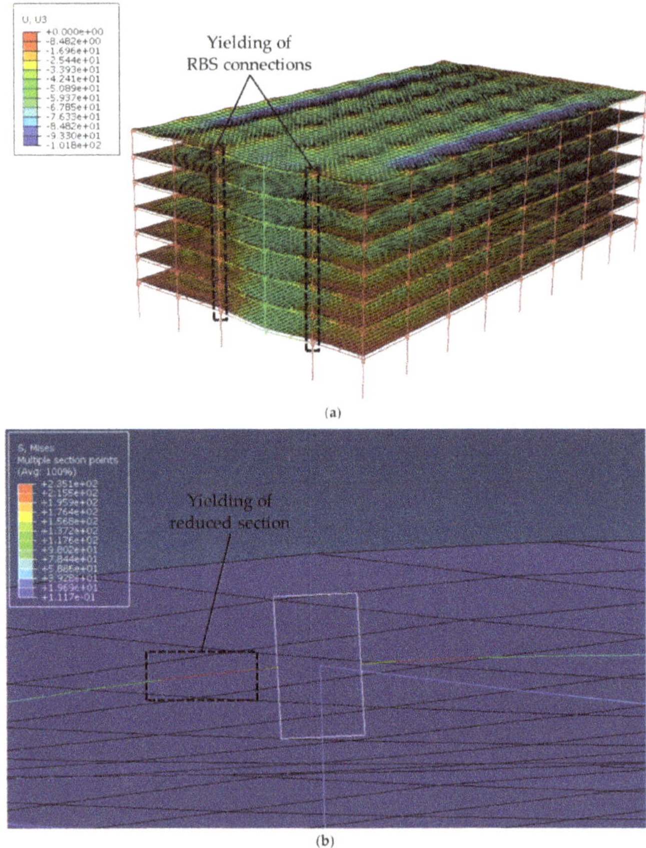

Figure 23. Elicited results from the building with RBS connections in accordance to the removal scenario of column A4. (**a**) deformation of structure for λ = 5.36; (**b**) von Mises stress for first yielding of the RBS connection at the roof floor.

Figure 24 shows the load factor-vertical displacement curves after the application of the energy-based approximate analysis introduced in Section 2.3 to the nonlinear static analysis results in Figure 21. These results, accounting for the dynamic effects associated with sudden column loss, can be used to evaluate the progressive collapse resistance performance using the load combination of the dynamic analysis. Table 7 shows the vertical displacements and the chord rotation under the load combination $1.0DL + 0.25LL$ specified in GSA 2003 [2]. The maximum vertical displacement δ_{GSA} = 40 mm of the two buildings was obtained from the elicited results of the removal scenario of column E4, while chord rotation was calculated to be equal to 0.0050 rad. These results showed that the chord rotation of these connections were within the range of 14~20% of the limit values (WUF-B = 0.025 rad and RBS = 0.035 rad) specified in GSA 2003 [2] and that the two buildings individually designed with different connections satisfied all of the collapse resistance performance criteria. It is evident that the composite slabs designed to have more than 3.3 times the required capacity improved the stiffness of the entire structure and induced very small bay deflection at the removed column. Therefore, it appears that the structures designed in this study will hardly fail due to progressive collapse.

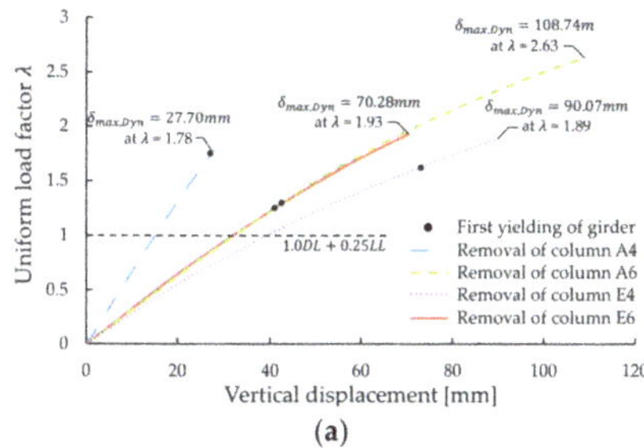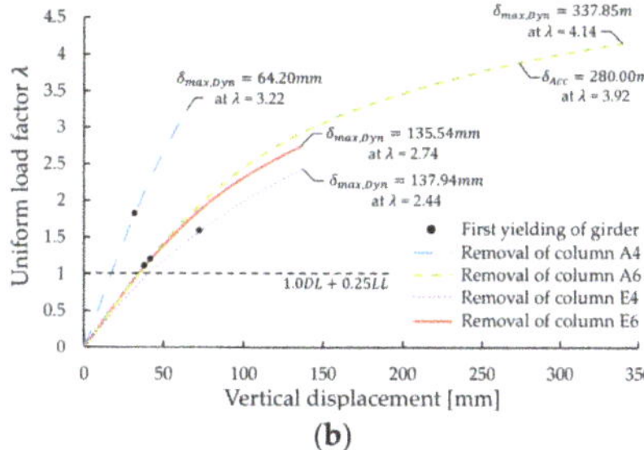

(a) **(b)**

Figure 24. Load factor-vertical displacement for column removal scenario using energy-based approximate analysis. (**a**) building with WUF-B connection; (**b**) building with RBS connection.

Table 7. Vertical displacement and chord rotation under the combination load of $1.0DL + 0.25LL$.

Removed Column	Building with WUF-B Connections		Building with RBS Connections	
	Vertical Displacement [mm]	Rotation [% rad]	Vertical Displacement [mm]	Rotation [% rad]
A4	14.59	0.0018	13.88	0.0017
A6	32.61	0.0041	38.58	0.0048
E4	39.92	0.0050	39.97	0.0050
E6	34.07	0.0038	34.43	0.0038

5. Evaluation of Structural Robustness and Sensitivity for Progressive Collapse

For the two buildings considered previously, the structural robustness for progressive collapse was evaluated using Equation (7) in Section 2.4. This study was based on the evaluation criteria for the WUF-B and RBS connections in GSA 2003 [2]. Therefore, required capacity D was defined as the load combination $G_0 = 1.0DL + 0.25LL$ that satisfies the maximum chord rotation, i.e., $\theta_{max} < \theta_{acc}$. Correspondingly, the maximum resistance C of the two buildings was defined as the maximum dynamic load $P_{Dyn,u}$ that satisfies the maximum chord rotation, i.e., $\theta_{max} < \theta_{acc}$. As shown in Table 8, from the analytical results of Section 4, in the column removal scenario of building with the WUF-B connections, since the maximum chord rotations do not exceed the limit chord rotation of 0.025 rad up to the point where the failure load factor is reached, the structural robustness r based on Equation (17) is determined by the failure load factor $r_{WUF-B} = 1.78$, corresponding to the removal of column A4. In the case of the building with the RBS connections, the failure load factor was calculated as $r_{RBS} = 3.92$ and decreased by 5.31% due to the maximum chord rotation that exceeded the allowable chord rotation limit of 0.035 rad. However, this value is 61% larger than the failure load factor estimated in the scenario of the sudden loss of column E4. Thus, the structural robustness is $r_{RBS} = 2.44$.

$$r = \frac{C}{D} = \frac{min P_{Dyn,u}}{1.0DL + 0.25LL} = min(\lambda_{max} \ for \ all \ column \ removal \ cases) \tag{17}$$

Table 8. Maximum chord rotation for $P_{Dyn,u}$.

Removed Column	Building with WUF-B Connections		Building with RBS Connections	
	Load Factor λ_{max}	Rotation [% rad]	Load Factor λ_{max}	Rotation [% rad]
A4	1.78	0.0035	3.22	0.0080
A6	2.63	0.0136	4.14 (3.92)	0.0422 (0.0350)
E4	1.89	0.0113	2.44	0.0172
E6	1.93	0.0078	2.74	0.0151

Figure 25 shows the comparison of the load factor and chord rotation relations in terms of structural robustness of the two buildings. The GSA 2003 [2] acceptance criterion indicates that the building with a smaller chord rotation has reliable collapse resistance. The building with the WUF-B connections was calculated to have a 59% smaller chord rotation than the other structure under a combination load of $1.0DL + 0.25LL$. This was because the WUF-B connections without defects of beams improved the stiffness of the structure in comparison with the RBS connections, resulting in a relatively small deformation. With regard to structural robustness, the building with the RBS connections was superior to the other building by 37% better performance than the counterpart.

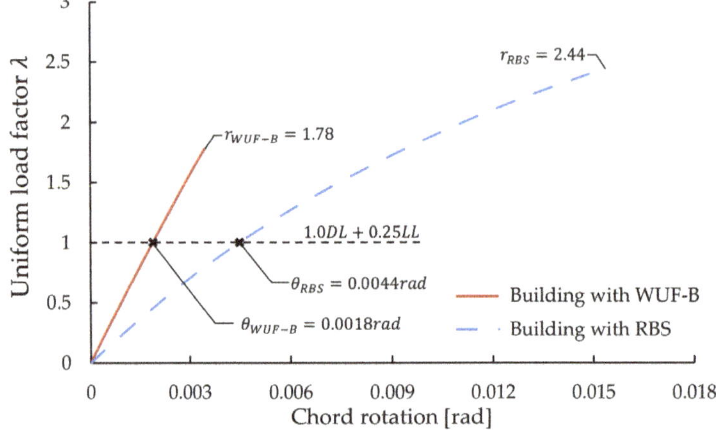

Figure 25. Comparison of load factor and chord rotation relations in terms of structural robustness of the two buildings.

The structural sensitivity S for the column removal scenario of the two buildings was evaluated based on Equation (8). The robustness index r_0 of the initial structure with an undamaged column was calculated by applying the nonlinear static analysis introduced previously. Correspondingly, all the seismically designed connections yielded in the roof floor, and the analysis was terminated due to the instability of the structure (Figure 26). The maximum failure load factor obtained from the analysis was $r_{0,WUF-B} = 5.15$ for the building with the WUF-B connections and $r_{0,RBS} = 5.34$ for the building with the RBS connections.

Consequently, the sensitivity index S of the two buildings estimated using the above results are presented in Table 9. For the abnormal loads specified in GSA 2003 [2], the building with the RBS connections was most sensitive to the removal of column E4, and corresponded to 33% of the initial bearing capacity. However, when column A4 was removed, the bearing capacity of the building with the WUF-B connections was reduced to 19% of the initial capacity. This is the most sensitive case in comparison with all the other cases. Therefore, the reduced bearing capacity of the building with the WUF-B connections was 14% higher than that of the building with the RBS connections, and it was thus considered to be a relatively more sensitive structure for progressive collapse.

Table 9. Evaluation of structural sensitivity.

Removed Column	Sensitivity Index S	
	Building with WUF-B Connections	Building with RBS Connections
A4	0.19	0.51
A6	0.39	0.72
E4	0.21	0.33
E6	0.22	0.40

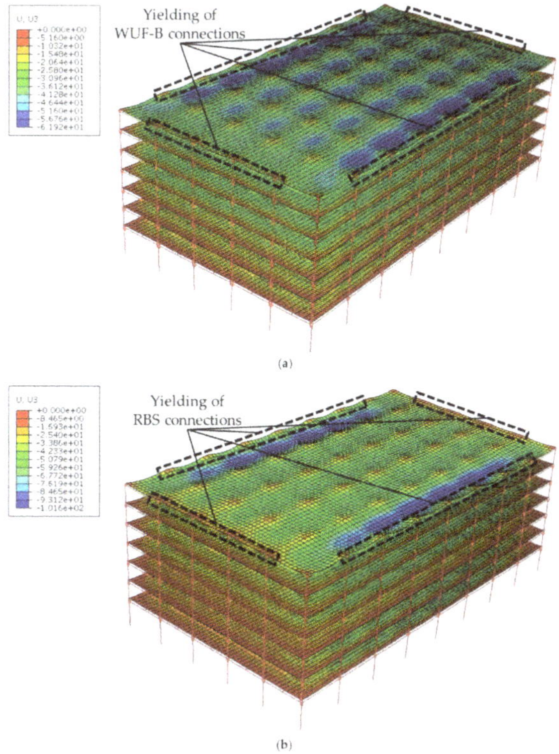

Figure 26. Failure mode of the initial structure. (**a**) building with WUF-B connection; (**b**) building with RBS connection.

6. Conclusions

This study evaluated the progressive collapse resistance performance of steel moment frame systems under column loss scenarios. For reliability, convenience, and efficiency of modeling and analysis, the reduced models consisted of 1-D and 2-D elements and the energy-based approximate analysis with dynamic effects were used. Based on these approaches, the structural robustness and sensitivity for progressive collapse was evaluated in column removal scenarios.

The comparison between the numerical results of the reduced models and the experimental results indicated that the computational models of the composite slab and the seismically designed connections employed in this study were suitable for the progressive collapse analysis. Additionally, they indicated that the Krawinkler model typically used in seismic research does not properly simulate the elicited behavior against gravity load. The progressive collapse resistance performance for the column removal scenario of the steel moment frames individually designed with the WUF-B and the RBS connections satisfied all the criteria specified in GSA 2003 [2]. These results were because the composite slabs designed to have more than 3.3 times the required capacity improved the stiffness of the entire structure. In the case of the building with the WUF-B connections, the reduction of the bearing capacity under the loss of a column at the short side of the plan was largest, but the structure was able to resist loads that were equal to 1.8 times the gravity load specified in the guide. Furthermore, the building with the RBS connections elicited the greatest reduction of bearing capacity when the interior column was removed, but it yielded a 14% higher bearing capacity than the building with the WUF-B connections and was considered to be less sensitive for progressive collapse. The results of this study indicated that the RBS connections showed better resistance performance against progressive collapse compared to the WUF-B connections, since the prototype structure was designed using a general steel structural plan and employed different connection details based on the standard.

In this study, the sensitivity for progressive collapse under sudden loss of a column was evaluated for seismically designed steel ordinary moment frames. Future research will focus on the analysis and comparison of the structural sensitivity of steel moment frames individually designed with different structural systems (e.g., intermediate moment frames and special moment frames) and of structures with multiple columns removed based on the guide.

Author Contributions: All authors contributed substantially to all aspects of this article.

Funding: This research received no external funding.

Conflicts of Interest: The authors declare no conflict of interest.

References

1. ASCE 7. *Minimum Design Loads for Buildings and Other Structures*; American Society of Civil Engineers: Reston, VA, USA, 2010.
2. GSA. *Progressive Collapse Analysis and Design Guidelines for New Federal Office Building and Major Modernization Project*; U.S. General Services Administration: Washington, DC, USA, 2003.
3. DoD (Department of Defense). *Design of Buildings to Resist Progressive Collapse*; Unified Facilities Criteria 4-023-03: Washington, DC, USA, 2009.
4. BS (British Standard) 5950. *Structural Use of Steelwork in Buildings-Part 1: Code of Practice for Design-Rolled and Welded Sections*; British Standard Institution: London, UK, 2000.
5. Korea Ministry of Land, Infrastructure and Transport. *Design Guidelines for Terror Prevention in Buildings*; Korea Ministry of Land, Infrastructure and Transport: Sejong, Korea, 2010.
6. Marjanishvili, S.M. Progressive Analysis Procedure for Progressive Collapse. *J. Perform. Constr. Facil.* **2004**, *18*, 79–85. [CrossRef]
7. Sadek, F.; Main, J.A.; Lew, H.S.; El-Tawil, S. Performance of Steel Moment Connections under a Column Removal Scenario. II: Analysis. *J. Struct. Eng.* **2013**, *139*, 108–119. [CrossRef]
8. Bao, Y.; Lew, H.S.; Kunnath, S.K. Modeling of Reinforced Concrete Assemblies under a Column Removal Scenario. *J. Struct. Eng.* **2014**, *140*, 04013026. [CrossRef]
9. Main, J.A. Composite Floor Systems under Column Loss: Collapse Resistance and Tie Force Requirements. *J. Struct. Eng.* **2014**, *140*, A4014003. [CrossRef]
10. Sadek, F.; El-Tawil, S.; Lew, H.S. Robustness of Composite Floor Systems with Shear Connections: Modeling, Simulation, and Evaluation. *J. Struct. Eng.* **2008**, *134*, 1717–1725. [CrossRef]
11. Alashker, Y.; El-Tawil, S.; Sadek, F. Progressive Collapse Resistance of Steel-Concrete Composite Floors. *J. Struct. Eng.* **2010**, *136*, 1187–1196. [CrossRef]
12. Izzuddin, B.A.; Vlassis, A.G.; Elghazouli, A.Y.; Nethercot, D.A. Progressive collapse of multi-storey buildings due to sudden column loss-Part I: Simplified assessment framework. *Eng. Struct.* **2008**, *30*, 1308–1318. [CrossRef]
13. Main, J.A.; Liu, J. Robustness of Prototype Steel Frame Buildings against Column Loss: Assessment and Comparisons. In Proceedings of the 2013 Structures Congress, Pittsburgh, PA, USA, 2–4 May 2013.
14. Bao, Y.; Main, J.A.; Noh, S.Y. Evaluation of Structural Robustness against Column Loss: Methodology and Application to RC Frame Buildings. *J. Struct. Eng.* **2017**, *143*, 04017066. [CrossRef]
15. Starossek, U.; Haberland, M. Approaches to measures of structural robustness. *J. Struct. Infrastruct. Eng.* **2011**, *7*, 625–631. [CrossRef]
16. Noh, S.Y.; Park, K.H.; Hong, S.C.; Lee, S.Y. Evaluation of Progressive Collapse Resistance of Steel Moment Frame with WUF-B Connection and Composite Slab using Equivalent Energy-based Static Analysis. *J. Arch. Inst. Korea Struct. Constr.* **2018**, *34*, 19–28.
17. KBC (Korean Building Code). *Chapter 3-Design Loads*; Korea Ministry of Land, Infrastructure and Transport: Sejong, Korea, 2016.
18. Elnashai, A.S.; Di Sarno, L. *Fundamentals of Earthquake Engineering*; John Wiley & Sons: West Sussex, UK, 2008; ISBN 978-0-470-02483-6.

19. Kim, G.D.; Huh, C.; Kim, J.H.; Yoon, M.H.; Lee, M.J.; Moon, T.S.; Kim, G.S.; Kim, D.J.; Kim, D.K. A Study on the Structural Behavior of the Composite Slabs with the Composite Metal Deckplate. *Mag. Korean Soc. Steel Constr.* **1996**, *8*, 273–282.
20. Hognestad, E. *A Study of Combined Bending and Axial Load in Reinforced Concrete Members*; Bulletin Series No. 399; University of Illinois Engineering Experimental Station: Urbana, IL, USA, 1951.
21. Mondal, T.G.; Prakash, S.S. Effect of Tension Stiffening on the Behaviour of Reinforced Concrete Circular Columns under Torsion. *Eng. Struct.* **2015**, *92*, 186–195. [CrossRef]
22. FEMA 350. *Recommended Seismic Design Criteria for New Steel Moment-Frame Buildings*; SAC Joint Venture and Federal Emergency Management Agency: Washington, DC, USA, 2000.
23. AISC 341. *Seismic Provisions for Structural Steel Buildings*; American Institute of Steel Construction: Chicago, IL, USA, 2016.
24. FEMA 355F. *State of the Art Report on Performance Prediction and Evaluation of Steel Moment-Frame Buildings*; SAC Joint Venture and Federal Emergency Management Agency: Washington, DC, USA, 2000.
25. Krawinkler, H. Shear Design of Steel Frame Joints. *Eng. J.* **1978**, *15*, 82–91.
26. Main, J.A.; Sadek, F. Modeling and Analysis of Single-Plate Shear Connections under Column Loss. *J. Struct. Eng.* **2014**, *140*, 04013070. [CrossRef]
27. AISC 360. *Specification for Structural Steel Buildings*; American Institute of Steel Construction: Chicago, IL, USA, 2016.
28. FEMA 355D. *State of the Art Report on Connection Performance*; SAC Joint Venture and Federal Emergency Management Agency: Washington, DC, USA, 2000.
29. Lee, C.H. Lateral Stiffness of Steel Moment Frames Having Dogbone Seismic Connection. *J. Comput. Struct. Eng. Inst. Korea* **2002**, *15*, 639–648.
30. KSSDC. *Korea Steel Structure Design Code-Load and Resistance Factored Design*; Korea Ministry of Land, Infrastructure and Transport: Sejong, Korea, 2016.
31. Astaneh-Asl, A.; Madsen, E.A.; Noble, C.; Jung, R.; McCallen, D.B.; Hoehler, M.S.; Li, W.; Hwa, R. *Use of Catenary Cables to Prevent Progressive Collapse of Buildings*; Report No. UCB/CEE-STEEL-2001/02; University of California: Berkeley, CA, USA, 2001.
32. Kim, S.W.; Lee, C.H.; Lee, K.K. Effects of Composite Floor Slab on Progressive Collapse Resistance of Steel Moment Frames. *J. Arch. Inst. Korea Struct. Constr.* **2014**, *30*, 3–10.
33. AISC 358. *Prequalified Connections for Special and Intermediate Steel Moment Frames for Seismic Applications*; American Institute of Steel Construction: Chicago, IL, USA, 2016.

Article

Post-Earthquake Restoration Simulation Model for Water Supply Networks

Jeongwook Choi [1], Do Guen Yoo [2] and Doosun Kang [3,*]

[1] Department of Civil Engineering, Kyung Hee University, Yongin-si, Gyeonggi-do 17104, Korea; cjw4859@naver.com
[2] Department of Civil Engineering, The University of Suwon, Hwaseong-si, Gyeonggi-do 18323, Korea; dgyoo411@suwon.ac.kr
[3] Department of Civil Engineering, Kyung Hee University, 1732 Deogyeong-daero, Giheung-gu, Yongin-si, Gyeonggi-do 17104, Korea
* Correspondence: doosunkang@khu.ac.kr; Tel.: +82-31-201-2513

Received: 7 September 2018; Accepted: 8 October 2018; Published: 10 October 2018

Abstract: A computer-based simulation model was developed to quantify the seismic damage that may occur in water supply networks and to suggest restoration strategies after such events. The model was designed to produce probabilistic seismic events and determine the structural damage of facilities. Then, the model numerically quantifies the system restoration rate over time by connecting it with a hydraulic analysis solver. The model intends to propose superb restoration plans by performing sensitivity analyses using several restoration scenarios. The developed model was applied to an actual metropolitan waterworks system currently operating in South Korea and successfully suggested the most efficient restoration approaches (given seismic damage) to minimize the complete recovery time and suspension of water service. It is expected that the proposed model can be utilized as a decision-making tool to determine prompt system recovery plans and restoration priorities in the case of an actual seismic hazard that may occur in water supply networks.

Keywords: seismic damage; simulation model; system restoration; water supply networks

---

1. Introduction

A water supply network is a human-made infrastructure supplying drinking water. If it partially or completely loses water supply function because of an unforeseen disaster, massive social and economic damage may result. Therefore, consistent maintenance to prevent such a setback in water service is critical. In general, there are several factors that inhibit such function, including the physical deterioration of network facilities (pipelines in particular), human-made hazards resulting in pipeline damage during ground construction, and natural disasters such as earthquakes and floods. Among them, natural disasters are beyond human control, and the extent of damage can be catastrophic. Thus, natural disasters require preemptive maintenance compared to other factors. Among other types of natural disasters, earthquakes have the most significant and direct impact on the water supply network because most components are buried underground.

In general, the approaches used to reduce seismic damage can be largely divided into two branches. The first one is reinforcing system durability to prevent damage or to minimize the immediate deterioration of a system. Because it is difficult to predict the exact timing of an earthquake event, it would be beneficial to develop an earthquake-resistant system through advanced reinforcement or system design [1,2]. The second approach is to enhance the resiliency of the system by post-earthquake restoration. Because the natural disasters are not entirely preventable, improving system resiliency in terms of post-event restoration is essential to avoid long-term losses.

Several representative seismic damage quantification models of water supply systems include (1) HAZUS developed by the Federal Emergency Management Agency [3]; (2) MAEviz [4] developed at the Mid-America Earthquake Center; (3) GIRAFFE (Graphical Iterative Response Analysis of Flow Following Earthquakes) introduced by Shi et al. [5]; and (4) REVAS.NET (Reliability EVAluation model for seismic hazard for water supply NETwork) created by Yoo [6]. HAZUS and MAEviz are computer models that quantify the damage occurred in social infrastructures in terms of societal and economic value and do not include restoration measures nor simulation of system performance. GIRAFFE and REVAS.NET present the extent of earthquake damage on water supply networks by quantifying the water supply capacity or system serviceability. These two models are capable of hydraulic analysis via linkage to a hydraulic solver; however, GIRAFFE lacks of hydraulic module to depict a detailed hydraulic behavior of the damaged water networks and REVAS. NET is mainly intended to estimate system reliability against seismic damage. They are not equipped to simulate the detailed post-earthquake restoration process.

Other studies that quantified the earthquake damage to water supply networks include Adachi and Ellingwood [7], and Yoo et al. [8]. Adachi and Ellingwood [7] claimed that water pipe networks and power systems should be evaluated jointly for more accurate assessment of system serviceability when seismic damage occurred to water pipe networks. Yoo et al. [8] quantified the serviceability of water supply system during an earthquake and proposed optimal design measures to reinforce the system. Those studies quantified the serviceability of water supply systems under seismic damage, but no applications for the post-event restoration simulation were provided.

In addition, Davenport et al. [9], Tabucchi et al. [10], and Luna et al. [11] published studies related to the reinforcement of post-earthquake system stability. These studies suggested various methods of restoring water pipe networks to a normal (or close to normal) state after an earthquake, but did not include any hydraulic analysis modules to simulate water supplies within an actual water pipe network in the event of a disaster. Recently, Tabucchi et al. [12] modeled a series of processes where various types of recovery support staffs were assigned to repair the system, and evaluated hydraulic performance of a system using a GIRAFFE model. The developed model was verified by simulating the 1994 Northridge earthquake that occurred in Los Angeles, California. The study demonstrated that the sequence and phases of a restoration process can be estimated accurately through modeling but did not attempt to propose and evaluate diverse recovery measures.

Lately, Davis et al. [13], Davis [14], and Davis and Giovinazzi [15] subdivided the serviceability of a water pipe network into several factors, such as "water delivery," "quality," "quantity," "fire protection," and "functionality." They showed the quantified recovery phases of each item and analyzed the phases by comparing them with the case of the Northridge earthquake recovery.

Most recently, Klise et al. [16,17] introduced the Water Network Tool for Resilience (WNTR), a new open source Python package designed to help water utilities investigating water service availability (WSA) and recovery time based on earthquake magnitude, location, and repair strategy. Their study mainly focused on introducing the developed model, yet not conducted simulations to compare and analyze diverse restoration strategies suitable to a damaged network.

In summary, the previous studies have limitations in that they lack hydraulic simulations required to describe the hydraulic behavior of damaged water networks and post-event restoration simulations for prompt system recovery. This study focuses on efforts to recover water system service after seismic damage in an efficient way and attempts to simulate the real-life restoration process. To that end, a computer simulation model was developed to simulate diverse damage recovery approaches in a virtual and risk-free environment. The developed model was applied to an actual metropolitan water supply network operating in South Korea and was used to evaluate various recovery strategies and suggest the most efficient scheme for the application network. This paper consists of four sections. Section 2 covers methodologies such as the earthquake and recovery simulations of the model, Section 3 introduces the applied network and the scenario-setting for a sensitivity analysis and compares and analyzes the results of each scenario. Lastly, Section 4 contains the conclusions of the study and provides future research directions.

2. Methodology

2.1. Model Overview

The proposed model was developed using MATLAB [18] in conjunction with a hydraulic solver, EPANET2 [19]. EPANET2 was used to perform hydraulic analyses of water pipe networks, and MATLAB was used to simulate seismic damage of the network and the post-event system restoration processes. The water network components simulated in the model include the pipelines, water storage tanks, and pump facilities. Note that the seismic damages to valves and other facilities were excluded from our simulation. The developed model has six simulation steps as shown in Figure 1, and a brief explanation of each step is provided below.

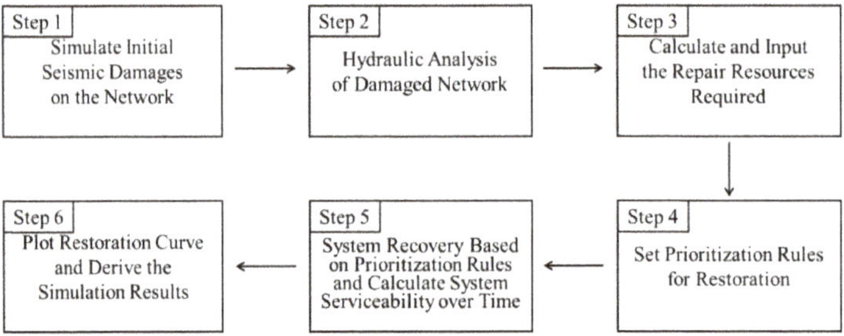

Figure 1. Model simulation flow chart.

Step 1. Simulate the initial seismic damage of the network immediately after the earthquake. The damage state of each component (pipelines, pumps, and tanks) is determined based on the seismic location and magnitude.

Step 2. The damage state of the components from Step 1 is provided as the input data of the hydraulic solver (EPANET2), and the hydraulic analysis is conducted to estimate the water supplying condition prior to a system restoration.

Step 3. After the water supply condition is estimated from Step 2, the required recovery resources are calculated and input. Recovery resources include repair crews, equipment, and restoration materials. In this study, only repair personnel were simulated.

Step 4. After entering the available recovery resources, the user sets the recovery priority rules. For example, a pipe carrying higher flow gets priority, or a pipe near a water source (water treatment plant, reservoir, or tank) would be fixed first.

Step 5. The system restoration simulation starts according to the pre-determined recovery plans from Step 4. The system restoration process is monitored over time, and the results are saved in the database. The restoration process is continued until the urgent recovery is completed. Here, the urgent recovery means the repair of the broken pipes. The detection and repair of the leaking pipes occurs subsequently and is not simulated in this study.

Step 6. Once urgent recovery is completed, the recovery strategies are quantitatively evaluated based on various simulation results, such as system restoration curves and recovery crew activity statistics.

2.2. Model Simulation Process

2.2.1. Earthquake Simulation and Damage Determination

(1) Determining Damage to Tanks and Pumps

The model determines the seismic damage to each component by considering the attenuation of the seismic waves, which is estimated based on the distance from the earthquake epicenter and magnitude of the earthquake. The earthquake occurrence time and the location/magnitude are randomly generated and applied. Damage to facilities is caused by the seismic waves after an earthquake occurs, and such waves imply the movement of vibrations from the earthquake, which generally decrease as the distance from the epicenter increases. Such seismic waves affect the estimation of peak ground acceleration (PGA). PGA is an index of how strongly the ground shakes, which directly influences the seismic damage condition. To estimate the PGA, a formula suggested by Baag et al. [20] was implemented as seen in Equation (1).

$$\ln PGA = 0.40 + 1.2M - 0.76 \ln R - 0.0094R \tag{1}$$

Here, PGA = peak ground acceleration (cm/s^2), M = magnitude of earthquake, and R = distance (km) from the epicenter assuming a focal depth of 10 km.

The facility damage caused by an earthquake can be estimated using the fragility curve. The fragility curve shows the probability that the extent of a facility's damage is beyond a certain level depending on the PGA. The fragility curves presented in the Seismic Fragility Formulations for Water Systems Part 1 Guidelines [21] were applied to determine the damage status of tanks and pumps, as shown in Figure 2. The y-axis indicates the probability that the facility incurs damage when the PGA value of the seismic wave corresponds to the *x*-axis. The "on-ground anchored concrete tank" and a "small-scale plant" were assumed for the tank and pump types in our model, respectively.

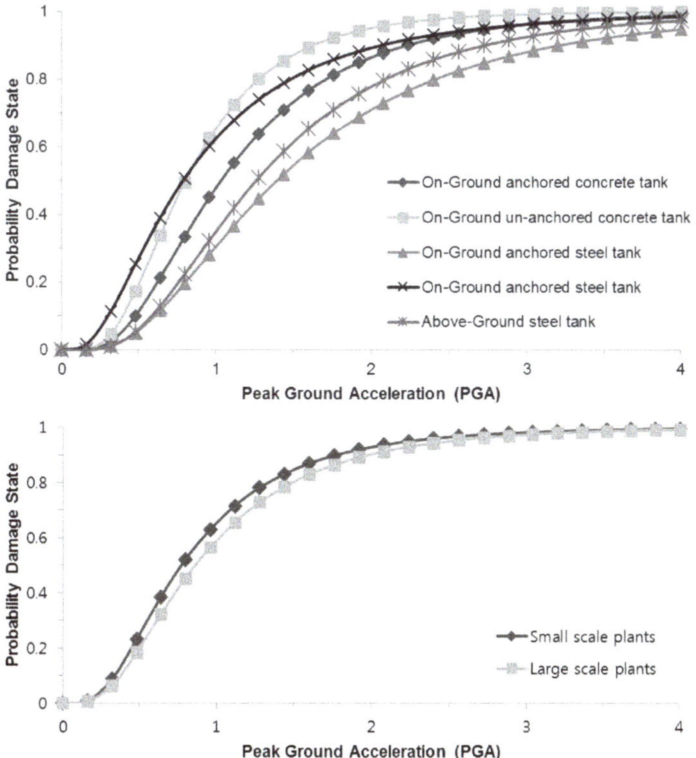

Figure 2. Fragility curves for tank (**top**) and pump (**bottom**) (data from ALA, 2001 [21]).

The model classifies the damage of the tank and pump into two states: damaged (function stopped) or normal. The facility damage estimation steps are as follows.

(1) Calculate the PGA of seismic waves that reached each facility using the seismic wave attenuation equation (Equation (1)).
(2) Calculate the damage probability of each facility using the obtained PGA value and the fragility curves (Figure 2).
(3) Generate a random number between 0 and 1 for each facility.
(4) If the random number is smaller than the damage probability derived in step 2, the facility is defined as "damaged", otherwise, it is "normal".

The repair time for the damaged facilities was estimated by a random number following a uniform distribution. Note the repair time of a tank is uniformly selected from 24 to 36 h, and that of a pump takes from 8 to 12 h.

(2) Determining Damage to Pipes

In general, damage to a pipeline can be determined based on the concept of repair rate (RR). The repair rate is defined as the number of repairs per unit length of a pipeline and is generally calculated by peak ground velocity (PGV) under seismic event. The American Lifelines Alliance (ALA) [21] suggested the repair rate of a pipe as expressed in Equation (2). The equation was derived based on 81 pipe failure data.

$$RR = K \times 0.00187 \times PGV \tag{2}$$

Here, RR = repair rate (no. of repairs/1000ft), K = modification factor

Isoyama [22] has provided various correction factors (C) representing the network information. The correction factors were estimated using the historical seismic data and represent four characteristics, such as pipe diameter, pipe material, topography and soil liquefaction. In this study, Equation (2) was adapted by implementing the correction factors (C) suggested by Isoyama [22] and finally expressed in Equation (3).

$$RR = C_1 \times C_2 \times C_3 \times C_4 \times 0.00187 \times PGV \tag{3}$$

Here, C_1, C_2, C_3 and C_4 represent the correction factors for the pipe diameter, pipe material, topography, and liquefaction, respectively.

To determine the pipe damage status, the Poisson distribution is widely utilized as noted by ALA [21]. To determine the probability of failure of an individual pipeline with a length of L, the Poisson distribution is expressed as:

$$P(x = k) = (\lambda L)^k e^{-\lambda L} / k! \tag{4}$$

Here, x is a random variable denotes the number of events (i.e., pipe breaks) and λ is an average rate of breaks (or repairs) per unit length of pipe (i.e., RR of a pipe).

Since a single break in a pipe takes the entire pipeline out of service, the probability of break of an individual pipeline can be easily calculated by setting k = 0 in Equation (4) and subtracting it from 1. Thus, the probability of pipe failure can be calculated as expressed in Equation (5).

$$P_b = 1 - e^{-RR \times L} \tag{5}$$

Here, P_b = probability of pipe breakage and L = pipe length.

In the model, the seismic damage to pipelines is divided into breakage and leakage. If a pipe is broken, the water flow through the pipe is completely suspended, while a pipe with leaks still conveys water with potential loss of flow and pressure. Here, the pipe damage condition was simulated through the following procedure:

(1) Calculate the PGV value of the seismic wave arriving at each pipe.

(2) Calculate the repair rate of the pipes using Equation (3).

(3) Estimate the distance between repair points of each pipe (L1 = 1/RR) and compare it with the actual pipe length (L2) to determine whether earthquake damage has occurred in the pipe. That is, if L1 is larger than L2, no damage occurs in the pipe.

(4) For the damaged pipe with L1 < L2, a random number between 0 and 1 is generated and compared with the pipe failure probability (Equation (5)). If the random number is smaller than the breakage probability, the pipe is considered to be broken, otherwise it is leaking.

(5) Using this procedure, all the pipes in the network are classified into either 'no damage', 'leakage' or 'breakage'.

The repair time of a broken pipe was estimated using the empirical formula suggested by Chang et al. [23] as expressed in Equation (6). The equation was derived by the Korea Water Resources Corporation (K-water) based on a number of field data. Although the repair time may be affected by the accessibility to the repair site, it was found that the repair time is generally proportional to the pipe diameter with a preparation time (2 h) as shown in Equation (6).

$$t_b = \frac{D}{100} + 2 \tag{6}$$

Here, t_b = repair time for a broken pipe (h) and D = pipe diameter (mm).

2.2.2. Hydraulic Analysis and Recovery Simulation

After determining the seismic damage of each component, hydraulic analysis was conducted to quantitatively estimate the water supply ability of the network as the restoration progresses. The hydraulic simulation of each system component is described as follows.

(1) Hydraulic Simulation of Damaged Tanks and Pumps

As shown in Figure 3, the model controls the front and rear pipelines directly connected to the damaged components to deactivate the water inflow and outflow to simulate the damaged tanks and pumps in EPANET. If a tank is damaged, the model closes the front and rear pipelines of the tank as described in Figure 3a. If a pump is connected to the failed tank, the pump operation is also suspended in the hydraulic simulation. If a pump failed, the status of the pump and discharge pipeline is set to "closed" so that there would be no active inflow to the downstream tank as illustrated in Figure 3b. Once the repair of the damaged tank or pump is completed, the closed pipelines and pumps would be set to "open" to simulate recovery of the facility.

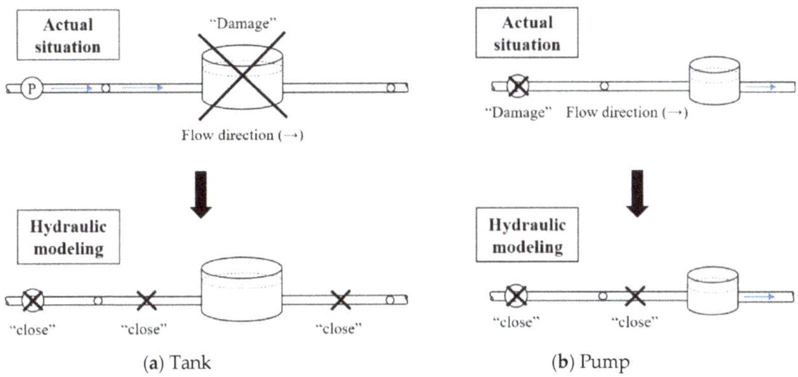

Figure 3. Hydraulic modeling of damaged tanks and pumps.

(2) Hydraulic Simulation of Damaged Pipes

Water loss through a damaged pipe (broken or leaking) is a pressure-dependent flow expressed as Equation (7), which was proposed by Puchovsky [24]. In the model, the EPANET emitter option was utilized to calculate the water loss from the damaged pipes. Note the emitter option is assigned at a node in the EPANET simulation.

$$Q = C_d P^{0.5} \tag{7}$$

Here, Q = water loss through leaks/breaks; C_d = discharge coefficient or emitter coefficient in EPANET ($= \left(\frac{2g}{\gamma_w}\right)^{0.5} A$, in which g = gravitational acceleration, γ_w = specific weight of water, A = opening area of the damaged pipe), and P = operational pressure. Note the total opening area (A) of the leaks is assumed to be equivalent to 10% of the entire cross-sectional pipe area.

For the damaged pipelines, the discharge coefficient (C_d) is calculated first. If a pipe is considered to be broken, C_d is assigned to the upper node of the flow direction. Then, the pipe status is set to "closed" and it does not function. Meanwhile, a pipe tagged as a leak may partially lose its function but still conveys water, and C_d is assigned to the downstream node of the flow direction.

The repair of broken pipes begins after all the information related to damaged pipes is provided in the hydraulic analysis model (EPANET). Figure 4 illustrates the recovery steps of a broken pipe. Note that we assumed that shut-off valves exist at both ends of the damaged pipeline. Step 1 indicates that the water loss through the break is allocated to the upstream node for the hydraulic simulation. In step 2, the water loss through the broken pipe is stopped by closing the valves, and pipe repair begins. In the hydraulic simulation, the discharge coefficient (C_d) assigned at the upper node is set to zero to stop the water loss, and the pipe status is set to closed. After the pipe is isolated, the demand of nodes connected to the isolated pipe is modified to reflect water supply suspension during the pipe repair. In the model, the demands of both end nodes of the isolated pipe are reduced by the degrees of node (DoN), which is defined as the number of links that a node contains. As an example, the water supply to the upstream and downstream nodes are reduced by 50% respectively, since 1/DoN = 1/2 (①) as shown in Figure 4. In step 3, the pipe repair is completed, and the pipe status is set to "open" to enable water flow through the pipeline (②). Finally, the reduced water supply at the connected nodes recovers to the initial value (③).

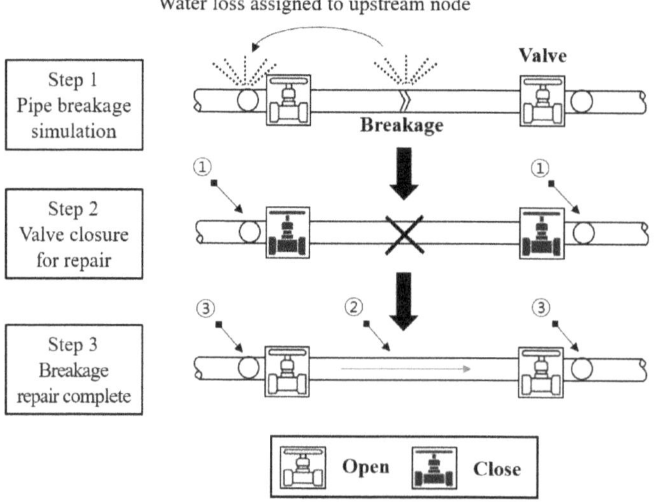

Figure 4. Pipe breakage repair simulation steps.

2.2.3. Quantification of System Recovery

(1) Hydraulic Analysis under Pressure-Deficient Condition

The developed model used a quasi-PDA (quasi-pressure driven analysis) for hydraulic simulation of the seismic-damaged network in which a pressure-deficient condition may occur. The quasi-PDA treats the negative pressure by repeatedly performing DDA (demand driven analysis) and the procedure is briefly described as follows. First, given seismic-damaged condition, initial EPANET analysis (i.e., DDA) is performed. If negative pressures occur in the network, the base demands of the negative pressure nodes are set to zero (meaning that no water can be supplied to the nodes). DDA is then repeated. This process is repeated until no negative pressure appears in the network. A detail procedure of the quasi-PDA for the treatment of pressure-deficient condition is described in Yoo et al. [1].

(2) Calculation of System Serviceability

Once seismic damage occurs in a water network, the overall water supply capacity is degraded due to water losses through damaged pipes resulting in insufficient water pressure at demand nodes. To quantify system performance and recovery over time, an indicator called "system serviceability" was utilized. The system serviceability (S_s) represents the ability of a network to supply water and is calculated as the ratio of the actual supply to the required demand of the entire network. The system serviceability (S_s) can be used as an index to evaluate the water supply capacity of the system under a seismic damage and is calculated as shown in Equation (8) (Shi [25]; Wang [26]).

$$S_s = \frac{\sum_{i=1}^{N} Q_{avl,i}}{\sum_{i=1}^{N} Q_{req,i}} \tag{8}$$

Here, S_s = system serviceability index, $Q_{avl,\,i}$ = available (or serviceable) demand at node i, $Q_{req,\,i}$ = required demand at node i, and N = total number of nodes.

The available nodal demand (Q_{avl}) is estimated based on the nodal pressure as expressed in Equation (9). If the water pressure at a node satisfies the minimum required pressure head ($P_i \geq P_{min}$), then the required demand is fully supplied. If the water pressure at a node does not reach the minimum required pressure head ($0 < P_i < P_{min}$), the required demand is partially supplied depending on the nodal pressure. Lastly, no water can be supplied for a node with zero pressure.

$$Q_{avl,i} = \begin{cases} 0 & when\ P_i = 0 \\ Q_{req,i} \times \sqrt{\frac{P_i}{P_{min}}} & when\ 0 < P_i < P_{min} \\ Q_{req,i} & when\ P_i \geq P_{min} \end{cases} \tag{9}$$

Here, P_i = nodal pressure at node i and P_{min} = minimum required pressure head for full water supply.

(3) System Restoration Curve

A system restoration curve is generated by estimating and plotting the system serviceability index over time as facility restoration proceeds as shown in Figure 5. Before an earthquake occurs, it is assumed that the system serviceability equals 1.0 (Point A) and the required water demand is fully supplied. An earthquake occurs at time T_0, and the water serviceability drops to Point B. Subsequently, the water supply improves through restoration work, and the system state returns to normal at time T_1 (Point C). Let us call the restoration process following an [A → B → C] curve a basic strategy. If a more effective recovery approach is applied to the damaged network, the whole restoration curve might move from [A → B → C] to [A → B → D]. This implies that establishing an effective recovery strategy may help to reduce the restoration complete time ($T_1 → T_2$) and also improve system serviceability during the restoration phase. The shaded area above the restoration curve in Figure 5 reflects both the depth and duration of the water supply shortage and is quantified in time (hours). Smaller curve areas indicate more effective restorations. Therefore, the serviceability index (S_s) and the area of the restoration curve can be utilized as indicators for quantitative comparison of different recovery strategies.

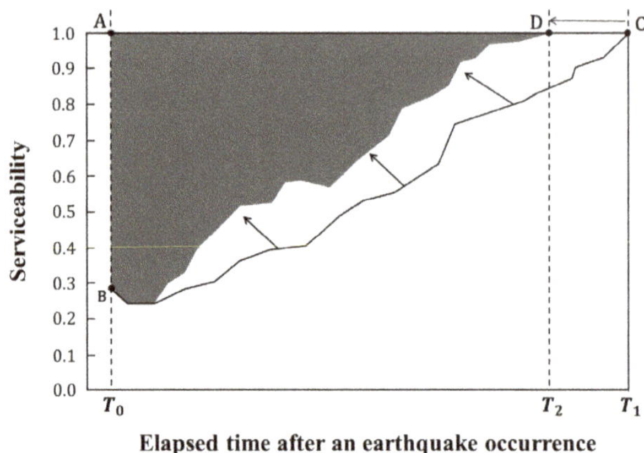

Figure 5. System restoration curve.

2.3. Summary of Assumptions and Simplifications

The assumptions used and simplifications made in the development and application of the model are summarized as follows. (1) All recovery personnel are available immediately after an earthquake and no shifts are considered. (2) The recovery equipment (crane, truck, etc.) and material usage (mechanical couplings, pipe sections, repair champs, etc.) were not considered in the simulation, only recovery personnel activity was simulated. (3) The application of the model is limited to the water pipe network; thus, interconnections with other lifeline networks (electric power, transportation, telecommunication, etc.) were not considered. (4) The transportation speed limit of the recovery personnel is set to 10 km/h by assuming that road conditions are not normal due to seismic damage. (5) The recovery simulation only considered the repairs of broken pipes. Locating and repairing the leaks will be more challenging and time-consuming compared to breaks and were excluded from the simulation. As a result, the system serviceability may not reach 100% even after the emergency repairs are completed due to water loss through unrepaired leaks. (6) Each pipeline has two shutoff valves at each end; thus, the pipe isolation for repair work only affects the customers connected to the isolated pipe.

3. Applications and Results

3.1. Application Network

As a real-life demonstration, the developed model was applied to a P-city water supply network that is currently operating in South Korea (Figure 6). The study network consists of one water treatment plant (WTP), four ground tanks, four pumping stations, 8949 pipelines, 8381 nodes, and 18 in-line pumps. The diameter of the pipelines ranges from a minimum of 25 mm to a maximum of 1100 mm, and pipes with diameters of less than 100 mm account for 51.4% of the total system. The four large pumping stations lift water to the four high-elevation ground tanks, from which water is fed by gravity. The service area is primary residential and spans approximately 32 km by 34 km in west/east and north/south directions, respectively. It is composed of five administrative districts as seen in Figure 6.

The MS area receives water directly from the WTP, while four other areas (JS, BW, GT and TH) are served from the ground storage tanks. Note that the total water demand is about 77,910 m^3/day.

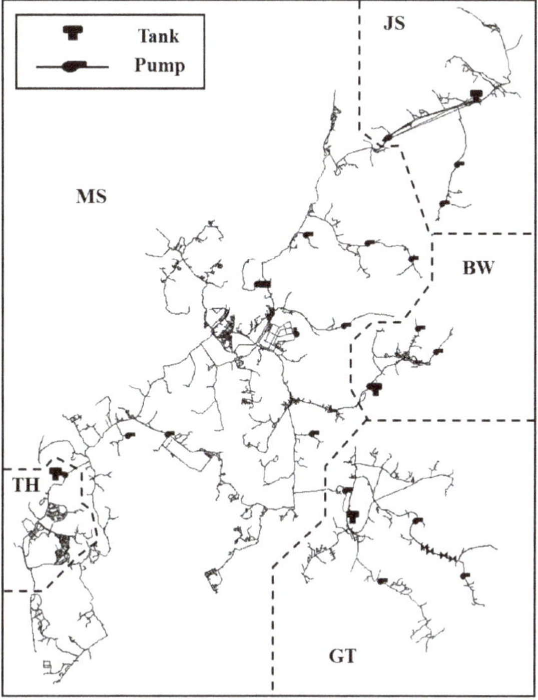

Figure 6. P-city water supply network layout.

3.2. Recovery Scenarios

First, a virtual earthquake was simulated near the study area. The epicenter was located 16 km east of the network center with a depth of 10 km and an assumed magnitude of 5.0 (M5.0). The system restoration performance can vary greatly depending on the repair strategies. Here, diverse recovery rules were applied and evaluated using the model to deduce the most efficient recovery strategy applicable for the study network. The applied six recovery strategies (recovery priorities) are summarized in Table 1. First, the cases were mainly categorized depending on whether a network is divided into repair zones. The network zoning was applied to possibly shorten the repair team's travel distance (the cases with the letter 'B'). That is, the study network was grouped into five zones that correspond to the five administrative districts shown in Figure 6, and repair teams were deployed based on the repair zones. If the repair work is completed in a designated area, the repair team can travel to support other areas where the repair is still in progress. For cases without zoning (the cases with the letter 'A'), all the repair teams would travel the entire network based on the repair priority without being allocated to a specific zone. Next, three specific recovery rules were suggested. Rule 1 states that the pipes delivering higher flows (generally those with a larger diameter) get the higher priority for repair. That is, the main transmission pipelines delivering higher flows, if damaged, will cause water suspension to a wide range of demand areas and need to be repaired first. Rule 2 indicates that the pipelines near water sources (WTP, ground tanks) should get priority. If the pipes near the water sources are damaged, there will be significant water loss in downstream areas, thus should be repaired first. Rule 3 suggests that the repair crews should move to the nearest repair location from the current repair-completed position, regardless of the flow rate or the distance from the sources. Rule 3 is likely to shorten the travel distance of the repair crews and consequently reduce the system recovery completion time. The recovery simulation was carried out hourly, the system restoration status was quantified, and the repair crew activity was recorded over time. For the study network, 50 repair personnel were available, and they were grouped into 10 teams (i.e., five personnel for each team). Note that for the cases with zoning (Cases B1~B3), two repair teams were allocated to each repair zone.

Table 1. System restoration strategies.

Case	Zoning	Rule	Description
A1	No	1	Pipes carrying higher water flow get higher repair priority
A2	No	2	Pipes closer to water sources get higher repair priority
A3	No	3	Pipes nearest to a current repair point get priority
B1	Yes	1	Pipes carrying higher water flow get higher priority within a zone
B2	Yes	2	Pipes closer to water sources get higher priority within a zone
B3	Yes	3	Pipes nearest to a current repair point get priority within a zone

3.3. Simulation Results

3.3.1. Comparison of Restoration Strategies

First, the restoration curves for each strategy were compared, as shown in Figure 7. Note the curves in general increase as the recovery progresses. However, at some time interval, the serviceability may decrease due to pipe isolation required for repair work. A sharp increase in serviceability was observed for all cases around 24–32 h of repair because the damaged tanks in BW, GT, TH areas were completely repaired around that time span. By resuming the water supply from the repaired ground tanks, the system serviceability increased significantly, indicating that the ground storage tanks play a significant role in the water supply of the studied network. We also observed that the system serviceability never reaches 100% since pipe leaks were not repaired. Note that the repair completion times are different for individual strategies. Comparing the six recovery scenarios, Case A1 seems the most efficient with the highest serviceability over time, while it takes the longest time to complete the repairs. Case A1 suggests that high flow pipes be repaired first regardless of the travel distance. Usually, when a main transmission line is damaged, the water supply is greatly affected. Thus, it is beneficial to repair such a pipeline first. The simulation results also reveal that network zoning is beneficial to reduce the repair completion time; that is, the repair completion times of B1~B3 are commonly shorter than those of A1~A3 as seen in Figure 7.

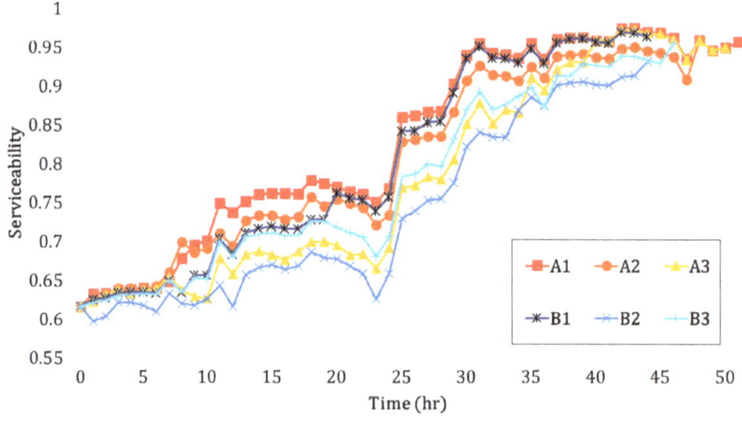

Figure 7. System restoration curves of six recovery strategies.

The system recovery curve allows intuitive and visual comparisons of the cases, but is not appropriate for quantitative comparisons. The curve area of each case was calculated for numerical comparisons. The curve area indicates the upper part of the restoration curve, and a smaller value denotes higher serviceability and shorter restoration times. The curve areas of each case are summarized in Table 2. Among the six applied cases, Case A1 showed the lowest curve area followed by Case B1. The results confirm that repairing the main pipelines first is advantageous in terms of restoration efficiency. Another important factor in evaluating the restoration strategy is to compare the recovery completion time, since delayed repairs may result in large socioeconomic losses. The total recovery time was measured as the time when the repairs of the broken pipes, damaged tanks and pumps were completed. Note the repair of pipe leaks was excluded in this study. As listed in Table 2, Case B1 was the quickest for completing the recovery. Overall, the cases with zoning (B1~B3) outperform the cases without zoning (A1~A3), showing a 15% reduction in total repair time.

Table 2. Restoration curve area and repair completion time.

Case	Curve Area (h)	Completion Time (h)
A1	8.9	53
A2	9.8	52
A3	11.1	53
B1	9.3	45
B2	12.0	46
B3	10.7	48

Finally, it would be better to determine the most efficient restoration strategy by considering both the curve area and total repair time. A rank was assigned to each recovery strategy using the two evaluation criteria separately, and the results are summarized in Table 3. Case B1 was selected as the most efficient recovery strategy for the studied network (second best for curve area and the best for repair time). Therefore, in case of the study network, the most efficient restoration strategy was to dispatch the repair teams for each zone and primarily repair the main transmission lines.

Table 3. Restoration total rank.

Case	Rank for Curve Area	Rank for Repair Time	Total Rank Sum
A1	1	5	6
A2	3	4	7
A3	5	5	10
B1	2	1	3
B2	6	2	8
B3	4	3	7
Best case/Total rank sum		Case B1/3	

3.3.2. Spatiotemporal Restoration Pattern

The restoration curve was used to understand the system-wide average recovery progress over time. In addition, it is required to monitor the spatial recovery conditions over the network as the restoration work progresses over time. Figure 8 shows the spatiotemporal distribution of system-wide serviceability illustrating the evolution of system restoration. Note that the system recovery simulation was performed by applying Case B1, which was determined to be the best recovery strategy in the previous analysis. Analyzing the snapshots, users can identify the locations of facility damage, the repairs in progress, and current conditions of water supply in an intuitive way. Figure 8a shows an initial damage state immediately after the earthquake. Among the four ground tanks, three (BW, GT, TH with 'X' mark) were damaged; thus, the downstream areas being supplied by the damaged tanks have significant effects of almost zero serviceability indicating no water to be supplied. Although the

JS tank still operates without damage, the JS area is significantly damaged since the area is relatively close to the epicenter. After 24 h, TH tank was repaired, and the serviceability increased to 0.756, but two other tanks were still in repair (Figure 8b). After 36 h, all the damaged tanks were completely repaired, and the system recovered a function higher than 90% (Figure 8c). Finally, urgent repairs were completed after 44 h, and the serviceability reached 0.963 (Figure 8d). Note the system serviceability did not reach 100% because of unrepaired leaks throughout the network.

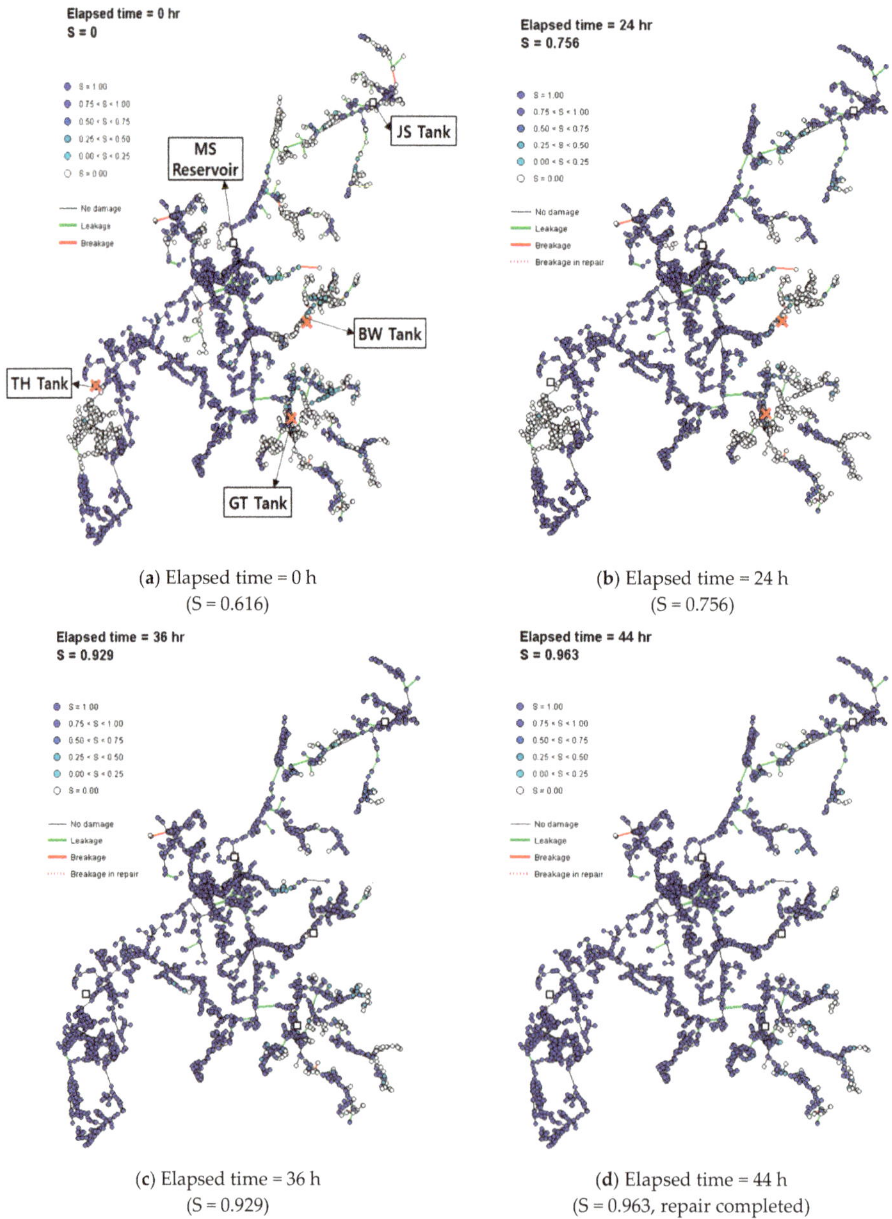

(**a**) Elapsed time = 0 h
(S = 0.616)

(**b**) Elapsed time = 24 h
(S = 0.756)

(**c**) Elapsed time = 36 h
(S = 0.929)

(**d**) Elapsed time = 44 h
(S = 0.963, repair completed)

Figure 8. Spatiotemporal distribution of system serviceability over time.

3.3.3. Repair Crew Activity

Estimating the repair team activity (including total number of repaired pipes, hourly work status, accumulated travel distance, and total working hours) would be helpful for both comparing recovery strategies and for estimating the required task force given seismic damage. Figure 9 illustrates the repair activities of individual teams over time when the restoration scenario of Case B1 was applied. Here, the "work" status means crew under repair, "move" indicates crew in travel, and "idle" implies the "on call" status because there is no remaining part to be repaired. The graph reveals that each team repaired approximately five to six pipes, and repairing a single damaged pipe took about six hours on average. We can also compare restoration strategies using the durations required for "work", "move", and "idle" status. Table 4 compares the average working, moving, and waiting hours of ten repair teams for the applied six restoration cases. Overall, we found that the restoration cases with zoning (B1~B3) required fewer hours for repair and travel compared to the cases without zoning (A1~A3). That is, on average 7 h (5 h for repair and 2 h for travel) were saved by district-based repairs. Note that the waiting hours were similar for both case groups. Using this statistical analysis of the repair crew activities, the work intensity of each team can be estimated, the restoration strategies can be assessed, and finally repair crews can be assigned in a more efficient way.

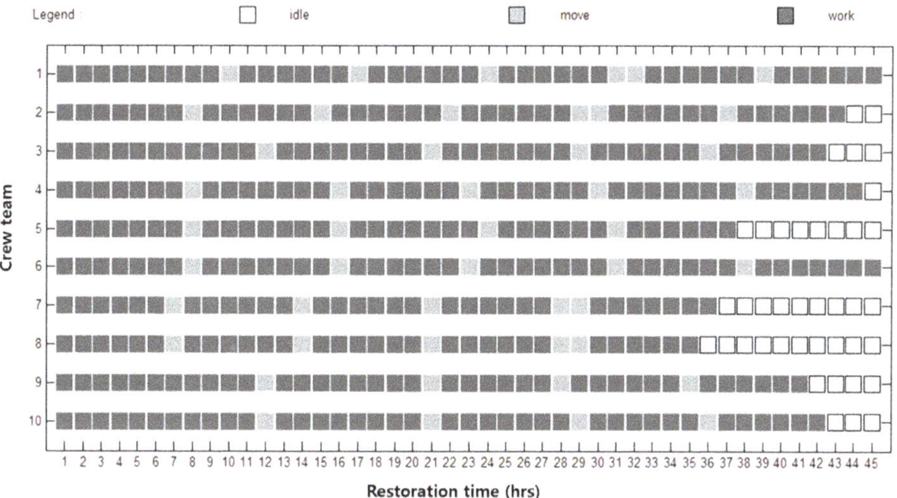

Figure 9. Repair crew activity chart.

Table 4. Average repair crew activity information for different recovery strategies.

Case	Average Time for Repair (h)	Average Time for Travel (h)	Average Time for Wait (h)
A1	41.3	7.7	4.0
A2	41.3	6.5	5.1
A3	41.4	7.1	4.5
B1	36.2	4.8	4.0
B2	36.2	5.2	4.6
B3	37.0	5.3	5.7

3.3.4. Impact on Tank Water Level

A ground storage tank stores and distributes a large amount of water. Thus, it plays a significant role in a water distribution system. When a WTP cannot supply water temporarily due to an emergency (e.g., power outage, pump failure, etc.), the storage tank can supply downstream service area for a certain period to mitigate the damage. Figure 10 illustrates the fluctuations of tank water level after an earthquake. Here, the solid line indicates the water level while the repair was in progress and the dotted line implies the tank water level after the emergency repair was completed (48 h later, approximately). Note the JS tank continued the service without the seismic damage, while other three tanks were damaged (operation ceased for repair) and recovered. The water level of the damaged tank is set to its initial value assuming no inflow and outflow from the tank until the repair is completed. It is observed that the tank water level declines and reaches a minimum operational depth even after the emergency repair has been completed. This is because of additional water loss through the unrepaired pipe leaks over the network. Therefore, to return to its normal operating condition, prompt leakage detection and repair would be required throughout the network.

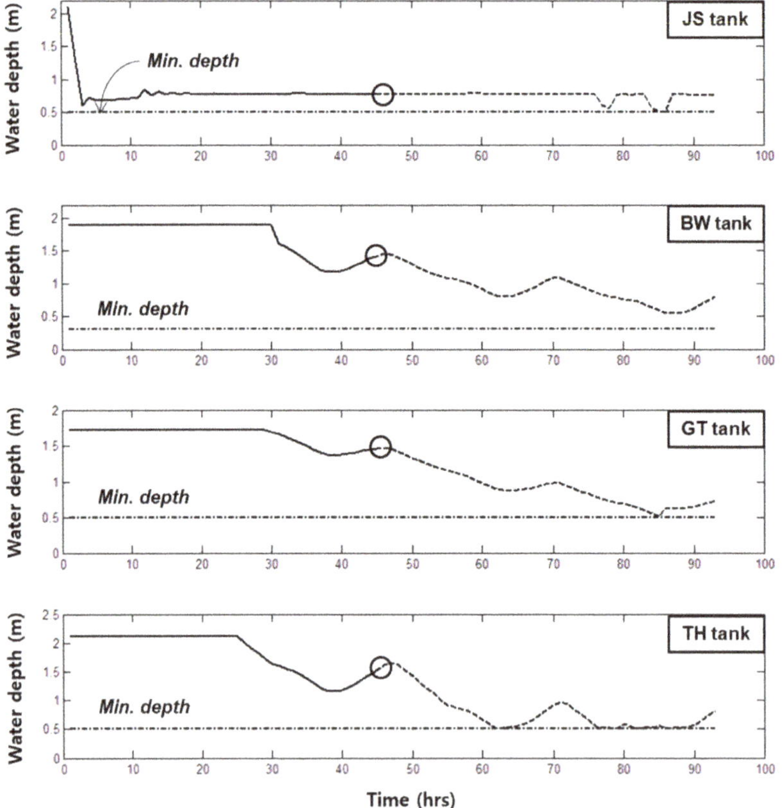

Figure 10. Tank water level fluctuation.

4. Conclusions

A computer-based simulation model was developed for the post-earthquake restoration simulation of water supply networks. The main purpose of the model was to derive a system-specific recovery strategy to enhance the resilience of the damaged network in terms of post-earthquake restoration. The model can reflect various seismic damage scenarios (epicenter location and seismic magnitude) to determine the potential damage to facilities. Given damage conditions (either from the simulation or collected information from the field), the model determines the appropriate recovery schemes, including the repair sequence rule and crew allocation decisions. Once the system recovery starts based on the user defined rules, the model provides the recovery progress information, such as the required repair completion time, required recovery resources, spatiotemporal restoration progress, and repair crew activity information for analysis. Using the information, the model can serve as a decision support tool for rapid and efficient system remedy.

The proposed model was applied to a P-city water supply network currently operating in South Korea as a real-life demonstration. It was possible to obtain the most effective scheme to recover the damaged network by applying six potential restoration strategies. We utilized the system serviceability index and the restoration curve to quantitatively evaluate the applied restoration strategy and intuitively assess the recovery process. The application results revealed that the most effective restoration strategy for the study network was 'Case B1', which separately dispatch the repair teams based on the pre-assigned repair zones and primarily repair the main transmission lines that deliver higher flows. The post-simulation analyses included further analysis of the spatiotemporal distribution of serviceability in graphical form, in which users can identify the locations of facility damage, repairs in progress, and current conditions of water supply. In addition, the statistical analysis of the repair crew activities revealed the work intensity of each repair team, from which system managers can determine alternative resources dispatch plans. The proposed model depicts a real-life process, thus can be used by system operators, engineers, and decision makers to better serve their community in an urgent situation.

Since the seismic damage is unpredictable and hard to simulate, the model results may have limited accuracy. To advance the developed model, we are planning to resolve the assumptions and simplifications embedded in the current model. First, the actual travel route of repair crews should be estimated based on the road network in conjunction with the water network for more accurate estimates of the required travel distance. For an interconnected simulation, it would be better to utilize a geographic information system (GIS) for both water-pipe and road networks. Second, the shut-off valve locations should be considered to identify the actual suspended area (depicted by a segment) while the pipe repairs are undertaken. Third, a full-PDA solver should be coupled with the model for accurate hydraulic analysis of pressure-deficient conditions. These are fruitful and encouraging topics to pursue in the future to advance the current model.

Author Contributions: J.C. carried out the analysis of proposed method, model simulations and drafted the manuscript. D.G.Y. surveyed the previous studies and provided conceptual ideas of the proposed model. D.K. conceived the original idea of the study and finalized the manuscript.

Funding: This research was supported by (1) Basic Science Research Program through the National Research Foundation of Korea (NRF) funded by the Ministry of Education, Science and Technology (NRF-2016R1A2B4014273) and (2) Korea Agency for Infrastructure Technology Advancement (KAIA) grant funded by the Ministry of Land, Infrastructure and Transport (Grant 18AWMP-B083066-05).

Conflicts of Interest: The authors declare no conflicts of interest.

References

1. Yoo, D.G.; Jung, D.; Kang, D.; Kim, J.H.; Lansey, K. Seismic hazard assessment model for urban water supply networks. *J. Water Resour. Plan. Manag.* **2015**, *142*, 04015055. [CrossRef]
2. Yoo, D.G.; Kang, D.; Kim, J.H. Optimal design of water supply networks for enhancing seismic reliability. *Reliab. Eng. Syst. Saf.* **2016**, *146*, 79–88. [CrossRef]
3. Federal Emergency Management Agency (FEMA). *HAZUS97 Technical Manual*; FEMA: Washington, DC, USA, 1997.
4. Mid America Earthquake (MAE) Center. MAEviz Software. Available online: http://mae.cee.illinois.edu/software/software_maeviz.html (accessed on 10 October 2018).
5. Shi, P.; O'Rourke, T.D.; Wang, Y. Simulation of earthquake water supply performance. In Proceedings of the 8th National Conf. on Earthquake Engineering, San Francisco, CA, USA, 18–22 April 2006.
6. Yoo, D. Seismic Reliability Assessment for Water Supply Networks. Ph.D. Thesis, Korea University, Seoul, Korea, 2018.

7.   Adachi, T.; Ellingwood, B.R. Serviceability of earthquake-damaged water systems: Effects of electrical power availability and power backup systems on system vulnerability. *Reliab. Eng. Syst. Saf.* **2008**, *93*, 78–88. [CrossRef]

8.   Yoo, D.G.; Jung, D.; Kang, D.; Kim, J.H.; Lansey, K. Seismic Reliability–Based Multiobjective Design of Water Distribution System: Sensitivity Analysis. *J. Water Resour. Plan. Manag.* **2016**, *143*, 06016005. [CrossRef]

9.   Davenport, P.N.; Lukovic, B.; Cousins, W.J. Restoration of earthquake damaged water distribution systems, in "Remembering Napier 1931". Paper No. 44. In Proceedings of the 2006 Conference of the New Zealand Society for Earthquake Engineering, Napier, New Zealand, 10–12 March 2006.

10.   Tabucchi, T.; Davidson, R.; Brink, S. Restoring the Los Angeles water supply system following an earthquake. In Proceedings of the 14th World Conference on Earthquake Engineering, Beijing, China, 12–17 October 2008.

11.   Luna, R.; Balakrishnan, N.; Dagli, C.H. Poste-arthquake recovery of a water distribution system: Discrete event simulation using colored petri nets. *J. Infrastr. Syst.* **2011**, *17*, 25–34. [CrossRef]

12.   Tabucchi, T.; Davidson, R.; Brink, S. Simulation of post-earthquake water supply system restoration. *Civ. Eng. Environ. Syst.* **2010**, *27*, 263–279. [CrossRef]

13.   Davis, C.A.; O'Rourke, T.D.; Adams, M.L.; Rho, M.A. Case study: Los Angeles water services restoration following the 1994 Northridge earthquake. In Proceedings of the 15th World Conference on Earthquake Engineering, Lisbon, Portugal, 24–28 September 2012.

14.   Davis, C.A. Water system service categories, post-earthquake interaction, and restoration strategies. *Earthq. Spectra* **2014**, *30*, 1487–1509. [CrossRef]

15.   Davis, C.A.; Giovinazzi, S. Toward Seismic Resilient Horizontal Infrastructure Networks. In Proceedings of the 6th International Conference on Earthquake Geotechnical Engineering, Christchurch, New Zealand, 2–4 November 2015.

16.   Klise, K.A.; Hart, D.; Moriarty, D.; Bynum, M.L.; Murray, R.; Burkhardt, J.; Haxton, T. *Water Network Tool for Resilience (WNTR) User Manual*; EPA/600/R-17/264; U.S. Environmental Protection Agency: Washington, DC, USA, 2017. Available online: https://cfpub.epa.gov/si/si_public_record_report.cfm?dirEntryId=337793 (accessed on 6 June 2017).

17.   Klise, K.A.; Bynum, M.; Moriarty, D.; Murray, R. A software framework for assessing the resilience of drinking water systems to disasters with an example earthquake case study. *Environ. Modell. Softw.* **2017**, *95*, 420–431. [CrossRef]

18.   *MATLAB User's Manual*; The MathWorks, Inc.: Natick, MA, USA, 2000.

19.   Rossman, L.A. *EPANET 2: Users Manual*; U.S. Environmental Protection Agency (EPA): Cincinnati, OH, USA, 2000.

20.   Baag, C.E.; Chang, S.J.; Jo, N.D.; Shin, J.S. Evaluation of seismic hazard in the southern part of Korea. In Proceedings of the 2nd International Symposium on Seismic Hazards and Ground Motion in the Region of Moderate Seismicity, Seoul, Korea, 1 November 1998; pp. 31–50.

21.   Eidinger, J. *Seismic Fragility Formulations for Water Systems*; American Lifelines Alliance, G&E Engineering Systems Inc.: Olympic Valley, CA, USA, 2001.

22.   Isoyama, R.; Ishida, E.; Yune, K.; Shirozu, T. Seismic damage estimation procedure for water supply pipelines. *Water Supply* **2000**, *18*, 63–68.

23.   Chang, Y.-H.; Kim, J.-H.; Jung, K.-S. A Study on the design and evaluation of connection pipes for stable water supply. *J. Korean Soc. Water Wastewater* **2012**, *26*, 249–256. [CrossRef]

24.   Puchovsky, M.T. (Ed.) *Automatic Sprinkler Systems handbook*; National Fire Protection Association (NFPA): Quincy, MA, USA, 1999.

25.   Shi, P.; O'Rourke, T.D. *Seismic Response Modeling of Water Supply Systems*; MCEER: Buffalo, NY, USA, 2006.

26.   Wang, Y.; O'Rourke, T.D. *Seismic Performance Evaluation of Water Supply Systems*; MCEER: Buffalo, NY, USA, 2006.

Article

Simplified Analytical Model and Shaking Table Test Validation for Seismic Analysis of Mid-Rise Cold-Formed Steel Composite Shear Wall Building

Jihong Ye [1,*] and Liqiang Jiang [2]

[1] Jiangsu Key Laboratory Environmental Impact and Structural Safety in Engineering, China University of Mining and Technology, Xuzhou 211116, China

[2] School of Mechanics, Civil Engineering and Architecture, Northwestern Polytechnical University, Xi'an 710072, China; 230149388@seu.edu.cn

* Correspondence: jhye@cumt.edu.cn

Received: 10 August 2018; Accepted: 5 September 2018; Published: 6 September 2018

Abstract: To develop the cold-formed steel (CFS) building from low-rise to mid-rise, this paper proposes a new type of CFS composite shear wall building system. The continuous placed CFS concrete-filled tube (CFRST) column is used as the end stud, and the CFS-ALC wall casing concrete composite floor is used as the floor system. In order to predict the seismic behavior of this new structural system, a simplified analytical model is proposed in this paper, which includes the following. (1) A build-up section with "new material" is used to model the CFS tube and infilled concrete of CFRST columns; the section parameters are determined by the equivalent stiffness principle, and the "new material" is modeled by an elastic-perfect plastic model. (2) Two crossed nonlinear springs with hysteretic parameters are used to model a composite CFS shear wall; the Pinching04 material is used to input the hysteretic parameters for these springs, and two crossed rigid trusses are used to model the CFS beams. (3) A linear spring is used to model the uplift behavior of a hold-down connection, and the contribution of these connections for CFRST columns are considered and individually modeled. (4) The rigid diaphragm is used to model the composite floor system, and it is demonstrated by example analyses. Finally, a shaking table test is conducted on a five-story 1:2-scaled CFS composite shear wall building to valid the simplified model. The results are as follows. The errors on peak drift of the first story, the energy dissipation of the first story, the peak drift of the roof story, and the energy dissipation of the whole structure's displacement time–history curves between the test and simplified models are about 10%, and the largest one of these errors is 20.8%. Both the time–history drift curves and cumulative energy curves obtained from the simplified model accurately track the deformation and energy dissipation processes of the test model. Such comparisons demonstrate the accuracy and applicability of the simplified model, and the proposed simplified model would provide the basis for the theoretical analysis and seismic design of CFS composite shear wall systems.

Keywords: cold-formed steel structure; cold-formed steel composite shear wall building; mid-rise; simplified modeling method; seismic analysis; shaking table test

---

1. Introduction

Cold-formed steel (CFS) composite shear wall buildings are widely used around the world due to their advantages, such as: light weight, recyclable materials, high assembly, short construction time cycle, etc. Such buildings were commonly used in low-rise villas or apartments within three stories. In China, this type of building system has attracted increasing attention from structural engineers. However, researchers [1] believe that mid-rise CFS buildings would be more appropriate for Chinese civil constructions, because China has insufficient land for its extensive people. In fact, the United States (US) National Science Foundation (NSF) and the American Iron and Steel Institute (AISI) funded the Cold-formed Steel-National Science Foundation Network for Earthquake Engineering Simulation (CFS-NEES) research project, and the main objective of the project is searching constructional technologies and perform-based seismic design methodologies for mid-rise CFS structures [2]. Thus, it can be seen that the technologies in mid-rise CFS structures are a common issue for the CFS researchers allover the world.

In the past three decades, researchers contributed lots of studies on the seismic performance and seismic design methods on low-rise CFS structures. Serrete et al. [3] discussed the CFS shear wall sheathed with different kinds of wallboard, including plywood, oriented strand board (OSB), gypsum, and fiberboard. Such works were adopted by the AISI S213 (AISI 2007). Then, Rogers et al. developed new types of CFS shear wall, such as a CFS wall with an X-shape strap [4] and a CFS steel-sheathed shear wall [5], and proposed two-stage modeling and seismic design methods for the steel-sheathed CFS shear walls [6]. Ronagh et al. [7] conducted experimental and numerical studies on the braced CFS shear walls. Yu et al. contributed several works on the seismic tests and design methods of steel-sheathed CFS shear walls [8]. Landolfo et al. [9–12] conducted cyclic tests on CFS wood-sheathed shear walls and fasteners, and CFS shear walls sheathed with nailed gypsum panels. Besides, Landolfo et al. tested a two-story CFS building to propose a perform-based seismic design method of CFS buildings [13,14]. Bourahla et al. [15,16] developed a deteriorating hysteresis model and seismic design procedure for CFS structures. Dubina et al. developed design methods for CFS walls sheathed with wood and gypsum [17], and generated a numerical model for seismic analysis for CFS buildings [18]. Schafer et al. conducted a shaking table test of a two-story full-scale CFS-framed office building, and discussed the system-level [19] and the subsystem-level [20] responses of the building. The modeling methodology on the test building was also proposed [21], and a reliability assessment was performed on the CFS buildings as well [22].

In addition to the research works on low-rise CFS structures, researchers have focused on realizing the construction of mid-rise CFS buildings through the following manners: (1) the hot-rolled section steel frame is proposed to be used as the main bearing component, and the CFS walls are filled in the steel frames; (2) new types of CFS shear walls with superior seismic performance are developed to satisfy the seismic demand of mid-rise CFS structures, such as a CFS shear wall sheathed with double-layer wallboards [23] and braced CFS shear walls [7]. Furthermore, various types of mid-rise CFS structural systems were also adopted by the structural engineers [2]. Ye et al. proposed a new type of CFS shear wall [24,25], in which the CFS concrete-filled steel tube (CFRST) was proposed as the end stud of the shear wall. Both a cyclic test and an analytical model were conducted to analyze the seismic performance and collapse mechanism of the proposed CFS shear walls [26–28]. The following researchers finished many experimental (including shaking table tests and cyclic tests), numerical, and theoretical works for the CFS structures, and many design guidelines were also adopted to design the CFS structures, such as AISI S400 (AISI 2015) [29]. However, as stated by Schafer [2], these works were mostly focusing on the component level; the system-level investigations of CFS structures are still insufficient, especially for the mid-rise CFS structures. The most effective way to understand the system-level responses of mid-rise CFS structures is developing shaking table tests and an accurate and reliable numerical modeling method for mid-rise CFS structures [2].

Therefore, this paper proposes a simplified numerical modeling methodology for the seismic analysis of mid-rise CFS structures based on a shaking table test of a 1:2 scaled five-story CFS composite shear wall building. The research steps can be summarized as follows. (1) In the second section of this paper, a brief introduction of the key construction technologies for the mid-rise CFS building system is made through comparison with the low-rise building system. (2) In the third section of this paper, the detailed modeling processes of the parts of the mid-rise building are described, including the CFRST columns, CFS composite shear walls, hold-down connections, and composite floor systems.

(3) In the fourth section of this paper, the results, including the natural frequency, structural drift, and cumulative energy obtained from the simplified model are validated by the test results.

2. A Five-Story Shaking Table Test Model

A 1:2 scaled five-story CFS composite shear wall building was designed and tested, as shown in Figure 1a. Such a building was designed according to the scale similarity. The prototype building was built in the a high seismic zone (eight-degree seismic fortification zone) in China, and it was designed according to the Chinese Code for the Seismic Design of Buildings GB 50011-2010 [30] and Chinese Technical Code of Cold-Formed Thin-Walled Steel Structures JGJ227-2011 [31]. The total height of the prototype building was 15 m, and it was 3 m for each story. The span was 3.6 m. The height of the scaled building was 7.5 m, and the span was 1.8 m. The scaled building was designed according to the geometrical, load, mass, stiffness, and initial condition similarities, and the scale similarity was shown in Table 1. The length and the width of the shaking table were 6 m and 4 m, respectively. Unidirectional earthquake was input by the Mechanical Testing System (MTS) actuator, and the layout of the test model on the shaking table is shown in Figure 1b. The bearing capacity of the shaking table is 25 t, and the maximum displacement of it is ±250 mm. The test model used the new scheme to realize the mid-rise construction, which was different from the low-rise construction scheme in the following two ways. (1) The continuous CFRST was used as end studs along the height, and the CFRST columns included "+"-shape inner columns, "T"-shape side columns, and "L"-shape corner columns, as shown in Figure 1b. (2) The CFS joists and the Autoclaved Lightweight Concrete (ALC) boards were used as a composite floor system, and the case-in-place concrete was placed on the composite floor to enhance the integrality of the floor system, as shown in Figure 2a. The details of the composite floor system, CFS studs placement, and the beam–column joints are shown in Figure 2.

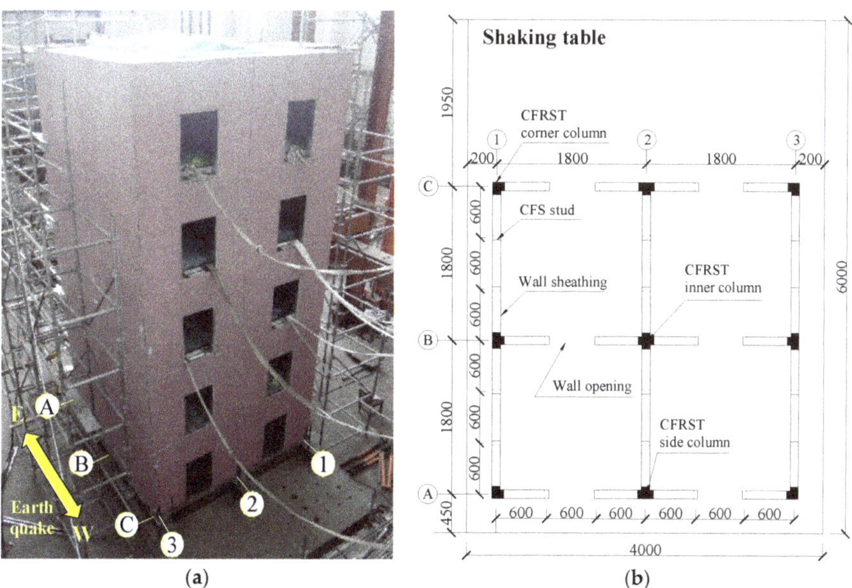

(a)   (b)

Figure 1. Layout and dimension of the test model. (**a**) Three-dimensional (3D) view; (**b**) Plan view.

The CFS composite shear walls used in the test model include the shear walls along the direction of earthquake (one-axial, two-axial, and three-axial), and the shear walls without earthquake input (A-axial, B-axial, and C-axial), as shown in Figures 1 and 2. All of the shear walls were sheathed with 12-mm gypsum wallboard at both sides, and were connected with the CFS frames through screws.

A 0.9-m height and 0.6-m width door opening was placed on each shear wall without earthquake input, as shown in Figure 1b. The CFS composite floor was placed between each story, and it included build-up I-shape CFS beams that were 150 mm in height (it is built by U150 × 50 × 0.8, unit: mm), 50-mm thick ALC boards, and a 30-mm thick concrete floor, as shown in Figure 2c. Two C89 studs (C89 × 50 × 13 × 0.8, unit: mm) were included in each shear wall, and the distance of these studs was 600 mm, as shown in Figure 1b. Hold-downs were placed at the ends of the CFRST columns, and the height of them was 270 mm. Such hold-downs were connected with the CFRST columns through screws.

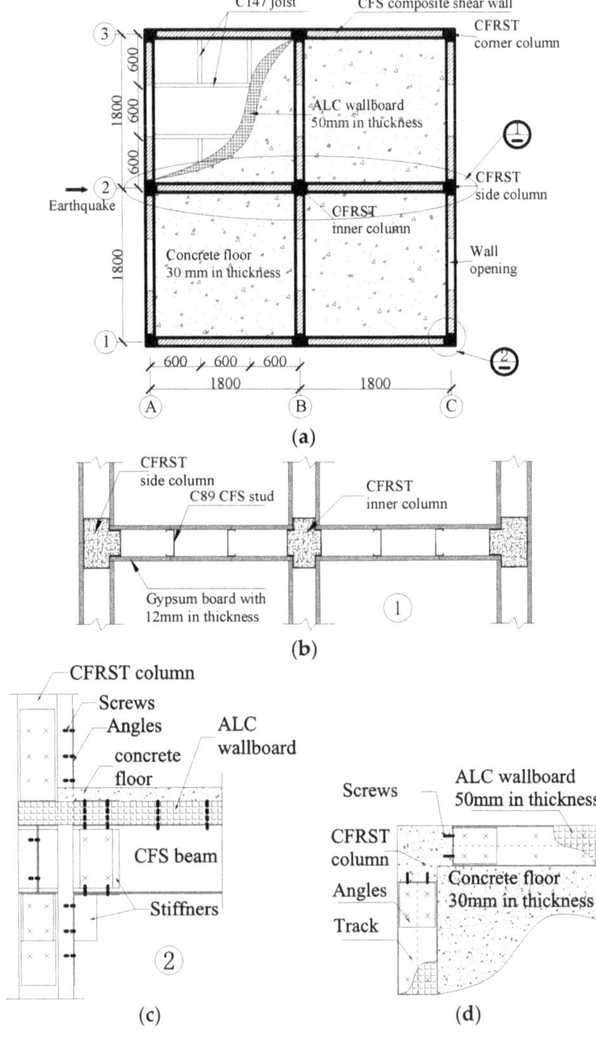

Figure 2. Details of the floor, shear wall, and beam–column joint (Unit: mm). (**a**) Layout of the floor system; (**b**) Layout of the partial shear wall ①; (**c**) Elevation of the corner column joint ②; (**d**) Planar of the corner column joint ②.

In total, five displacement sensors and five acceleration sensors were placed on each story to measure the lateral displacement and acceleration of the test model subject to the earthquake, as shown in Figure 3. The El Centro earthquake along the north–south direction was used as the prototype earthquake, as shown in Figure 4. The earthquakes were input on the test model consequently according to the peak acceleration on a multiple of 100 gal, and the vibration results of the test model were recorded.

Table 1. Scale similarity between the prototype and the test models.

Items	Parameters	Symbols	Similarity Factors
Geometric parameters	Length	S_L	1/2
	Area	S_A	1/4
	Drift ratio	S_a	1
Material parameters	Strain	S_ε	1
	Elastic modulus	S_E	1
	Stress	S_σ	1
	Poisson's ratio	S_μ	1
	Mass density	S_ρ	2
Load parameters	Concentrated force	S_P	1/4
	Area load	S_q	1
Dynamic parameters	Period	S_T	$1/\sqrt{2}$
	Frequency	S_f	$\sqrt{2}$
	Acceleration	S_a	1
	Gravity acceleration	S_g	1
	Damping ratio	S_C	$1/2^{1.5}$

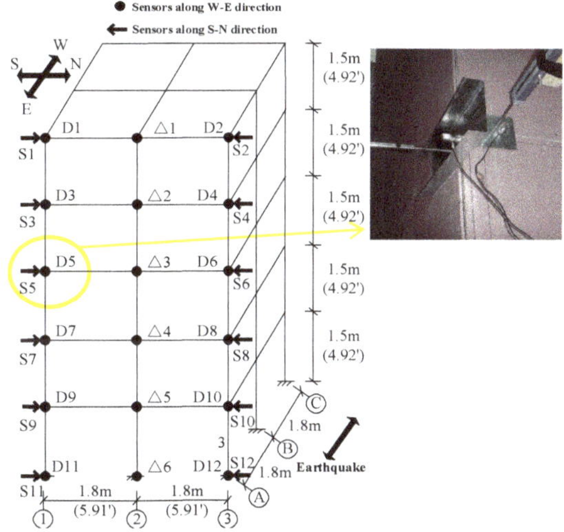

Figure 3. Measurements of the shaking table test model.

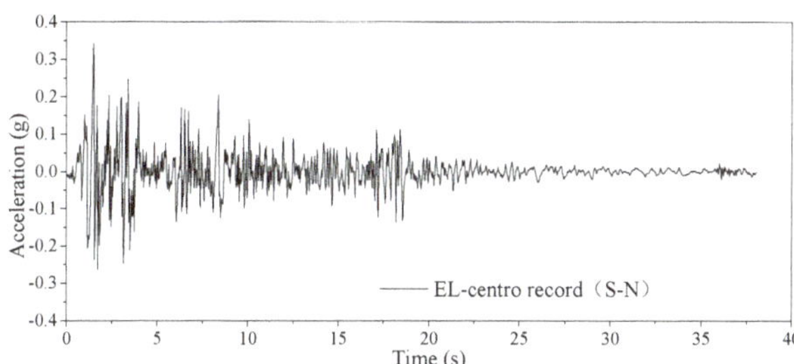

Figure 4. Prototype earthquake record of the shaking table test (El Centro in S–N direction).

3. Simplified Analytical Model of the Test Specimen

The simplified analytical model of the test specimen is presented in Figure 5. Since the two-axial is the symmetry axial for the test model, only the simplified models for the one-axial and two-axial shear walls are shown in Figure 5. Such simplified models are realized by the *Open System of Earthquake Engineering Simulation* software OpenSees [32]. The simplified model follows the following estimations.

(1) A build-up section beam is idealized as two crossed rigid links, and such links are pinned with CFRST columns. The CFS composite floor system is idealized as a rigid plane, and the rigid plane is pinned with the CFRST columns.

(2) A CFS composite shear wall (including the sheathing wallboards and the CFS studs) is idealized as two crossed nonlinear springs, and the hysteretic characteristics of the composite shear wall are represented by the nonlinear springs.

(3) The hold-down connections at the ends of the CFRST columns are idealized as rigid connections, and the uplift behavior of the hold-down connections is modeled by an axial linear spring according to suggestions of previous study [26]. Thus, in the simplified model, three rotational freedoms and two translation freedoms are restrained, and the axial translation freedom is restrained by the axial linear spring.

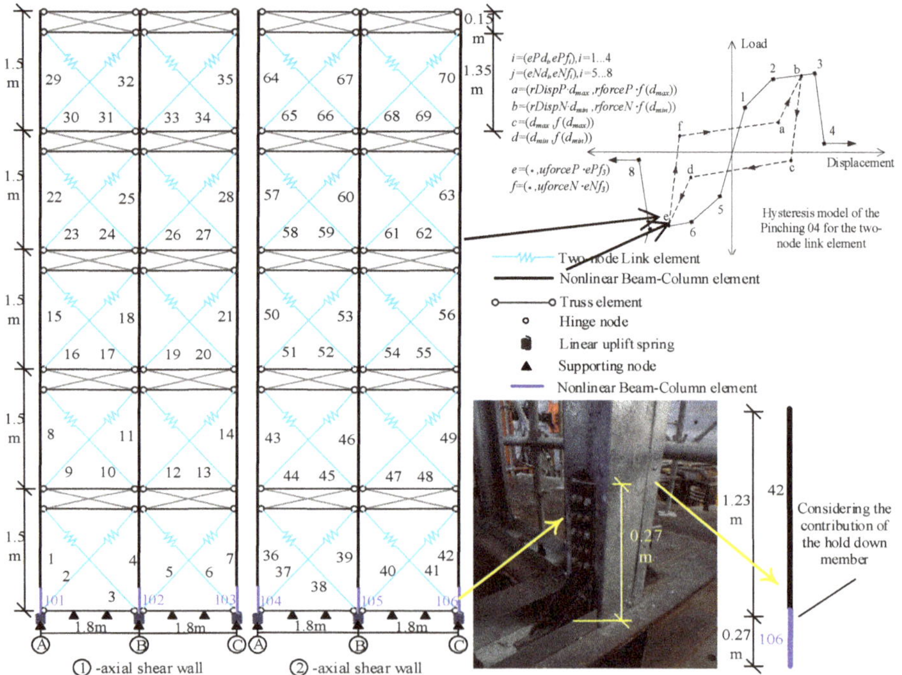

Figure 5. Simplified analytical model for the shaking table test specimen.

3.1. Modeling the CFRST Column

According to the failure modes of the shaking table test model, the CFRST columns buckled under the combination of gravity loads and earthquakes; thus, the nonlinear behavior of the CFRST columns should be considered in the simplified model. In previous studies, the end studs were simplified as a linear truss [6], or a CFRST column was simplified as a linear beam–column element and a linear spring [26]. However, such simplifications cannot capture the nonlinear behavior of the

CFRST columns. Padilla-Llano [33] proposed a complete nonlinear hysteretic model for each CFS stud; however, as stated by Leng et al. [21], such a model would result in excessive computational cost or computational non-convergence. Therefore, this paper simplifies the CFRST columns as a build-up section with "new material", and the columns are modeled by the nonlinear beam–column elements of OpenSees. The material properties of the build-up section are listed in Table 2 according to the mechanical interaction between the CFS and infilled concrete, including axial force, moment, and torque.

Table 2. Definition of the cold-formed steel (CFS) concrete-filled tube (CFRST) column in the simplified model.

Items	Constitutive	Stiffness	Load Capacity
Axial force	Elastic-perfect plastic	EA	P
Strong axis moment	Elastic-perfect plastic	EI_x	M_{nx}
Weak axial moment	Elastic-perfect plastic	EI_y	M_{ny}
Torque	Elastic	GJ	∞

The section parameters of the build-up section are determined according to the equivalent bending stiffness of the CFRST columns, the resistance of which is determined according to the American Institute of Steel Construction- Load and Resistance Factor Design (AISC-LRFD) [34,35]:

$$N_{AISC} = A_s(0.658^{\lambda_c^2})F_{my}; \ \lambda_c = \frac{kL}{\pi}\sqrt{\frac{F_{my}}{E_m}};$$
$$F_{my} = f_y + 0.85f'_c(A_c/A_s); \ E_m = E_s + 0.4E_c(A_c/A_s) \tag{1}$$

where E_s, E_c, and E_m are the elastic modulus for the CFS members, the infilled concrete, and the "new material", respectively; A_s and A_c are the area of the CFS members and concrete in CFRST columns, respectively; f_y and f_c are the yield strength and compressive strength of the CFS and concrete, respectively; and N_{AISC} is the axial compressive capacity of the CFRST columns. Since the elastic-perfect plastic model is used for the build-up section in the cases of axial force and moment, thus, the yield strength of the "new material" is simplified as N_{AISC}/A_e, where A_e is the area of the build-up section.

Due to the CFRST columns being designed as "+"-shape, "T"-shape, and "L"-shape sections for the test model, as shown in Figure 6, it is thus necessary to calculate the section and material parameters for them; the results are shown in Table 3. In Table 3, the bottom region of the CFRST columns is the region considering the contribution of the hold-down members, which can be seen in Figure 5. In the simplified model, the contributions on bending stiffness and axial compressive stiffness of the hold-down members are considered; such regions are individually modeled, and the section and material parameters of these regions are listed in Table 3.

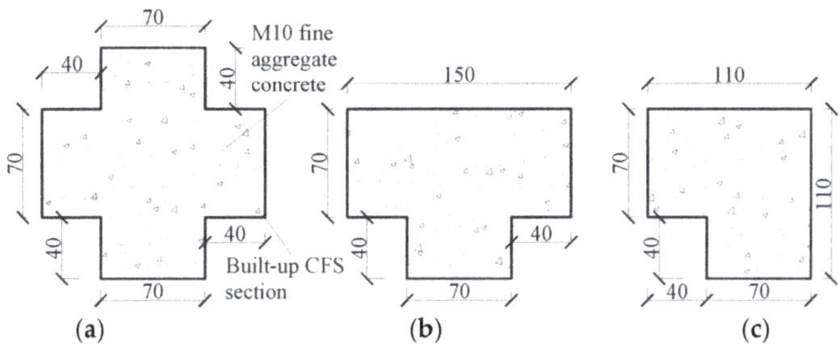

Figure 6. Section of the CFRST columns (unit: mm, thickness of CFS members is 0.8 mm). (a) Inner column; (b) Side column; (c) Corner column.

Table 3. Parameters of the simplified model for the CFRST columns in the test specimen.

Items		Middle Region of the Column				Bottom Region of the Column		
		E_{m}/MPa	A_{e}/mm^2	f_{ey}/MPa	I_{ex}/mm^4	E_{m}/MPa	A_{es}/mm^2	I_{esx}/mm^4
Inner column		4.26×10^5	565.7	411.1	1.27×10^6	4.26×10^5	952.6	2.74×10^6
Side column	Strong axial	3.92×10^5	528.1	369.3	1.04×10^6	3.92×10^5	818.3	2.63×10^6
	Weak axial	3.92×10^5	528.1	383.0	1.30×10^6	3.92×10^5	818.3	2.24×10^6
Corner column		3.76×10^5	473.4	324.5	0.73×10^6	3.76×10^5	666.7	1.71×10^6

Note: E_{m}, A_{e}, f_{ey}, and I_{ex} are the elastic modulus, area, yield strength, and inertia moment along the direction of the earthquake, respectively; the bottom region of the column is the region strengthened by the hold-down (the height is 270 mm, as shown in Figure 5); A_{es}, f_{esy}, and I_{esx} are the equivalent area, yield strength, and inertia moment of the CFRST column at the strengthened region.

3.2. Modeling the CFS Composite Shear Wall

The CFS composite shear wall includes build-up I-shape CFS beams, gypsum wallboards, and CFS studs. Two crossed rigid trusses are used to model a build-up I-shape CFS beam, because no obvious damage and deformation were observed on these CFS beams, according to the shaking table test. Two crossed two-node link elements are used to model the nonlinear behavior of a composite shear wall, and Pinching04 material is used to represent the hysteretic behavior of these two-node link elements. The hysteretic parameters for the Pinching04 material are always determined from the cyclic test results of the prototype CFS composite shear walls. As the composite shear wall is simplified as two crossed elements; thus, the relation between the load–displacement curve of the shear wall and the load–displacement curve of the simplified element is following the geometric relationship, which is also presented in Figure 7:

$$F' = F/2\cos\theta \tag{2}$$

$$\Delta'_{\mathrm{w}} = \Delta_{\mathrm{w}} \cdot \cos\theta \tag{3}$$

where θ is the angle between the simplified element and the top track of the shear wall; F and Δ_{w} are the load and the displacement of the shear wall, respectively; and F' and Δ'_{w} are the load and the displacement of the simplified element, respectively.

However, there is not cyclic test data for the 1:2 scaled CFS composite shear wall constructed in the test model of this paper. Thus, this paper determines the hysteretic parameters for the scaled composite shear wall according to the fastener-based model, as shown in Figure 8a; such a model was proposed by the CFS-NESS team [36]. In the fastener-based model, the wallboard is estimated as a rigid plane, a screw is idealized as a nonlinear spring with hysteretic parameters, and the hysteretic parameters are obtained from the cyclic tests on fasteners. Such a model was verified by cyclic test results [21]; the authors' research team validated such a model through cyclic tests on single-story and double-story CFS composite shear walls [26], and the results of specimen W1 are depicted in Figure 8b. Such a model can be realized by the following steps.

(1) Based on the fastener-based model, the hysteretic curves for the target CFS composite shear walls can be obtained. (2) The hysteretic parameters for the target shear wall can be obtained from the hysteretic curves, as shown in Figure 9. (3) The hysteretic parameters of the nonlinear springs for the target shear wall can be determined according to Equations (1) and (2), and such parameters can be used in the simplified model. Thus, the hysteretic parameters for the composite shear walls that are used in the test model are listed in Table 4.

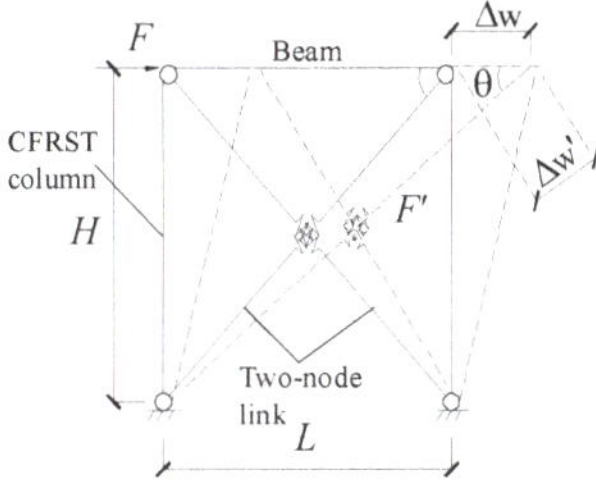

Figure 7. Mechanical and deformation diagrams of the shear wall subject to the horizontal load.

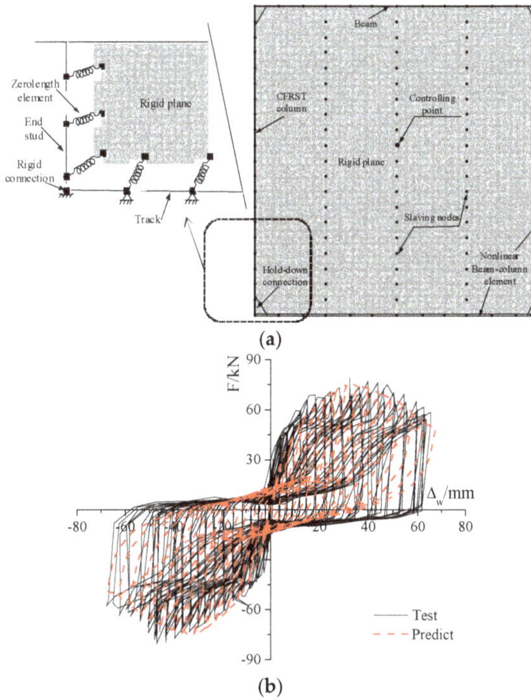

(a)

(b)

Figure 8. Test validation of the fastener-based model for numerical analysis of the shear wall. (**a**) Fastener-based model; (**b**) Test validation.

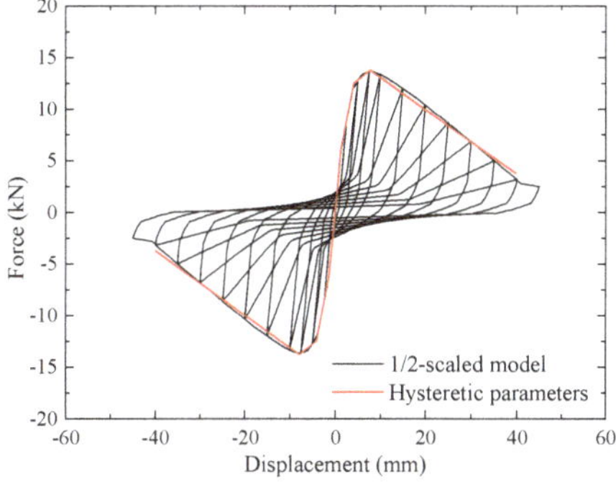

Figure 9. Hysteretic parameter determination of the half-scaled shear wall.

Table 4. Hysteretic parameters of the shear walls in the simplified model for the test specimen.

							$a_{\text{Klimit}} = 0.5$		$a_{\text{Dlimit}} = 0.2$		$a_{\text{Flimit}} = 0.05$	
	Item	epd_i/mm $i = 1\text{–}4$	epf_i/kN $i = 1\text{–}4$	$rDispP$	$rforceP$	$uforceP$	$a_{\text{K1,2}}$	$a_{\text{K3,4}}$	$a_{\text{D1,2}}$	$a_{\text{D3,4}}$	$a_{\text{F1,2}}$	$a_{\text{F3,4}}$
Equivalent members of the shear walls	Axial of 1, 2, 3	1.96, 4.15, 8.95, 50.52	5.88, 8.14, 9.96, 2.44	0.3	0.3	0.05	0.5	1.5	0.4	1.5	0.4	1.5
	Axial of A, B, C	1.67, 5.02, 10.56, 45.30	4.23, 5.68, 7.84, 2.79	0.3	0.3	0.05	0.5	1.5	0.4	1.5	0.4	1.5

Note: epd_1–epd_4, epf_1–epf_4, $rDispP$, $rforceP$, and $uforceP$ are the hysteretic parameters of the shear walls, as shown in Figure 5; a_{Klimit}, a_{Dlimit}, and a_{Flimit} are the damage factors to describe the hysteretic characteristics of the shear walls, which can be determined by the stiffness and strength of the loading and unloading stages for the Pinching04 model in Figure 5.

3.3. Modeling the Hold-Down Connections

Due to the restraining effects of hold-down connections on CFRST columns, the uplift behavior might occur on these connections [6,21,26]. Thus, the deformation of the uplift behavior should be considered in the simplified model. According to the shaking table test results, there were not obvious damages and deformation observed on the hold-down connections; thus, such connections were estimated as working in an elastic stage according the suggestions of cyclic tests on composite shear walls by the authors' research team [26]. Therefore, a linear spring is used to model the uplift behavior of a hold-down connection, according to the suggestions of Shamim and Rogers [6] and Wang et al. [26], and such a spring considers the tensile deformation, but does not consider the compressive deformation.

Due to the same size and same type of hold-down connection being used in the shaking table test model of this paper and the cyclic test model W89 shear wall of the authors' previous study [26], the stiffness of the linear spring was 39.2 kN/mm for two hold-down connections. In the shaking table test model, two, three, and four hold-down connections were used for the corner column, side column, and inner column, respectively. Thus, the stiffness of these springs is $k_{\text{corner}} = 39.2$ kN/mm, $k_{\text{side}} = 58.8$ kN/mm, and $k_{\text{mid}} = 78.4$ kN/mm for the corner column, side column, and inner column, respectively.

3.4. Modeling the Composite Floor System

Leng et al. [21] proposed rigid and semi-rigid diaphragm models for the composite floor system in the three-dimensional (3D) numerical model. The semi-rigid diaphragm model determined the shear deformation response by using engineering judgment, and it was found that the semi-rigid diaphragm was more appropriate for the CFS composite floor system. However, a new ALC wallboard CFS composite floor system is proposed and used for the shaking table test model by Ye et al. [37], as shown in Figure 2. The new floor system is different from the one used by Leng et al. [21]: the CFS joists and ALC wallboard (thickness of 50 mm to 100 mm) are used as the composite floor, and the cast-in-place concrete (thickness of 30 mm to 50 mm) is then poured on the composite floor to enhance its integrality. Among the load-carrying capacity, stiffness, fire resistance, and integrality would be effectively improved for the covering of the ALC wallboard and cast concrete [37].

Due to the complicate calculating processes and the engineering-based judgment of the semi-rigid diaphragm model proposed by Leng et al. [21], this paper adopts the rigid diaphragm model in the simplified model for the new composite floor system. To valid the simplified model, both semi-rigid and rigid diaphragm models are used to analyze the roof displacement time–history curves of the shaking table test model in the cases of 200 gal and 800 gal. In the simplified model, the RigidPlane element is used for the rigid diaphragm model, and crossed truss elements are used for the semi-rigid diaphragm model, according to the principle of equivalent stiffness. The results are shown in Figure 10, which demonstrate that the roof displacement time–history curves for rigid diaphragm model and semi-rigid diaphragm model closely resemble both the 200-gal case and the 800-gal case. Therefore, the rigid diaphragm model is used to model the new composite floor system in the shaking table test model.

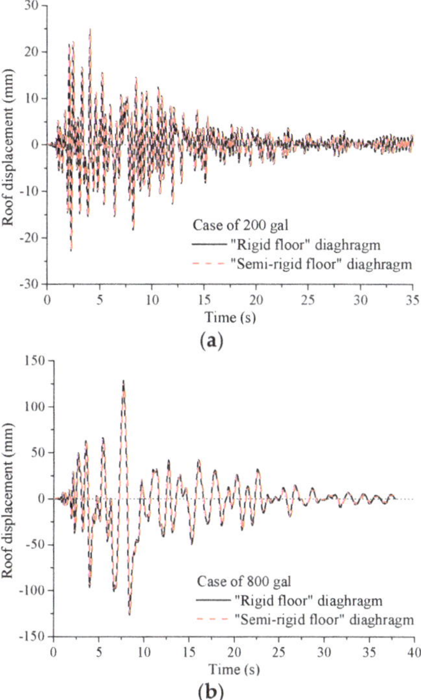

Figure 10. Comparison of roof displacement time–history curves between "Rigid floor" and "Semi-rigid floor" diaphragms. (**a**) Case of 200 gal; (**b**) Case of 800 gal.

3.5. Mass and Damping Ratio of the Building

The mass of a story is directly input on the rigid plane of this story, and it would be evenly distributed to the floor, CFRST columns, and studs through the rigid plane. The mass of a story includes the parts of the composite floor, half of the CFS shear walls and CFRST columns in the upper and lower stories, and the additional mass for modeling the live loads. The mass is 3753 kg and 1362 kg for the standard floor (first story to fourth story) and roof story (fifth story), respectively. Besides, the P-delta effect is also considered in the simplified model. Firstly, nine individual regions are divided for the floor according to the action area of the nine CFRST columns, and the gravity load of each region is input on the joint between the floor and the CFRST column corresponding to the region. The gravity load is directly input on the intersecting joint, and the position of the gravity load would be changed following the lateral deformation of the joint of the CFRST column due to the earthquake. Therefore, the P-delta effect can be considered in the simplified model.

The Rayleigh damping ratio is used in the simplified model. Shamim and Rogers [6] stated that the Rayleigh damping ratio showed significant influence of the numerical results for the CFS buildings, and that a 4% to 5% damping ratio was suitable for the CFS building through trial calculation. It was also found that such a value was larger than the commonly used value of steel structures (2% damping ratio), because the friction behavior would occur in the fabricated screw connections in the shear walls and the high-strength bolt connections at the column bases, and such behavior would increase the damping ratio of the CFS buildings. In the meanwhile, a 5% damping ratio was also proposed by Leng et al. [21] for CFS buildings, and the shaking table test results on a two-story full-scale CFS building were used to valid the value of the damping ratio. Therefore, the Rayleigh damping ratio of 5% is used in the simplified model of this paper.

4. Validation of Simplified Models by Shaking Table Test

To validate the simplified analytical model, a five-story CFS 1:2 scaled composite shear wall building was tested and compared with the analytical results. Due to the relative slight residual deformation (less than 0.5 times of the value of maximum drift) of the test model that was subjected to different earthquake cases, the numerical analysis is individually proceeded for each earthquake case (including 300 gal, 500 gal, and 800 gal cases, which are used to valid the simplified models) according to the similar founding in the two-story full-scaled shaking test model by Peterman et al. [17,18].

To start with, white noise analysis with 100-gal peak acceleration was performed to obtain the natural frequency of the test model, which was 4.98 HZ along the earthquake direction. The first natural frequency of the simplified analytical model was 5.07 HZ along the same direction, and the error between the test and analytical models was 1.8%. The reason for the error can be concluded as: the boundary condition at the column base is considered as a rigid connection in the simplified model, and such an estimation would overestimate the lateral stiffness of the test model; thus, the natural frequency of the analytical model is larger than the value of the test model.

Figures 11–13 show the comparisons on first-story displacement time–history curves, the energy dissipation of the shear wall in the first story, the roof displacement time–history curves, and energy dissipation of whole structure between the shaking table test and analytical results corresponding to the 300-gal, 500-gal, and 800-gal cases, respectively. It can be seen that the simplified analytical model captures the dynamic responses of the shaking table model, and predicts the time–history processes and energy dissipation of the test model within different earthquake cases effectively. Such comparisons demonstrate the validity of the proposed simplified analytical model. By comparing the test and analytical results in the 300-gal, 500-gal, and 800-gal cases, the error among the peak drift of the first story, cumulative energy of the first story, the peak drift of the roof story, and the cumulative energy of the whole structure are no more than 20.8%, and most of the errors are about 10%, as shown in Table 5. Such comparisons are also demonstrating the high precision of the simplified analytical model.

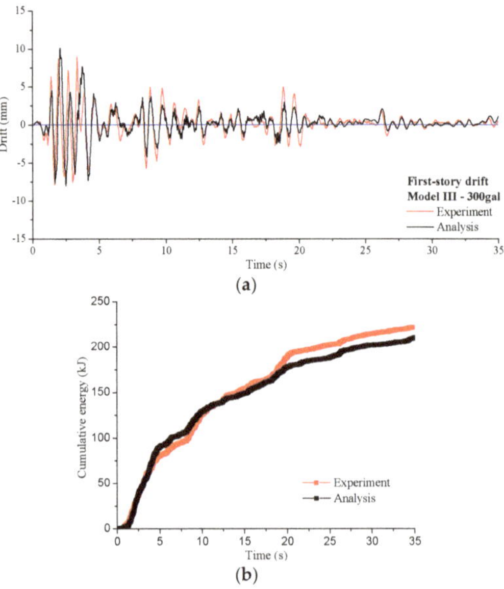

Figure 11. *Cont.*

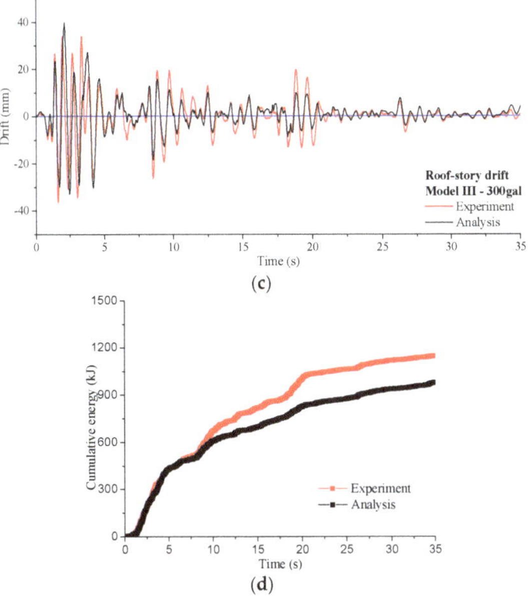

Figure 11. Comparison of time–history curves and cumulative energy between the test and numerical results in the case of 300 gal. (**a**) Story drift of the first story; (**b**) Cumulative energy of shear walls in the first story; (**c**) Roof drift; (**d**) Cumulative energy of the whole structure.

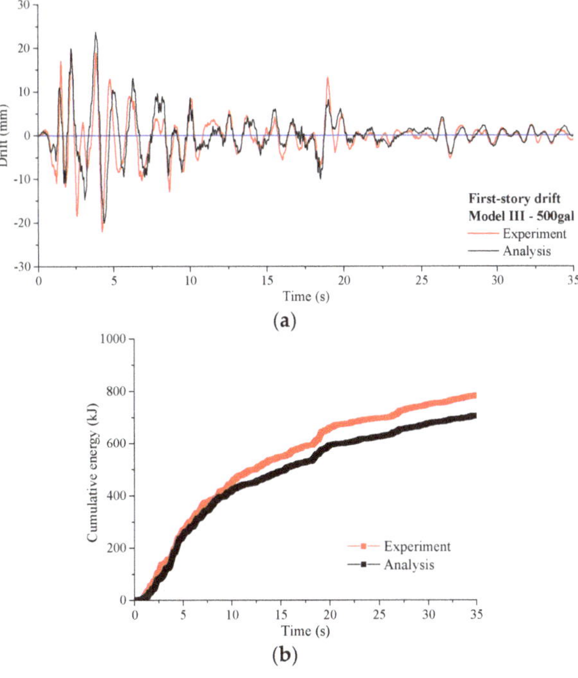

Figure 12. *Cont.*

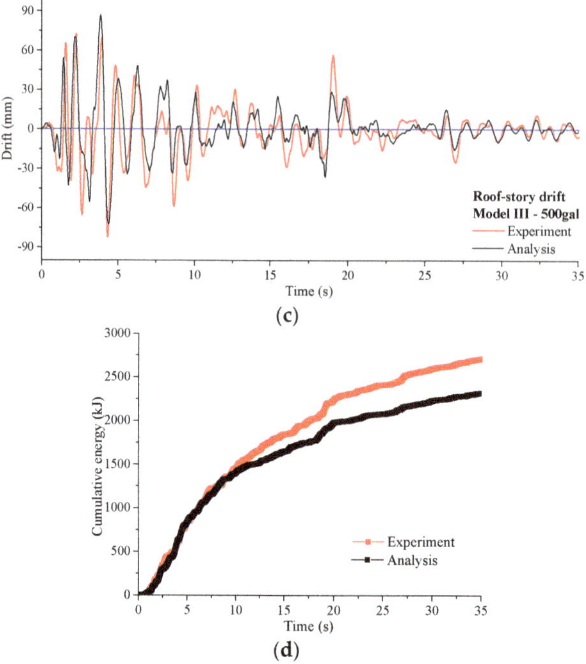

(c)

(d)

Figure 12. Comparison of time–history curves and cumulative energy between the test and numerical results in the 500-gal case. (**a**) Story drift of the first story; (**b**) Cumulative energy of the shear walls in the first story; (**c**) Roof drift; (**d**) Cumulative energy of the whole structure.

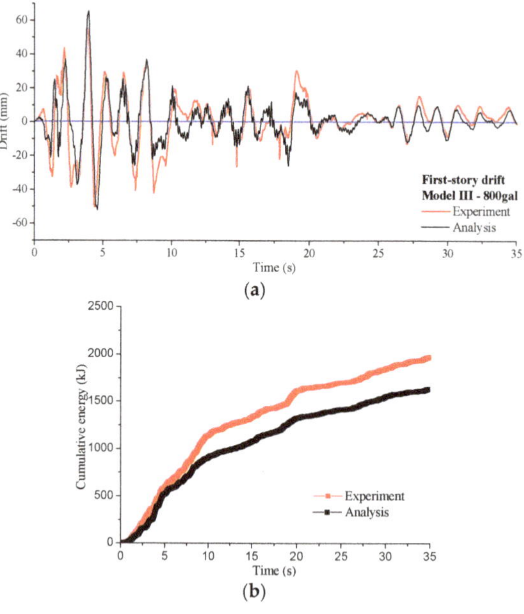

(a)

(b)

Figure 13. *Cont.*

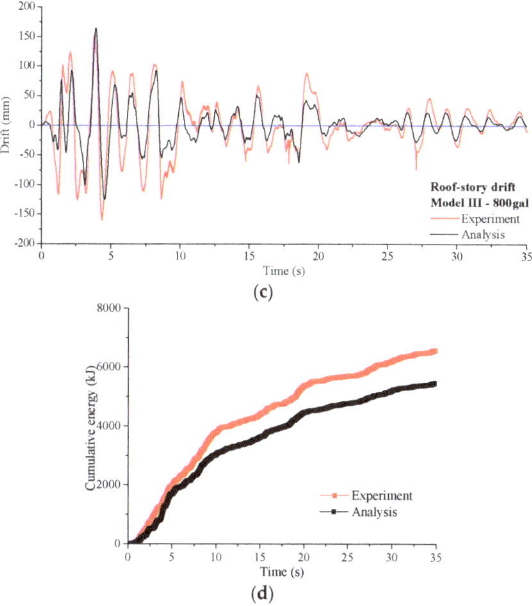

Figure 13. Comparison of time–history curves and cumulative energy between the test and numerical results in the 800-gal case. (**a**) Story drift of the first story; (**b**) Cumulative energy of shear walls in the first story; (**c**) Roof drift; (**d**) Cumulative energy of the whole structure.

Table 5. Comparison of the results between the test and numerical analysis.

Cases	Items	Maximum Drift of the First Story (mm)	Cumulative Energy of the Shear Wall in the First Story (kJ)	Maximum Roof Drift (mm)	Cumulative Energy of the Whole Structure (kJ)
300 gal	Test	8.98	221.72	36.76	1146.33
	Analysis	10.09	210.15	39.65	981.60
	Error	12.3%	5.2%	7.9%	14.4%
500 gal	Test	22.20	783.43	82.09	2703.41
	Analysis	23.52	705.19	86.35	2317.30
	Error	5.9%	10.0%	5.2%	14.3%
800 gal	Test	55.66	1969.88	165.05	6559.92
	Analysis	65.41	1631.09	153.47	5457.08
	Error	17.5%	20.8%	7.0%	16.8%

Note: The maximum acceleration of the output record from the shaking table is 305 gal, 514 gal, and 821 gal for the 300-gal, 500-gal, and 800-gal cases, respectively.

From Figures 11–13, it can be found that the cumulative energies both of the first story and the whole structure of the test model are larger than the analytical model, and the distance increases with the increasing peak acceleration of the earthquake. The reason can be drawn as follows. (1) The analytical model underestimates the energy dissipation of the structure, because the Pingching04 hysteretic model stipulated that the unloading stiffness was lower than the loading stiffness, but the unloading stiffness of the test shear wall was larger than the loading stiffness, as shown in Figure 8b (the opposition friction force derived from the pretension force of the screws between the CFS studs and gypsum wallboard). (2) The connections between the CFS beams and CFRST columns are idealized as pin connections in the analytical model; thus, the energy dissipation of these connections cannot be considered. Furthermore, both the peak drift of the first story and the peak drift of the roof story of the analytical results are larger than the values of the test, and the errors between them are not related to the peak acceleration of earthquake. Since the bond curve of the Pinching04 hysteretic model is divided into four segments, thus, the consecutive degradation relationship of the stiffness and strength of the composite shear walls is simplified as piecewise functions. This is the reason why there is an error between the tested and analyzed results.

5. Conclusions

This paper proposes a simplified analytical numerical model for the seismic analysis of mid-rise CFS composite shear wall building systems. The simplified model considers the mechanical behaviors of CFRST columns, composite CFS shear walls, hold-down connections, and composite floor systems, and detailed modeling methods are described for this simplified model. Finally, shaking table test results on a five-story 1:2-scaled CFS composite shear wall building are used to valid the simplified model, and the following conclusions are drawn:

1. The nonlinear mechanical behavior of the CFRST columns is considered in the simplified model, including the bucking behavior and the yielding of materials. A build-up section with "new material" is proposed to model the CFS tube and infilled concrete, and the equivalent stiffness principle is used to determine the section parameters. The material property of the "new material" is modeled by an elastic-perfect plastic model, and the equivalent yield strength is determined by AISC-LRFD guidance. Besides, the contribution of the hold-down connections on the lateral stiffness and axial strength of the column base of the CFRST columns is also considered in the simplified model, and the strengthened region (270 mm in height) is separately modeled with the CFRST column. Among the "+"-shape inner CFRST columns, the "T"-shape side CFRST columns, and the "L"-shape corner CFRST columns, their strengthened regions are modeled individually in the simplified model.

2. Two crossed nonlinear springs with hysteretic parameters are used in the simplified model to model the hysteretic behavior of a composite CFS shear wall subjected to earthquakes, and such behaviors are modeled by Pinching04 material. Two crossed rigid trusses are used to model a CFS beam. The fastener-based modeling method is used to determine the hysteretic parameters of the 1:2-scaled composite shear walls due to no cyclic test data for them.

3. A linear spring is used to model the uplift behavior of a hold-down connection in the simplified model, and the stiffness of this linear spring is determined by the cyclic test results of the composite shear walls. The stiffness of this linear spring is determined according to the numbers of hold-down connections for the CFRST inner columns, side columns, and corner columns, respectively.

4. To improve the computational efficiency of the simplified model, the rigid diaphragm method is used to model the composite floor system, and such a method is demonstrated by example analyses.

To sum up, the simplified model of the shaking table test model is built by OpenSees software, according to the above methodologies. By comparing the peak drift of the first story, the energy dissipation of the first story, the peak drift of the roof story, and the energy dissipation of the whole structure of the displacement curves between the simplified model and the test model, it is found that the errors between them are about 10%, and the largest one of these errors is 20.8%. It is also found that the rules of change of the time–history curves and cumulative energy curves obtained from the simplified model align closely with the measured results of the test model. The simplified model accurately tracks the whole deformation and energy dissipation processes of the test model. These demonstrations indicate that the simplified model exhibits high computational accuracy. Such works provide the basis for the theoretical analysis and seismic design of mid-rise CFS composite shear wall buildings.

Author Contributions: All authors contribute equally to this paper.

Acknowledgments: This research is sponsored by the National Key Program Foundation of China (51538002) and the Priority Academic Program Development of Jiangsu Higher Education Institutions, Scientific Innovation Research Foundation of Jiangsu, China (KYLX15_0090).

Conflicts of Interest: The authors declare no conflict of interest.

References

1. Ye, J.H. An introduction of mid-rise thin-walled steel structures: Research progress on cold-formed steel-framed composite shear wall systems. *J. Harbin Inst. Technol.* **2016**, *48*, 1–9.
2. Schafer, B.W.; Ayhan, D.; Leng, J.; Liu, P.; Padilla-Llano, D.; Perterman, K.D.; Stehman, M.; Buonopane, S.G.; Eatherton, M.; Madsen, R.; et al. Seismic responses and engineering of cold-formed steel framed building. *Structures* **2016**, *8*, 197–212. [CrossRef]
3. Serrette, R.; Encalada, J.; Juadines, M.; Nguyen, H. Static racking behavior of plywood, OSB, gypsum, and fiberboard walls with metal framing. *J. Struct. Eng.* **1997**, *123*, 1079–1086. [CrossRef]
4. Al-Kharat, M.; Rogers, C.A. Inelastic performance of cold-formed steel strap braced walls. *J. Constr. Steel Res.* **2007**, *63*, 460–474. [CrossRef]
5. Balh, N.; DaBreo, J.; Ong-Tone, C.; El-Saloussy, K.; Yu, C.; Rogers, C.A. Design of steel sheathed cold-formed steel framed shear walls. *Thin-Walled Struct.* **2014**, *75*, 76–86. [CrossRef]
6. Shamim, I.; Roger, C.A. Steel sheathed/CFS framed shear walls under dynamic loading: Numerical modelling and calibration. *Thin-Walled Struct.* **2013**, *71*, 57–71. [CrossRef]
7. Zeynalian, M.; Ronagh, H.R. Seismic performance of cold formed steel walls sheathed by fibre-cement board panels. *J. Constr. Steel Res.* **2015**, *107*, 1–11. [CrossRef]
8. Yu, C. Shear resistance of cold-formed steel framed shear walls with 0.686 mm, 0.762 mm, and 0.838 mm steel sheet sheathing. *Eng. Struct.* **2010**, *32*, 1522–1529. [CrossRef]
9. Landolfo, R.; Fiorino, L.; Della Corte, G. Seismic behavior of sheathed cold-formed structures: Physical tests. *J. Struct. Eng.* **2006**, *132*, 570–581. [CrossRef]
10. Fiorino, L.; Della Corte, G.; Landolfo, R. Experimental tests on typical screw connections for cold-formed steel housing. *Eng. Struct.* **2007**, *29*, 1761–1773. [CrossRef]
11. Macillo, V.; Fiorino, L.; Landolfo, R. Seismic response of CFS shear walls sheathed with nailed gypsum panels: Experimental tests. *Thin-Walled Struct.* **2017**, *120*, 161–171. [CrossRef]
12. Fiorino, L.; Shakeel, S.; Macillo, V.; Landolfo, R. Seismic response of CFS shear walls sheathed with nailed gypsum panels: Numerical modelling. *Thin-Walled Struct.* **2018**, *122*, 359–370. [CrossRef]
13. Fiorino, L.; Macillo, V.; Landolfo, R. Shake table tests of a full-scale two-story sheathing-braced cold-formed steel building. *Eng. Struct.* **2017**, *151*, 633–647. [CrossRef]
14. Fiorino, L.; Iuorio, O.; Macillo, V.; Landolfo, R. Performance-based design of sheathed CFS buildings in seismic area. *Thin-Walled Struct.* **2012**, *61*, 248–257. [CrossRef]
15. Kechidi, S.; Bourahla, N. Deteriorating hysteresis model for cold-formed steel shear wall panel based on its physical and mechanical characteristics. *Thin-Walled Struct.* **2016**, *98*, 421–430. [CrossRef]
16. Kechidi, S.; Bourahla, N.; Castro, J.M. Seismic design procedure for cold-formed steel sheathed shear wall frames: Proposal and evaluation. *J. Constr. Steel Res.* **2017**, *128*, 219–232. [CrossRef]
17. Fülop, L.A.; Dubina, D. Design criteria for seam and sheeting-to-framing connections of cold-formed steel shear panels. *J. Struct. Eng.* **2006**, *132*, 582–590. [CrossRef]
18. Dubina, D. Behavior and performance of cold-formed steel-framed houses under seismic action. *J. Constr. Steel Res.* **2008**, *64*, 896–913. [CrossRef]
19. Peterman, K.D.; Stehman, M.J.; Madsen, R.L.; Buonopane, S.G.; Nakata, N.; Schafer, B.W. Experimental seismic responses of a full-scale cold-formed steel-framed building. I: System-level response. *J. Struct. Eng.* **2016**, *142*, 04016127. [CrossRef]
20. Peterman, K.D.; Stehman, M.J.; Madsen, R.L.; Buonopane, S.G.; Nakata, N.; Schafer, B.W. Experimental seismic responses of a full-scale cold-formed steel-framed building. II: Subsystem-level response. *J. Struct. Eng.* **2016**, *142*, 04016128. [CrossRef]
21. Leng, J.Z.; Peterman, K.D.; Bian, G.B.; Buonopane, S.G.; Schafer, B.W. Modeling seismic responses of a full-scale cold-formed steel-framed building. *Eng. Struct.* **2017**, *153*, 146–165. [CrossRef]
22. Smith, B.H.; Arwade, S.R.; Schafer, B.W.; Moen, C.D. Design component and system reliability of a low-rise cold formed steel framed commercial building. *Eng. Struct.* **2016**, *127*, 434–446. [CrossRef]
23. Wang, X.; Pantoli, E.; Hutchinson, T.C.; Restrepo, J.I.; Wood, R.L.; Hoehler, M.S.; Grezesik, P.; Sesma, F.H. Seismic performance of cold-formed steel wall systems in a full-scale building. *J. Struct. Eng. ASCE* **2015**, *141*, 04015014. [CrossRef]

24. Ye, J.H.; Wang, X.X.; Jia, H.Y.; Zhao, M.Y. Cyclic performance of cold-formed steel shear walls sheathed with double-layer wallboards on both sides. *Thin-Walled Struct.* **2015**, *92*, 146–159. [CrossRef]

25. Wang, X.X.; Ye, J.H. Reversed cyclic performance of cold-formed steel shear walls with reinforced end studs. *J. Constr. Steel Res.* **2015**, *113*, 28–42. [CrossRef]

26. Wang, X.X.; Ye, J.H.; Yu, Q. Improved equivalent bracing model for seismic analysis of mid-rise CFS structures. *J. Constr. Steel Res.* **2017**, *136*, 256–264. [CrossRef]

27. Ye, J.H.; Jiang, L.Q.; Wang, X.X. Seismic failure mechanism of reinforced cold-formed steel shear wall based on structural vulnerability analysis. *Appl. Sci.* **2017**, *7*, 182. [CrossRef]

28. Ye, J.H.; Jiang, L.Q. Collapse mechanism analysis of a steel-moment frame based on structural vulnerability theory. *Arch. Civ. Mech. Eng.* **2018**, *18*, 833–843.

29. American Iron and Steel Institute (AISI). *North American Standard of Cold-Formed Steel Framing–Lateral Design*; AISI S400; AISI: Washington, DC, USA, 2015.

30. *Code for Seismic Design of Buildings (GB 50011-2010)*; China Architecture & Building Press: Beijing, China, 2016. (In Chinese)

31. *Technical Specification for Low-Ries Cold-Formed Thin-Walled Steel Buildings (JGJ 227-2011)*; China Environmental Science Press: Beijing, China, 2011. (In Chinese)

32. Pacific Earthquake Engineering Research Center. *Open System for Earthquake Engineering*; Pacific Earthquake Engineering Research Center, University of California: Berkley, CA, USA, 2006.

33. Padilla-Llano, D.A. *A Framework for Cyclic Simulation of Thin-Walled Cold-Formed Steel Members in Structural Systems*; Virginia Polytechnic Institute and State University: Blacksburg, VA, USA, 2015.

34. AISC-LRFD. *Load and Resistance Factor Design Specification for Structural Steel Buildings*; American Institute of Steel Construction: Chicago, IL, USA, 1999.

35. Tao, Z.; Uy, B.; Liao, F.Y.; Han, L.H. Nonlinear analysis of concrete-filled square stainless steel stub columns under axial compression. *J. Constr. Steel Res.* **2011**, *67*, 1719–1732. [CrossRef]

36. Buonopane, S.G.; Bian, G.; Tun, T.H.; Schafer, B.W. Computationally efficient fastener-based models of cold-formed steel shear walls with wood sheathing. *J. Constr. Steel Res.* **2015**, *110*, 137–148. [CrossRef]

37. Ye, J.H.; Chen, W.; Wang, Z. Fire-resistance behavior of a newly developed cold-formed steel composite floor. *J. Struct. Eng.* **2017**, *143*, 04017018. [CrossRef]

Article

Probabilistic Generalization of a Comprehensive Model for the Deterioration Prediction of RC Structure under Extreme Corrosion Environments

Xingji Zhu [1], Zaixian Chen [1,*], Hao Wang [2] , Yabin Chen [1] and Longjun Xu [1,3]

[1] Department of Civil Engineering, Harbin Institute of Technology at Weihai, Weihai 264209, China; zhuxingji@hit.edu.cn (X.Z.); 14B933032@hit.edu.cn (Y.C.); xulongjun80@163.com (L.X.)
[2] Department of Civil Engineering, Southeast University, Nanjing 210096, China; wanghao1980@seu.edu.cn
[3] Cooperative Innovation Center of Engineering Construction and Safety in Shandong Blue Economic Zone, Qingdao 266033, China
* Correspondence: zaixian_chen@sina.com; Tel.: +86-0631-568-7845

Received: 3 July 2018; Accepted: 25 August 2018; Published: 28 August 2018

Abstract: In some extreme corrosion environments, the erosion of chloride ions and carbon dioxide can occur simultaneously, causing deterioration of reinforced concrete (RC) structures. This study presents a probabilistic model for the sustainability prediction of the service life of RC structures, taking into account that combined deterioration. Because of the high computational cost, we also present a series of simplifications to improve the model. Meanwhile, a semi-empirical method is also developed for this combined effect. By probabilistic generalization, this simplified method can swiftly handle the original reliability analysis which needs to be based on large amounts of data. A comparison of results obtained by the models with and without the above simplifications supports the significance of these improvements.

Keywords: reinforced concrete; corrosion; chloride ingress; carbonation; probabilistic; sustainability prediction

---

1. Introduction

The initiation of corrosion is generally due to the penetration of free chloride ions, carbonation, or their combined effect [1–6]. Meanwhile, previous studies also showed that the effect of temperature and humidity on the transport of harmful ions was significant [7–9]. Some works which studied the whole process of reinforcement corrosion indicate that the load-capacity of RC structures would decrease rapidly after steel reinforcements get depassivated [10–14]. As we know, the load-carrying capacity of the entire RC structure reduces seriously only after the extent of corrosion exceeds a certain limit. This characteristic can be easily represented by defining a probability threshold. Therefore, probabilistic analysis for the deterioration of corroded RC structures is very useful for service life sustainability predictions.

The reliability of chloride-induced corrosion was studied in many papers [15–25]. However, the penetration of harmful ions was only considered with a simple diffusion equation, for the sake of simplicity. All of these works did not consider the effect of chloride binding capacity, convection, temperature, and humidity. Meanwhile, the boundary concentration of chloride was only considered as a normal random variable. This means that none of the above-mentioned probabilistic models can be used for RC deterioration assessments under a de-icing salt environment. In that particular case, the chloride concentration on the surface of the concrete should be considered a piecewise function. Therefore, it is more natural and appropriate to use stochastic processes to model the uncertainties of time-dependent boundary conditions in probabilistic analyses. Meanwhile, the probability of carbonation-induced corrosion was studied in [26–40]. However, all of these works just predicted the depth of carbonation based on the empirical formula alone.

As we know, chlorides are the dominant deterioration mechanisms in most cases. However, in some extreme corrosion environments, such as at the entrance and exit of various types of offshore tunnels, carbon dioxide pollution and salt spray intrusion are both very serious [31–33]. At some moments of heavy traffic, the rate of carbonation of concrete will be ten times faster than normal [31]. Another more common case occurs on the concrete bridges in cities. Due to the influence of vehicle exhaust, the concrete pavement generally suffers from severe carbonation-induced deterioration. At this point, once de-icing salt is used heavily, severe combined corrosion of concrete will occur. Our previous observations indicated that the service-life of RC structures under these extreme environments will be shortened to less than one-third of the design life [32]. At this time, it is very important to study the deterioration of concrete structures under these special conditions.

Many experiments have shown that the penetration of free chloride ions and moisture are influenced significantly by carbonation [34–43]. However, most chloride- and carbonation-induced corrosion is considered independently in all existing probabilistic analyses [15–30], including the widely recognized FIB model code for service life design [44]. Therefore, we need a more accurate probabilistic model to calculate the reliability of RC structures with combined deterioration under extreme environments. The interaction mechanism of carbonation and chloride ingress was studied recently in other papers of Zhu et al. [32,33]. The influence mechanism can be summarized as: (1) the reduction in the chloride binding capacity of concrete by carbonation; (2) change in the critical radius of pores and porosity of concrete; (3) release of free chloride ions from bound chloride; and (4) the change in the threshold chloride content. Appendix A shows the mineralogical mechanism of carbonation-induced change of chloride binding capacity and the release of free chloride ions form bound chloride.

A probabilistic generalization for this combined action was presented based on that theoretical model [45]. However, the complexity of that model causes increased computational cost, meaning that the probabilistic study is expensive, and only the Monte Carlo method is applicable. Therefore, the theoretical model will be further simplified, and a series of approaches is presented to simplify the solution procedure and the limit state functions. Meanwhile, a new semi-empirical method is developed specially to swiftly calculate the probability of deterioration under the combined action, when large amounts of data are needed.

Following this introduction, the reliability model for the combined deterioration is reviewed in Section 2. Then, Section 3 provides further simplification by combining the limit state functions to be one for the original probabilistic model. Two simplified approaches for the solution of the comprehensive probabilistic model are given, too. A new semi-empirical method for the approximate reliability study of this combined deterioration is proposed in Section 4. An example is discussed in Section 5, and the conclusion follows in Section 6.

2. Review of the Reliability Model of Combined Deterioration

2.1. Governing Equation and Limit State Functions

In previous studies [1,46–49], the transport of free chloride ions in concrete is considered as meeting Fick's second law of diffusion. The one-dimensional solution of the partial differential diffusion with Crank's error is widely used. Similarly, the depth of carbonation was interpreted to be linearly proportional to the square root of time. Then, it was used to predict the probability of the carbonation-induced corrosion initiation by comparing with the thickness of the concrete cover [26–30]. However, none of these models can consider the coupling effect of carbonation and chloride attack.

To improve the shortcoming of the analytical one-dimensional probabilistic analysis, a comprehensive probabilistic model of reinforcement corrosion is developed in [45]. The governing equation of the transport of free chloride ions with carbonation can be given by [32]

$$\frac{\partial C_{fc}}{\partial t} = \nabla \cdot \left(D_{fc}^{car} \nabla C_{fc} \right) + \nabla \cdot \left(C_{fc} D_h^{car} \nabla h \right) + Q_{rc} \tag{1}$$

$$\frac{\partial C_{CO2}}{\partial t} = \nabla \cdot \left(D_{CO2} \nabla C_{CO2} \right) - Q_h \tag{2}$$

where C_{fc} is the content of free chloride ions (\% binder wt.), Q_{rc} is the release rate of free chloride ions at the carbonation interface, h is the humidity, C_{CO2} is the CO_2 concentration in the pore of concrete, Q_h is the consumption rate of hydration product in the reaction process of carbonation, involved $Ca(OH)_2$ and CSH et al. [32], D_{fc}^{car}, D_h^{car} and D_{CO2} are the apparent diffusion coefficients of free chloride ions, humidity and CO_2 with the influence of carbonation.

Then, two limit state functions are introduced to calculate the probability of corrosion initiation due to the combined action of chloride ingress and carbonation, i.e.,

$$g_1(x, t) = \eta(x, t) C_{fc,th} - C_{fc}(x, t) \tag{3}$$

$$g_2(x, t) = pH(x, t) - pH_{th} \tag{4}$$

where x is a vector reflecting the random variables $C_{fc,th}$ is the threshold chloride content; $\eta(x,t) = C_{ch,d}/[C_{ch,d}]_0$ is a correction coefficient depending on the degree of carbonation; $C_{ch,d}$ and $[C_{ch,d}]_0$ are the instantaneous and initial concentration of dissolved calcium hydroxide during the process of carbon dioxide intrusion, respectively; $pH(x,t)$ is the instantaneous pH value; and pH_{th} is the critical pH value which can cause the initiation of reinforcement corrosion.

Because the individual action of both agents can cause corrosion initiation, the failure domain under this combined action should be defined as the union of the corrosion domain due to chloride ingress and the corrosion domain due to carbonation, as shown in Figure 1. Then, the probability of corrosion initiation under the combined action $p_{f,comb}$ at time t can be given by:

$$p_{f,comb}(t) = \int_{g_1(x,t)<0 \text{ or } g_2(x,t)<0} f(x, t) dx \tag{5}$$

The stochastic processes with Karhunen-Loeve expansion are used to model the random nature of the environmental parameters varies with exposure time, including, boundary temperature T_b, boundary humidity h_b, boundary chloride concentration $C_{fc,b}$ and boundary carbon dioxide $C_{CO2,b}$. For detailed information, refer to [45–47].

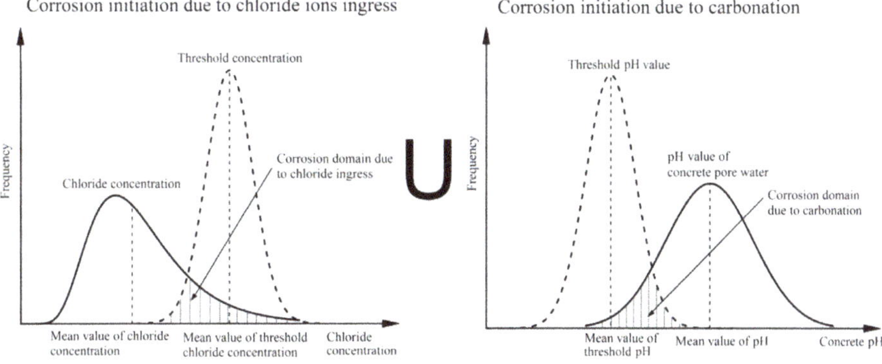

Figure 1. Prediction of corrosion initiation under the combined action.

2.2. Reduced Number of Random Variables

However, the comprehensive coupling model which simulates the combined action of chloride ingress and carbonation is very complex; it contains nine important control equations and more than one hundred parameters [32]. Therefore, the computational cost of that theoretical model is very high. Due to the complexity of the basic theoretical model, each deterministic case that is solved consumes a lot of time in the Monte Carlo method, for an actual problem.

For instance, a beam with fly ash concrete is taken as an example to study its corrosion-induced failure probability in Section 5. That model is divided into 461 2D quadrilateral elements with 536 nodes and 21,516 degrees of freedom. We use a commercial finite element software COMSOL Multiphysics to solve each sample of the deterministic theoretical model. Then, by linking with Matlab, the numerical results obtained by COMSOL are used for the probability calculation.

In a personal computer, it takes approximately 70 s per computer run of the original basic theoretical model. When we adopt the conventional solution procedure based on the Monte Carlo method, even if only eight of the most sensitive parameters are selected as independent random variables, and all of them are divided as five equal intervals, a computer still needs a very long time to solve this 2D reliability model. Therefore, it is necessary to further simplify the original basic theoretical model and develop some simplified approaches for the solution process of the probabilistic model.

Note that not all of the parameters meaningfully influence the results of the calculation in actual problems. To reduce the computational cost of the probabilistic model, we also reasonably reduce the random variables. Here, a sensitivity analysis is conducted for the 24 most important parameters. The specific analysis process, environmental condition, and concrete composition and reference value are introduced in our previous work [45]. With each key parameter increased and decreased by 20% of their original value, the tornado diagram of the changes in the initiation time of corrosion is shown in Figure 2.

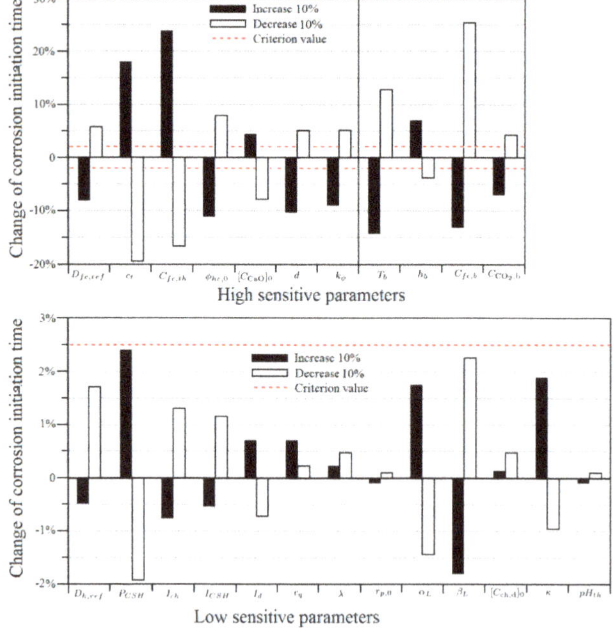

Figure 2. Change of initiation time of corrosion if each key parameter is increased and decreased by 20%.

The parameters are sorted from high to low sensitivity as follows: $C_{fc,th}$, c_t, $\varphi_{hc,0}$, d, k_φ, $D_{fc,ref}$, $[C_{CaO}]_0$, P_{CSH}, β_L, α_L, κ, $D_{h,ref}$, I_{ch}, I_{CSH}, I_d, c_q, λ_q, $[C_{ch,d}]_0$, $r_{p,0}$, and pH_{th}. These all are the parameters in our proposed comprehensive theoretical model for predicting the combined deterioration of chloride ingress and carboantion. The meaning and explanation for those parameters can be found in the literature [32]. Parameters $C_{fc,b}$, $C_{CO2,b}$, T_b, and h_b are used to determine the boundary conditions in all the cases; their sensitivity is not involved in the comparison and sorting.

In this study, only seven key parameters are selected as random variables for probabilistic analysis. These seven were selected based on the test results for one-way sensitivity, including $C_{fc,th}$, c_t, $\varphi_{hc,0}$, d, k_φ, $D_{fc,ref}$, $[C_{CaO}]_0$. Compared to a case that considers all 20 parameters to be random variables, it can be found that computational cost based on the conventional Monte Carlo method is reduced by a factor greater than 200 million.

3. Other Further Improvements for the Probabilistic Model of Combined Durability

3.1. Single Limit State Function

Because there are two limit state functions in the original probabilistic model, we cannot use other simple solution methods to solve the original model aside from the complicated Monte Carlo method. Here, a step function $\beta(x,t)$ is introduced to reflect the contribution of carbonation on the corrosion initiation, instead of a separate limit state function. Then, Equations (3) and (4) are integrated to be one limit state function, i.e.,

$$g_3(x,t) = \beta(x,t)\eta(x,t)C_{fc,th} - C_{fc}(x,t) \tag{6}$$

$$\beta(x,t) = \begin{cases} 1 & \text{for} \quad pH(x,t) - pH_{th} > 0 \\ 0 & \text{for} \quad pH(x,t) - pH_{th} \leq 0 \end{cases} \tag{7}$$

Furthermore, we found that the value of $\beta(x,t)$ is highly coincident with the value of $\eta(x,t)$. Therefore, $\beta(x,t) \times \eta(x,t)$ can be directly replaced by $\eta(x,t)$, for simplicity. Then, the failure probability under the combined deterioration can be further simplified and adjusted as:

$$p_{f,comb}(t) = \int_{g_1(x,t)<0} f(x,t)dx \tag{8}$$

With this simplification, it is possible to use of other analytical methods, such as the response surface method, to solve this probabilistic model.

3.2. Simplification for the Algorithm of Conventional Solution Procedure

The computational cost of the previous method is still very expansive, even when the random variables were reduced and the original two limit state functions were combined. Here, a simplification method for the solution procedure of the probabilistic model is also provided. This method was firstly proposed by Bazant in [48]. In the process of Latin hypercube sampling, all of the random variables are divided as n intervals. During the solving procedure, only one sample is randomly selected from the sample array of each random variable and used in one computer run. In this simplified method, the total times of runs for the theoretical deterministic models are the divided intervals of the random variables n.

3.3. Response Surface Method

In some actual problems, maybe there are only a few random variables have different mean values and variances, such as some concrete components with different mix proportions in the same environment. Then, it is not necessary to run the expansive Monte Carlo method to solve individual problems. Instead, the response surface model can be adopted. For a set of models in which only a few random variables differ, this response surface method will greatly simplify the calculation process.

According to the sampling points of primary random variables, the Monte Carlo result can be fitted and converted to the sample space of a response surface model. Based on this response surface result, the new probabilities of corrosion initiation in other cases can be easily obtained by changing the mean values. The calculation procedure is as follows:

1. The probability of corrosion initiation of a reference specimen is calculated using the Monte Carlo method.
2. The Monte Carlo result for the above reference specimen is used to build a response surface.
3. By changing the mean value of reach random variable, the probabilities of corrosion initiation in other specimens can be calculated by the above mentioned response surface.

4. Semi-Empirical Approach for the Probabilistic Study of Combined Durability

4.1. Concentration Distribution of Free Chloride ions Considering Carbonation

Here, we also provide a simple semi-empirical method for this combined deterioration problem. Except for the reliability analysis, its simple equations are also particularly suitable for some special circumstances of the application, such as the compilation of design specifications. According to the coupling mechanism of carbonation and chloride ingress [32] and the chemical reaction mechanism of carbonation [49,50], we believe that the influence of chloride ions concentration for the rate of carbonation reaction is negligible. The classic expression of carbonation depthwith $\sqrt{t}$ is still applicable in this combined durability problem.

Here, a partial carbonated depth (carbonation-process zone) is introduced to optimize Papadakis' analytical formula [49], i.e.,

$$c_{car} = \begin{cases} 0 & \text{for} \quad t \le \frac{c_{par}^2 [C_{CaO}]_0}{2 D_{CO2} C_{CO2,b}} \\ \sqrt{\frac{2 D_{CO2} C_{CO2,b} \cdot t}{[C_{CaO}]_0}} & \text{for} \quad t > \frac{c_{par}^2 [C_{CaO}]_0}{2 D_{CO2} C_{CO2,b}} \end{cases} \tag{9}$$

where c_{car} and c_{par} are the fully and partial carbonated depth.

Previous studies have shown that the reaction rate of carbon dioxide with $Ca(OH)_2$ is much faster than with CSH and Friedel's salt. That means that the release phenomenon of free chloride ions from the Friedel's salt should be mainly after $Ca(OH)_2$ is completely consumed by carbon dioxide. Therefore, the critical degree of carbonation corresponding to this moment is significantly meaningful for the combined deterioration caused by chloride with carbonation. Here, the partial carbonated zone is defined as the degrees of carbonation in the following range:

$$1 - P_{CSH} \le \alpha_c < 1 \tag{10}$$

By analyzing the numerical results and experimental data in [32,34,36,38], it shows that the value of c_{par} depends on the diffusion coefficient of carbon dioxide. This is determined by the mechanism of this partial carbonation phenomenon. If carbon dioxide is transported too fast, the calcium hydroxide and CSH at the reaction interface cannot absorb it completely in time. Then, part of the carbon dioxide passes through the interface and continues to transport to the interior of the concrete. Its formula can be fitted as a cubic polynomial function, i.e.,

$$c_{par} = 0.0158 \overline{D}^3 - 0.3702 \overline{D}^2 + 36656 \overline{D} + 5.945 \quad \text{for} \quad 0.1 \le \overline{D} \le 10 \tag{11}$$

where $\overline{D} = D_{CO2} / D_{CO2}^{ref}$ is a dimensionless parameter, $D_{CO2}^{ref} = 1 \times 10^{-8} \text{ m}^2/s$ is a reference value for the diffusion coefficient of CO_2.

Then, we can decouple the combined effect of carbonation and chloride ingress with a semi-empirical model. The transport of free chloride ions is considered as a separate diffusion phenomenon. The completely and partial carbonated zones of concrete are both removed, and the front surface of carbonation is reset as the new boundary for the transport of chloride ions into the internal concrete.

Note that the value of the new boundary condition is also changed because of the carbonation-induced release of free chloride ions. It can be regarded as a linear relationship with the depth of carbonation. The slope of this line can be determined by the mixture proportions of concrete. The theoretical mechanism of that semi-empirical approach is as follows: generally, the depth of carbonation interface is very small and close to the concrete boundary; once a large number of free chloride ions are released at the carbonation interface, a rapid reverse diffusion process will be formed from the carbonation interface to the concrete boundary; and, the peak value of chloride concentration at the carbonation interface is only related to the mineralogical characteristic of concrete and the diffusion rate of chloride ions.

According to large amounts of data, the slope of the linear change of chloride concentration in the carbonated concrete zone, β_c, can be fitted as:

$$\beta_c = -0.004 f_{car}^3 + 0.048 f_{car}^2 - 0.147 f_{car} + 0.201 \quad \text{for} \quad 1.5 \le f_{car} \le 5 \tag{12}$$

for the ordinary Portland Concrete, and

$$\beta_c = -0.009 f_{car}^3 + 0.089 f_{car}^2 - 0.340 f_{car} + 0.399 \quad \text{for} \quad 1.5 \le f_{car} \le 5 \tag{13}$$

for fly ash concrete; where f_{car} is the comprehensive influence function of carboantion for the apparent diffusion coefficient of free chloride ion, and its estimated method can be found in [32]. Here, the unit of carbonation depth should be taken as centimeters (cm).

Then, a semi-empirical formula for predicting the chloride concentration considering the effect of carbonation can be expressed as:

$$C_{fc}(y) = (1 + \beta_c \cdot y)C_{fc,b} \tag{14}$$

for $y \le c_{car} + c_{par}$; and

$$C_{fc}(y) = C_{fc,0} + [(1 + \beta_c \cdot y)C_{fc,b} - C_{fc,0}]\left[1 - \operatorname{erf}\left(\frac{y - c_{car} - c_{par}}{2\sqrt{D_{fc,app} \cdot t}}\right)\right] \tag{15}$$

for $y > c_{car} + c_{par}$; where y shows the position of the calculating point, and $C_{fc,0}$ is the initial condition for the chloride concentration.

The advantage of this semi-empirical method is that the formula of chloride concentration distribution considering the effect of carbonation can be obtained directly, instead of solving the complex and complicated partial differential equations. This will enable the engineers and designers to directly use the simple expression function to predict the combined deterioration of structures.

4.2. Experimental Verification and Comparison with the Comprehensive Combined Model for the Semi-Empirical Solution

Here, an alternating test of carboantion and chloride penetration is used to verify this proposed semi-empirical solution [38]. In this test, three Ordinary Portland concrete with different mix proportions are subjected to carbonation for three days by placing in a carbonation chamber. These specimens are named as D-1, D-2, and D-3. For D-1, D-2, and D-3, the vaules of w/b are 0.45, 0.50, 0.55, respectively; and the vaules of a/b are 0.45, 0.50, 0.55, respectively. Before the alternating test, three Ordinary Portland concrete specimens have been used to test the diffusion coefficient of free chloride ions without carbonation. For D-1, D-2, and D-3, their apparent chloride diffusion coefficients are 3.8×10^{-12}, 4.5×10^{-12}, and 5.6×10^{-12} m^2/s, respectively.

Then, the volume fraction of CO_2 was set to be 5%, and the temperature and humidity were also set to be constants, 20 °C and 65%, respectively, for the testing chamber. After three days of carbonation, those specimens are then placed in a NaCl solution with 0.5 M. A year later, these two procedures were alternately carried out 120 times.

The distribution of free chloride ions from the alternating test, the comprehensive combined model, and the proposed semi-empirical solution, are shown and compared in Figure 3. It can be found that the peak value of chloride ions content appeared on the carbonation interface, rather than the boundary surface, which is due to the carboantion-induced release of free chloride ions from the bound chloride. The results calculated by the semi-empirical solution are close to the alternating experimental results and the comprehensive combined numerical results.

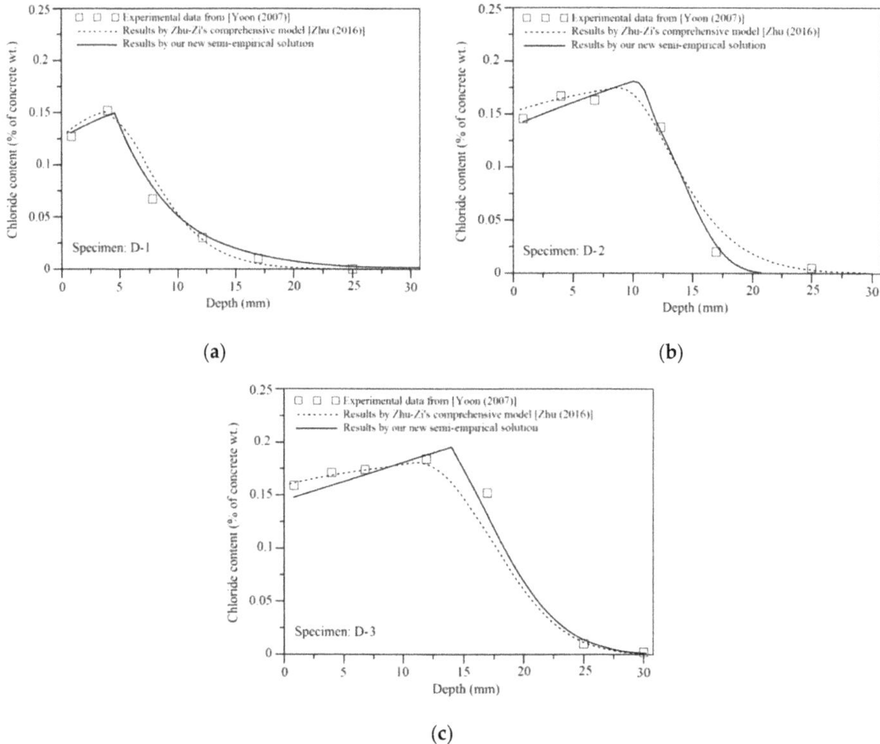

Figure 3. Distribution of the chloride ions for (**a**) D-1, (**b**) D-2, and (**c**) D-3.

4.3. Probability Calculation

According to above-mentioned semi-empirical formulae, the beginning time of corrosion is defined as when the chloride content around the steel rebar exceeds the threshold concentration, or the length of $c_{car} + c_{par}$ larger than the thickness of concrete cover. Then, the limit state functions is formulated by:

$$g_4(x, t) = \gamma(x, t)C_{fc,th} - C_{fc}(x, t) \tag{16}$$

$$g_5(x, t) = c_t - c_{car}(x, t) - c_{par}(x, t) \tag{17}$$

where $\gamma(x,t)$ is a dimensionless parameter reflecting the carbonation-induced decrease of the threshold value of chloride concentration.

As we know, the new value of chloride threshold content depends on the content of calcium hydroxide. The front of the partial carbonated zone should be taken as the interface. In the range of carbonation-process zone, the change of $\gamma(x,t)$ is assumed as linear, for simplicity. Then, $\gamma(x,t)$ can be formulated as:

$$\gamma(x,t) = \begin{cases} 0 & \text{for} & y \leq c_{car}(x,t) \\ \frac{1}{c_{par}(x,t)}[y - c_{car}(x,t)] & \text{for} & c_{car}(x,t) < y \leq c_{car}(x,t) + c_{par}(x,t) \\ 1 & \text{for} & y > c_{car}(x,t) + c_{par}(x,t) \end{cases} \tag{18}$$

Similar to the discussion in Section 3.2, the contribution of carbonation on the initiation of corrosion also has been reflected in the parameter $\gamma(x,t)$. Then, $g_5(x,t)$ can be ignored. The probability of this combined deterioration should be approximately given as:

$$p_{f,comb}(t) = \int_{g_4(x,t)<0} f(x,t)\mathrm{d}x \tag{19}$$

5. Illustrative Example and Parametric Study

5.1. Problem Description

A concrete beam with 300 mm $\times$ 500 mm is used for the parametric study. For the concrete, we assume that the $a/b = 4$, $w/b = 0.45$, and $FA = 10\%$. Four steel reinforcements with 22 mm are set in the bottom of this beam. It is assumed that this beam is under severe salt fog environment. $T_{b,max} = 30$ °C and $T_{b,min} = -10$ °C are defined as the maximum and minimum mean values of environmental temperature, with the correlation length 1 year. $h_{b,max} = 80\%$ and $h_{b,min} = 70\%$ are defined as the maximum and minimum mean values of environmental relative humidity, with the correlation length 0.001 year. The effect of saturation on the transport rates of chloride and carbon dioxide is taken into account in their diffusion coefficients by a correction function of relative humidity. Table 1 summarizes the parameter characteristics of random variables for this illustrative example, as suggested by our previous work [32,45].

Table 1. Parameter characteristics of random variables for this illustrative example.

Variable	Mean Value	COV	Distribution
$C_{fc,b}$	0.75% binder w. t.	0.2	Log-Normal
$C_{CO2,b}$	0.02 mol/m^3 of pore air	0.1	Normal
$C_{fc,th}$	0.4% binder w. t.	0.2	Normal
c_t	40 mm	0.2	Normal
$D_{fc,ref}$	6.5×10^{-12} m^2/s	0.2	Log-Normal
$\varphi_{hc,0}$	0.5	0.3	Normal
$[C_{CaO}]_0$	3000 mol/m^3 of concrete	0.05	Normal
d	0.9	0.05	Normal
k_φ	0.2	0.1	Normal

For the prediction of corrosion initiation of steel reinforcement, previous study indicated that numerous parameters can affect the threshold concentration of chloride ions, and many of them are interrelated [51]. This leads to that the overall trends are not visible for the changing of threshold chloride content. Among so many factors, the influence of the steel-concrete interface is considered the most significant [52–56]. In this example, the threshold chloride content is defined as the chloride content required for depassivation of the steel. The ribbed steel bars are taken as the reinforcements. Some pores exist between reinforcement and concrete due to the low workability. The reinforcement and the surrounding concrete are always in the tension state. All these assumptions mean that the threshold chloride content should take a small value, as given in Table 1.

5.2. Result Discussion

5.2.1. Probability

The failure probabilities of corrosion-induced deterioration due to the combined action, penetration of chloride and carbonation calculated by our model can be found in Figure 4. A comparison of Figure 4a with Figure 4b,c shows that the corrosion-induced failure probability considering the influence of carbonation is significantly larger than the probability without considering carbonation.

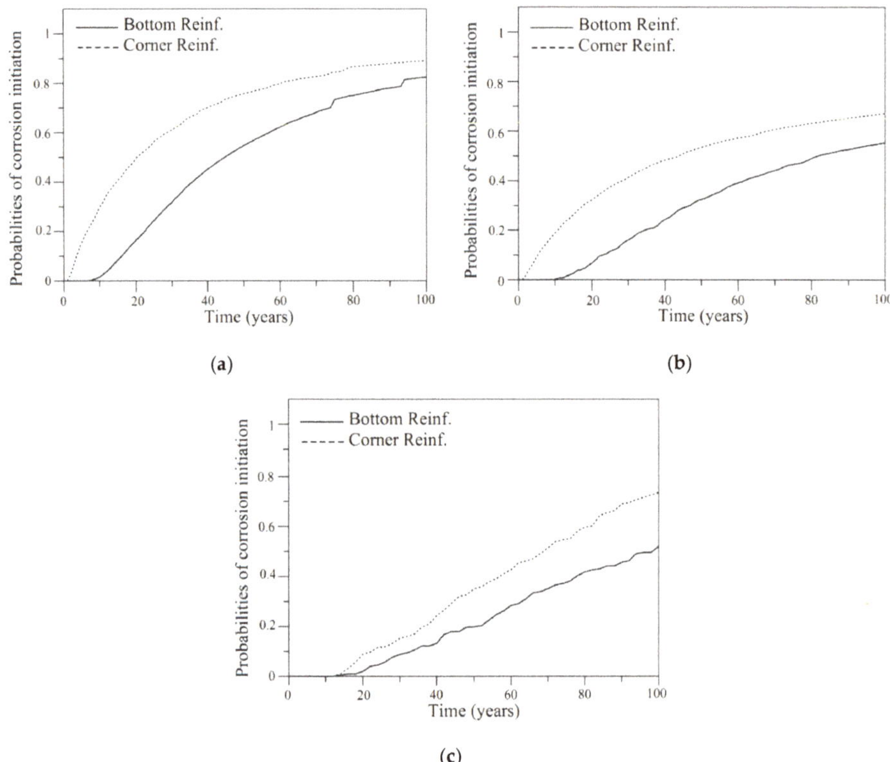

Figure 4. Probability of corrosion initiation calculated by our model (**a**) due to the combined action of chloride penetration and carbonation, and (**b**) only due to the ingress of chloride, and (**c**) only due to the carbonation.

5.2.2. Effect of Reducing Limit State Functions

The probabilities with and without reducing the limit state functions are both shown in Figure 5. It can be seen that the results considering this simplification method are almost identical to the original results. This indicates that the method provided in Section 3.1 is feasible for this comprehensive probabilistic model.

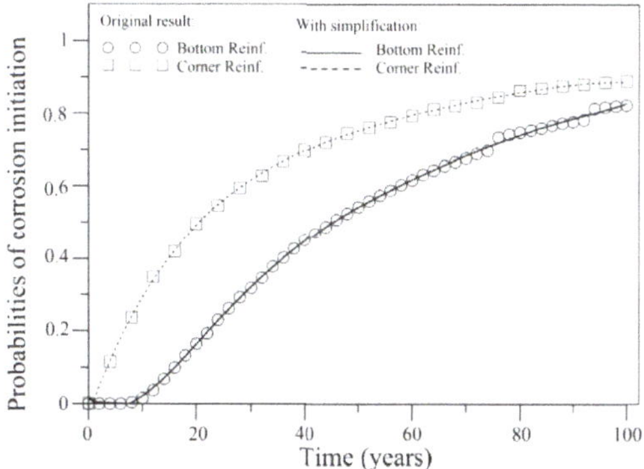

Figure 5. Probabilities with and without the reduction of limit state functions for the bottom reinforcement and the corner reinforcement.

5.2.3. Effect of Simplified Solution Procedure

The comparison of corrosion initiation probabilities with and without the simplified solution procedure are shown in Figure 6. In the simplified approach, all of the random variables is divided into 1000 intervals by the Latin hypercube sampling. It was found that the corrosion-induced failure probabilities calculated by the conventional solution procedure, and the method with simplification, are close. Here, an average error, e, for checking the difference between corrosion probabilities with and without this simplification is introduced here, and its calculation equation is defined in Appendix B. It is found that the average error is 0.0124 and 0.0254 for the bottom reinforcement and the corner reinforcement, respectively. This accuracy is acceptable for the reliability analysis of a concrete durability problem.

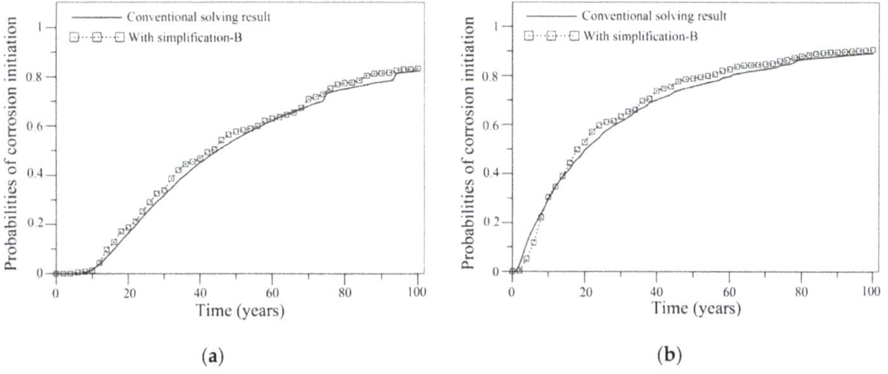

Figure 6. Corrosion-induced failure probabilities with and without the simplification for solution procedure for the (a) bottom and (b) corner reinforcement.

5.2.4. Application of the Response Surface Method

Here, three fly ash concrete and three ordinary Portland concrete sections are used to explain the application of the above mentioned response surface method in the Section 3.3, and verify its accuracy. Their mix proportions are given in Table 2. They are applied to the estimation of parameters for the theoretical model. The mean values of random variables of each specimen are calculated according to [32]. The results from the FAC-1 specimen calculated by the Monte Carlo method are taken to build the reference response surface.

The probabilities of the corrosion initiation obtained by the Monte Carlo method and the response surface method are shown and compared in Figure 7. It shows that the results calculated by these two methods closely concur. The average error is 0.0120, 0.0139, 0.0343, 0.0345, and 0.0285 for specimens OPC-1, OPC-2, OPC-3, FAC-2 and FAC-3, respectively. However, the amount of calculation will be cut by several orders of magnitude if this response surface method is obtained instead of the conventional Monte Carlo method.

Table 2. Mix proportions of concretes.

Mix Proportions	Specimen					
	FAC-1 (Ref.)	OPC-1	OPC-2	OPC-3	FAC-2	FAC-3
w/b	0.45	0.35	0.35	0.55	0.45	0.45
a/b	4	4	6	4	4	4
FA	0.1	0	0	0	0.2	0.3

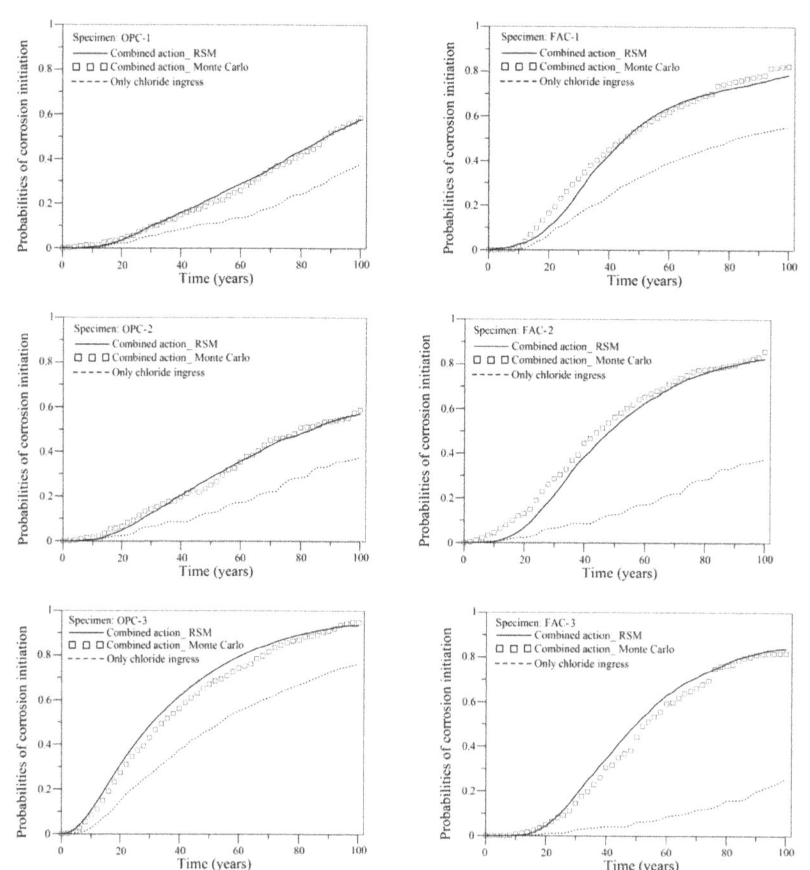

Figure 7. Comparison of the Monte Carlo result and the response surface result for concrete specimens with different mix proportions.

5.2.5. Discussion for the Semi-Empirical Probabilistic Model

The accuracy and error calculated by the semi-empirical probabilistic method is discussed in this section. The parameter characteristics of the selected random variables in the study with semi-empirical method is shown in Table 3. Here, the COVs of β_c and c_{par} are fitted by the numerical results of the comprehensive model. The range of each random variable is also divided as five intervals for Latin hypercube sampling.

The probabilities obtained by the comprehensive model and the proposed semi-empirical method are compared in Figure 8. It shows that the probability curves obtained by these two approaches are very close. The average error is about 0.049; this can meet the needs of the accuracy for the actual project.

Table 3. Parameter characteristics of the selected random variables in semi-empirical probabilistic study.

Variable	Mean Value	COV	Distribution
$C_{fc,b}$	0.75% binder w. t.	0.2	Log-Normal
$C_{CO2,b}$	0.02 mol/m^3 of pore air	0.1	Normal
$C_{fc,th}$	0.4% binder w. t.	0.2	Normal
c_t	40 mm	0.2	Normal
$D_{fc,ref}$	6.5×10^{-12} m^2/s	0.2	Log-Normal
D_{CO2}	3×10^{-8} m^2/s	0.2	Log-Normal
$[C_{CaO}]_0$	3000 mol/m^3 of concrete	0.05	Normal
β_c	0.15	0.1	Normal
c_{par}	15 mm	0.05	Normal

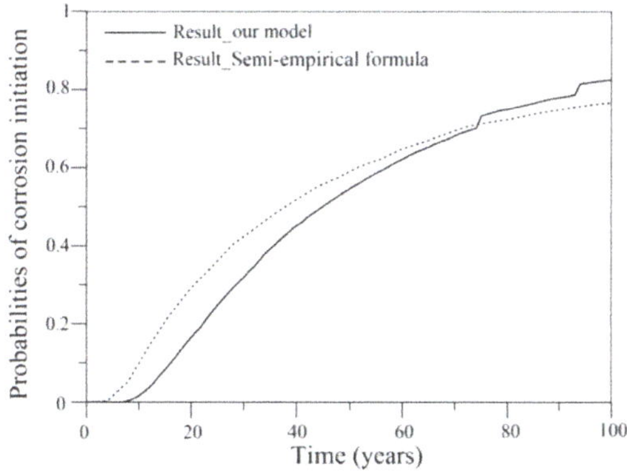

Figure 8. Corrosion-induced failure probabilities calculated by the comprehensive model and the proposed semi-empirical method.

5.2.6. Comparison with the Conventional One-Dimensional Solution

The failure probabilities of corrosion-induced deterioration obtained by our model and the widely used analytical one-dimensional method are compared in Figure 9. We can find that the difference of these probability results are significant. This indicates that the influence of the uncertainty of chloride binding capacity, convection, temperature, and humidity should be considered in the probabilistic analysis for a more accurate result. Only using the analytical one-dimensional simplified method to estimate the reliability of the durability of reinforced concrete structure is not rigorous for some cases.

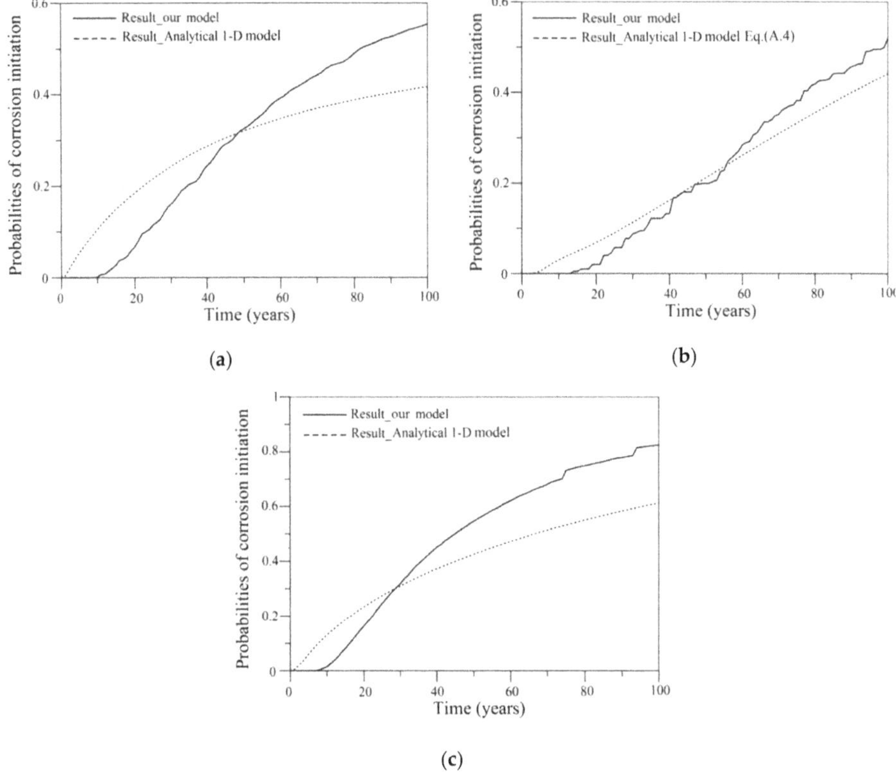

Figure 9. Corrosion-induced failure probabilities obtained by the comprehensive model and the conventional one-dimensional solution under the (**a**) penetration of free chloride ions, (**b**) carbonation, and (**c**) their combined effect.

6. Conclusions

1. A comprehensive model is reviewed to calculate the reliability of corrosion deterioration due to carbonation, chloride penetration, and their combined effect.
2. Two simplified approaches for the solution procedure of this probabilistic model are provided.
3. A semi-empirical method is developed to approximately predict the concentration distribution of chloride ions, and swiftly calculate the probability of combined deterioration.
4. Comparison to the original non-simplified model shows that all of these improved approaches can greatly simplify the computational cost, and that the accuracies are acceptable.

Author Contributions: Conceptualization and Methodology, X.Z.; Validation, X.Z. and L.X.; Resources, Z.C.; Writing-Original Draft Preparation, X.Z. and Z.C.; Writing-Review & Editing, Y.C. and H.W.

Funding: This work was financially supported by the Natural Science Foundation of Shandong Province (No. ZR2018BEE044), and the National Natural Science Foundation of China (No. 51678199, 51578151, 51678208). We also appreciates the support of the Major Program of Mutual Foundation of Weihai City with Harbin Institute of Technology (Weihai).

Appendix A. Mechanism of Chemical Link between Carbonation and Chloride Movement

With the ingress of free chloride ions, the Friedel's salt (bound chloride) is formed in concrete. That chemical reaction process can be depicted by

$$Ca_4AlH_6 + CaCl_2 + H_2O \rightarrow 3CaO{\cdot}Al_2O_3{\cdot}CaCl_2{\cdot}10H_2O$$

Noted that the process of carbonation is much slower than the movement of chloride. Once carbon dioxide comes into contact with the formed Friedel's salt, free chloride ions will be released at the carbonation interface with the following chemical reaction

$$3CaO{\cdot}Al_2O_3{\cdot}CaCl_2{\cdot}10H_2O + 3CO_2 \rightarrow 3CaCO_3 + 2Al(OH)_3 + CaCl_2 + 7H_2O$$

Appendix B. Calculation of Average Error *e*

The average error e within y years is given by

$$e = \sum_{t_i=1}^{y} \frac{\left| p_{f,sim}(t_i) - p_{f,0}(t_i) \right|}{y} \tag{A1}$$

where y is an integer showing the number of exposure years, $p_{f,sim}(t_i)$ is the probability of corrosion initiation at different time nodes t_i after simplification and $p_{f,0}(t_i)$ is the initial probability of corrosion initiation at different time nodes t_i without any simplification.

References

1. Böhni, H. (Ed.) *Corrosion in Reinforced Concrete Structures*; Woodhead Publishing: Oxford, UK, 2005.
2. Bazant, Z.P. Physical model for steel corrosion in concrete sea structures-theory. *ASCE J. Struct. Div.* **1979**, *105*, 1137–1153.
3. Safehian, M.; Ramezanianpour, A.A. Prediction of RC structure service life from field long term chloride diffusion. *Comput. Concr.* **2015**, *15*, 589–606. [CrossRef]
4. Kwon, S.J.; Kim, S.C. Concrete mix design for service life of RC structures exposed to chloride attack. *Comput. Concr.* **2012**, *10*, 587–607. [CrossRef]
5. Ho, D.W.S.; Lewis, R.K. Carbonation of concrete and its prediction. *Cem. Concr. Res.* **1987**, *17*, 489–504. [CrossRef]
6. Demis, S.; Papadakis, V. Software-assisted comparative assessment of the effect of cement type on concrete carbonation and chloride ingress. *Comput. Concr.* **2012**, *10*, 391–407. [CrossRef]
7. Na, O.; Cai, X.; Xi, Y. Corrosion prediction with parallel finite element modeling for coupled hygro-chemo transport into concrete under chloride-rich environment. *Materials* **2017**, *10*, 350. [CrossRef] [PubMed]
8. Isteita, M.; Xi, Y. The effect of temperature variation on chloride penetration in concrete. *Constr. Build. Mater.* **2017**, *156*, 73–82. [CrossRef]
9. Homan, L.; Ababneh, A.; Xi, Y. The effect of moisture transport on chloride penetration in concrete. *Constr. Build. Mater.* **2016**, *125*, 1189–1195. [CrossRef]
10. Ozbolt, J.; Orsanic, F.; Balabanic, G.; Kuster, M. Modeling damage in concrete caused by corrosion of reinforcement: Coupled 3D FE model. *Int. J. Fract.* **2012**, *178*, 233–244. [CrossRef]
11. Ozbolt, J.; Balabanic, G.; Kuster, M. 3D Numerical modelling of steel corrosion in concrete structures. *Corros. Sci.* **2011**, *53*, 4166–4177. [CrossRef]
12. Cao, C.; Cheung, M.; Chan, B. Modelling of interaction between corrosion-induced concrete cover crack and steel corrosion rate. *Corros. Sci.* **2013**, *69*, 97–109. [CrossRef]
13. Pantazopoulou, S.; Papoulia, K. Modeling cover-cracking due to reinforcement corrosion in RC structures. *J. Eng. Mech.-ASCE* **2001**, *127*, 342–351. [CrossRef]
14. Zhu, X.; Zi, G. A 2D mechano-chemical model for the simulation of reinforcement corrosion and concrete damage. *Constr. Build. Mater.* **2017**, *137*, 330–344. [CrossRef]
15. Saassouh, B.; Lounis, Z. Probabilistic modeling of chloride-induced corrosion in concrete structures using first-and second-order reliability methods. *Cem. Concr. Comp.* **2012**, *34*, 1082–1093. [CrossRef]

16. Shin, K.; Kim, J.; Lee, K. Probability-based durability design software for concrete structures subjected to chloride exposed environments. *Comput. Concr.* **2011**, *8*, 511–524. [CrossRef]
17. Kong, J.; Ababneh, A.; Frangopol, D.; Xi, Y. Reliability analysis of chloride penetration in saturated concrete. *Probabl. Eng. Mech.* **2002**, *17*, 305–315. [CrossRef]
18. Stewart, M.; Mullard, J. Spatial time-dependent reliability analysis of corrosion damage and the timing of first repair for RC structures. *Eng. Struct.* **2007**, *29*, 1457–1464. [CrossRef]
19. Stewart, M.; Suo, Q. Extent of spatially variable corrosion damage as an indicator of strength and time-dependent reliability of RC beams. *Eng. Struct.* **2009**, *31*, 198–207. [CrossRef]
20. Ryan, P.; O'Connor, A. Probabilistic analysis of the time to chloride induced corrosion for different self-compacting concretes. *Constr. Build. Mater.* **2013**, *47*, 1106–1116. [CrossRef]
21. Nogueira, C.; Leonel, E. Probabilistic models applied to safety assessment of reinforced concrete structures subjected to chloride ingress. *Eng. Fail. Anal.* **2013**, *31*, 76–89. [CrossRef]
22. Strauss, A.; Wendner, R.; Bergmeister, K.; Costa, C. Numerically and experimentally based reliability assessment of a concrete bridge subjected to chloride-induced deterioration. *J. Infrastruct. Syst.* **2012**, *19*, 166–175. [CrossRef]
23. Val, D.; Trapper, P. Probabilistic evaluation of initiation time of chloride-induced corrosion. *Reliab. Eng. Syst. Saf.* **2008**, *93*, 364–372. [CrossRef]
24. Saetta, A.; Scotta, R.; Vitaliani, R. Analysis of chloride diffusion into partially saturated concrete. *ACI Mater. J.* **1993**, *90*, 441–451.
25. Suo, Q.; Stewart, M. Corrosion cracking prediction updating of deteriorating RC structures using inspection information. *Reliab. Eng. Syst. Saf.* **2009**, *94*, 1340–1348. [CrossRef]
26. Bastidas-Arteaga, E.; Schoefs, F.; Stewart, M.; Wang, X. Influence of global warming on durability of corroding RC structures: A probabilistic approach. *Eng. Struct.* **2013**, *51*, 259–266. [CrossRef]
27. Duprat, F.; Vu, N.; Sellier, A. Accelerated carbonation tests for the probabilistic prediction of the durability of concrete structures. *Constr. Build. Mater.* **2014**, *66*, 597–605. [CrossRef]
28. Duprat, F.; Vu, N. Probabilistic threshold for the onset of carbonation-induced corrosion. *Eur. J. Environ. Civ. Eng.* **2013**, *17*, 478–495. [CrossRef]
29. Lollini, F.; Redaelli, E.; Bertolini, L. Analysis of the parameters affecting probabilistic predictions of initiation time for carbonation-induced corrosion of reinforced concrete structures. *Mater. Corros.* **2012**, *63*, 1059–1068. [CrossRef]
30. Mai-Nhu, J.; Sellier, A.; Duprat, F.; Rougeau, P.; Caprad, B.; Hyvert, N.; Francisco, P. Probabilistic approach for durable design of concrete cover: Application to carbonation. *Eur. J. Environ. Civ. Eng.* **2012**, *16*, 264–272. [CrossRef]
31. Huang, T. The Experimental Research on the Interaction between Concrete Carbonation and Chloride Ingress under Loading. Master's Thesis, Zhejiang University, Hangzhou, China, 2013.
32. Zhu, X.J.; Zi, G.; Cao, Z.; Cheng, X. Combined effect of carbonation and chloride ingress in concrete. *Constr. Build. Mater.* **2016**, *110*, 369–380. [CrossRef]
33. Zhu, X.J.; Kim, S.; Kwak, D.; Bea, K.; Zi, G. Parametric analysis for the simultaneous carbonation and chloride ion penetration in reinforced con crete sections. *J. Korea Inst. Struct. Maint. Insp.* **2016**, *20*, 66–74.
34. Lee, M.; Jung, S.; Oh, B. Effects of carbonation on chloride penetration in concrete. *ACI Mater. J.* **2013**, *110*, 559–566. [CrossRef]
35. Backus, J.; Mcpolin, D.; Basheer, M.; Long, A.; Holmes, N. Exposure of mortars to cyclic chloride ingress and carbonation. *Adv. Cem. Res.* **2013**, *25*, 3–11. [CrossRef]
36. Tumidajski, P.; Chan, G. Effect of sulfate and carbon dioxide on chloride diffusivity. *Cem. Concr. Res.* **1996**, *26*, 551–556. [CrossRef]
37. Yoon, I. Simple approach to calculate chloride diffusivity of concrete considering carbonation. *Comput. Concr.* **2009**, *6*, 1–18. [CrossRef]
38. Yoon, I. Deterioration of concrete due to combined reaction of carbonation and chloride penetration: Experimental study. *Key Eng. Mater.* **2007**, *348–349*, 729–732. [CrossRef]
39. Yoon, I. Simple approach for computing chloride diffusivity of (non)carbonated concrete. *Key Eng. Mater.* **2008**, *385–387*, 281–284. [CrossRef]
40. Yuan, C.; Niu, D.; Luo, D. Effect of carbonation on chloride diffusion in fly ash concrete. *Disaster Adv.* **2012**, *5*, 433–436.

41. Delnavaz, A.; Ramezanianpour, A. The assessment of carbonation effect on chloride diffusion in concrete based on artificial neural network model. *Mag. Concr. Res.* **2012**, *64*, 877–884. [CrossRef]

42. Ngala, V.; Page, C. Effects of carbonation on pore structure and diffusional properties of hydrated cement pastes. *Cem. Concr. Res.* **1997**, *27*, 995–1007. [CrossRef]

43. CEB-FIP. *Fib Model Code for Concrete Structures*; Wiley-VCH Verlag GmbH: Weinheim, Germany, 2010.

44. Zhang, S.; Zhao, B. Research on chloride ion diffusivity of concrete subjected to CO_2 environment. *Comput. Concr.* **2015**, *15*, 589–606.

45. Zhu, X.J.; Zi, G.; Lee, W.; Kim, S.; Kong, J. Probabilistic analysis of reinforcement corrosion due to the combined action of carbonation and chloride ingress in concrete. *Constr. Build. Mater.* **2016**, *124*, 667–680. [CrossRef]

46. Ghanem, R.; Spanos, P. *Stochastic Finite Elements: A Spectral Approach*; Springer: New York, NY, USA, 1991; pp. 15–36.

47. Xiu, D. *Numerical Methods for Stochastic Computations: A Spectral Method Approach*; Princeton University Press: Princeton, NJ, USA, 1991; pp. 33–45.

48. Bazant, Z.P.; Liu, K. Random creep and shrinkage in structures: Sampling. *J. Struct. Eng. ASCE* **1985**, *111*, 1113–1134. [CrossRef]

49. Papadakis, V.; Vayenas, C.; Fardis, M. Fundamental modeling and experimental investigation of concrete carbonation. *ACI Mater. J.* **1991**, *88*, 363–373.

50. Papadakis, V.; Vayenas, C.; Fardis, M. Experimental investigation and mathematical modeling of the concrete carbonation problem. *Chem. Eng. Sci.* **1991**, *46*, 1333–1338. [CrossRef]

51. Angst, U.; Elsener, B.; Larsen, C.; Vennesland, Ø. Critical chloride content in reinforced concrete—A review. *Cem. Concr. Res.* **2009**, *39*, 1122–1138. [CrossRef]

52. Söylev, T.; François, R. Corrosion of reinforcement in relation to presence of defects at the interface between steel and concrete. *J. Mater. Civ. Eng.* **2005**, *17*, 447–455. [CrossRef]

53. Shi, J.; Ming, J. Influence of defects at the steel-mortar interface on the corrosion behaviour of steel. *Constr. Build. Mater.* **2017**, *136*, 118–125. [CrossRef]

54. Horne, T.; Richardson, I.; Brydson, R. Quantitative analysis of the microstructure of interfaces in steel reinforced concrete. *Cem. Concr. Res.* **2007**, *37*, 1613–1623. [CrossRef]

55. Angst, U.; Elsener, B.; Larsen, C.; Vennesland, Ø. Chloride induced reinforcement corrosion: Electrochemical monitoring of initiation stage and chloride threshold values. *Corros. Sci.* **2011**, *53*, 1451–1464. [CrossRef]

56. Zhang, R.; François, R.; Castel, A. Influence of steel-concrete interface defects owing to the top-bar effect on the chloride-induced corrosion of reinforcement. *Mag. Concr. Res.* **2011**, *63*, 773–781. [CrossRef]

Article

Brazier Effect of Thin Angle-Section Beams under Bending

Zhiguang Zhou [1,2] **, Liuyun Xu** [1] **, Chaoxin Sun** [3] **and Songtao Xue** [1,*]

[1] Research Institute of Structural Engineering and Disaster Reduction, Tongji University, Shanghai 200092, China; zgzhou@tongji.edu.cn (Z.Z.); 1832641@tongji.edu.cn (L.X.)
[2] State Key Laboratory of Disaster Reduction in Civil Engineering, Tongji University, Shanghai 200092, China
[3] General Contracting Department, Shanghai Construction Group, Shanghai 200036, China; suncx1991@163.com
* Correspondence: xue@tongji.edu.cn; Tel.: +86-21-6598-1033

Received: 30 July 2018; Accepted: 22 August 2018; Published: 27 August 2018

Abstract: Thin-walled section beams have Brazier effect to exhibit a nonlinear response to bending moments, which is a geometric nonlinearity problem and different from eigenvalue problem. This paper is aimed at investigating the Brazier effect in thin-walled angle-section beams subjected to pure bending about its weak axis. The derivation using energy method is presented to predict the maximum bending moment and section deformation. Both numerical analyses and experimental results were used to show the validity of the proposed formula. Numerical results show that the boundary condition can influence the results due to the end effect, and that the influence tends to be negligible when the length of angle beam goes up to 30 times as the length of beam side. When the collapse in experiments is governed by Brazier flattening, the moment vs. curvature curve deviates significantly from the linear beam theory, but coincides well with the proposed formula in consideration of the restraint due to limited span of experimental setup. It can be concluded that the proposed formula shows good agreement with numerical results and experimental results.

Keywords: Brazier effect; angle section; Brazier flattening; variational method; numerical simulation; beam

---

1. Introduction

When an initially straight thin-walled circular cylindrical shell is subjected to bending, there is a tendency for the cross section to become progressively more oval as the curvature increases. The moment–curvature response deviates significantly from the linear beam theory. This nonlinear bending response phenomenon is due to geometric nonlinearity and was first investigated by Brazier [1] in 1927, thus it is called Brazier effect or Brazier flattening. Brazier flattening is not an eigenvalue problem, which is different from buckling. The main characteristic of Brazier flattening is the reduction of flexural stiffness of the shell with the increase of curvature. Furthermore, under steadily increasing curvature, the bending moment has a maximum value that is defined as the instability critical moment, which is given by employing the well-known energy method. Reissner [2] reconsidered Brazier's theory and demonstrated that Brazier's solution is a first-order approximate solution. Reissner also derived the second-order approximate solution and found the maximum moment is 8% lower than the value Brazier had gotten. Either is a geometric nonlinearity problem based on elastic material.

Aksel'rad [3] developed the application of Brazier's theory to finite length shells. Gerber [4] studied Brazier effect by taking the plastic properties into consideration. Tomasz and Sinmao [5] proposed a theoretical model to predict the bending moment and section deformation of cylindrical tubes under pure plastic bending. Li and Kettle [6] investigated the nonlinear bending response of finite length cylindrical shells with stiffening rings using a modified Brazier approach. Poonaya [7] developed a close-form solution of thin-walled circular tube subjected to bending using a rigid plastic mechanism analysis. Besides circular sections, studies also have been carried out on the analysis of other shapes. The deformation of angle-section beams has been analyzed by Yu [8], while it is not based on Brazier's energy principle. The Brazier effect in elliptical cross sections, box sections, rectangular sections, airfoil sections, and wind turbine blades have been analyzed by Huber [9], Rand [10], Paulsen and Welo [11], Cecchini and Weaver [12] and Jensen [13], respectively. In addition, the Brazier effect of elastic pipe beams, single- and double-walled elastic tubes, and multilayered cylinders have been studied by Luongo [14], Sato [15], and Shima [16], respectively. The studes on the other shapes of sections demonstrated that all long thin-walled beams exhibit a nonlinear response to bending moments. However, the Brazier effect of angle-section beams is not yet studied on the basis of the variational method as Brazier have done for cylindrical tubes

A cylinder under bending, however, can also take place another instability phenomenon which is called bifurcation instability, i.e. buckling. The main characteristic of buckling is a form of longitudinal wavy type 'wrinkles' caused by the increased axial stress at the compression side due to ovalization. The interaction between Brazier flattening and bifurcation buckling has been investigated by many researchers such as Stephens [17], Fabian [18], Libai [19] and Karamanos [20]. Karamanos's study shows that the interaction between the two instability modes depends on the value and the sign of the initial tube curvature. Tatting [21,22] investigated the local buckling behavior of composite shells associated with Brazier effect. Angle-section beams can firstly buckle and then suddenly collapse under additional bending moment and Brazier flattening. This phenomenon was found in the 1995 Hyogoken Nanbu earthquake [23], as illustrated in Figure 1.

Figure 1. Diagonal brace collapsed in the 1995 Hyogoken Nanbu earthquake [23].

In this paper, the Brazier effect in thin-walled angle-section beams subjected to pure bending about its weak axis is investigated. It is aimed to predict the maximum bending moment and section deformation under extreme loads, so as to study the sustainable load capacity of the angle steel. Theoretical analysis on the basis of the variational method was performed [1], and a formula to predict the maximum bending moment and section deformation was proposed. Both numerical results and experimental results were used to show the validity of the proposed theoretical formula. A number of beam section sizes were used for comparison. The least length of angle beam is also recommended to eliminate the end effect, i.e., the restraint due to limited beam length of the numerical model and experimental setup. The ratio of width to thickness to make sure that Brazier effect does not occur within elastic range is derived for angle beams. Furthermore, the bending moment of thin-walled angle beam considering elasto-plastic material is discussed as well.

2. Theoretical Derivation

2.1. Elastic Theoretical Analysis Based on Variational Method

Angle beam section is illustrated in Figure 2. The variational method can be applied to examine the equilibrium of the flange when the angle beam is bent by pure bending moment. Referring to Figure 3, a system of coordinates x, y, z, and of displacements u, v, w, is described. The axis of x is parallel to the beam axis and the axis of z is normal to the flange. The thickness of the flange is defined as t, the length of side as b, and the angle between flange and the symmetry axis of the section as α.

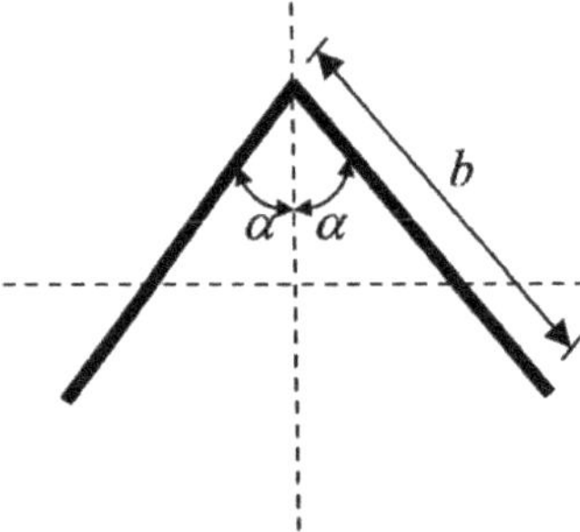

Figure 2. Angle beam section.

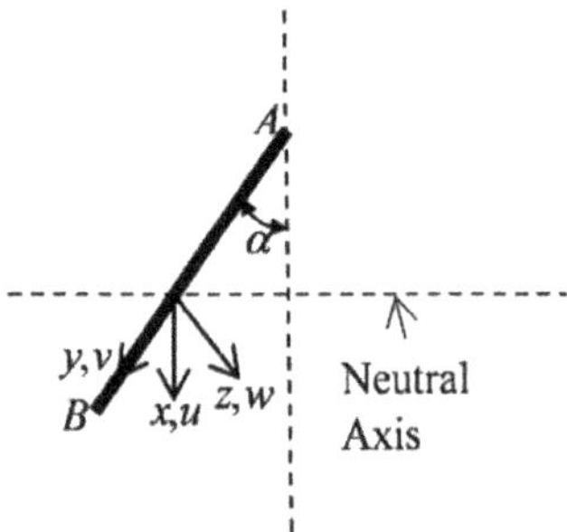

Figure 3. System of coordinate.

It is supposed that the beam is subjected to a constant axis curvature ϕ and that the stress at any point is proportional to the distance from the neutral axis. The flange is allowed to take up an inextensible system of deformation w, v. The condition of inextensibility is

$$\frac{dv}{dy} = 0 \tag{1}$$

Because of the geometrical symmetry of the section, the half section is analyzed. It can get that $v = 0$. The change of curvature at a point can be obtained

$$\Delta\phi_L = |w''| = \left|\frac{d^2w}{dy^2}\right| \tag{2}$$

The resultant distance from the neutral axis is

$$d = \begin{cases} -y\cos\alpha - w\sin\alpha & (y \le 0) \\ y\cos\alpha + w\sin\alpha & (y > 0) \end{cases} \tag{3}$$

After neglecting squares and products of the small quantities w with respect to y, it thus can be obtained for the total strain energy between point A and point B per unit length of the beam the expression

$$U = \frac{1}{2}\left[\frac{Et^3}{12(1-v^2)}\int_{-\frac{b}{2}}^{\frac{b}{2}}(w'')^2 dy + t\phi^2\int_{-\frac{b}{2}}^{\frac{b}{2}}(y^2\cos^2\alpha + yw\sin 2\alpha)dy\right] \tag{4}$$

In Equation (4), E is Young's modulus and v is Poisson ratio. If this is to be a minimum then according to the calculus of variations w must satisfy the following equation

$$\begin{aligned}
\delta U &= \frac{E}{2}\left[\frac{t^3}{12(1-v^2)}\int_{-\frac{b}{2}}^{\frac{b}{2}}2w''\,\delta w''\,dy + t\phi^2\int_{-\frac{b}{2}}^{\frac{b}{2}}y\sin 2\alpha\,\delta w\,dy\right]\\
&= \frac{E}{2}\left[\frac{t^3}{12(1-v^2)}(\int_{-\frac{b}{2}}^{\frac{b}{2}}2w''\,d(\delta w')) + t\phi^2\int_{-\frac{b}{2}}^{\frac{b}{2}}y\sin 2\alpha\,\delta w\,dy\right]\\
&= \frac{E}{2}\left[\frac{t^3}{12(1-v^2)}(2w''\,\delta w'|_{-\frac{b}{2}}^{\frac{b}{2}} - \int_{-\frac{b}{2}}^{\frac{b}{2}}2\delta w'\,w'''\,dy) + t\phi^2\int_{-\frac{b}{2}}^{\frac{b}{2}}y\sin 2\alpha\,\delta w\,dy\right]\\
&= \frac{E}{2}\left[\frac{t^3}{12(1-v^2)}(2w''\,\delta w'|_{-\frac{b}{2}}^{\frac{b}{2}} - 2w'''\,\delta w|_{-\frac{b}{2}}^{\frac{b}{2}} + \int_{-\frac{b}{2}}^{\frac{b}{2}}2\delta w\cdot w^{(4)}dy) + t\phi^2\int_{-\frac{b}{2}}^{\frac{b}{2}}y\sin 2\alpha\,\delta w\,dy\right] = 0\\
&\quad\therefore w^{(4)} = -\frac{6\phi^2(1-v^2)\sin 2\alpha}{t^2}y\\
&\quad\quad \frac{d^4w}{dy^4} = C_1 y
\end{aligned} \tag{5}$$

where

$$C_1 = -\frac{6\phi^2(1-v^2)\sin 2\alpha}{t^2} \tag{6}$$

The solution of Equation (5) is

$$w = \frac{1}{120}C_1 y^5 + C_2 y^3 + C_3 y^2 + C_4 y + C_5 \;(C_2, C_3, C_4, C_5 \text{ are constants}) \tag{7}$$

Considerations of symmetry and of the free end effect require that Equations (8)–(11) are satisfied.

$$w(y)\big|_{y=-\frac{b}{2}} = 0 \tag{8}$$

$$\frac{dw}{dy}\bigg|_{y=-\frac{b}{2}} = 0 \tag{9}$$

$$\frac{d^2w}{dy^2}\bigg|_{y=\frac{b}{2}} = 0 \tag{10}$$

$$\frac{d^3w}{dy^3}\bigg|_{y=\frac{b}{2}} = 0 \tag{11}$$

Solving the equations, we can get

$$w = C_1\left(\frac{1}{120}y^5 - \frac{1}{48}b^2 y^3 + \frac{1}{48}b^3 y^2 + \frac{13}{384}b^4 y + \frac{3}{320}b^5\right) \tag{12}$$

If these expressions are substituted in Equation (4) and the integrations effected, we can obtain

$$U = \frac{E}{2}\left[\frac{1}{12}\phi^2\cos^2\alpha b^3 t - \frac{13\phi^4\sin^2 2\alpha(1-v^2)b^7}{1680t}\right] \tag{13}$$

The moment transmitted at a section of the beam is given by

$$M = \frac{dU}{d\phi} = \frac{1}{12} Eb^3 t \left[\cos^2 \alpha \phi - \frac{13 \sin^2 2\alpha (1 - v^2) b^4}{70 t^2} \phi^3 \right]$$ (14)

This is a maximum when

$$\phi = \sqrt{\frac{70}{39}} \frac{\cos \alpha \cdot t}{\sin 2\alpha \sqrt{1 - v^2} b^2}$$ (15)

at which point the moment is

$$M_{br} = \frac{1}{18} \sqrt{\frac{70}{39}} \frac{E \cos^3 \alpha \cdot bt^2}{\sin 2\alpha \sqrt{1 - v^2}}$$ (16)

Because of symmetry, the moment of the whole section is $2M_{br}$, that is

$$M_t = 2M_{br} = \frac{1}{9} \sqrt{\frac{35}{78}} \frac{E \cos^2 \alpha \cdot bt^2}{\sin \alpha \sqrt{1 - v^2}}$$ (17)

If Equation (15) is substituted in Equation (12), the extreme displacement w_{ext} at free end at this instant can be obtained,

$$w_{ext} = -\frac{49}{312} b \cdot \cot \alpha$$ (18)

so that the approximations based on the smallness of w are justified to this extent. The deformation of the cross section at this point is shown at Figure 4.

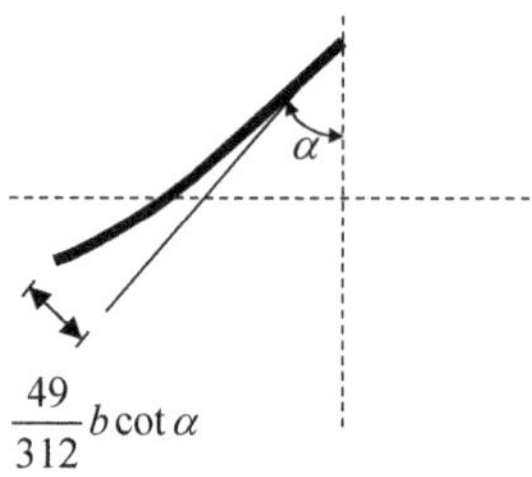

Figure 4. Deformation of angle beam.

2.2. Threshold Ratio of Width to Thickness

The occurrence order of Brazier flattening and yielding is governed by the ratio of width to thickness of the angle-section beam. The upper limit ratio of width to thickness in order to make sure that Brazier flattening does not occur within elastic range is here called threshold ratio of width to thickness. In another words, the threshold ratio of width to thickness is the lower limit ratio to make sure that Brazier flattening occurs within elastic range. Yield moment M_y can be determined by Equation (19) using the initial shape of cross section.

$$M_y = \sigma_y \frac{I}{h} = \sigma_y \frac{\frac{tb^3}{6} \cos^2 \alpha}{\frac{b}{2} \cos \alpha} = \sigma_y \frac{tb^2}{3} \cos \alpha$$ (19)

According to $M_y = M_t$, threshold ratio of width to thickness can be derived as

$$\left(\frac{b}{t} \right)_{cr} = \frac{E}{\sigma_y} \cot \alpha \sqrt{\frac{35}{702(1 - v^2)}}$$ (20)

Considering a general steel, E = 205,000 N/mm^2, v = 0.3, σ_y = 235 N/mm^2, the threshold ratio of width to thickness is 354, 204 and 118 for α being 30°, 45°, and 60°, respectively. These ratio values are relatively large compared to the critical ratio for general local plate buckling. For general structural angle-section beams with the ratio of width to thickness being about 10, Brazier flattening will not occur within elastic limit. In other words, Brazier flattening within elastic limit is restricted to angle-section beams with large ratio of width to thickness. Accordingly, Brazier flattening is expected to couple with local buckling in a test as Brazier had met in his tests [1].

3. Finite Element Analysis

3.1. Elastic Finite Element Analysis

In order to test the validity of the proposed formula, static elastic analyses considering geometric nonlinearity have been performed using finite element code ABAQUS. The angle-section beam model is discretized by four node-reduced integration-hourglass control shell elements (S4R) as shown in Figure 5. In this model, b = 100 mm, t = 2.5 mm, α = 45°, E = 205,000 MPa, v = 0.3. Considering the symmetry of the model, a half angle beam is meshed for analysis, with the symmetrical boundary condition enforced at the left end of the model. The bending moment is exerted through a rigid plate glued at the right end of the model. As the theoretical derivation is based on elastic material assumption, the material is firstly assumed to be elastic in order to observe the geometric nonlinear response caused by Brazier effect. Figure 6 shows a partial illustration of the superimposed undeformed and deformed shape of the model. Figure 7a,b illustrate the contours of Mises stress and rotational moment. Figure 8 shows the normalized numerical maximum moment (M_f/M_t) with the change of beam length. It can be seen from the figure that the maximum moments calculated from finite element analyses decrease with the beam length and tend to be converging when the beam length becomes longer. When the length l of angle beam goes up to 30 times as the length of edge length, the influence of the end effect tends to be negligible, and the pure bending condition is approached. Thus, this length is used for the analyses.

Figure 5. Finite element model.

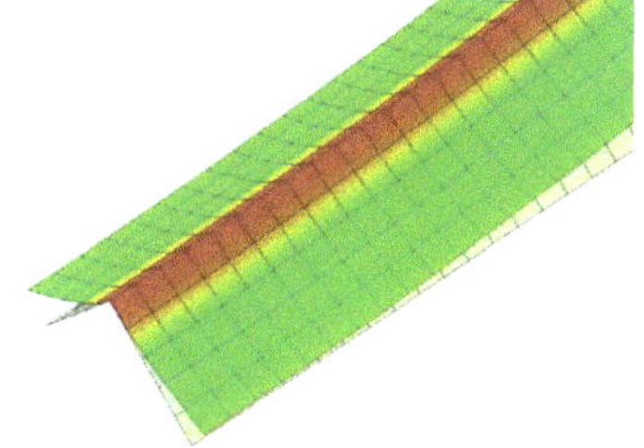

Figure 6. Defomed and undeformed shape.

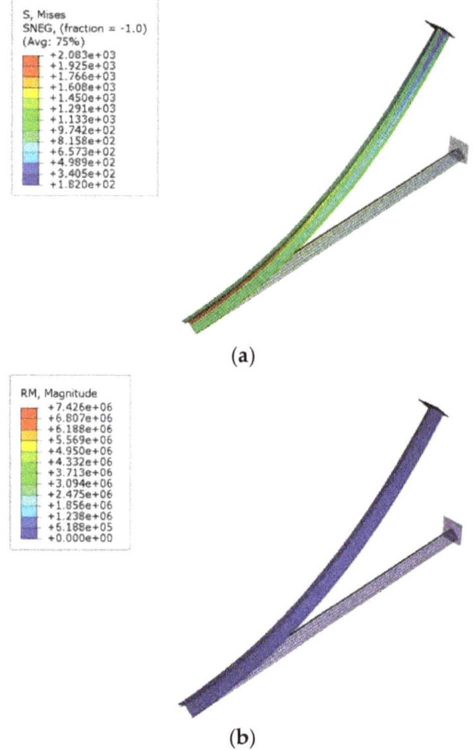

(a)

(b)

Figure 7. FEM Results: (**a**) Mises stress; (**b**) rotational moment.

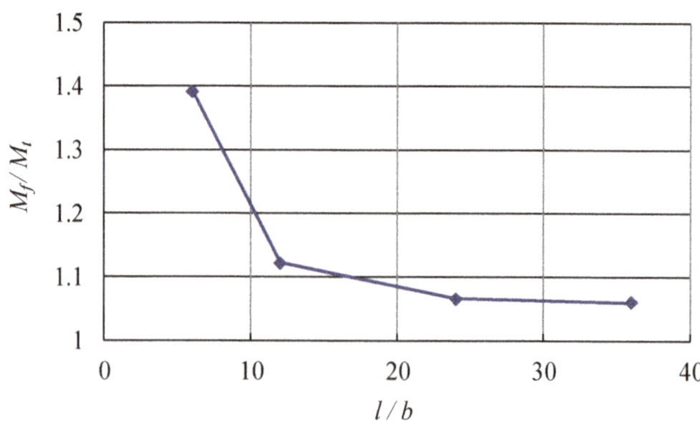

Figure 8. Normalized numerical maximum moment versus the ratio of beam length to edge length.

Comparison of numerical and theoretical elastic maximum moments is carried out using a number of beam section sizes, as illustrated in Table 1. Numerical values show agreement with these theoretical values, with the relative difference being -2.19–7.57%. It can be concluded that elastic theoretical analysis based on variational method is validated by finite element analyses.

Table 1. Comparison of numerical and theoretical elastic maximum moments

Size			Numerical Value	Theoretical Value	Relative
b (mm)	b/t	α	(N × mm)	(N × mm)	Difference (%)
50			5.97×10^5	5.66×10^5	+5.59
100	50	45°	4.82×10^6	4.52×10^6	+6.45
200			3.80×10^7	3.62×10^7	+4.86
	40		7.55×10^6	7.07×10^6	+6.75
100	50	45°	4.82×10^6	4.52×10^6	+6.45
	60		3.33×10^6	3.14×10^6	+5.82
		30°	9.39×10^6	9.60×10^6	−2.19
100	50	45°	4.82×10^6	4.52×10^6	+6.45
		60°	1.99×10^6	1.85×10^6	+7.57

3.2. Elasto-Plastic Finite Element Analysis

The above theoretical derivation is based on an elastic energy approach, which does not consider the plastic property of material. Herein, elasto-plastic finite element analysis is performed to show the influence of plasticity. The yielding strength σ_y is assumed as 325 MPa and the ultimate strength σ_u is assumed as 554 MPa. Then finite element analyses considering geometric nonlinearity and material nonlinearity were performed using ABAQUS. Comparison of the maximum moment for different analyses is shown in Table 2. It can be seen that the plastic property will significantly decrease the maximum moment of the angle-section beam when ratio of width to thickness is smaller than the critical value $\left(\frac{b}{t}\right)_{cr}$, which is 204 for the cases in Table 2. Plastic deformation will occur before Brazier flattening for these cases. The FEM value from the elasto-plastic analyses is approximate to the plastic moment, which can be calculated from Equation (21).

$$M_p = 0.5tb^2 \cos \alpha \sigma_y \tag{21}$$

Table 2. Comparison of maximum moments for elastic and elasto-plastic analyses

Size			FEM Value (N × mm)		Elastic	Relative Difference (%)	
b (mm)	b/t	α	Elastic	Elasto-Plastic	Theoretical Value (N × mm)	FEM Elastic	FEM Elasto-Plastic
	40		7.55×10^6	2.63×10^6	7.07×10^6	6.75	−62.80
100	50	45°	4.82×10^6	2.04×10^6	4.52×10^6	6.45	−54.88
	60		3.33×10^6	1.65×10^6	3.14×10^6	5.82	−47.45

4. Comparison with Experimental Results

The result of a symmetrical four-point bending test was used to compare with theoretical results. The illustration of the experiment setup is shown in Figure 9. The total length of the thin-walled angle beam is 1800 mm. Three displacement meters were set to measure the vertical displacement at the load points and the center point. The material properties of the specimen are shown in Table 3.

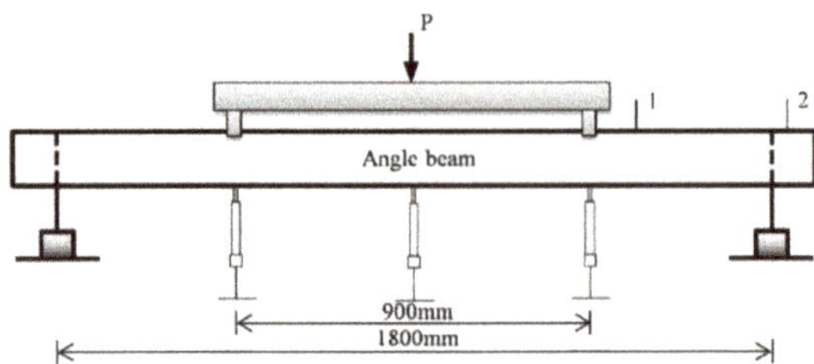

Figure 9. Experiment setup of the four-point bending test [24].

Table 3. Material properties of the specimen [24].

Material	Flange Thickness t (mm)	Young's Modulus E (GPa)	Poisson's Ratio v	Yielding Strength σ_y (MPa)
Steel	0.324	199,000	0.3	300

Two specimen types—i.e., specimen type A and B with individual kind of section sizes—were tested. The numerical, theoretical, and experimental results are shown in Table 4. As for the FEM analyses, both elastic and elasto-plastic material are considered. During the experiment, it was demonstrated that the wave of plate buckling was generated and developed, which obviously decreased the bending stiffness of the beam. Nevertheless, the collapse is not controlled by the plate buckling, but by Brazier flattening, load-point crippling, or ridge-line buckling. The collapse mode of specimen type A is Brazier flattening. When the collapse was governed by flattening, the moment–curvature curve is different from the linear beam theory, but well coincides with the flattening theory in consideration of the end effect. For specimen type A which shows Brazier flattening collapse, the theoretical maximum moment is 17.97% smaller than the experimental maximum moment and close to the numerical values. Substituting $\alpha = 60°$ and the parameter in Table 3 to Equation (20), the threshold ratio of width to thickness gets the value of 117, which is bigger than the ratio of width to thickness of specimen type A. This means that yielding happened in the specimen before Brazier flattening. The picture of Brazier flattening of specimen A3 is shown in Figure 10. However, for specimen type B, load-point crippling or ridge-line buckling occurred before Brazier flattening happens. Thus, the experimental max moments are far below theoretical value, Brazier flattening is unlikely to happen in specimen type B.

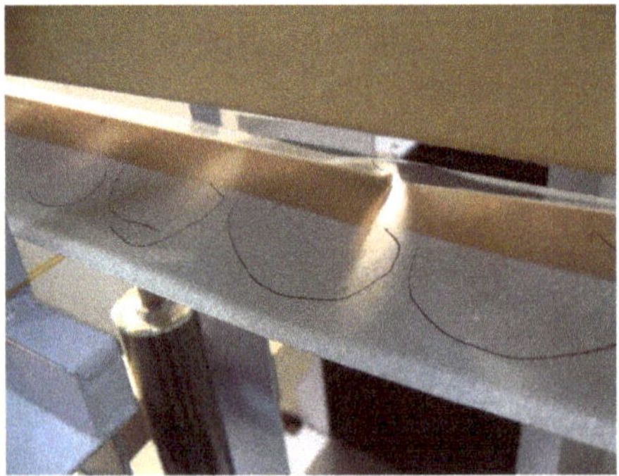

Figure 10. Brazier flattening of specimen A3.

Table 4. Comparison of numerical, theoretical, and experimental results

Specimen No.		A1	A2	A3	B1	B2
Size	L (mm)		1800		1800	
	α		60°		45°	
	b (mm)		30		60	
Experimental max moment (N × mm)		1.61×10^4	1.60×10^4	1.58×10^4	4.83×10^4	5.07×10^4
Collapse mode			Brazier Flattening		Load-point crippling	Ridge-line buckling
Elastic theoretical max moment (N × mm)			1.31×10^4		6.41×10^4	
FEM max moment (N × mm)	Elastic		1.51×10^4		7.84×10^4	
	Elasto-plastic		1.31×10^4		7.13×10^4	
Relative difference to experimental value (%)	Theoretical value		-17.97		29.49	
	Elastic		-5.44		58.38	
	Elasto-plastic		-17.97		44.04	

5. Conclusions

This paper mainly aims at presenting a theoretical analysis for Brazier effect in thin-walled angle-section beams. The variational method is applied to derive formulas to predict the maximum bending moment and section deformation of angle-section beams under pure bending about its weak axis. Both numerical analyses and experimental results were used to show the validity of the proposed formula. Numerical results show that the end effect can influence the numerical results, but the end effect can be eliminated by making the length of angle beam to thirty times as the length of beam side. During the experiment, it was demonstrated that the wave of plate buckling was generated and developed, which obviously decreased the bending stiffness of the beam. Nevertheless, the collapse is not controlled by the plate buckling, but by Brazier flattening, load-point crippling, or ridge-line buckling. When the collapse was governed by Brazier flattening, the moment–curvature curve is different from the linear beam theory, but coincides well with the flattening theory in consideration of the end effect. It can be concluded that the proposed formula shows good agreement with numerical results and experimental results.

Author Contributions: Z.Z. wrote the paper and directed the study. L.X. analyzed the data and revised the paper. C.S. performed the numerical analyses. S.X. conceived the idea, provided valuable discussions, and revised the paper. He took responsibility of the corresponding work.

Acknowledgments: The authors gratefully acknowledge the financial supports from the National Natural Science Foundation of China (No. 51478357) and a collaborative research project under International Joint Research Laboratory of Earthquake Engineering (ILEE) (No. ILEE-IJRP-P1-P3-2016).

Conflicts of Interest: The authors declare no conflict of interest.

References

1. Brazier, L.G. On the flexure of thin cylindrical shells and other thin sections. *Proc. R. Soc. Lond. Ser. A* **1927**, *116*, 104–114. [CrossRef]
2. Reissner, E. On finite bending of pressurized tubes. *Trans. ASME* **1959**, *26*, 386–392.
3. Akselrad, E.L. Pinpointing the upper critical bending load of a pipe by calculating geometric nonlinearity. *Izv. Akad. Nauk SSR. Mekh.* **1965**, *4*, 133–139.
4. Gerber, T.L. Plastic Deformation of Piping Due to Pipe-Whip Loading. 1974. Available online: http://jglobal.jst.go.jp/en/public/201002025727941512 (accessed on 22 August 2018).
5. Tomasz, W.; Sinmao, M.V. A simplified model of Brazier effect in plastic bending of cylindrical tubes. *Int. J. Press. Vessels Pip.* **1997**, *71*, 19–28.
6. Li, L.Y.; Kettle, R. Nonlinear bending response and buckling of ring-stiffened cylindrical shells under pure bending. *Int. J. Solids. Struct.* **2002**, *39*, 765–781. [CrossRef]

7. Poonaya, S.; Teeboonma, U.; Thinvongpituk, C. Plastic collapse analysis of thin-walled circular tubes subjected to bending. *Thin Wall. Struct.* **2009**, *47*, 637–645. [CrossRef]

8. Yu, T.X.; Teh, L.S. Large plastic deformation of beams of angle-section under symmetric bending. *Int. J. Mech. Sci.* **1997**, *39*, 829–839. [CrossRef]

9. Huber, M.T. The Bending of Curved Tube of Elliptic Sections. In *Proceedings of the Seventh International Congress for Applied Mechanics*; Levy, H., Ed.; Her Majesty's Stationery Office: London, UK, 1948; pp. 322–328.

10. Rand, O. In-Plane Warping Effects in Thin-Walled Box Beams. *AIAA J.* **2000**, *38*, 542–544. [CrossRef]

11. Paulsen, F.; Welo, T. Cross-Sectional Deformations of Rectangular Hollow Sections in Bending: Part II, Analytical Models. *Int. J. Mech. Sci.* **2001**, *43*, 131–152. [CrossRef]

12. Cecchini, L.S.; Weaver, P.M. Brazier effect in multibay airfoil sections. *AIAA J.* **2005**, *43*, 2252–2258. [CrossRef]

13. Jensen, F.M.; Weaver, P.M.; Cecchini, L.S. The Brazier effect in wind turbine blades and its influence on design. *Wind Energy* **2012**, *15*, 319–333. [CrossRef]

14. Luongo, A.; Zulli, D.; Scognamiglio, I. The Brazier effect for elastic pipe beams with foam cores. *Thin-Wall Struct.* **2018**, *124*, 72–80. [CrossRef]

15. Sato, M.; Ishiwata, Y. Brazier effect of single- and double-walled elastic tubes under pure bending. *Struct. Eng. Mech.* **2015**, *53*, 17–26. [CrossRef]

16. Shima, H.; Sato, M.; Park, S.J. Suppression of Brazier Effect in Multilayered Cylinders. *Adv. Condens. Matter. Phys.* **2014**, *10*, 11–55. [CrossRef]

17. Stephens, W.B.; Starnes, J.J. Collapse of long cylindrical shells under combined bending and pressure loads. *AIAA J.* **1975**, *13*, 5–20.

18. Fabian, O. Collapse of cylindrical, elastic tubes under combined bending, pressure and axial loads. *Int. J. Solids Struct.* **1977**, *13*, 1257–1270. [CrossRef]

19. Libai, A.; Bert, C.W. A mixed variational principle and its application to the nonlinear bending problem of orthotropic tubes-II. Application to nonlinear bending of circular cylindrical tubes. *Int. J. Solids Struct.* **1994**, *31*, 1019–1033. [CrossRef]

20. Karamanos; Spyros, A. Bending instabilities of elastic tubes. *Int. J. Solids Struct.* **2002**, *39*, 2059–2085. [CrossRef]

21. Tatting, B.F.; Gürdal, Z. Nonlinear shell theory solution for the bending response of orthotropic finite length cylinders including the Brazier effect. In Proceedings of the 36th Structures, Structural Dynamics and Materials Conference, New Orleans, LA, USA, 10–13 April 1995.

22. Tatting, B.F.; Gürdal, Z.; Vasiliev, V.V. The Brazier effect for finite length composite cylinders under bending. *Int. J. Solids Struct.* **1997**, *34*, 1419–1440. [CrossRef]

23. Makoto, W.; Gakkai, N.H. *Preliminary Reconnaissance Report of the 1995 Hyogoken-Nanbu Earthquake*; The Architectural Institute of Japan: Tokyo, Japan, 1995.

24. Kuwamura, H. Flattening and buckling of thin angle-section beams. In Proceedings of the Fifth International Conference on "Coupled Instabilities in Metal Structures" CIMS2008, Sydney, Australia, 23–25 June 2008; pp. 101–108.

Article

Nonlinear Error Propagation Analysis of a New Family of Model-Based Integration Algorithm for Pseudodynamic Tests

Bo Fu [1], Huanjun Jiang [2] and Tao Wu [1,*]

[1]   School of Civil Engineering, Chang'an University, Xi'an 710061, China; 90_bofu@chd.edu.cn
[2]   College of Civil Engineering, Tongji University, Shanghai 200092, China; jhj73@tongji.edu.cn
*   Correspondence: wutao@chd.edu.cn; Tel.: +86-139-9132-2194

Received: 29 June 2018; Accepted: 7 August 2018; Published: 10 August 2018

Abstract: Error propagation properties of integration algorithms are crucial in conducting pseudodynamic tests. The motivation of this study is to investigate the error propagation properties of a new family of model-based integration algorithm for pseudodynamic tests. To develop the new algorithms, two additional coefficients are introduced in the Chen-Ricles (CR) algorithm. In addition, a parameter—i.e., degree of nonlinearity—is applied to describe the change of stiffness for nonlinear structures. The error propagation equation for the new algorithms implemented in a pseudodynamic test is derived and two error amplification factors are deduced correspondingly. The error amplification factors for three structures with different degrees of nonlinearity are calculated to illustrate the error propagation effect. The numerical simulation of a pseudodynamic test for a two-story shear-type building structure is conducted to further demonstrate the error propagation characteristics of the new algorithms. It can be concluded from the theoretical analysis and numerical study that both nonlinearity and the two additional coefficients of the new algorithms have great influence on its error propagation properties.

Keywords: integration algorithm; pseudodynamic test, earthquake; nonlinearity; model-based

1. Introduction

In the field of civil engineering, experimental studies are crucial to investigate and enhance the sustainability and resilience of civil infrastructures in the event of extreme loads, such as earthquakes. The experimental approaches of studying the seismic performance of building structures include the quasi-static tests, shake table tests, pseudodynamic tests, and real-time hybrid simulations. Among them, the pseudodynamic tests, also known as on-line tests or hybrid simulations, have been paid great attention for decades [1,2].

In conducting a pseudodynamic test, two types of errors—i.e., displacement control error and restoring force measurement error—introduced in each time step will be propagated and accumulated owing to a feedback procedure [1,2]. This feedback procedure roots from the use of an integration algorithm to conduct the step-by-step integration during the pseudodynamic test. Therefore, it is crucial to investigate the error propagation of an integration algorithm because the errors from the pseudodynamic test must be controlled within certain limits to obtain reliable test results.

An integration algorithm can be labeled as explicit or implicit depending on its formulation of displacement. Traditional explicit algorithms, such as the Newmark explicit algorithm, are always conditionally stable, which requires a very severe limitation on the integration time step if the tested structure system has high frequency modes [3]. For the past decade, researchers developed several innovative integration algorithms which are both explicit and unconditionally stable.

The representative algorithms include the Chang algorithm [4], the Chen-Ricles (CR) algorithm [5], and the Kolay-Ricles-α (KR-α) algorithm [6]. These algorithms are nominated as model-based integration algorithms [7]. When using these algorithms, the algorithmic parameters, which are functions of the complete model of the system, have to be calculated. The model-based algorithms can achieve unconditional stability within the framework of an explicit formulation. On the contrary, there is no need to calculate the model-based parameters for the conventional integration algorithms such as the well-known Newmark algorithms, so they are called model-independent algorithms. Although a number of studies [8–15] have been conducted for the error propagation of the model-independent integration algorithms, there is very limited research with regard to the error propagation of the model-based integration algorithms.

In addition, most of the previous work [8–12] on the error propagation of the integration algorithms were conducted for linear elastic systems. Only a few research [13–15] were aimed at nonlinear structures. It should be noted that most of the pseudodynamic tests were carried out in nonlinear range of behavior as pseudodynamic tests are mainly used to investigate the nonlinear behavior of the structures subjected to seismic excitations [2,16]. Therefore, it is much more meaningful to perform the nonlinear error propagation analysis. To take the nonlinearity into consideration, the degree of nonlinearity, which was adopted in [13–15], is introduced in this study. Furthermore, the influence of viscous damping, which is a critical structural parameter, on the error propagation characteristics have not been considered in the previous studies [13–15].

The motivation of this study is to investigate the error propagation properties of a new family of model-based integration algorithm for pseudodynamic tests of nonlinear systems by considering the influence of viscous damping. The new family of model-based integration algorithms is firstly introduced in Section 2. The procedure of using the new algorithms in pseudodynamic tests is then presented in Section 3. The error propagation equation and two error amplification factors for the new algorithms are derived in Section 4 (Appendix Figure A1 shows flowchart of calculating the error amplification factors). To further illustrate the influences of the degree of nonlinearity, viscous damping, and two additional coefficients of the new algorithms, the error amplification factors derived from Section 4 for three structures with different degrees of nonlinearity are obtained to demonstrate the error propagation effect in Section 5. Finally, the numerical simulations of pseudodymamic tests for a two-story shear-type building are also conducted to further illustrate the error propagation properties of the new algorithms in Section 6.

2. A New Family of Model-Based Integration Algorithm

In implementing a pseudodynamic test, the structure must be idealized as a discrete model, whose equation of motion can be written as

$$M\ddot{X}_{i+1} + C\dot{X}_{i+1} + KX_{i+1} = F_{i+1} \tag{1}$$

where M, C, and K are mass, damping and stiffness matrices, respectively; X_{i+1}, $\dot{X}_{i+1}$, and $\ddot{X}_{i+1}$ are displacement, velocity, and acceleration vectors at the $(i+1)$th time step, respectively; and F_{i+1} is the external force vector at the $(i+1)$th time step.

The integration algorithms can be used to solve the equation of motion (Equation (1)) in conducting a pseudodynamic test. Recently, the authors [17] proposed a new family of explicit integration algorithm, which was a generalized version of the CR algorithm [5], which was proposed by Chen and Ricles and has been successfully implemented in a series of real-time hybrid simulation tests. The new family of algorithms is named 'generalized CR algorithm' and is abbreviated as 'GCR algorithm' in this paper. The formulation of the GCR algorithm is inherited from the CR algorithm and expressed as

$$\dot{X}_{i+1} = \dot{X}_i + \Delta t\alpha_1 \ddot{X}_i \tag{2a}$$

$$X_{i+1} = X_i + \Delta t\dot{X}_i + \Delta t^2\alpha_2 \ddot{X}_i \tag{2b}$$

where Δt is the time step; α_1 and α_2 are the integration parameter matrices

$$\alpha_1 = \left[M + \kappa_1 \Delta t C + \kappa_2 \Delta t^2 K\right]^{-1} M; \alpha_2 = (1/2 + \kappa_1)\alpha_1 \tag{3}$$

where κ_1 and κ_2 are two additional coefficients controlling the numerical characteristics of the GCR algorithm. The CR algorithm is the special case of the GCR algorithm with $\kappa_1 = 1/2, \kappa_2 = 1/4$.

It can be concluded from the preliminary study [17] that the subfamily of the GCR algorithm with $\kappa_1 = 1/2, 2\kappa_2 \geq \kappa_1$ has no numerical damping, while the subfamily of the GCR algorithm with $\kappa_1 > 1/2, \kappa_2 \geq (\kappa_1 + 1/2)^2/4$ has numerical damping. The GCR algorithm is superior to commonly used integration algorithms due to its explicit expressions and unconditionally stability. In addition, the numerical characteristics of the GCR algorithm are identical with the well-known Newmark algorithm. Therefore, the GCR algorithm can be easily adopted in solving the equation of motion in the pseudodynamic tests and other structural dynamic problems.

For the brevity of mathematics, the following derivations are only considered for a single-degree-of-freedom (SDOF) system. As for the SDOF system, all the matrices and vectors in Equations (1)–(3) become scalars.

The degree of nonlinearity used in [13–15] is introduced to evaluate the change of stiffness for the nonlinear system and defined as

$$\delta_{i+1} = \frac{k_{i+1}}{k_i} \tag{4}$$

It should be noted that $\delta_{i+1} = 1$ means the stiffness remains unchanged during the $(i + 1)$th time step, while $\delta_{i+1} > 1$ and $0 < \delta_{i+1} < 1$ can be used to represent the nonlinear system with stiffness hardening and the stiffness softening during the $(i + 1)$th time step, respectively.

3. Pseudodynamic Test Procedure

Because there are possible variations in the changes of stiffness and the resulting complicated eigendecomposition, it is almost impossible to conduct the error propagation analysis for the whole step-by-step integration procedure. Therefore, only a specific time step is analyzed for the error propagation analysis of a nonlinear system [13].

It should be noted that the GCR algorithm is a dual-explicit algorithm, which means it is explicit for both displacement and velocity, so the pseudodynamic test procedure is much easier and more direct compared to the pseudodynamic test using the Newmark explicit algorithm [13], the subfamily of Hilber-Hughes-Taylor-α (HHT-α) algorithm [14] or the constant average acceleration method [15].

The step-by-step integration procedure for the GCR algorithm can be written in a recursive matrix form. It is

$$X_{i+1} = A_{i+1}X_i + L_{i+1}F_{i+1} \tag{5}$$

where $X_{i+1} = [x_{i+1}, \Delta t \dot{x}_{i+1}, \Delta t^2 \ddot{x}_{i+1}]^T$. The amplification matrix A_{i+1} and the load vector L_{i+1} for the $(i + 1)$th time step are

$$A_{i+1} = \begin{bmatrix} 1 & 1 & \alpha_2 \\ 0 & 1 & \alpha_1 \\ -\Omega_{i+1}^2 & -(\Omega_{i+1}^2 + 2\zeta\Omega_{i+1}) & -(\alpha_2\Omega_{i+1}^2 + 2\alpha_1\zeta\Omega_{i+1}) \end{bmatrix} \tag{6}$$

$$L_{i+1} = [0, 0, \Delta t^2/m]^T \tag{7}$$

where $\Omega_{i+1} = \omega_{i+1}\Delta t$ and $\omega_{i+1} = \sqrt{k_{i+1}/m}$ is the natural frequency of the system at the end of the $(i + 1)$th time step. $\zeta = c/(2m\omega_0)$ is the equivalent viscous damping ratio, where ω_0 is the initial natural frequency of the system. Differing from the work conducted by other researchers [13–15], the influence of viscous damping is also taken into consideration in this study.

After the eigendecomposition of the matrix A_{i+1}, its characteristic equation can be obtained by the equation of $|A_{i+1} - \lambda I| = 0$ and expanded as

$$|A_{i+1} - \lambda I| = \lambda\left(\lambda^2 - 2A_1\lambda + A_2\right) = 0 \tag{8}$$

where I is the identity matrix of dimension 3×3; λ is the eigenvalue of the matrix A_{i+1}; $A_1 = (2 - \alpha_2\Omega_{i+1}^2 - 2\alpha_1\xi\Omega_{i+1})/2$; $A_2 = (\alpha_1 - \alpha_2)\Omega_{i+1}^2 - 2\alpha_1\xi\Omega_{i+1} + 1$. It is apparent that there are three eigenvalues and one of them is zero, so the above equation can be further simplified as

$$\lambda^2 - 2A_1\lambda + A_2 = 0 \tag{9}$$

There are two eigenvalues $\lambda_{1,2} = A_1 \pm \sqrt{A_1^2 - A_2}$. According to [17], the condition of the unconditionally stable GCR algorithm can be derived

$$2\kappa_2 \geq \kappa_1 \geq 1/2 \tag{10}$$

Therefore, the error propagation characteristics of the subfamily of the unconditionally stable GCR algorithm is investigated in the following content.

4. Derivation of Error Propagation Equation

In conducting a pseudodynamic test, control and measurement errors are unavoidable and are introduced at each time step. Firstly, it is very difficult to exactly impose the computed displacement on the specimen because of the displacement control error. Secondly, the restoring force actually developed in the specimen may not be correctly measured. These errors are carried over to the subsequent time steps and exhibit a cumulative effect from the beginning to the end of the test.

The equation of motion of a nonlinear SDOF system is

$$m\ddot{x}_{i+1} + c\dot{x}_{i+1} = F_{i+1} - r_{i+1} \tag{11}$$

where r_{i+1} is the restoring force at the $(i + 1)$th time step. The displacement and velocity calculated from the SDOF formulation of Equation (2) are accurate values, and the corresponding restoring force obtained by state determination is also accurate and without error.

The below two equations define the actual displacement and restoring force during a pseudodynamic test [13]:

$$x_{i+1}^a = x_{i+1}^e + e_{i+1}^x \tag{12}$$

$$r_{i+1}^a = r_{i+1}^e + e_{i+1}^r \tag{13}$$

where x_{i+1}^a and r_{i+1}^a are the actual displacement and restoring force at the $(i + 1)$th time step, respectively, including the effects of errors at previous steps and the current step; x_{i+1}^e and r_{i+1}^e are the exact displacement and restoring force at the $(i + 1)$th time step, respectively, including the effects of errors at previous steps; e_{i+1}^x and e_{i+1}^r are the displacement error and the restoring force error at the $(i + 1)$th time step, respectively.

The restoring force error at the $(i + 1)$th time step e_{i+1}^r can be further expressed as [13]

$$e_{i+1}^r = k_{i+1}e_{i+1}^{rx} \tag{14}$$

where e_{i+1}^{rx} is the displacement error corresponding to the restoring force error e_{i+1}^r. According to the above definition and the formulations of the GCR algorithm, the following relationships can be obtained

$$\dot{x}_{i+1}^e = \dot{x}_i^e + \Delta t\alpha_1\ddot{x}_i^e \tag{15a}$$

$$x_{i+1}^e = x_i^e + \Delta t\dot{x}_i^e + \Delta t^2\alpha_2\ddot{x}_i^e + e_i^x \tag{15b}$$

$$m\ddot{x}^e_{i+1} + c\dot{x}^e_{i+1} + r^e_{i+1} + e^r_{i+1} = F_{i+1} \tag{16}$$

It should be noted that $r^e_{i+1} = k_{i+1}x^e_{i+1}$, so the aforementioned equations can be reformulated in a recursive matrix form [13]

$$X^e_{i+1} = A_{i+1}X^e_i + L_{i+1}F_{i+1} + M_{i+1}e^x_i - N_{i+1}e^{rx}_{i+1} \tag{17}$$

where $X^e_{i+1} = \left[x^e_{i+1}, \Delta t\dot{x}^e_{i+1}, \Delta t^2\ddot{x}^e_{i+1}\right]^T; M_{i+1} = \left[1, 0, -\Omega^2_{i+1}\right]^T; N_{i+1} = \left[0, 0, \Omega^2_{i+1}\right]^T$.

Subtracting Equation (5) from Equation (16), the following error propagation equation can be obtained

$$X^e_{i+1} - X_{i+1} = A_{i+1}(X^e_i - X_i) + M_{i+1}e^x_i - N_{i+1}e^{rx}_{i+1} \tag{18}$$

Define $\varepsilon_{i+1} = X^e_{i+1} - X_{i+1}$, then the error propagation equation can be rewritten as

$$\varepsilon_{i+1} = A_{i+1}\varepsilon_i + M_{i+1}e^x_i - N_{i+1}e^{rx}_{i+1} \tag{19}$$

The error cumulative vector ε_{i+1} can be explicitly expressed as

$$\varepsilon_{i+1} = \begin{bmatrix} \varepsilon^1_{i+1} \\ \varepsilon^2_{i+1} \\ \varepsilon^3_{i+1} \end{bmatrix} = \begin{bmatrix} x^e_{i+1} - x_{i+1} \\ \Delta t\left(\dot{x}^e_{i+1} - \dot{x}_{i+1}\right) \\ \Delta t^2\left(\ddot{x}^e_{i+1} - \ddot{x}_{i+1}\right) \end{bmatrix} \tag{20}$$

It is obvious that $\varepsilon^1_{i+1} = x^e_{i+1} - x_{i+1} = e^x_{i+1}$ is the cumulative displacement error for the $(i+1)$th time step. When the specimen enters into a nonlinear range, the amplification matrix A_{i+1} and vectors M_{i+1} and N_{i+1} are varied for each time step, so it is almost impossible to calculate the cumulative displacement error e^x_{i+1} for a complete pseudodynamic test such as that developed by Shing and Mahin [11] for a linear elastic system. Even so, if the error propagation for a specific time step is thoroughly studied, some useful information can still be acquired. Therefore, the error propagation from the previous one and two time steps to the current time step is derived as follows.

After the repeated substitutions of ε_{i-1} and ε_i into ε_{i+1} using Equation (20), the following cumulative equation can be acquired

$$\begin{aligned} \varepsilon_{i+1} = {} & A_{i+1}A_iA_{i-1}\varepsilon_{i-2} + \left(A_{i+1}A_iM_{i-1}e^x_{i-2} + A_{i+1}M_ie^x_{i-1} + M_{i+1}e^x_i\right) \\ & - \left(A_{i+1}A_iN_{i-1}e^{rx}_{i-1} + A_{i+1}N_ie^{rx}_i + N_{i+1}e^{rx}_{i+1}\right) \end{aligned} \tag{21}$$

It is assumed that ε_{i+1} is mainly affected by ε_i and ε_{i-1}, while less affected by ε_{i-2}, then the item of $A_{i+1}A_iA_{i-1}\varepsilon_{i-2}$ can be omitted and the cumulative equation can be further simplified as

$$\varepsilon_{i+1} = \varepsilon^x_{i+1} - \varepsilon^r_{i+1} \tag{22a}$$

$$\varepsilon^x_{i+1} = A_{i+1}A_iM_{i-1}e^x_{i-2} + A_{i+1}M_ie^x_{i-1} + M_{i+1}e^x_i \tag{22b}$$

$$\varepsilon^r_{i+1} = A_{i+1}A_iN_{i-1}e^{rx}_{i-1} + A_{i+1}N_ie^{rx}_i + N_{i+1}e^{rx}_{i+1} \tag{22c}$$

where ε^x_{i+1} and ε^r_{i+1} are the cumulative error vector caused from the displacement feedback errors and restoring force feedback errors.

Substituting the amplification matrix into Equation (22), the cumulative displacement error for the $(i+1)$th time step can be expressed as

$$e^x_{i+1} = \left(D_ie^x_i + D_{i-1}e^x_{i-1} + D_{i-2}e^x_{i-2}\right) - \left(R_ie^{rx}_i + R_{i-1}e^{rx}_{i-1}\right) \tag{23}$$

As the coefficients D_i, D_{i-1}, D_{i-2}, R_i, and R_{i-1}, of Equation (23) are so complicated, they are not exhibited in the paper. However, it is very convenient to numerically calculate them by using MATLAB [18]. Equation (23) can be further expressed as

$$e_{i+1}^x = E^x \sum_{k=i}^{i-1} \cos\left[(i-k)\overline{\Omega}_{i+1} + \alpha_{i+1}\right] e_k^x + D_{i-2} e_{i-2}^x - E^r \sum_{k=i}^{i-1} \sin\left[(i-k)\overline{\Omega}_{i+1} + \beta_{i+1}\right] e_k^{rx} \qquad (24)$$

where α_{i+1} and β_{i+1} are the phase angles, E^x and E^r are the displacement and restoring force error amplification factors arising from the previous one and two steps to the cumulative displacement error, respectively. These two amplification factors are all dimensionless and widely used to evaluate the error propagation properties of the integration algorithms in pseudodynamic tests [10–15]. The expressions of the two factors for the GCR algorithm are

$$E^x = \sqrt{(D_i)^2 + \left(\frac{D_{i-1} - D_i \cos \overline{\Omega}_{i+1}}{\sin \overline{\Omega}_{i+1}}\right)^2} \qquad (25)$$

$$E^r = \sqrt{(R_i)^2 + \left(\frac{R_{i-1} - R_i \cos \overline{\Omega}_{i+1}}{\sin \overline{\Omega}_{i+1}}\right)^2} \qquad (26)$$

5. Examples for Error Propagation Effect

The aforementioned two error amplification factors, i.e., the displacement error amplification factor E^x (Equation (25)) and the restoring force error amplification factor E^r (Equation (26)), are used to illustrate the error propagation effect of the GCR algorithm. The degree of nonlinearity is used to define three cases as:

(1) Case 1: nonlinear structure with stiffness softening, $\delta_{i-1} = \delta_i = \delta_{i+1} = 0.5$
(2) Case 2: linear elastic structure, $\delta_{i-1} = \delta_i = \delta_{i+1} = 1.0$
(3) Case 3: nonlinear structure with stiffness hardening, $\delta_{i-1} = \delta_i = \delta_{i+1} = 1.5$

Figures 1 and 2 show the two error amplification factors, i.e, E^x and E^r, with respect to the initial $\Omega_0 = \omega_0 \Delta t$, respectively.

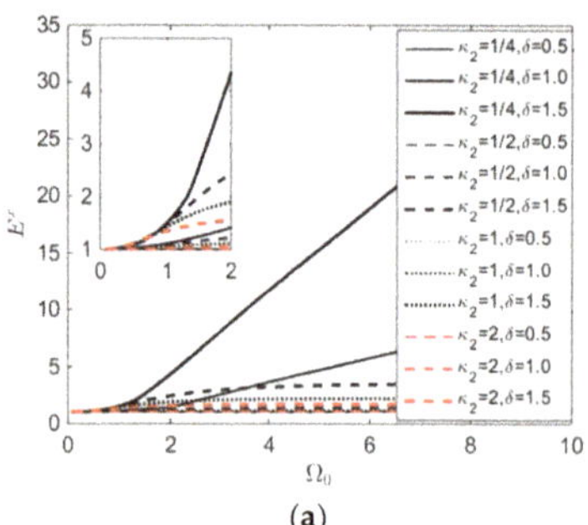

(a)

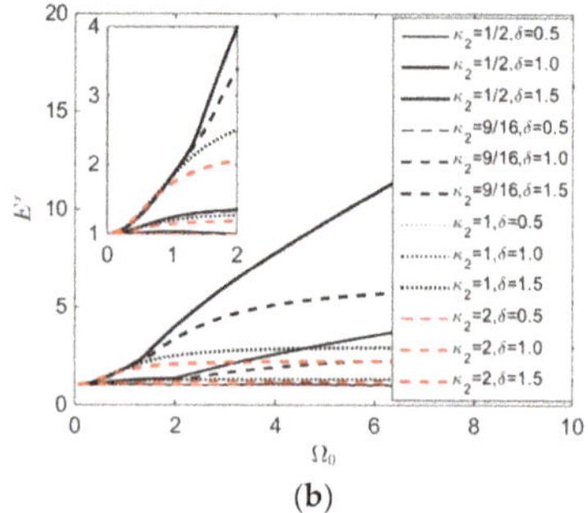

(b)

Figure 1. *Cont.*

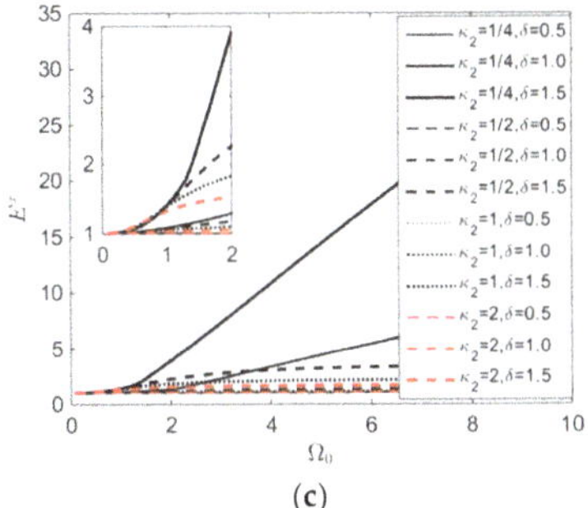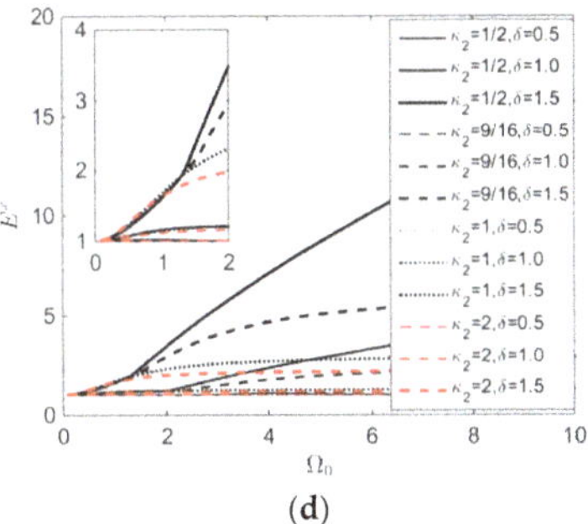

Figure 1. Variations of displacement error amplification factor E^x with respect to Ω_0: (**a**) $\kappa_1 = 1/2$, $\xi = 0$; (**b**) $\kappa_1 = 1$, $\xi = 0$; (**c**) $\kappa_1 = 1/2$, $\xi = 0.1$; (**d**) $\kappa_1 = 1$, $\xi = 0.1$.

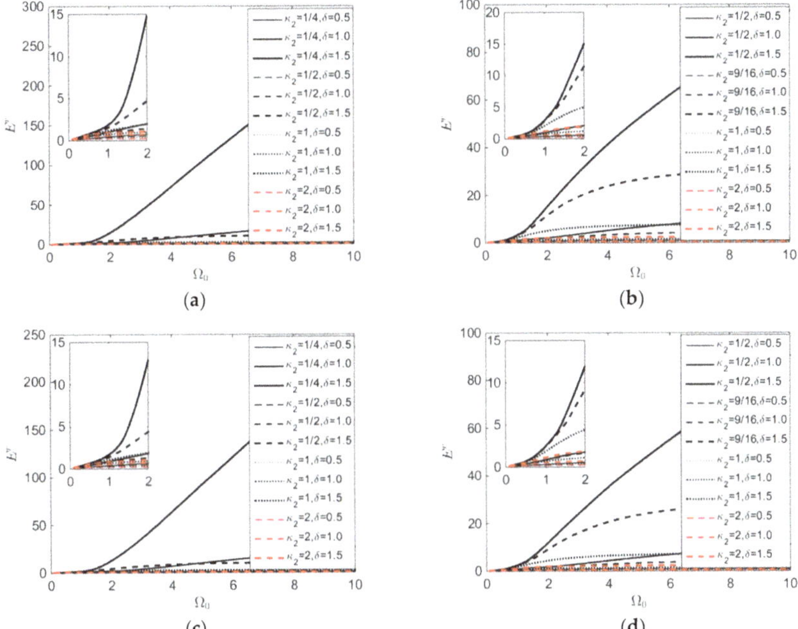

Figure 2. Variations of restoring force error amplification factor E^r with respect to Ω_0: (**a**) $\kappa_1 = 1/2$, $\xi = 0$; (**b**) $\kappa_1 = 1$, $\xi = 0$; (**c**) $\kappa_1 = 1/2$, $\xi = 0.1$; (**d**) $\kappa_1 = 1$, $\xi = 0.1$.

It can be seen from Figures 1 and 2 that both amplification factors E^x and E^r increase with Ω_0. This result is in consistent for those of other integration algorithms [10–15]. Besides, both amplification factors increase with the degree of nonlinearity δ, which is in line with the observations from [13–15]. It indicates that the degree of nonlinearity has a significance influence on the error propagation properties of the GCR algorithms. More specifically, the nonlinear structures with stiffness hardening possess the most errors, while the nonlinear structures with stiffness softening have the lowest errors. As for the viscous damping, it has a positive effect on reducing the errors. Therefore, it is conservative to ignore the influence of viscous damping. Both the two additional coefficients—i.e., κ_1 and κ_2—of the GCR algorithms have great impact on the error propagation property. With the increasing of κ_1, the errors increase, while both amplification factors E^x and E^r increase with the decrease of κ_2. In terms of error propagation, it is recommended to select the GCR algorithms with a small value of κ_1 and a large value of κ_2. For instance, it is superior to use the GCR algorithm with $\kappa_1 = 1/2, \kappa_2 = 1/2$ than the original CR algorithm with $\kappa_1 = 1/2, \kappa_2 = 1/4$ in a pseudodynamic test, especially for the nonlinear structures with stiffness hardening. Taking the structure of Case 3 without viscous damping ($\xi = 0$) as an example, the displacement amplification factors E^x of the GCR algorithm with $\kappa_1 = 1/2, \kappa_2 = 1/2$ and the original CR algorithm with $\kappa_1 = 1/2, \kappa_2 = 1/4$ are 2.43 and 4.33 (=1.78 $\times$ 2.43), respectively, for $\Omega_0 = 2$, and are 3.56 and 32.91 (=9.24 $\times$ 3.56), respectively, for $\Omega_0 = 10$ (Figure 1a). The restoring force amplification factors E^r of the GCR algorithm with $\kappa_1 = 1/2, \kappa_2 = 1/2$ and the original CR algorithm with $\kappa_1 = 1/2, \kappa_2 = 1/4$ are 4.69 and 14.76 (= 3.15 $\times$ 4.69), respectively, for $\Omega_0 = 2$; are 11.55 and 251.80 (= 21.80 $\times$ 11.55), respectively, for $\Omega_0 = 10$ (Figure 2a). It can be clearly seen that the differences between the two specific algorithms rapidly increase with Ω_0.

6. Numerical Simulation of Pseudodynamic Testing

A two-story shear-type building structure is taken as the numerical simulation example of the pseudodynamic testing. The lumped masses of the two stories are $m_1 = 10^3$ kg and $m_2 = 5 \times 10^2$ kg, respectively. The initial stiffness of the two stories are $k_{1,0} = 10^7$ N/m and $k_{2,0} = 10^4$ N/m, respectively. The damping matrix for the substructure is based on using Rayleigh proportional damping [19], with a damping ratio of 0.02 in both the first and second modes. The natural initial circular frequencies for the first and second modes are found to be $\omega_{1,0} = 4.47$ rad/s and $\omega_{2,0} = 100.05$ rad/s, respectively. The first two order of initial frequencies are $f_{1,0} = 0.7114$ Hz and $f_{2,0} = 15.9235$ Hz, respectively. The structure is subjected to the 1940 El Centro NS earthquake ground motion.

In order to take the influence of error into consideration, displacement error is introduced to the displacement response at each story. Assume the displacement error is a random variable which obeys the truncated normal distribution with the probability density function as [20]

$$f(x) = \frac{1.00135}{\sigma\sqrt{2\pi}} \exp\left[-\frac{(x-\mu)^2}{2\sigma^2}\right], x \in (\mu - 3\sigma, \mu + 3\sigma) \tag{27}$$

where μ and σ are the mean value and the standard deviation of the displacement error x, respectively. In conducting the pseudodynamic testing, it is assumed that $\mu = 0$ and $\sigma = tol/3 = 10^{-5}$ m. *tol* is the tolerance error and assigned as 3×10^{-5} m in this paper. It denotes that the maximum displacement error of each time step is 3×10^{-5} m. By using the MATLAB software [18], a series of random numbers conforming the aforementioned distribution can be generated and used as the displacement errors.

Numerical simulation of two situations—i.e., without displacement error and with random displacement error—using the GCR algorithm are conducted. The time step Δt is assigned as 0.01 s. As a result, $\Omega_{1,0} = 0.04$ and $\Omega_{2,0} = 1$. Figures 3–5 show the displacement response, displacement error at each time step, and cumulative displacement error of each story, respectively. The numerical results using the Newmark explicit algorithm with the time step of 0.001 s are used as references.

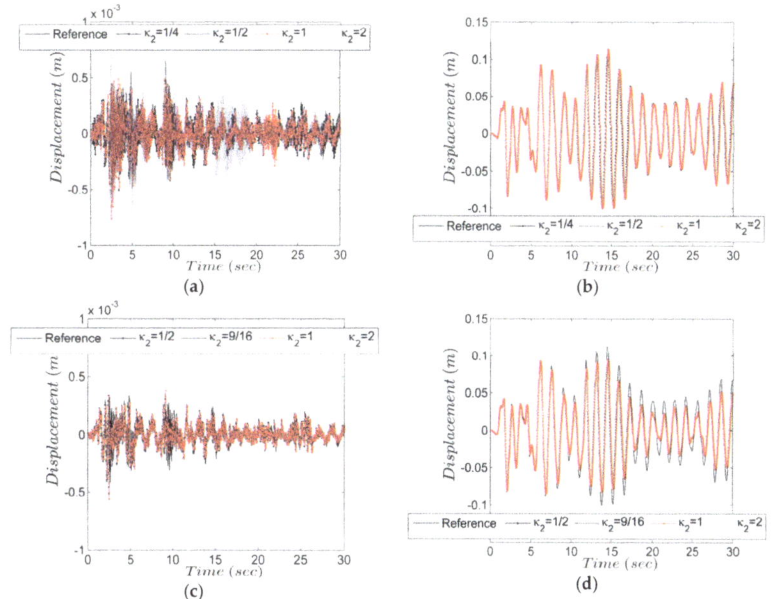

Figure 3. Displacement response: (**a**) $\kappa_1 = 1/2$, first story; (**b**) $\kappa_1 = 1/2$, second story; (**c**) $\kappa_1 = 1$, first story; (**d**) $\kappa_1 = 1$, second story.

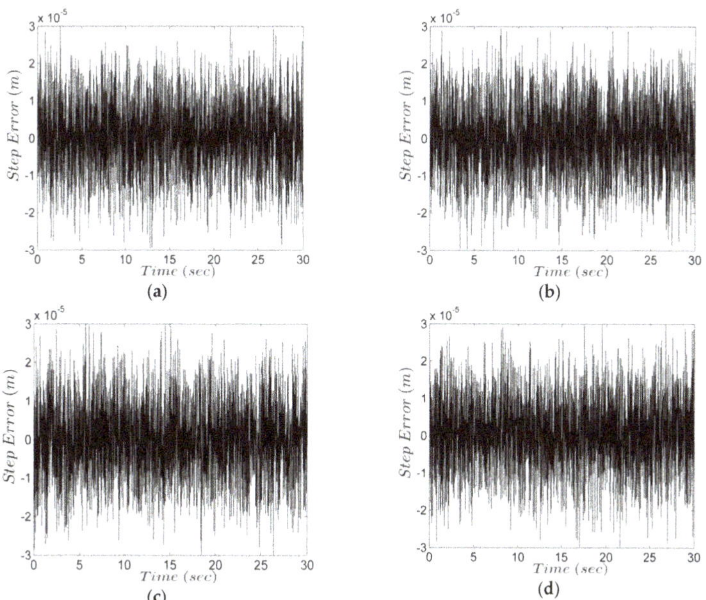

Figure 4. Displacement error at each time step: (**a**) $\kappa_1 = 1/2$, first story; (**b**) $\kappa_1 = 1/2$, second story; (**c**) $\kappa_1 = 1$, first story; (**d**) $\kappa_1 = 1$, second story.

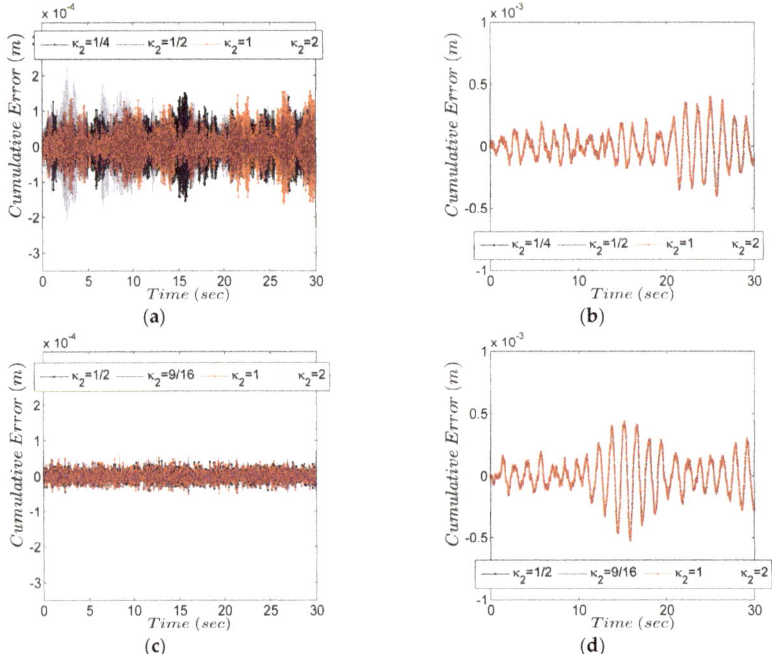

Figure 5. Cumulative displacement error: (**a**) $\kappa_1 = 1/2$, first story; (**b**) $\kappa_1 = 1/2$, second story; (**c**) $\kappa_1 = 1$, first story; (**d**) $\kappa_1 = 1$, second story.

It can be seen from Figure 3 that the displacement response at the first story is significantly disturbed by the error. On the contrary, the influence of error on the displacement response at the second story is inconspicuous. The main reason is that the displacement at the second story is mainly contributed by the first order modal response, while the displacement at the first story is, to some extent, contributed by the second modal response, whose value of Ω is larger than that of the first modal response and therefore exhibits a higher error propagation result.

Although the cumulative displacement error can be significantly reduced with the increase of κ_2, the accurate displacement response at first story is still overlapped by the error (Figure 5a,c). As for $\kappa_2 = 1/2$, the displacement at the second story is in good agreement with the reference value and less affected by the displacement error (Figure 3b). As for $\kappa_2 = 1$, the displacement at the second story is also less affected by the displacement error, but the accuracy of the numerical result is impaired due to the existing of numerical damping of the integration algorithm (Figure 3d). The maximum displacement (0.094 m) at the second story calculated using the GCR algorithms with $\kappa_2 = 1$ are 83% of the reference value (0.113 m).

Figure 6 shows the spectral characteristic of the displacement by conducting fast Fourier transform (FFT) of the displacement response shown in Figure 3.

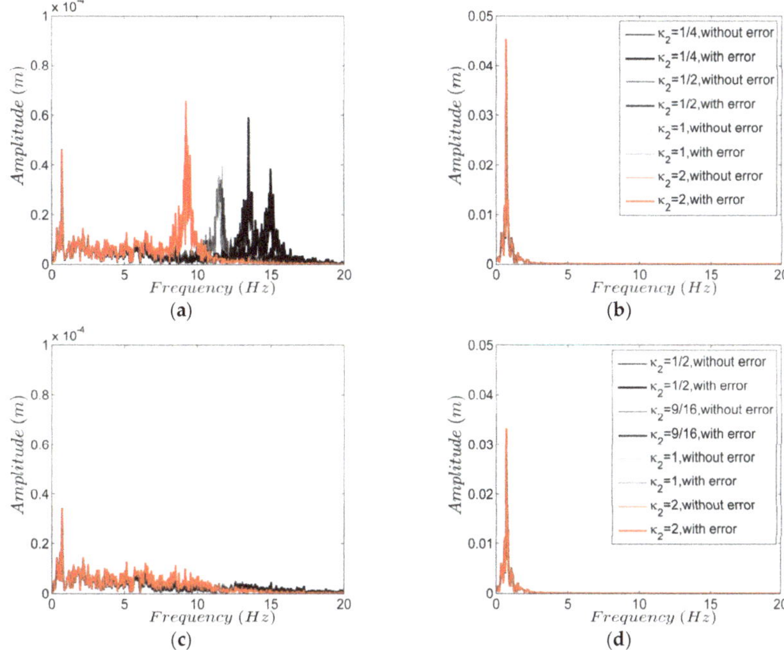

Figure 6. Spectral characteristic of displacement response at each story: (**a**) $\kappa_1 = 1/2$, first story; (**b**) $\kappa_1 = 1/2$, second story; (**c**) $\kappa_1 = 1$, first story; (**d**) $\kappa_1 = 1$, second story. Note: The legends in Figure 6a are identical with those in Figure 6b; the legends in Figure 6c are identical with those in Figure 6d.

Figure 6a implies that the displacement at the first story is contributed by the first and second order of modal responses. The first order frequency obtained by several GCR algorithms are very close and approximately 0.7060 Hz, which is slightly smaller than the first order frequency of the structure, i.e., 0.7114 Hz. This is due to period elongation [19] of the integration algorithm. However, there exists obvious difference of the second order frequency between the calculated value by using several GCR algorithms and the actual frequency as period elongation increases with the increase of Ω. In Figure 6c, the first order response is obvious, while the second order response is less obvious. The reason for this is the numerical damping and the random displacement error. By comparing Figure 6b,d, the displacement at the second story is only contributed by the first order modal response, which further explains the phenomenon shown in Figure 3. In addition, due to the existing of numerical damping, the spectral amplitude of $\kappa_2 = 1$ is less than that of $\kappa_2 = 1/2$, which is consistent with the displacement response indicated in Figure 3.

7. Conclusions

This paper presents a new family of model-based GCR integration algorithm. The error propagation equation for the GCR algorithms implemented in a pseudodynamic test is derived. The error amplification factors for three structures with different degrees of nonlinearity are calculated to illustrate the error propagation effect. The numerical simulation of a pseudodynamic test for a two-story shear-type building structure is conducted to further demonstrate the error propagation characteristics of the new algorithms. The original contributions of this study are to investigate the error propagation properties of the GCR algorithms for pseudodynamic tests of nonlinear systems by considering the influence of viscous damping. The following conclusions can be drawn from the examples and numerical simulations in Sections 5 and 6:

(1) Both amplification error factors, i.e., E^x and E^r, increase with $\Omega_0 = \omega_0 \Delta t$.

(2) Both amplification factors, i.e., E^x and E^r, increase with the degree of nonlinearity δ, so it is crucial to take the degree of nonlinearity δ into consideration.

(3) As for the viscous damping, it has a positive effect on reducing the errors, so it is conservative to ignore the influence of viscous damping.

(4) Both the two additional coefficients, i.e., κ_1 and κ_2, of the GCR algorithms have great impact on the error propagation property. With the increasing of κ_1, the errors increase, while both amplification factors E^x and E^r increasing with the decrease of κ_2.

(5) The original CR algorithm with $\kappa_1 = 1/2, \kappa_2 = 1/4$ has a relatively large error propagation property, especially for the nonlinear structures with stiffness hardening. The two amplification error factors, i.e., E^x and E^r, of the GCR algorithm with $\kappa_1 = 1/2, \kappa_2 = 1/2$ are only 10.8% and 4.6% of those of the original CR algorithm with $\kappa_1 = 1/2, \kappa_2 = 1/4$ when $\Omega_0 = 10$ for the nonlinear structure with stiffness hardening of Case 3 without viscous damping. Therefore, the GCR algorithm with $\kappa_1 = 1/2, \kappa_2 = 1/2$ is a superior alternative of the original CR algorithm with $\kappa_1 = 1/2, \kappa_2 = 1/4$.

(6) The displacement response at the first story is significantly disturbed by the error, whereas the influence of error on the displacement response at the second story is inconspicuous. The main reason is that the displacement at the second story is mainly contributed by the first order modal response, while the displacement at the first story is, to some extent, contributed by the second modal response.

(7) The maximum displacement at the second story obtained by using the GCR algorithms with $\kappa_2 = 1$ are 83% of the reference value. It means that the accuracy of the numerical result is impaired by excessive numerical damping for the high frequency response, thus the numerical dissipation characteristics of the GCR algorithms should be optimized in the future.

Author Contributions: B.F. conducted the numerical simulations and wrote the paper. H.J. conceived the idea and revised the paper. T.W. provided valuable discussions and revised the paper. He took responsibility of the corresponding work.

Funding: Financial support from the National Natural Science Foundation of China (Grant No. 51578072, 51478354), Shaanxi Province Natural Science Foundation (Grant No. 2018JQ5078) and Program of Shanghai Subject Chief Scientist (18XD1403900) are gratefully acknowledged.

Conflicts of Interest: The authors declare no conflict of interest.

Appendix A: Notations

M: Mass matrix
C: Damping matrix
K: Stiffness matrix
X_{i+1}: Displacement vector at the $(i + 1)$th time step
$\dot{X}_{i+1}$: Velocity vector at the $(i + 1)$th time step
$\ddot{X}_{i+1}$: Acceleration vector at the $(i + 1)$th time step
F_{i+1}: External force vector at the $(i + 1)$th time step
Δt: Time step
α_1 and α_2: Integration parameter matrices
κ_1 and κ_2: Additional coefficients of the GCR algorithm
δ_{i+1}: Degree of nonlinearity
A_{i+1}: Amplification matrix at the $(i + 1)$th time step
L_{i+1}: Load vector at the $(i + 1)$th time step
ω_{i+1}: Natural frequency of the system at the $(i + 1)$th time step

ω_0: Initial natural frequency of the system

Ω_{i+1}: The product of natural frequency ω_{i+1} at the $(i + 1)$th time step multiplies time step Δt

Ω_0: The product of initial natural frequency ω_0 multiplies time step Δt

$\xi = c/(2m\omega_0)$: The equivalent viscous damping ratio

I: Identity matrix of dimension 3×3

λ: Eigenvalue of the matrix A_{i+1}

r_{i+1}: Restoring force at the $(i + 1)$th time step

x_{i+1}^a: Actual displacement at the $(i + 1)$th time step, including the effects of errors at previous steps and the current step

r_{i+1}^a: Restoring force at the $(i + 1)$th time step, including the effects of errors at previous steps and the current step

x_{i+1}^e: Exact displacement at the $(i + 1)$th time step, including the effects of errors at previous steps

r_{i+1}^e: Exact restoring force at the $(i + 1)$th time step, including the effects of errors at previous steps

e_{i+1}^x: Displacement error at the $(i + 1)$th time step

e_{i+1}^r: Restoring force error at the $(i + 1)$th time step

e_{i+1}^{rx}: Displacement error corresponding to the restoring force error e_{i+1}^r

ε_{i+1}: Error cumulative vector at the $(i + 1)$th time step

D_i, D_{i-1} and D_{i-2}: Coefficients for the displacement feedback errors

R_i and R_{i-1}: Coefficients for the restoring force feedback errors

α_{i+1} and β_{i+1}: Phase angles

E^x: Displacement error amplification factor

E^r: Restoring force error amplification factor

Appendix B

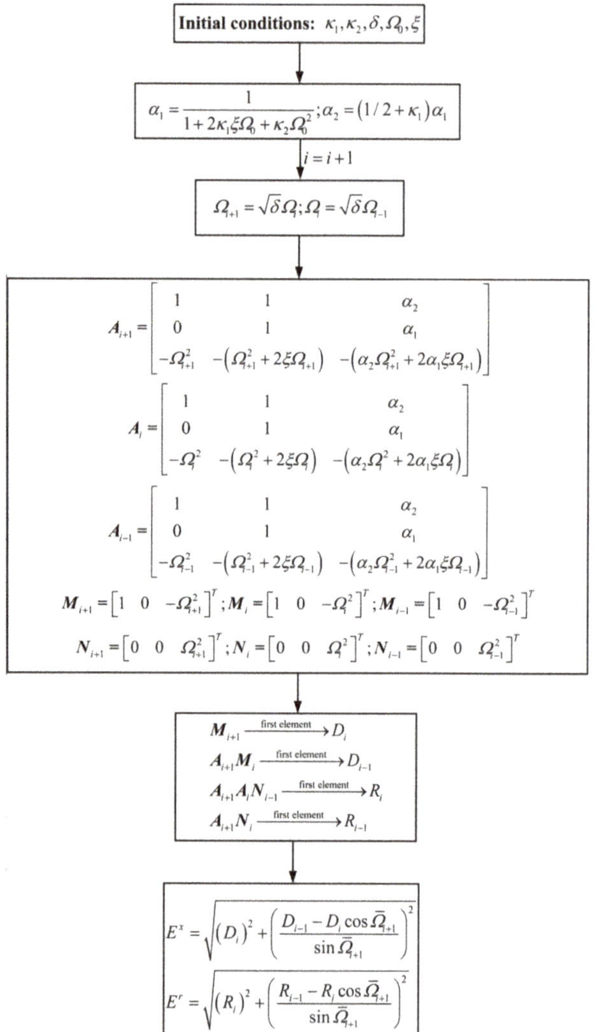

Figure A1. Flowchart of calculating the error amplification factors.

References

1. Takanashi, K.; Udagawa, K.; Seki, M.; Okada, T.; Tanaka, H. *Nonlinear Earthquake Response Analysis of Structures by a Computer-Actuator On-Line System*; University of Tokyo: Tokyo, Japan, 1975.
2. Abbiati, G.; Bursi, O.S.; Caperan, P.; Di Sarno, L.; Molina, F.J.; Paolacci, F.; Pegon, P. Hybrid simulations of a multi-span RC viaduct with plain bars and sliding bearings. *Earthq. Eng. Struct. Dyn.* **2015**, *44*, 2221–2240. [CrossRef]
3. Elnashai, A.S.; Di Sarno, L. *Fundamentals of Earthquake Engineering*; Wiley and Sons: Chichester, UK 2008.
4. Chang, S.Y. Explicit pseudodynamic algorithm with unconditional stability. *J. Eng. Mech.* **2002**, *128*, 935–947. [CrossRef]

5. Chen, C.; Ricles, J.M. Development of direct integration algorithms for structural dynamics using discrete control theory. *J. Eng. Mech.* **2008**, *134*, 676–683. [CrossRef]

6. Kolay, C.; Ricles, J.M. Development of a family of unconditionally stable explicit direct integration algorithms with controllable numerical energy dissipation. *Earthq. Eng. Struct. Dyn.* **2014**, *43*, 1361–1380. [CrossRef]

7. Kolay, C.; Ricles, J.M. Assessment of explicit and semi-explicit classes of model-based algorithms for direct integration in structural dynamics. *Int. J. Numer. Methods Eng.* **2016**, *107*, 49–73. [CrossRef]

8. Peek, R.; Yi, W.H. Error analysis for pseudodynamic test method. I: Analysis. *J. Eng. Mech.* **1990**, *116*, 1618–1637. [CrossRef]

9. Peek, R.; Yi, W.H. Error analysis for pseudodynamic test method. II: Application. *J. Eng. Mech.* **1990**, *116*, 1638–1658. [CrossRef]

10. Shing, P.B.; Mahin, S.A. Cumulative experimental errors in pseudodynamic tests. *Earthq. Eng. Struct. Dyn.* **1987**, *15*, 409–424. [CrossRef]

11. Shing, P.B.; Mahin, S.A. Experimental error effects in pseudodynamic testing. *J. Eng. Mech.* **1990**, *116*, 805–821. [CrossRef]

12. Shing, P.B.; Manivannan, T. On the accuracy of an implicit algorithm for pseudodynamic tests. *Earthq. Eng. Struct. Dyn.* **1990**, *19*, 631–651. [CrossRef]

13. Chang, S.Y. Nonlinear error propagation analysis for explicit pseudodynamic algorithm. *J. Eng. Mech.* **2003**, *129*, 841–850. [CrossRef]

14. Chang, S.Y. Error propagation of HHT-α method for pseudodynamic tests. *J. Earthq. Eng.* **2005**, *9*, 223–246. [CrossRef]

15. Chang, S.Y. Error propagation in implicit pseudodynamic testing of nonlinear systems. *J. Eng. Mech.* **2005**, *131*, 1257–1269. [CrossRef]

16. Bousias, S.; Sextos, A.; Kwon, O.H.; Taskari, O.; Elnashai, A.S.; Evangeliou, N.; Di Sarno, L. Intercontinental hybrid simulation for the assessment of a three-span R/C highway overpass. *J. Earthq. Eng.* **2017**. [CrossRef]

17. Fu, B. Substructure Shaking Table Testing Method Using Model-Based Integration Algorithms. Ph.D. Thesis, Tongji University, Shanghai, China, 2017. (In Chinese)

18. *MATLAB. Version 2014a*; The MathWorks Inc.: Natick, MA, USA, 2014.

19. Chopra, A.K. *Dynamic of Structures: Theory and Applications to Earthquake Engineering*, 4th ed.; Prentice Hall: Upper Saddle River, NJ, USA, 2012.

20. Chang, S.Y. The γ-function pseudodymamic algorithm. *J. Earthq. Eng.* **2000**, *4*, 303–320. [CrossRef]

 *sustainability*

Article

Substructure Hybrid Simulation Boundary Technique Based on Beam/Column Inflection Points

Zaixian Chen [1], Xueyuan Yan [2,*], Hao Wang [3] , Xingji Zhu [1,*] and Billie F. Spencer [4]

[1]   Department of Civil Engineering, Harbin Institute of Technology at Weihai, Weihai 264209, China; Zaixian_chen@sina.com
[2]   College of Civil Engineering, Fuzhou University, Fuzhou 350116, China
[3]   Department of Civil Engineering, Southeast University, Nanjing 210096, China; wanghao1980@seu.edu.cn
[4]   Department of Civil and Environmental Engineering, University of Illinois at Urbana-Champaign, Urbana, IL 61801, USA; bfs@illinois.edu
*   Correspondence: yxy820910@sina.com (X.Y.); zhuxingji@hit.edu.cn (X.Z.); Tel.: +86-591-2286-5364 (X.Y.)

Received: 2 July 2018; Accepted: 26 July 2018; Published: 28 July 2018

Abstract: Compatibility among substructures is an issue for hybrid simulation. Traditionally, the structure model is regarded as the idealized shear model. The equilibrium and compatibility of the axial and rotational direction at the substructure boundary are neglected. To improve the traditional boundary technique, this paper presents a novel substructure hybrid simulation boundary technique based on beam/column inflection points, which can effectively avoid the complex operation for realizing the bending moment at the boundary by using the features of the inflection point where the bending moment need not be simulated in the physical substructure. An axial displacement prediction technique and the equivalent force control method are used to realize the proposed method. The numerical simulation test scheme for the different boundary techniques was designed to consider three factors: (i) the different structural layers; (ii) the line stiffness ratio of the beam to column; and (iii) the peak acceleration. The simulation results for a variety of numerical tests show that the proposed technique shows better performance than the traditional technique, demonstrating its potential in improving HS test accuracy. Finally, the accuracy and feasibility of the proposed boundary technique is verified experimentally through the substructure hybrid simulation tests of a six-story steel frame model.

Keywords: substructure; boundary technique; inflection point; hybrid simulation; force-displacement control

---

1. Introduction

Hybrid simulation (HS), also known as pseudo-dynamic testing, was initially proposed in 1969 [1]. This approach [2–4], which is a combination of physical testing and numerical simulation, has been widely used to evaluate the seismic performance of a structure, because it is efficient and cost-effective. HS takes the critical parts of a structure, which cannot be simulated adequately with the numerical simulation, as the physical substructure to test in the lab. The rest of the structure is taken as the numerical substructure and simulated using appropriate finite element software. Simulation of the laboratory and computer parts of the structure are carried out simultaneously, exchanging information about the responses at the boundary at each time step. Furthermore, the HS method transforms a real-time dynamic test under earthquake loads into a static loading test by simulating the dynamic components of the structural response in the computer, thus reducing the requirement for the loading devices and making large-scale or even full-scale model tests possible.

Initial research on the HS was overwhelmingly focused in Japan and the United States. Beginning in the 1990s, development and application of HS in Europe, Asia, and other regions grew significantly [5–7]. For structural systems with rate-dependent components, real-time HS testing was proposed and developed [8–15]. Moreover, HS also allows testing to be geographically distributed in multiple testing facilities. Many countries have been begun to build networked structures laboratories and developed a variety of HS platforms [16–19] to allow performance assessment of modern building structure systems. The largest is the Network for Earthquake Engineering Simulation (NEES) supported by the United States National Science Foundation from 1999 to 2014 [20–23]. NEES featured 14 geographically distributed and shared-use laboratories to test modern structural system.

Although substantial advances have been made in HS, achieving equilibrium and compatibility of the boundaries between the two substructures remains problematic, possibly limiting the accuracy of HS tests. There are currently four Multi-Axis Substructure Testing MAST systems in the world including MAST at the University of Illinois at Urbana-Champaign (UIUC), the University of Minnesota (USA), Swinburne University of Technology (Australia) and Ecole Polytechnique Montreal (Canada) that have the capabilities of six-DOF switched and mixed displacement and force modes of control [24–26]. The MAST at Swinburne has already been used in such hybrid simulation tests. However, controlling and measuring the force and displacement in all DOF is challenging. The method we present is relatively simple and can be widely implemented.

The fact that it cannot consider the effect of bending moments has always been a flaw for the HS method. However, this method has also been widely used in various structures due to its effectiveness, not just limited to shear buildings, as described in the literature [27–35]. In these tests, the conventional method of placing the loading position at each layer was widely adopted without considering the effect of the bending moment.

This paper proposes and improves a novel boundary technique based on a beam's inflection point that can avoid the complex operation for realizing bending moments at the substructure boundary, based on previous a series of experimental techniques [34,36,37]. Following a description of the proposed boundary condition technique, a discussion of various implementation challenges is presented. The proposed method was validated through virtual HS, and experimental validation was carried out using a six-story steel frame model. The results demonstrate the efficacy of the proposed approach.

2. Boundary Technique for the HS

The compatibility and equilibrium conditions should be satisfied at the boundary between the physical and the numerical substructures during the HS. Consider the five-story example shown in Figure 1. Figure 1a shows the full model of the structure. Figure 1b illustrates the traditional boundary between two substructures, which is located at the top of the first story. This traditional boundary strategy mainly takes the structure model as an idealized shear model. The equilibrium of shear forces and the compatibility of the horizontal or shear displacement is satisfied by the hydraulic actuator. The equilibrium of the bending moment and axial force, as well as the compatibility axial and rotational deformations are typically neglected. Consequently, only two kinds of loading device are required at the boundary between the numerical substructure and the physical substructure: (i) one is the horizontal loading device to impose the shear force or displacement; and (ii) the other is the vertical loading device to simulate the gravity load due to upper stories, which is usually taken to be constant during the test. Therefore, several inevitable errors are introduced by not satisfying compatibility and equilibrium conditions.

To meet the equilibrium and compatibility in the axial and rotational direction at the boundary, more than two hydraulic actuators are required, as well as appropriate sensors to measure the rotational angle and the axial deflection, which can be challenging. To address the problems, a new boundary technique based on the inflection point is proposed, as shown in Figure 1c. The bending moments at the section of the inflection point in a structure is zero; therefore, this new boundary technique can effectively avoid the complex operation for realizing the bending moment at the boundary. However, two new problems arise in the proposed boundary technique: (i) how to calculate the horizontal displacement of the inflection point; and (ii) how to meet equilibrium and compatibility in the axial direction.

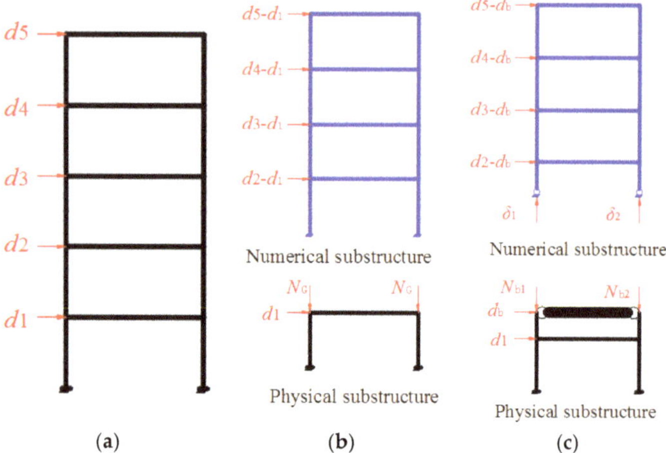

Figure 1. Schematic diagram of the different boundary techniques: (**a**) Full model; (**b**) traditional technique; and (**c**) proposed technique. d_i denotes the displacements of each level on the full models. i ∈ [1, 5]. d_b denotes the displacements at the inflection point. δ_1 and δ_2 denote the axial deformation on interfaces. N_G denotes the constant representative value of gravity load calculated by the numerical substructure. N_{b1} and N_{b2} denote the axial forces resulted from the overturning moments on physical substructure transmitted from numerical substructure on each interface.

For the first problem, assuming a uniform column between two floors, then the displacement with respect to the lower floor at the inflection point is a linear interpolation value of the interstory drift for the corresponding floor. Adding a new concentrated mass or a degree of freedom (DOF) on the inflection point of each story is the direct method to solve the first problem for the proposed boundary technique. However, the DOF of the equation of motion (EOM) will increase significantly so that the stability and efficiency of the HS is influenced. A linear interpolation technique is used to determine the displacement at the inflection point. The calculated displacement at the inflection point is from the interstory drift of the corresponding floor. Therefore, the solution of the EOM is not different from the traditional method.

For the second problem, the axial displacement prediction technique at the boundary is presented, because the axial displacement is usually too small to measure accurately using displacement sensors [25]. Moreover, the axial displacement at the boundary must be obtained to start the simulation of the numerical substructure for the next step, so that the axial forces can be obtained from the numerical substructure simulation results to apply to the boundary of the physical substructure by the hydraulic actuators. Traditionally, an iteration strategy should be introduced to calculate the axial displacement. However, such iteration strategies will result in inevitable measure errors, because the physical substructure will be loaded many times during the iteration. To avoid iteration, an axial displacement prediction technique is presented to predict the axial displacement using the loading axial force, the assumption of the elastic axial deformation during the test, and the equilibrium equation.

The proposed prediction method is given by the following equation

$$\delta_N(i) = \sum_{j=1}^{m} \frac{N_j(i)h_j}{E_j A_j} + \frac{N_b(i)h_{I,b}}{E_b A_b} \tag{1}$$

in which $\delta_N(i)$ is the prediction axial deformation at the step i; m is the number of stories in the physical substructure; $N_j(i)$ is the calculated axial force of the j-story at the step i; $N_b(i)$ is the calculated axial force of the boundary story, which is calculated from the numerical substructure; h_j is the height of the j-story of the physical substructure; $h_{I,b}$ is the height from the inflection point to the floor of the boundary story; $E_j A_j$ is the axial stiffness of the jth-story of the physical substructure, which is constant during the test due to the assumption of the elastic axial deformation; and $E_b A_b$ is the axial stiffness of the boundary story of the physical substructure. Here, $N_j(i)$ can be calculated by the equilibrium equation. For example, consider the top story of the physical substructure shown in Figure 2. Here, L is the span of the frame; point $O_{m,1}$ and $O_{m,2}$ are inflection points of the m story; $h_{I,m}$ is the height from the inflection point to the roof of the m story; R_b is the applied force by the actuator at the boundary; and R_m is the applied force by the actuator at the mth-story. Noting that the dynamics of the frame are represented in the computer and using the moment equilibrium equation, the axial force can be calculated by taking moments about point O_1 or O_2. In the same way, the axial force of each story of the physical substructure can be calculated from the results of the upper story. Then, the axial forces of the physical substructure can be obtained. Subsequently, the prediction axial displacement at the boundary can be calculated using Equation (1), i.e.,

$$N_{m1}(i) = \frac{N_{b1}(i)L - R_m(i)h_{I,m} - R_b(i)(h_{I,b} + h_{I,m})}{L} \tag{2}$$

$$N_{m2}(i) = \frac{N_{b2}(i)L + R_m(i)h_{I,m} + R_b(i)(h_{I,b} + h_{I,m})}{L} \tag{3}$$

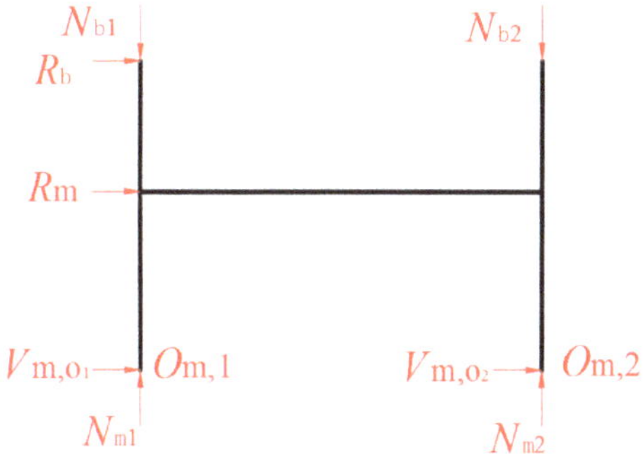

Figure 2. Calculation diagram for the upper story axial force of the physical substructure.

3. Implement of Boundary Technique

3.1. HS Component Interaction

Typically, the HS test includes three main components: the physical substructure, the numerical substructure, and the numerical integration routine. Early HS tests usually used an idealized mass-spring-damper model whose performance is known a priori to simulate the numerical substructure, allowing the software controlling the entire HS test to be realized as in a single program (e.g., written in FORTRAN or C). As HS testing has developed, researchers have sought to use more and more complicated models, in terms of both experimental and computational components. Computational components tailored for a particular strength are combined with general purpose finite element programs to achieve more accurate simulation. For example, a high-fidelity ABAQUS model was used [38], combined with the Matlab-based UI-SimCor [17], to investigate the seismic performance of semirigid steel joints. To address current research needs, HS testing software must accommodate different computational components.

Communication between the computational and experimental components must be considered to realize a HS test. Figure 3 shows the sketch of the HS component interaction. First, OpenSEES is used to simulate both the physical and the numerical substructure in the virtual HS test, while MATLAB is employed to integrate the EOM. To realize the required data transfer, a communication bridge between the two different computational components is constructed. Data communication will take place at each step; each independent unit writes data to the other units and waits for the data written by other units. Moreover, the waiting or calculating time at different step is not uniform. This paper employs the network socket technique as a bridge to communicate the data between the two different computational components. In the MATLAB component, the TCPIP and ECHOTCPIP commands are used to construct an ICPIP object associated with OpenSEES. Concurrently, the command SOCKET in OpenSEES program is used to communicate with the Matlab program. Subsequently, a data acquisition card of PCI 1712 manufactured by the Advantech is employed as the communication bridge between the MATLAB and the MTS loading device. The PCI-1712 card provides a total of 16 single-ended or eight differential A/D input channels or a mixed combination, two 12-bit D/A output channels, 16 digital input/output channels, and three 10 MHz 16-bit multifunction counter channels.

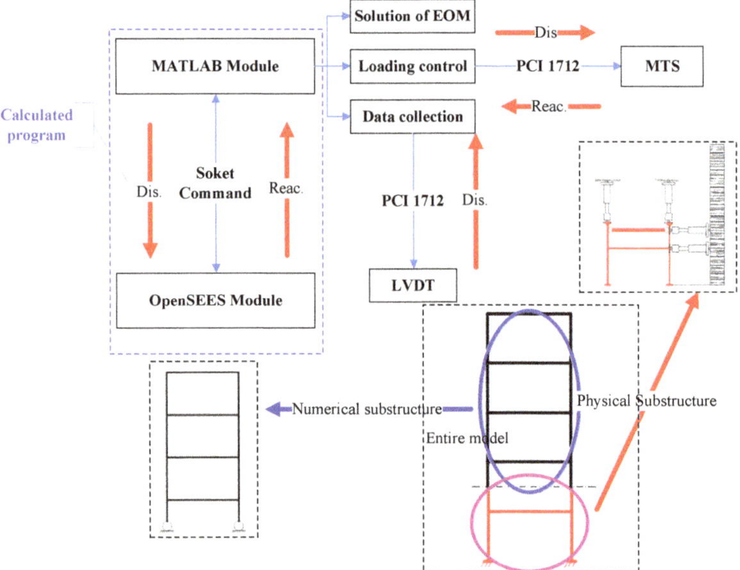

Figure 3. Sketch of the HS component interaction.

3.2. Equivalent Force Control Method

Research on various integration algorithms employed for HS testing is very active. For example, the CR and KR-α methods [10,35] which have unconditional stability for linear elastic and stiffness softening nonlinear systems, were presented. However, the implicit algorithms generally require iteration, which can result in inconsistent HS test results. To avoid iteration, an equivalent force control (EFC) method [39], which uses feedback control to replace numerical iteration for the real-time HS testing, was proposed. EFC method was used to a full-scale six-story frame-supported reinforced concrete masonry shear wall [39].

This proposed approach combines the EFC method with the proposed inflection point boundary condition technique to conduct HS tests. The EFC method is based on the feedback control method shown in Figure 4. Here, the $F_{EQ,i+1}$, which is used for the equivalent force (EF) command, can be calculated by the information before the $i + 1$ step. Therefore, the $F_{EQ,i+1}$ is available and remains unchanged at $i + 1$ step. The $F_{EQ,i+1}$, which is used for equivalent force (EF) feedback, is obtained by the three terms: (i) the reaction of the experimental substructure $R_E\left[d^c_{E,i+1}(t)\right]$, which is measured by the load cells; (ii) the reaction of the numerical substructure $R_N\left[d^c_{E,i+1}(t)\right]$, which is obtained from the results of the numerical simulation; and (iii) the pseudo-dynamic force, which is linearly related to displacement by K_{PD}. The displacement command, $d^c_{i+1}(t)$, is determined from the force signal using the force-displacement conversion matrix C_F. The equivalent force controller is used to make certain that the EF feedback can track the EF command quickly without overshoot. This paper uses a PD controller as the equivalent force controller. The parameters of the PD controller are obtained as discussed in Reference [39].

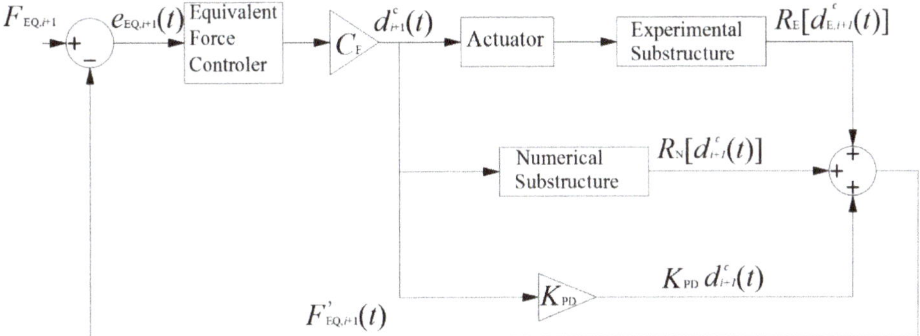

Figure 4. The block diagram of EFC method [39].

3.3. Basic Process

Figure 5 shows the flow diagram for the proposed HS for both the virtual and the physical HS. The only difference between the virtual and the physical HS is that the physical substructure is modeled computationally. For the physical HS, the results of the physical substructure can be measured by the sensors from the model in the lab. For the virtual HS, the results of the physical substructure can be obtained from the results of an additional numerical simulation using the OpenSEES. In Figure 5, *istep* is the *i*th step of the earthquake records, and *n* corresponds the total number of the earthquake records. Matlab is used to integrate the EOM for calculating the responses of the structure. OpenSEES is used for simulation of the numerical substructure. The basic detailed process is as follows:

- Step 1: The initial stiffness matrix, the initial mass matrix, the initial damping matrix, other initial parameters, and the ground motion record are obtained. The ground motion record may be scaled in amplitude to accommodate the goal of different HS tests.
- Step 2: The EF command is calculated. The displacement command is then calculated using the force-displacement conversion matrix and the EF controller.
- Step 3: The pseudo-dynamic force is calculated using the pseudo-dynamic stiffness K_{PD}, which can be determined from the initial parameters. Concurrently, the predicted axial deformation is obtained using Equation (1). Before the next step, all calculations are realized using Matlab. After this step, the calculated displacement command of the numerical substructure and the prediction axial deformation at the boundary is applied to the numerical substructure modeled by OpenSEES.
- Step 4: The calculated displacement command of the numerical substructure and the prediction axial deformation at the boundary are sent to the numerical substructure. Subsequently, the reaction force at the boundary is calculated. Similarly, the data calculated by OpenSEES

are sent to the Matlab program and subsequently wait to receive the data calculated by Matlab for the next step.

- Step 5: The calculated reaction force in the vertical direction and the displacement command in the horizontal direction are sent to the actuator controllers for the physical substructure. Then, the reaction of the physical substructure is obtained by: (i) the sensors in the associated loading devices; or (ii) OpenSEES for the virtual test.

- Step 6: The EF feedback is obtained by summing up the pseudo-dynamic force, and the reaction of both the numerical substructure and the physical substructure. Then, compare the EF feedback with the EF command to obtain the error between them. If the error between the EF feedback with the EF command is less than the specified tolerance, go the next step. If not, repeat the steps from the third step to the sixth step.

- Step 7: The responses of the entire structure are obtained. Then, repeat the above-mentioned steps until the entire ground motion record has been processed.

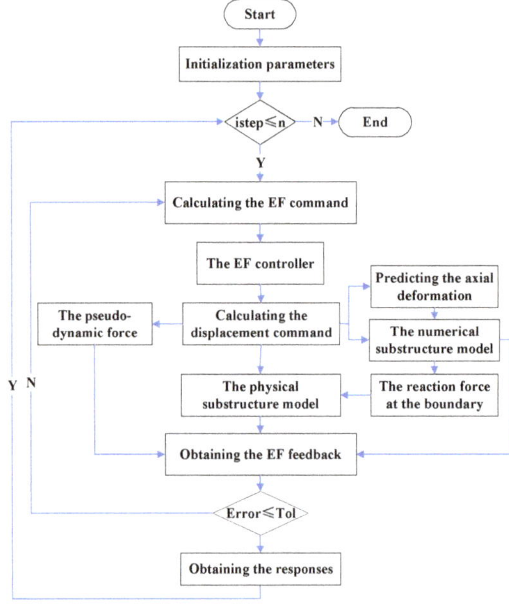

Figure 5. Flow diagram of substructure boundary simulation for the HS testing method.

After the details of proposed HS, a series of numerical simulation is carried out though the virtual HS in the next section. Then, experimental validation is accomplished using a six-story steel frame model. The El Centro (N-S, 1940) is taken as the ground motion record for both the virtual and the physical HS tests.

4. Numerical Simulation

4.1. Numerical Simulation Model Based on the Uniform Design

To illustrate the proposed approach, a series of steel frame structures with 3–17 stories were considered. A nonlinear Beam–column element, which has a hardening bilinear constitutive model, was chosen in OpenSEES to model the frame. The yield strength is 235 Mpa. The elastic modulus is 200,000 Mpa. The post-yield to pre-yield stiffness ratio is 0.02. The mass matrix was calculated using provision 5.1.3 of the Chinese Code for Seismic Building Design GB50011-2010. Rayleigh Damping was adopted with a damping ratio of 0.05 in the first two modes. The number of the stories in the building, the beam–column's linear stiffness ratio (LSR), and the peak ground acceleration (PGA) of the ground motion record were taken as parameters for the numerical study. Each story height is 4 m, and the span of the steel frame structure is 6 m.

This experimental design method proposed by Wang and Fang [40] was used to increase efficiency and reduce the required number of tests by focusing on the uniformity of the test matrix. For example, assume that the m parameters are chosen in an experiment. Let the number of values of each parameter to be considered be n. The number of experiments is m^n for a full factorial design. The number of samples is about m^2 for an orthogonal design, which takes both the orthogonality and uniformity of the design space into consideration. However, the number of experiments is about m for the uniform design, which only highlights the uniformity of the design space.

The uniform design table, $Un(q^s)$, which is similar to the orthogonal design table, was constructed in advance, where the symbol U is the label for the uniform design, n is the number of test, s is the maximum number of parameters, and q is the number of levels for each parameter. The deviation D of the uniform design table is taken as the index to evaluate the uniformity of the table. The smaller is the value of D, the better is the uniformity of the design table. By considering the number of stories, the beam–column's linear stiffness ratio, and the peak ground acceleration (PGA) of the ground motion record, the uniform design table $U^*_{15}(15^7)$ was chosen. The symbol * means that the design table has more uniformity [40]. The deviation D of the selected uniform table is 0.1361, which meets the precision required. More detailed parameters are listed in Table 1.

Table 1. Structural parameter information.

Test Number	Number of Story	LSR of Beam–Column	PGA (cm/s^2)	Number of Story for the Physical Substructure	Beam Section (mm)	Column Section (mm)
1	3	0.9	540	1	300 × 550 × 18 × 12	400 × 400 × 21 × 13
2	4	1.4	420	2	300 × 600 × 24 × 14	400 × 400 × 21 × 13
3	5	1.9	300	1	300 × 700 × 24 × 12	400 × 400 × 21 × 13
4	6	0.8	180	1	300 × 500 × 20 × 14	400 × 400 × 21 × 13
5	7	1.3	55	1	300 × 600 × 22 × 14	400 × 400 × 21 × 13
6	8	1.8	580	3	300 × 700 × 22 × 12	400 × 400 × 21 × 13
7	9	0.7	460	1	300 × 500 × 17 × 12	400 × 400 × 21 × 13
8	10	1.2	340	3	350 × 550 × 22 × 14	400 × 400 × 21 × 13
9	11	1.7	220	1	450 × 800 × 28 × 16	500 × 500 × 28 × 18
10	12	0.6	100	1	450 × 550 × 22 × 14	500 × 500 × 28 × 18
11	13	1.1	620	3	450 × 650 × 28 × 18	500 × 500 × 28 × 18
12	14	1.6	500	2	450 × 800 × 26 × 16	500 × 500 × 28 × 18
13	15	0.5	380	1	400 × 500 × 24 × 16	500 × 500 × 28 × 18
14	16	1	260	1	450 × 600 × 31 × 18	500 × 500 × 28 × 18
15	17	1.5	140	1	450 × 800 × 24 × 16	500 × 500 × 28 × 18

4.2. Numerical Simulation Results Analysis

This section provides a comparison of the numerical simulation for each model listed in Table 1 for the traditional technique and the proposed elastic inflection point technique. The simulation result of entire structure using OpenSEES is considered to be the "exact" solution and used as a point of reference. The errors in the numerical simulation for the different boundary simulation techniques at the ith step, as compared to the reference solution, are designated as e_i. Then, the cumulative error in the numerical simulation for the different boundary simulation techniques is given by

$$\delta = \sqrt{\sum_i e_i{}^2} \tag{4}$$

Due to space limitation, only representative results are given for three of these buildings: tests No. 2, No. 8, and No. 13.

4.2.1. Displacements Analysis

Figure 6 shows the time history of the errors e_i for tests No. 2, No. 8, and No. 13 for the traditional and proposed boundary simulation techniques. In Figure 6, we can see clearly that the error in the displacement for the proposed boundary technique is smaller than the traditional technique. The simulation results for the other numerical tests shown in Table 1 followed similar trends. Additionally, the maximum displacement errors and the cumulative errors for the different boundary techniques were calculated for the 15 tests shown in Table 1. The mean of the maximum displacement error and the mean of the cumulative error over all 15 tests was calculated for both the traditional technique and the proposed technique. Both error measures were found to be an order of magnitude smaller for the proposed approach, as compared with the traditional technique. The maximum displacement error ratio between two different techniques had an average of 9.18 with a maximum of 13.00 and a minimum of 2.66. The cumulative error ratio between two different techniques had an average of 9.87 with a maximum of 18.04 and a minimum of 3.12.

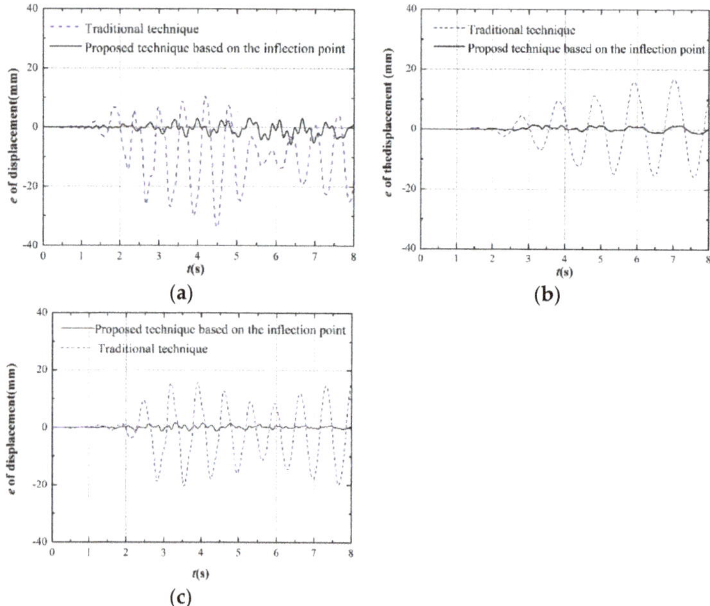

Figure 6. The time history of the displacement errors for the different boundary techniques at the first story: (**a**) Test 2; (**b**) Test 8; and (**c**) Test 13.

4.2.2. Internal Force Analysis

Figures 7–9 show the error time history of the bending moment, shear force, and axial force, respectively, at the top of the column of the 1st story for the different boundary techniques. Here, the internal force (including bending moment, shear force, and axial force) errors have a similar tendency with the displacement errors in Section 4.2.1. The internal force errors of the proposed boundary technique based on the inflection point are the smaller than the traditional one.

The maximum internal forces (the bending moment, the shear force and the axial force) errors and the cumulative errors of different boundary techniques at the top of the first story column follow a similar tendency with the displacement. The means of maximum errors ratio and the cumulative errors ratio of the bending moment between the traditional and proposed technique are 4.86 and 9.42, respectively. Both the shear force and the axial force have a similar tendency but the means of the corresponding ratio are larger than the bending moment. The means of maximum errors ratio and the cumulative errors ratio of the shear force are 9.44 and 12.90, respectively. The means of maximum errors ratio and the cumulative errors ratio of the axial force are 17.62 and 23.65, respectively.

Therefore, these results demonstrate that the proposed technique based on the inflection point can improve substantially the test accuracy and show the better performance than the traditional method.

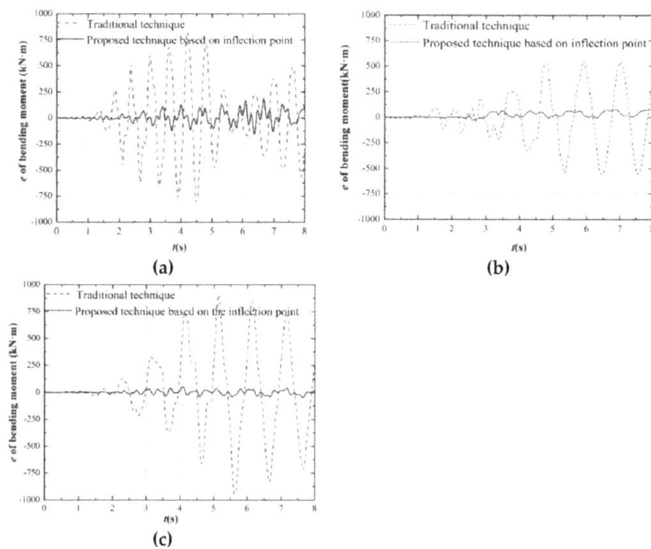

Figure 7. Time history of bending moment errors for different boundary technique at the top of the column of the first story: (**a**) Test 2; (**b**) Test 8; and (**c**) Test 13.

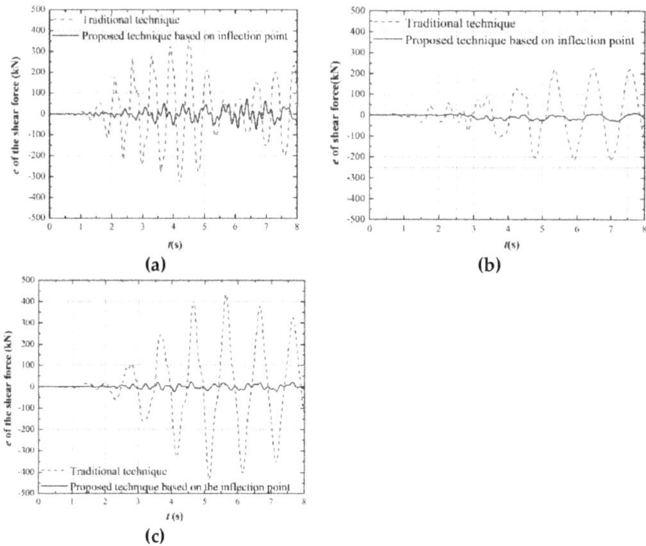

Figure 8. Time history of shear force errors for different boundary technique at the top of the column of the first story: (**a**) Test 2; (**b**) Test 8; and (**c**) Test 13.

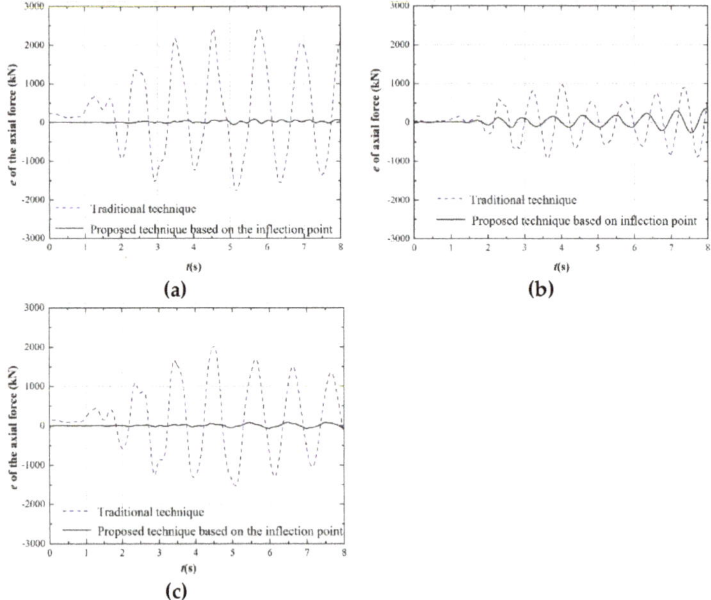

Figure 9. Time history of axial force errors for different boundary technique at the top of the column of the first story: (**a**) Test 2; (**b**) Test 8; and (**c**) Test 13.

5. Experimental Validation

To validate the proposed boundary method, a six-story steel frame structure, in which the first story was taken as the physical substructure and the rest of the structure as the numerical substructure, was designed as the HS testing model. Ground motions with PGAs of 310, 620, 800, 1200, and 1600 cm/s^2 were sequentially applied to the test structure.

5.1. Test Model

The six-story frame structure was designed based on both Chinese Code GB50017-2003 for Design of Steel Structures and Chinese Code for Seismic Building Design GB50011-2010. The height of the first story is 4200 mm and others are 4000 mm. The span of the bay for the test model is 8000 mm. The sections of the columns are HW600 × 600 × 20 × 30, and the section of the beams are HW500 × 300 × 14 × 22 (with units of mm). Q235 steel is used. Because of laboratory limitations, a 1/2-scale test model was manufactured in the Seismic Experimental Center of Civil Engineering at the Harbin Institute of Technology.

The first story, along with the columns of the second story up to the inflection point, was taken as the physical substructure for the proposed boundary technique, as illustrated in Figure 10. A transfer-structure, which is hinge connected with the inflection point of both columns by the high-strength bolts M10.9, was designed to realize the synchronous horizontal displacement by using the two-equilateral angle steel 100 × 10 (mm). The physical experiment is shown in Figure 11, and Table 2 provides the test results for the performance of the steel material. The mass of the first–fifth stories m_{1-5} = 88,000 kg, and the mass of the topper story m_6 = 84,000 kg. The record of the ground motion is the El Centro. Rayleigh Damping is chosen, with parameters a is 0.115 and b is 0.0024, corresponding to 5% damping in the first two modes.

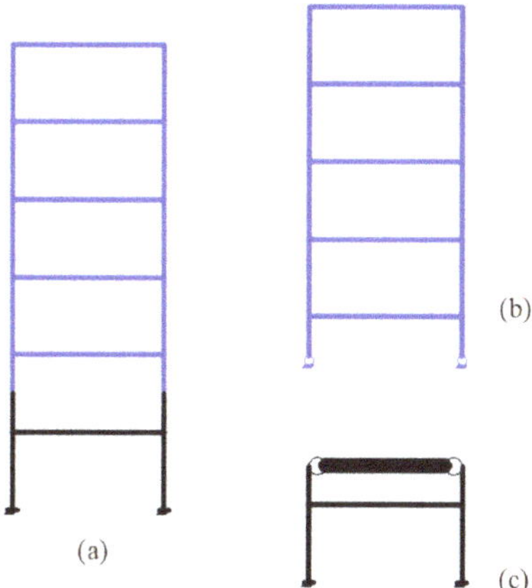

Figure 10. Sketch of the 6-story steel frame: (**a**) entire frame; (**b**) numerical substructure; and (**c**) physical substructure.

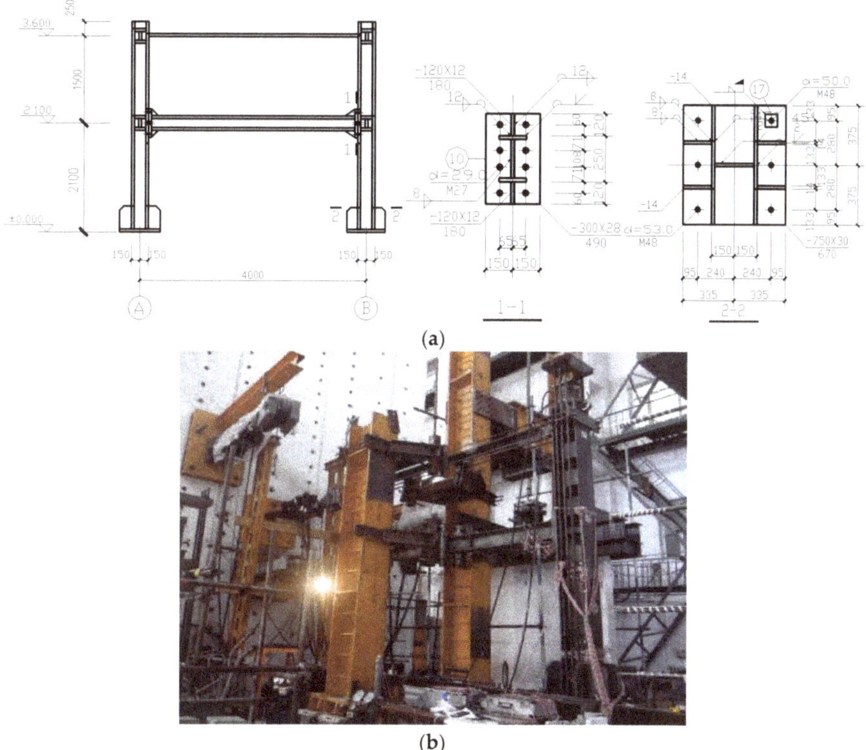

Figure 11. The physical substructure test model of the steel frame structure: (**a**) the design drawing of the physical substructure test model; and (**b**) the manufactured test model at the lab.

Table 2. The test results for the steel material performance.

Theoretical Thickness of Steel Plate (mm)	Real Thickness of Steel Plate (mm)	Yield Strength (Mpa)	Elasticity Modulus (10^5 Mpa)	Ultimate Strength (Mpa)
7	6.8	220.5	2.02	390.4
10	9.85	202.3	1.90	323.2
12	10.5	216.0	2.04	345.6
16	15.7	206.5	1.91	339.6

5.2. Displacement–Force Mixed Control Technique

The traditional HS test, which usually focuses on the compatibility and equilibrium only in the horizontal direction, is conducted using the displacement control. The calculated displacements obtained by solving the EOM were applied to the physical substructure test model. Then, the reaction forces corresponding to the command displacements were measured and fed back to the EOM for the calculation of the next displacement increment. The stiffness of the columns in the vertical (axial) direction is typically high, consequently making displacement control challenging to implement, primarily because the resolution of the displacement transducer used for feedback is inadequate. Therefore, force control was used to realize the axial loading in the proposed approach. To address the problem of measuring the axial displacement of the column, the technique proposed in Section 3.2 for predicting the axial displacement was employed.

To realize the displacement–force mixed control, the arrangement of the vertical and horizontal loading devices was designed as shown in Figure 12. Two hydraulic jacks were chosen as the vertical loading devices while two hydraulic servo actuators as the horizontal loading devices.

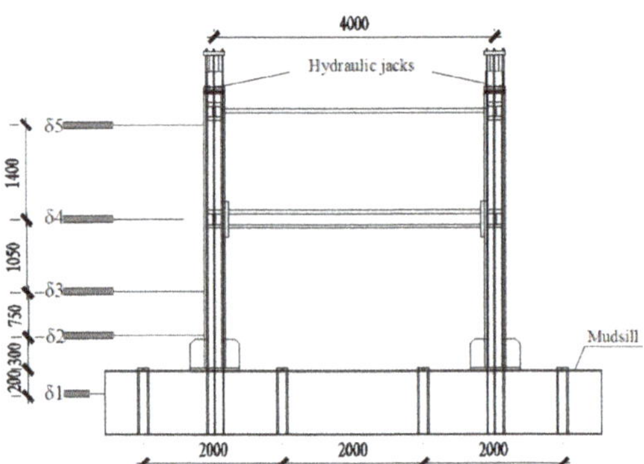

Figure 12. The arrangement of the loading device.

5.2.1. Vertical Loading Devices

Figure 13a shows the vertical loading devices which are two hydraulic jacks. The hydraulic jacks were arranged at the boundary of the physical substructure model where the self-balancing device (Figure 13a) was designed to realize the axial load at the boundary. The self-balancing device includes the 20 mm-thick steel plate which is placed on the top of the hydraulic jack and four 25 mm-diameter steel bars which connects the 20 mm-thick steel plate and the mudsill of the physical substructure model. The loading capacity of the hydraulic jacks is 1000 kN. The loading errors are 1% of the loading capacity.

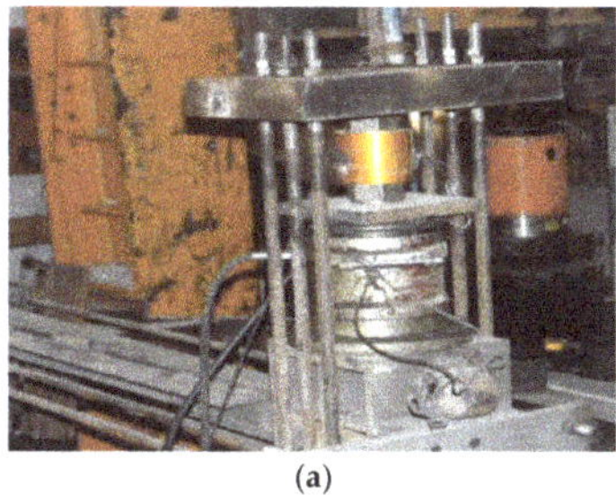
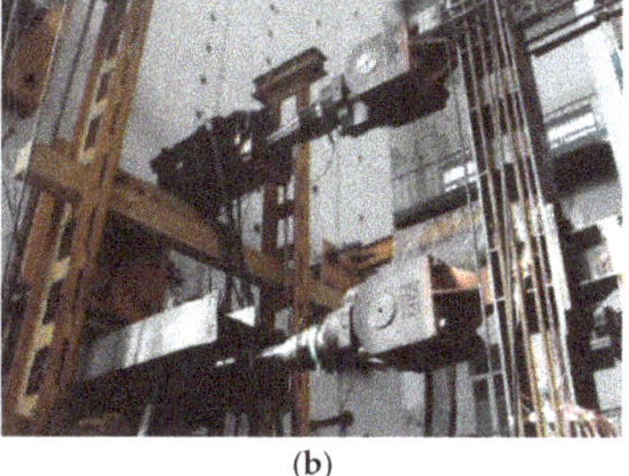

(**a**) (**b**)

Figure 13. Loading device: (**a**) vertical hydraulic jacks; and (**b**) horizontal hydraulic servo actuators.

5.2.2. Horizontal Loading devices

Figure 13b shows the horizontal loading devices which are two hydraulic servo actuators. One end of the hydraulic servo actuators is fixed into the actuator wall by four 100 mm-diameter bolt bars. The other end of the hydraulic servo actuators connects with the physical substructure model, in which one is located at the beam–column joint of the first story while the other at the boundary of the physical substructure model. The force capacity of two hydraulic servo actuators is ±1000 kN, and the displacement capacity is ±100 mm. The errors of both force and displacement are the 0.1% of the corresponding capacity. To avoid the horizontal slip of the physical substructure model, two measures were designed: one is that the mudsill of the physical substructure model is fixed on the floor of the lab by using 100 mm-diameter bolts spaced on 1 m centers; and the other is that two hydraulic jacks are used on both sides of the mudsill to apply the equal and opposite forces (Figure 14). In addition, four steel rollers were used on the side of both the first beam and the truss at the boundary, to avoid the out-of-plane deformation of the steel frame (Figure 15).

Figure 14. The device for avoiding the horizontal slip.

Figure 15. The device for avoiding the out-of-plane deformation.

5.3. Arrangement of the Measurement Devices

5.3.1. Arrangement of Displacement Measuring Points

Five linear variable differential transformers (LVDTs) were employed to measure the physical displacements of the substructure model. One of the LVDTs is placed at the mudsill to measure the slip of the physical substructure model. Two of the LVDTs, which are 300 and 1050 mm above the top of the mudsill, are placed on the column. One of the LVDTs is placed at the beam–column joint of the first story, and one of the LVDTs is placed at the boundary of the physical substructure model. The stroke of all the used LVDTs is ±150 mm. The errors of the LVDTs are 0.1% of the corresponding capacity.

5.3.2. Arrangement of Strain Gauge

Figure 16a shows the arrangement of strain gauges. Each column has five sections which are chosen to paste the strain gauges. Both the top and bottom of the first column has two sections, respectively. The bottom of the second column has one section. Each end of the first beam has two sections which are chosen to paste the strain gauges. Each section has eight strain gauges to calculate the bending moment and axial force of the section (see Figure 16b).

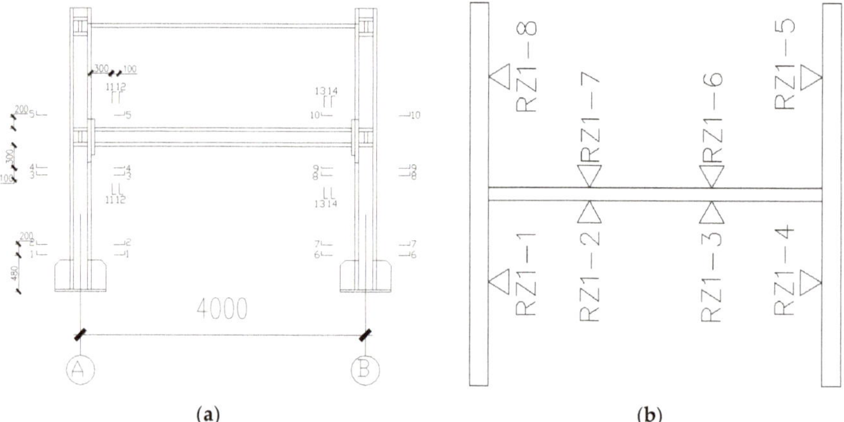

(a)   (b)

Figure 16. Arrangement of the strain gauges: (**a**) arrangement of the sections for pasting the strain gauge; and (**b**) strain gauges at each section.

5.4. Test Results and Analysis

Figure 17 shows the comparison curves of time history between the test results and the simulation results at the first story for different levels of PGA. The results in Figure 17 were obtained by simulating the entire structure in OpenSEES, based on the parameters in Section 5.1. The HS test results are close to the simulation results at the different PGA levels.

Figures 18 and 19 show the comparison curves of the internal force time history between the test results and the simulation results with the 310 and 620 cm/s^2, respectively. Note that, for PGA levels above 620 cm/s^2, the response of the HS test model is significantly nonlinear, resulting in failure of most of the strain gauges; as a result, the internal forces calculated using the stain response are not available above a PGA of 620 cm/s^2.

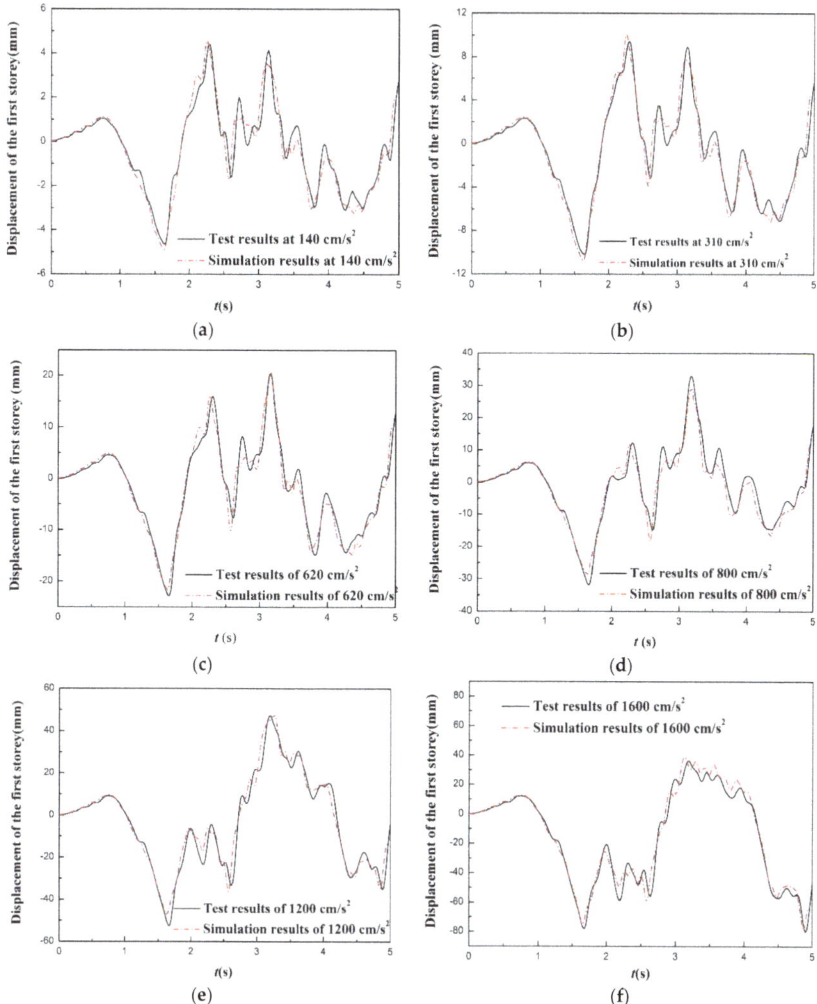

Figure 17. The comparison curves of the displacement time history between the test results and the simulation results with various PGA: (**a**) 140 cm/s² of the PGA; (**b**) 310 cm/s² of the PGA; (**c**) 620 cm/s² of the PGA; (**d**) 800 cm/s² of the PGA; (**e**) 1200 cm/s² of the PGA; and (**f**) 1600 cm/s² of the PGA.

The shear force in Figures 18 and 19 can be obtained conveniently from the force sensor of the horizontal hydraulic actuators. The bending moment and axial force can be calculated from the stain gauges of the calculated section by the following equation:

$$N = \frac{EA}{2} \times \frac{\sum\limits_{i-1}^{8} \xi_i}{8} \tag{5}$$

$$M = \frac{EI}{2y} \times \left(\frac{\xi_1 + \xi_8}{2} - \frac{\xi_4 + \xi_5}{2} \right) \tag{6}$$

where ζ_i is measured by the stain gauge at the section, ζ_1 and ζ_8 are two compressed strain gauges at the flange of the section, ζ_4 and ζ_5 are two tensile strain gauges at the flange of the section, E is the elasticity modulus, I is the sectional inertia moment, A is the sectional area, y is the distance between the point for the stain gauge to the center of the section, N is the axial force on the section, and M is the bending moment. Because the measure of the stain gauge is only recorded at the step when the vertical force changes, Figures 18 and 19 give the calculated values of the bending moment and the axial force at the recorded step.

Figures 18 and 19 also show that the test results of internal forces are close to the simulation one. Table 3 gives the relative errors at the peak of the internal force. In Table 3, the maximum relative error is less than 10%, showing good performance of the proposed boundary technique based on the inflection point.

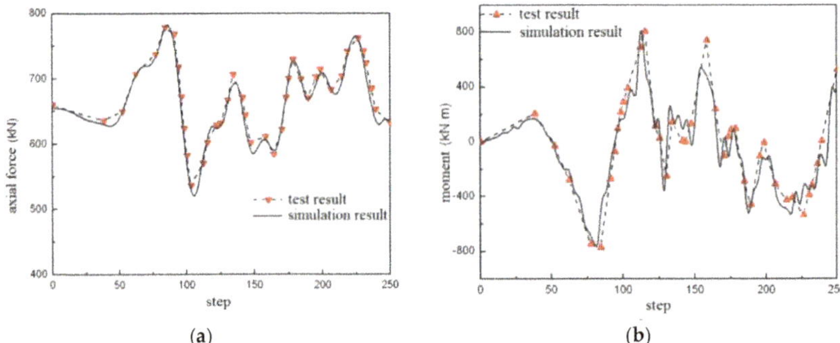

Figure 18. The comparison curves of the internal force time history between the test results and the simulation results at the 310 cm/s² of PGA: (**a**) axial force at the bottom of column; and (**b**) bending moment at the bottom of column.

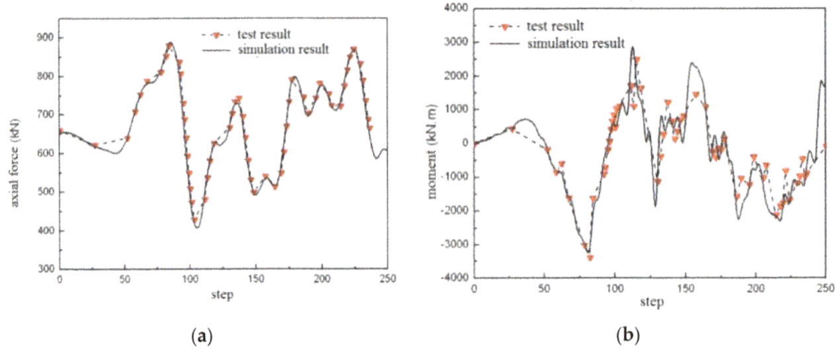

Figure 19. The comparison curves of the internal force time history between the test results and the simulation results at the 620 cm/s² of PGA: (**a**) axial force at the bottom of column; and (**b**) bending moment at the bottom of column.

Table 3. Relative errors of the internal force at the peak of with various peak accelerations.

Peak Acceleration (cm/s²)	Axial Force at the Bottom of the Column	Bending Moment at the Bottom of the Column	Bending Moment at the End of Beam
310	2.5%	3.7%	6.1%
620	6.7%	9.3%	4.0%

6. Conclusions

This paper proposes a new boundary technique based on the inflection point, where the bending moment is zero, to conveniently realize the equilibrium and compatibility at the boundary. The axial displacement prediction technique is used to meet the equilibrium and compatibility of the axial direction at the boundary. To realize the process of the HS based on the proposed boundary technique, the HS component interaction is used to build a bridge between the Matlab and the OpenSEES, and the equivalent force control method is employed for solution of the EOM. Comparison of the results of the numerical simulation and the physical test show that the proposed method has the better performance than the traditional boundary technique.

Fifteen numerical examples are presented to verify the accuracy and feasibility of the proposed boundary technique. The numerical simulation results show that the proposed technique based on the inflection point can improve tremendously the test accuracy over the traditional technique.

The systematic laboratory hybrid simulation tests of the six-story steel frame structure model were designed to verify the accuracy and feasibility of the proposed boundary technique. The displacement–force mixed control technique is proposed to realize the real hybrid simulating test based on the proposed boundary technique. The tests results show reliable and accurate performance of proposed boundary technique.

Author Contributions: All authors made substantial contributions to this study. Z.C. provided the concept and design of the study. X.Y. and X.Z. performed the experiment and collected the data. H.W. and B.F.S. wrote and revised the manuscript. Z.C., X.Y. and X.Z. analyzed the experimental data.

Funding: The research is financially supported by the Natural Science Foundation of China (51678199, 51578151, and 51161120360).

Acknowledgments: The authors gratefully acknowledge the financial support provided by the China Scholarship Council (CSC) at UIUC in 2016 (201606125079).

Conflicts of Interest: The authors declare no conflict of interest.

References

1. Hakuno, M.; Shidowara, M.; Haa, T. Dynamic destructive test of a cantilevers beam, Controlled by an analog-computer. *Trans. Jpn. Soc. Civ. Eng.* **1969**, *171*, 1–9. [CrossRef]
2. Takanashi, K.; Nakashima, M. Japanese activities on online testing. *J. Eng. Mech.* **1987**, *113*, 1014–1032. [CrossRef]
3. Mahin, S.A.; Shing, P.B.; Thewalt, C.R.; Hanson, R.D. Pseudodynamic test method current status and future directions. *J. Struct. Eng.* **1989**, *115*, 2113–2128. [CrossRef]
4. Shing, P.B.; Nakashima, M.; Bursi, O.S. Application of psuedodynamic test method to structural research. *Earthq. Spectr.* **1996**, *12*, 29–56. [CrossRef]
5. Saouma, V.; Sivaselvan, M. *Hybrid Simulation: Theory, Implementation and Applications*; Taylor & Francis: London, UK, 2008; ISBN 978-0-415-46568-7.
6. Bursi, O.S.; Wagg, D. *Modern Testing Techniques for Structural Systems: Dynamics and Control*; Springer: New York, NY, USA, 2008; ISBN 978-3-211-09445-7.
7. Wu, B.; Wang, Q.; Shing, P.B.; Ou, J. Equivalent force control method for generalized real-time substructure testing. *Earthq. Eng. Struct. Dyn.* **2007**, *36*, 1127–1149. [CrossRef]
8. Wallace, M.I.; Sieber, J.; Neild, S.A.; Wagg, D.J.; Krauskopf, B. Stability analysis of real-time dynamic substructuring using delay differential equation models. *Earthq. Eng. Struct. Dyn.* **2005**, *34*, 1817–1832. [CrossRef]
9. Ahmadizadeh, M.; Mosqueda, G.; Reinhorn, A.M. Compensation of actuator delay and dynamics for real-time hybrid structural simulation. *Earthq. Eng. Struct. Dyn.* **2008**, *37*, 21–42. [CrossRef]
10. Chen, C.; Ricles, J.M.; Marullo, T.; Mercan, O. Real-time hybrid testing using the unconditionally stable explicit CR integration algorithm. *Earthq. Eng. Struct. Dyn.* **2009**, *38*, 23–44. [CrossRef]
11. Amin, M.; Shirley, D.; Siamak, R.; Arun, P. Predictive stability indicator: A novel approach to configuring a real-time hybrid simulation. *Earthq. Eng. Struct. Dyn.* **2017**, *46*, 95–116.

12. Ou, G.; Ozdagli, A.I.; Dyke, S.J.; Wu, B. Robust integrated actuator control experimental verification and real-time hybrid simulation implementation. *Earthq. Eng. Struct. Dyn.* **2015**, *44*, 441–460. [CrossRef]

13. Phillips, B.M.; Takada, S.; Spencer, B.F.; Fujino, Y. Feedforward actuator controller development using the backwarddifference method for real-time hybrid simulation. *Smart Struct. Syst.* **2014**, *14*, 1081–1103. [CrossRef]

14. Friedman, A.; Dyke, S.J.; Phillips, B.; Ahn, R.; Dong, B.; Chae, Y.; Castaneda, N.; Jiang, Z.; Zhang, J.; Cha, Y.; et al. Large-scale real-time hybrid simulation for evaluation of advanced damping system performance. *J. Struct. Eng.* **2015**, *141*, 04014150. [CrossRef]

15. Gao, X.; Castaneda, N.; Dyke, S.J. Real-time hybrid simulation: From dynamic system, motion control to experimental error. *Earthq. Eng. Struct. Dyn.* **2013**, *42*, 815–832. [CrossRef]

16. Takahashi, Y.; Fenves, G. Software framework for distributed experimental-computational simulation of structural systems. *Earthq. Eng. Struct. Dyn.* **2006**, *35*, 267–291. [CrossRef]

17. Kwon, O.S.; Elnashai, A.S.; Spencer, B.F. UI-SIMCOR: A global platform for hybrid distributed simulation. In *Hybrid Simulation: Theory, Implementation and Applications*; CRC Press: Boca Raton, FL, USA, 2008; pp. 157–167.

18. Pan, P.; Tomofuji, H.; Wang, T.; Nakashima, M.; Ohsaki, M.; Mosalam, K.M. Development of peer-to-peer (P2P) internet online hybrid test system. *Earthq. Eng. Struct. Dyn.* **2006**, *35*, 867–890. [CrossRef]

19. Guo, Y.R.; Zhu, X.J.; Wang, Z.M.; Zhang, Z.X. A NetSLab Based Hybrid Testing Program for Composite Frame Structures with Buckling Restrained Braces. *Adv. Mater. Res.* **2013**, *639–640*, 1142–1147. [CrossRef]

20. Buckle, I.; Reitherman, R. The consortium for the George E. Brown J. network for earthquake engineering simulation. In Proceedings of the 13th World Conference on Earthquake Engineering, Vancouver, BC, Canada, 1–6 August 2004; p. 4016.

21. Stojadinovic, B.; Mosqueda, G.; Mahin, S.A. Event-driven control system for geographically distributed hybrid simulation. *J. Struct. Eng.* **2006**, *132*, 68–77. [CrossRef]

22. Kim, J.K. KOCED collaboratory program. In Proceedings of the Annual Meeting: Networking of Young Earthquake Engineering Researchers and Professionals (ANCER), Hawaii, HI, USA, 2004.

23. Ohtani, K.; Ogawa, N.; Katayama, T.; Shibata, H. Project 'E-Defense' (3-D Full-Scale Earthqyake Testing Facility). In Proceedings of the Joint NCREE/JRC Workshop on International Collaboration on Earthquake Disaster Mitigation Research, Taipei, Taiwan, 17–18 November 2003.

24. Elnashai, A.S.; Spencer, B.F.; Kuchma, D.A.; Yang, G.; Carrion, J.E.; Gan, Q.; Kim, S.J. The multi-Axial full-scale sub-structured testing and simulation (MUST-SIM) facility at the University of Illinois at Urbana-Champaign. In *Advances in Earthquake Engineering for Urban Risk Reduct*; Kluwer Academic Publishers: Alphen aan den Rijn, The Netherlands, 2006; pp. 245–260.

25. Fermandois-Cornejo, G.; Spencer, B.F. Framework development for multi-axial real-time hybrid simulation testing. In Proceedings of the 16th World Conference on Earthquake Engineering, Santiago, Santiago Metropolitan Region, Chile, 9–13 January 2017; p. 2770.

26. Hashemi, M.J.; Tsang, H.-H.; Al-Ogaidi, Y.; Wilson, J.L.; Al-Mahaidi, R. Collapse Assessment of Reinforced Concrete Building Columns through Multi-Axis Hybrid Simulation. *ACI Struct. J.* **2017**, *114*, 437–449. [CrossRef]

27. Gao, X.; Castaneda, N.; Dyke, S. Experimental Validation of a Generalized Procedure for MDOF Real-Time Hybrid Simulation. *J. Eng. Mech.* **2014**, *140*, 04013006. [CrossRef]

28. Murray, J.A.; Sasani, M. Collapse Resistance of a Seven-Story Structure with Multiple Shear-Axial Column Failures Using Hybrid Simulation. *J. Struct. Eng.* **2016**, *143*, 04017012. [CrossRef]

29. Kammula, V.; Erochko, J.; Kwon, O.; Christopoulos, C. Application of hybrid-simulation to fragility assessment of the telescoping self-centering energy dissipative bracing system. *Earthq. Eng. Struct. Dyn.* **2014**, *43*, 811–830. [CrossRef]

30. Chae, Y.; Ricles, J.M.; Sause, R. Large-scale real-time hybrid simulation of a three-story steel frame building with magneto-rheological dampers. *Earthq. Eng. Struct. Dyn.* **2014**, *43*, 1915–1933. [CrossRef]

31. Dong, B.; Sause, R.; Ricles, J.M. Accurate real-time hybrid earthquake simulations on large-scale MDOF steel structure with nonlinear viscous dampers. *Earthq. Eng. Struct. Dyn.* **2015**, *44*, 2035–2055. [CrossRef]

32. Abbiati, G.; Bursi, O.S.; Caperan, P.; Sarno, L.D.; Molina, F.J.; Paolacci, F.; Pegon, P. Hybrid simulation of a multi-span RC viaduct with plain bars and sliding bearings. *Earthq. Eng. Struct. Dyn.* **2015**, *44*, 2221–2240. [CrossRef]

33. Murray, J.A.; Sasani, M. Near-collapse response of existing RC building under severe pulse-type ground motion using hybrid simulation. *Earthq. Eng. Struct. Dyn.* **2016**, *45*, 1109–1127. [CrossRef]
34. Shao, X.; Pang, W.; Griffith, C.; Ziaei, E.; van de Lindt, J. Development of a hybrid simulation controller for full-scale experimental investigation of seismic retrofits for soft-story woodframe buildings. *Earthq. Eng. Struct. Dyn.* **2016**, *45*, 1233–1249. [CrossRef]
35. Kolay, C.; Ricles, J.M.; Marullo, T.M.; Mahvashmohammadi, A.; Sause, R. Implementation and application of the unconditionally stable explicit parametrically dissipative KR-alpha method for real-time hybrid simulation. *Earthq. Eng. Struct. Dyn.* **2015**, *44*, 735–755. [CrossRef]
36. Hashemi, M.J.; Mosqueda, G. Innovative substructuring technique for hybrid simulation of multistory buildings through collapse. *Earthq. Eng. Struct. Dyn.* **2014**, *43*, 2059–2074. [CrossRef]
37. Wang, T.; Mosqueda, G.; Jacobsen, A. Performance evaluation of a distributed hybrid test framework to reproduce the collapse behavior of a structure. *Earthq. Eng. Struct. Dyn.* **2012**, *41*, 295–313. [CrossRef]
38. Mahmoud, H.N.; Elnashai, A.S.; Spencer, B.F.; Kwon, O.S.; Bennier, D.J. Hybrid simulation for earthquake response of semirigid partial-strength steel frames. *J. Struct. Eng.* **2013**, *139*, 1134–1148. [CrossRef]
39. Chen, Z.X.; Xu, G.Sh.; Wu, B.; Sun, Y.; Wang, H.; Wang, F. Equivalent force control method for substructure pseudo-dynamic test of a full-scale masonry structure. *Earthq. Eng. Struct. Dyn.* **2014**, *43*, 969–983. [CrossRef]
40. Fang, K.T.; Ma, C.X. *The Uniform Design Method*; Science Press: Beijing, China, 2001; ISBN 7-03-009189-2.

Article

Numerical Simulation and In-Situ Measurement of Ground-Borne Vibration Due to Subway System

Jinping Yang [1], Peizhen Li [1,2] and **Zheng Lu [1,2,*]**

[1]   Research Institute of Structural Engineering and Disaster Reduction, Tongji University,
     Shanghai 200092, China; 1410231@tongji.edu.cn (J.Y.); lipeizh@tongji.edu.cn (P.L.)
[2]   State Key Laboratory of Disaster Reduction in Civil Engineering, Tongji University, Shanghai 200092, China
*   Correspondence: luzheng111@tongji.edu.cn; Tel.: +86-216-598-6186

Received: 29 June 2018; Accepted: 9 July 2018; Published: 12 July 2018

Abstract: A coupled finite element–infinite element boundary method performed by the harmonic analysis based on the commercial software ABAQUS was verified by comparing it with the thin layer method to address the issue of subway induced vibration, a major environmental concern in urban areas. In addition, an interface program was developed to automatically read the simulation result files in the harmonic analysis, and then put the data into MATLAB, achieving the frequency domain analysis. Moreover, a site measurement was performed on a practical engineering track bed-tunnel lining-surrounding formation located on Line 2, Shanghai Metro and rich vibration data were acquired. Then, the corresponding simulation model was established and the numerical results were compared with the measured data based on the developed program, which was verified applicable for the practical engineering of subway induced vibration on the soft site. The proposed prediction formula of the vibration level, by comparison with the measurement, is applicable for the prediction in subway induced vibration. The results show that there exists a vibration amplifying zone a certain distance to the tunnel under high frequency loads due to the wave propagation and reflection. Finally, a parametric study was conducted in an elastic half-space simulation to investigate the influence of model widths and depths with infinite element boundary on the numerical results. The higher performance of the combined finite element–infinite element boundary method, which can decrease the model sizes in widths and heights 50% effectively, was demonstrated. Consequently, the coupled finite element–infinite element boundary method and developed frequency analysis with interface program provide rational numerical methods for the models of subway induced vibration.

Keywords: subway induced vibration; numerical simulations; finite element; infinite element boundary; measurement

---

1. Introduction

In recent decades, with the rapid development of urban modernization, the impact of traffic system induced vibration on urban life and working environment has been brought to public attention since vibration has been listed as one of the seven major environmental hazards in the world [1]. Subways, key parts of the urban traffic, are highly demanding on the environment, particularly in major cities, e.g., Shanghai and Beijing, China. Therefore, the vibration induced by the passage of underground trains is a major environmental concern in urban areas [2].

The influences of subway induced vibration is mainly reflected in the following areas: the influence of vibration on the surrounding buildings, especially historical architecture, such as the collapse of the old church caused by vibration. Vibration may also interfere with sensitive equipment used in scientific research and high-tech industries [3]. Moreover, vibration and secondary noise affect people's work, life and even their health. Whole-body human exposure to vibration indoors has been evaluated in the frequency range 1–80 Hz [4], which is consistent with the range of subway induced vibration [2]. Therefore, it is imperative to focus on the subway induced vibration.

Great efforts have been made on the measurements of subway induced vibration to study the vibration characteristics, the vibration causes, the propagation path and the controlling means [5]. Vibration measurements are performed on the building foundation slab and in open fields adjacent to the building [3], the subway train-induced vibrations [6], viaduct track and nearby ground [7]. Vibration measurement systems are various, including a remote system for the online monitoring [8], real-time monitoring of adjacent cross operation subway [9], and field monitoring at permafrost testing section [10]. Moreover, effective measures to mitigate vibration are developed and suggested based on field or test results, such as a floating stab tracks [11], rail pads [12], and wave barrier backfill material [13]. Besides, track vibration isolation [14] and energy-absorbing rubber [15,16] are effective strategies to mitigate the rail-way infrastructure issues, and structural resonant behavior can be avoided [17–19]. Moreover, structural vibration control methods [20–22], such as tuned mass damper [23–25], nonlinear energy sink [26,27], particle damper [28–30], etc., may also be applied to suppress subway induced vibrations.

Some interrelated research, theoretical studies and numerical analysis of subway induced vibration have recently received attention [31]. Numerical modeling includes: two dimensional (2D) models [32], two-and-a-half-dimensional (2.5D) models [33] and three-dimensional (3D) models [34]. 2D analytical methods have the advantage of high computational efficiency and are suitable for simple geometries. 2.5D models have received wide concern and many studies have been performed, including on the 2.5D finite/infinite element approach [35] and the 2.5D finite element coupled boundary element BEM model [36]. Other novel numerical methods, such as a closed-form semi-analytical solution in a half-space [37], have also been developed for the prediction of subway induced vibrations. 3D models have high accuracy, however, are more computationally expensive [38].

It has proven effective to combine numerical studies together with experimental results in studies on the subway induced vibration. Chiacchiari [39] discussed measurement methods for rail corrugation and proposed an application method for rail irregularities. Kaynia [40] performed a measurement of the vibration induced railway embankment by high speed trains and developed a rigorous numerical model for the prediction of ground response.

It can be seen from the above studies that the issue of subway induced vibration is a sustainable topic, causing annoyance and adverse physical, physiological, and psychological effects on humans in dense urban environments [5]. Consequently, the adopted approaches and methods in the vibration mitigation must be sustainable to reduce damage and adverse effects on humans, and it is of great significance to focuses on methods to enhance the sustainability in design of newly built metro depots.

Based on the basic principles of the infinite element method, a coupled finite element–infinite element boundary method that combines the finite element method in the key areas and infinite element method in the far field is introduced in Section 2 to address the issue of subway induced vibration. Then, the finite element method performed by the harmonic analysis based on the commercial software ABAQUS is verified by comparing with the thin layer method. An interface program is developed to realize the frequency domain calculation on the MATLAB platform and it is verified by the transfer function method. In Section 3, a field measurement of practical engineering is performed on a track bed-tunnel lining-surrounding formation located on Line 2, Shanghai Metro, and the corresponding simulation model is established based on the coupled method. The numerical results of vibration level and spectrum characteristic, obtained by the developed interface programs, are discussed. Then, a simplified prediction formula of ground vibration level is proposed based on the measurement results. Finally, parametric study is conducted in an elastic half-space simulation to investigate the influence of model widths and depths with the combined method on the numerical results in Section 4.

2. Numerical Methods of Subway Induced Environmental Vibration

2.1. Basic Principles and Realization of Infinite Element Method in Programs

Infinite element, a special kind of finite element, is a unit that tends to infinity geometrically. It is an effective supplement to the finite element in dealing with the unbounded domain issues, and can be seamlessly connected to the finite element.

The directionality is the big difference between infinite and finite units and the two sides cannot intersect in the infinite direction. Therefore, the infinite element must be divergent in the infinite direction mathematically. Only under these conditions can the quantities be monotonic in the process of tending towards infinity. In the dynamic analysis of infinite large medium (e.g., foundations), artificial boundary is affected by the wave reflection, and therefore the energy is transmitted back to the analytical zone and cannot be propagated to infinity. Damping force is introduced to absorb the radiant energy of planar body wave in the infinite boundary.

The damping stress is introduced to the boundary of $x = L$ and can be expressed as follows:

$$\sigma_{xx} = -d_p \dot{u}_x, \sigma_{xy} = -d_s \dot{u}_y, \sigma_{xz} = -d_s \dot{u}_z \tag{1}$$

where $\dot{u}_x$, $\dot{u}_y$ and $\dot{u}_z$ are the vibration velocities in x, y and z direction, respectively.

The constants d_p and d_s, proportional to the velocity, are selected to avoid the energy reflection of longitudinal and shear waves to the infinite zones. They are generally expressed as:

$$d_p = \frac{\lambda + 2G}{c_p} = \rho c_p \tag{2}$$

$$d_s = \rho c_s \tag{3}$$

where c_p and c_s are the propagation velocities of longitudinal wave and transverse wave, respectively.

The infinite elements include plain strain element, plain stress element, axial symmetry element and three-dimensional infinite element in ABAQUS program. The first surface of the infinite element is required to be the interface between finite element and infinite element to ensure the direction of the infinite element extending from near field to far field. In addition, the infinite elements cannot be intersected with each other in the extension direction.

The coupled finite–infinite–infinite element boundary is performed in the numerical analysis. The focused areas of interest (near field) are simulated by finite elements and the far field by infinite elements [41]. Finite elements with certain nodes are firstly defined, and the unit type is then changed in INP file to accomplish the definition of infinite elements, which cannot be defined directly in ABAQUS program. The simulation model is conducted in the harmonic analysis based on the commercial software ABAQUS.

2.2. Verification of Infinite Element Method in Programs

It is an effective way to test the applicability of the presented model in different methods. The proposed finite element coupled infinite element boundary method in ABAQUS software is compared with the thin layer method (TLM), a semi-analytical and semi-numerical approach for the analysis of elastic wave propagation in stratified media [42].

An elastic half-space simulation is taken as an example in the subway induced vibration models. The basic parameters of the numerical model are as follows: the mass density ρ is 2000 kg/m^3; shear wave velocity Cs is 500 m/s; Poisson's ratio μ is 0.4; and the material damping ratio ζ of soil is 0.05. The depth of soil layer is 80 m and the base is assumed as fixed rock. The two-dimensional numerical model is built and half-structure is utilized based on symmetry in the present study. The cosine function $F = F_0 \cos(2\pi f t)$, with the amplitude F_0 5 kN, is applied to the models, and the applied frequency range f is from 1 Hz to 80 Hz according to the required frequency band under the subway-induced vibration. The ground evaluation points are within the range of 2–30 m from the distance of load application with total 15 points and 2 m interval and the vertical vibrating displacement of these points are obtained and discussed in this section.

The base of the numerical model is fixed and the horizontal degrees are constrained on the right side. The model can be expressed as L_H, where L denotes the width of the model and H is the height (depth). The model 1280_80 notes the model width is 1280 m and the depth is 80 m. ABAQUS is abbreviated to ABA in the numerical figures.

Figure 1 presents the comparison attenuation curves at the typical frequencies of TLM and ABAQUS results with element dimension 1 m, which can be drawn that the numerical results of the two methods are similar under the low frequency load, such as 8 Hz. There are differences under the high frequency load due to the larger element dimensions. However, the differences become smaller with the finer element dimensions [35,41] in ABAQUS software, as shown in Figure 2, indicating that the finite element method is verified applicable in ABAQUS software.

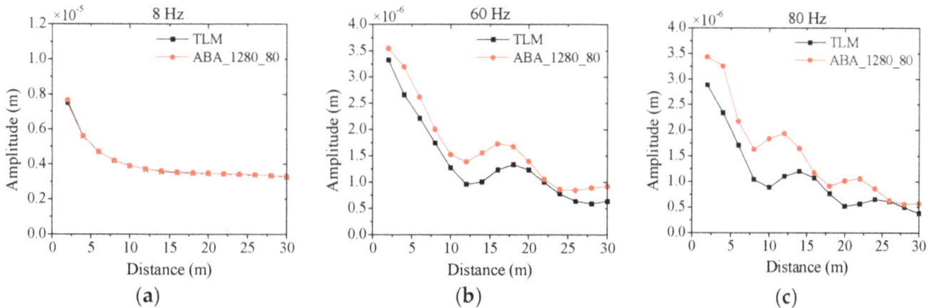

Figure 1. The attenuation curves of TLM and ABAQUS results under the excitation of: (**a**) 8 Hz; (**b**) 60 Hz; and (**c**) 80 Hz.

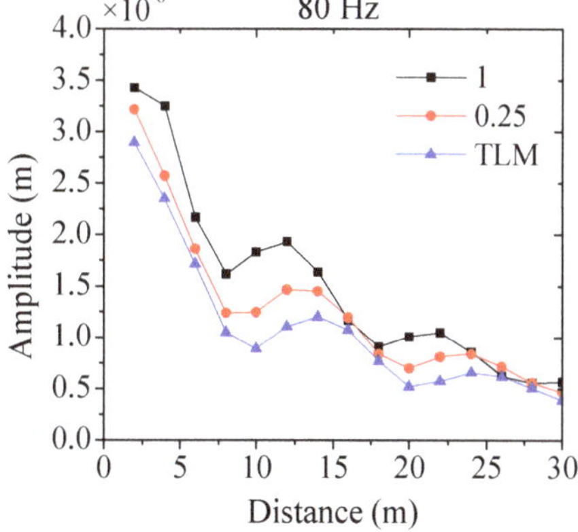

Figure 2. The attenuation curves of TLM results and ABAQUS results with two different element dimensions. Note: 1 and 0.25 mean the element dimensions are 1 m and 0.25 m, respectively. TLM is the thin layer method.

2.3. The Realization of Infinite Element Method in the Frequency Domain

To make the simulation results statistically significant, many excitations are applied to the two-dimensional, or more complex, models in the subway induced vibration, since the subway incentive is a random process. The required calculation process becomes time-consuming in the time domain analysis. Moreover, the frequency characteristics of the vibration is important in the subway induced vibration. Therefore, the numerical analysis conducted in the frequency domain is of great significance. The frequency analysis is a process of data operation, rather than the step-by-step solution process in the time history analysis. The calculation time is then saved effectively.

In this study, an interface program is developed in the commercial software ABAQUS environment. It can automatically read the simulation result files (dat file) in the ABAQUS harmonic analysis, and then the frequency domain calculation is realized by MATLAB. The main steps of frequency domain analysis could be summarized by the following:

Step 1. The coupled finite element–infinite element boundary simulation model is built based on the commercial software ABAQUS.

Step 2. The harmonic analyses are conducted based on the commercial software ABAQUS to obtain the amplitude and phase data of the evaluation measures on the ground in the subway induced vibration.

Step 3. The numerical results are saved as dat file type in ABAQUS, and then read automatically by the developed interface programs to MATLAB.

Step 4. Frequency domain analysis is performed based on MATLAB platform and the responses, such as acceleration and displacement time-histories of the evaluation points are obtained.

The simulation results acquired in Steps 1 and 2 are verified by TLM in Section 2.2. The time history with certain frequency is obtained based on the amplitude and phase, which are calculated in Step 2. Then, the total responses of evaluation points with each frequency are obtained based on the superposition principle. The frequency domain analysis in Step 4 is verified by the transfer function method, whose principle is expressed as follows:

If the transfer function $H(\omega)$ is known, then the responses of the system can be calculated by Equations (4) and (5).

$$Y(t) = \int_{-\infty}^{\infty} H(\omega)X(\omega)e^{i\omega t}d\omega \tag{4}$$

$$Y(\omega) = H(\omega)X(\omega) \tag{5}$$

where $X(\omega)$ is the frequency response functions and $Y(\omega)$ is the output Fourier transform.

Therefore, the responses of system can be calculated under any excitation $X(t)$, provided the transfer function $H(\omega)$ is known.

The correctness and the application of the program is verified by a numerical simulation, a track bed-tunnel lining-surrounding formation model, as shown in Figure 3a. Displacement load, as expressed in Equation (6), is applied to the center of the tunnel track-bed (A point) in the vertical direction. The applied displacement curve is shown in Figure 3b, and the displacement time-history of the ground surface B point is required to be calculated. Table 1 lists the soil parameters of different soil layers in this model.

$$x(t) = 1 \times 10^{-5}[\cos(2\pi t) - 2\sin(2\pi t) - 2\sin(4\pi t)] \tag{6}$$

Figure 4 shows the comparison displacement time history of Point B by the developed interface program in frequency domain analysis and the transfer function method. The two curves almost coincide with each other, indicating that the frequency analysis is applicable in the interface programs. There is little difference between the two results in the amplified graph, as shown in Figure 4b. The reason is that the sampling points are limited in the frequency analysis, resulting in frequency leakage and numerical calculation errors.

Table 1. Soil parameters of different soil layers.

Soil Layer	Thickness (m)	Depth (m)	Density (kg/m^3)	Shear Wave Velocity (m/s)	Poisson's Ratio
1	1.00	1.00	1890	74	0.35
2	3.30	4.30	1850	89	0.30
3	2.10	6.40	1850	85	0.30
4	11.26	17.66	1790	108	0.30
5	26.40	44.06	1820	220	0.25
6	2.00	46.06	1930	189	0.25
7	2.00	48.06	1940	191	0.25
8	3.40	51.46	2040	195	0.25
9	8.05	59.51	1920	230	0.25
10	16.96	76.47	1950	220	0.25
11	2.20	78.67	1920	263	0.25
12	3.78	82.45	1960	267	0.25
13	2.99	85.44	1920	272	0.25
14	6.65	92.09	1940	279	0.25
15	4.36	96.45	1930	287	0.25
16	3.55	100.00	1920	296	0.25

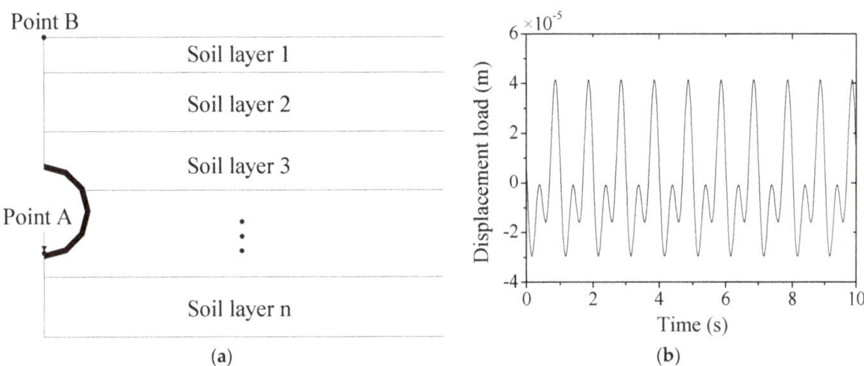

Figure 3. (**a**) The simulation model; and (**b**) the applied displacement load time-history at Point A.

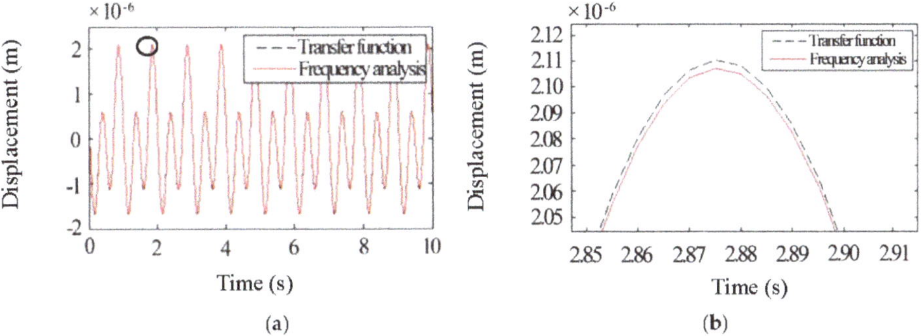

Figure 4. (**a**) Vertical displacement time-history; and (**b**) local displacement time-history of Point B.

3. Experimental Verification of the Coupled Model for Subway Induced Vibrations

3.1. Vibration Measurement

Field measurements were performed on a track bed-tunnel lining-surrounding formation in a certain park project of Line 2, Shanghai Metro, which is an ideal test site due to the few surroundings and buildings to study the vibration attenuation characteristics of the ground surface under the subway induced vibration. The acceleration time-histories of the central tunnel and the typical ground points were measured and acquired in this test. The track bed-tunnel lining-surrounding formation model and the arrangement of measuring points in lateral direction, which were 0 m, 5 m, 11 m, 15 m, 25 m and 40 m from the central tunnel, are shown in Figure 5.

The outer radius of the tunnel is 3.1 m and inner radius is 2.75 m. The tunnel depth, the distance from the top of the tunnel to the surface of the tunnel, is 13.8 m. The elastic modulus, the density and Poisson's ratio of concrete are 34,500 MPa, 2500 kg/m^3 and 0.25, respectively. The typical clay in Shanghai, not homogeneous in the depth direction, was employed in this field test and numerical simulation. The soil parameters of different soil layers are shown in Table 1 in Section 2.3.

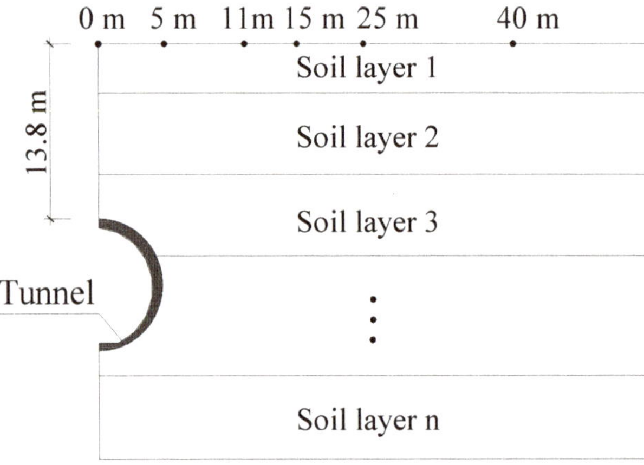

Figure 5. The schematic diagram of the model and the arrangement of measuring points.

The 31 accelerations of the evaluation points in the center of the track bed were measured and acquired in this field test, with total excitation time 10 s and sample frequency 200 Hz. The 31 accelerations correspond to 31 cases. The typical acceleration time history of the track bed, its corresponding Fourier spectra and 1/3 octave frequency are provided in Figure 6a–c, in which it can be observed that the dominant frequencies of the tunnel track bed induced by the subway are above 40 Hz. The measured results agree with other tests [43] that the vibration frequency induced by the subway is high. Vibration level of each recorded accelerations is shown in Figure 6d based on the Shanghai regulation, in which it can be seen that the vibration level is within 80–85 dB and the mean value is 82.4 dB.

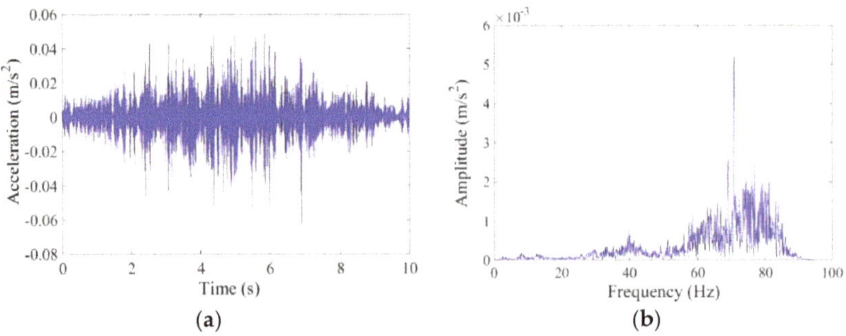

Figure 6. *Cont.*

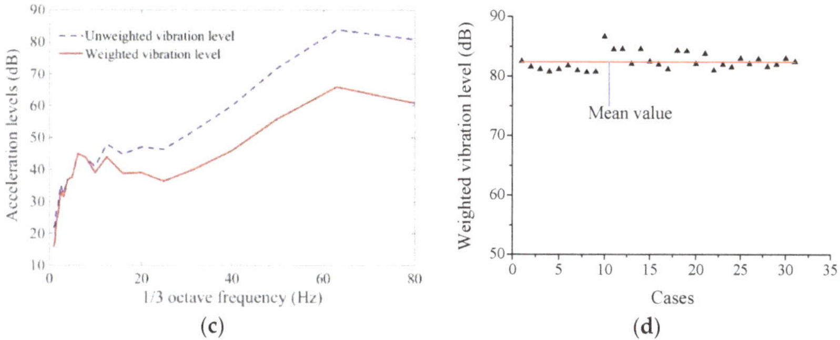

(c)

Figure 6. (a) The acceleration time-history of the track-bed; (b) the corresponding Fourier spectrum of the track-bed; (c) 1/3 octave frequency; and (d) acceleration levels of track-bed.

3.2. The Coupled Finite Element–Infinite Element Boundary Model

In this engineering, the minimum wave velocity of the soil is 100 m/s, the width of the finite element region is designed as 100 m, and the model depth is 100 m. Infinite element boundary is applied to the right side and the bottom of the models. The element size in this engineering along the horizontal direction and vertical direction within 50 m depth is designed as 0.2 m, while other regions are meshed with 0.4 m. The tunnel is analyzed with plane strain solid elements, with the element size 0.1 m × 0.1 m (width and height). The measured 31 accelerations of the evaluation points in the center of the track bed are applied vertically to this position in the simulation model. The simulation frequency is 0–100 Hz, with 0.1 Hz interval. The simulation points are within 0–40 m of the center of the tunnel, coincident to the measure points displayed in Figure 5. The meshed model of track bed-tunnel lining-surrounding formation is shown in Figure 7. The numerical results are obtained by the harmonic analysis in ABAQUS software and frequency analysis in MATLAB, connecting by the developed interface programs, as introduced in Section 2.3. Soil behaviors are simplified in this simulation model and soil properties are based on the practical engineering, as shown in Table 1.

Figure 7. The meshed model of track bed-tunnel lining-surrounding formation.

3.3. Numerical Results

The simulation attenuation of peak vibration levels in the vertical direction with respect to distance under the 31 measured accelerations excitations are shown in Figure 8a,b. In addition, the mean attenuations of peak vibration levels, which are obtained by the measurement and simulation, are compared in Figure 8c. The following conclusions can be drawn.

The attenuation curves of vibration levels generally decrease with the increasing distance away from the central tunnel. However, there exists an amplifying area of vibration that the vibration level at 10 m from the center is about 5 dB greater than that at 15 m.

This amplification effect due to the wave propagation and reflection is consistent with the results of measurement and numerical simulation obtained by other researchers [44]. If the reflection frequency is dominant and obvious in the applied frequency components, the overall vibration will reflect due to the wave superposition in this area. A part of the vibration wave is directly transmitted through the bound site, and part of the wave first spreads to the ground surface and then propagates from the ground to the lateral sides, forming a superposition effect in this area. This may result in a larger vibration in this field than that directly above the tunnel. The wave propagation phenomenon is clearly observed in the measured ground data by Takemiya [7].

Moreover, the amplification phenomenon is obvious in some cases, such as in Case 11, while some are not apparent, such as in Case 29, as shown in Figure 8b, indicating that the amplification has relationship with the spectrum of the applied excitation accelerations.

The numerical mean value of vibration level basically maintains within the measured range, which is a little smaller than the mean value of experimental results, as shown in Figure 8c. The amplification of vibration level is also obtained in the simulation, about 15 m away from the center of the tunnel.

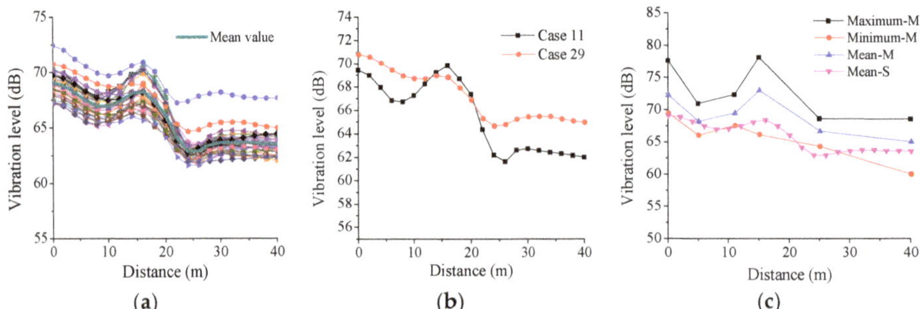

Figure 8. (**a**) Vibration levels with respect to distance under the 31 measured accelerations excitations; (**b**) vibration levels in Cases 11 and 29; and (**c**) the mean vibration levels of measurement and simulation. Note: In (**c**), M and S mean the measurement results and the simulation results.

Figure 9 plots the simulation and measurement frequency spectra 0 m, 5 m and 15 m from the center tunnel. The calculated and measured frequency spectra, with dominant frequency 40–90 Hz, are generally similar. The frequencies are consistent with the findings by Kaewunruen that most vibration issues are between 2 and 150 Hz for the neighborhood near railway lines in tunnels, on embankments, in a cutting or on the level ground [14]. In addition, there is a peak value within 5 Hz in the simulation, which is different from the measurement. The reason is that there is a fundamental frequency in the numerical model due to the limited element dimensions, and amplification effect generates as the excited frequency is close to the model's basic frequency.

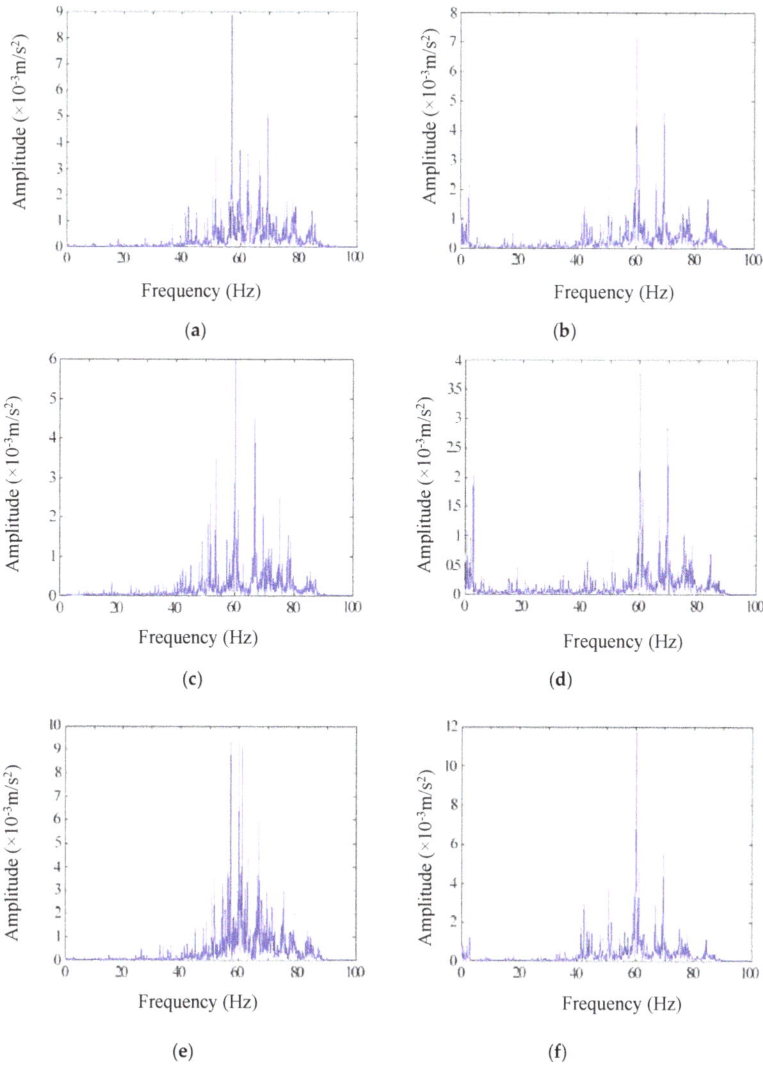

Figure 9. The Fourier spectrum under Case 1: (**a**) measurement at 0 m; (**b**) simulation at 0 m; (**c**) measurement at 5 m; (**d**) simulation at 5 m; (**e**) measurement at 15 m; and (**f**) simulation at 15 m.

Figure 10 displays the compassion 1/3 octave band frequency spectra of simulation with the letter S drawn by the blue lines and measurement with the letter M to show the trend of the vertical response at the distance of 0 m, 5 m, 11 m and 15 m to the central tunnel (as shown in Figure 5). In Figure 10, the calculated acceleration vibration levels are larger than that in the measurement as the 1/3 octave frequency is smaller than 5 Hz, which is mainly due to the amplification effect. On the other hand, the simulation vibration levels are consistent with the measurement in the higher frequencies, in which the trend of the 1/3 octave band frequency spectra by the observation and simulation results agree with the in-situ vibration measurements [7] and the observations on the waiting platform in Boston [45]. Generally, the directivity differences in the measurement and simulation frequencies are small, demonstrating that the coupled finite element–infinite element boundary method can better capture the vibration characteristics in practical engineering, since the frequency of subway-induced vibration is high [43].

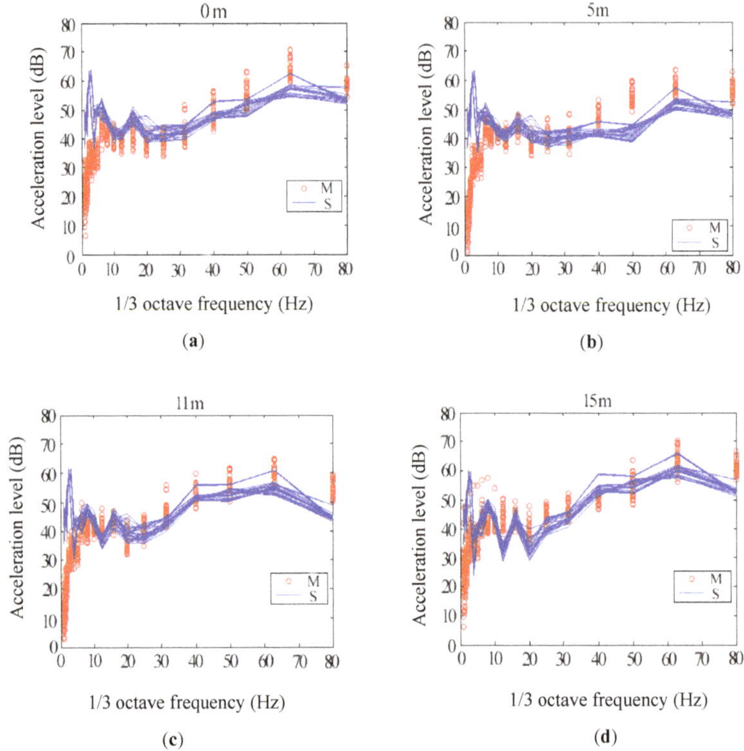

Figure 10. The measurement and simulation 1/3 octave frequency at: (**a**) 0 m; (**b**) 5 m; (**c**); 11 m; and (**d**) 15 m.

3.4. Simplified Prediction Formula of Ground Vibration Level

The ground vibration level is mainly influenced by the tunnel track-bed vibration and the tunnel depth. A simplified prediction formula of the ground vibration level of the track bed-tunnel lining-surrounding formation on this engineering certain soft site, based on the numerical simulation models with different tunnel depths and the track-bed accelerations, is presented as Equation (7).

$$VL_Z = VL_{ZS}(-0.0028x - 0.0033h + 0.942)$$ (7)

where VL_Z is the ground vertical acceleration level (dB), and VL_{ZS} is the tunnel track-bed vertical acceleration level (dB). x is the horizontal distance away from the center of the tunnel (m), and h is the tunnel depth (m).

Figure 11 plots the attenuation of peak vibration levels with respect to distance in this practical engineering. The predicted vibration levels are approximately similar to the measured results, with maximum error 3 dB. Therefore, the proposed prediction formula is acceptable and applicable for the prediction of vibration levels induced by subway in the practical engineering.

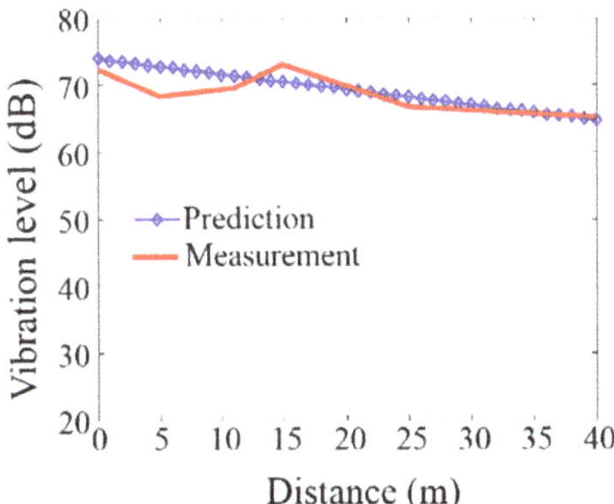

Figure 11. The comparison of vibration levels of the engineering project by the measurement and presented prediction formula.

4. Parametric Study of Coupled Finite Element-Infinite Element Boundary Method in the Subway Induced Vibration

The elastic half-space simulation introduced in Section 2.2 is used in this section to discuss the efficiency of the presented finite element method coupled infinite element boundary and the parametric study is conducted to investigate the influence of model widths and depths on the numerical results.

4.1. The Effect of Model Depths on the Simulation Results

Table 2 shows the numerical cases with different model heights. The model can be expressed as LIN_H, where L denotes the width of the model, H is the height and IN means that the infinite element is adopted to the lateral side of the model. The base of the model is fixed in the simulation. The dimension of the element is 1 m × 1 m. Plain strain element CPE4 is used as the finite element in this model. Different model depths are analyzed to obtain a stable solution. Figure 12 is the sketch of the model with infinite element. Symmetrical constraints are imposed on the left boundary and the infinite element is applied to the right side of the model.

Table 2. Numerical cases with different model depths.

Case	1	2	3	4	5
L × H (m × m)	240 × 80	240 × 160	240 × 320	240 × 640	240 × 1280
Expression	240IN_80	240IN_160	240IN_320	240IN_640	240IN_1280

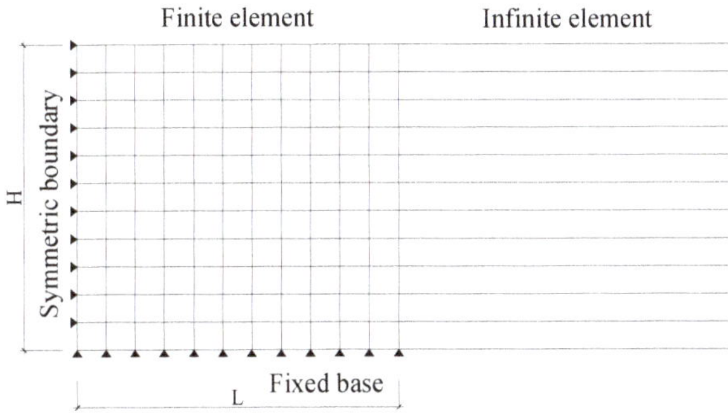

Figure 12. The diagram of finite element method coupled infinite element boundary

The attenuation curves of the model applied to lateral infinite element boundary with different heights under the excitations of different vibration frequencies are shown in Figure 13. The numerical results of models with different heights are similar under the high frequency load, such as 40 Hz and 80 Hz, while the results are obviously different under the low frequency load. The numerical errors compared to the benchmark model 240IN_1280 mainly depend on the vertical frequencies, using an equation $f_0 = c_p/4H$, with 3.8 Hz, 1.9 Hz, 0.95 Hz, 0.48 Hz and 0.24 Hz of the models in Cases 1–5, respectively. Therefore, the vertical frequencies of the models are small, resulting relative larger errors under low frequency load (close to the vertical fundamental frequency) than that under high frequency load. Vertical frequency, depending on the depth of the model with certain soil property, has great influence on the simulation results. Therefore, the depth of the model should be designed seriously to reduce the numerical errors due to the improper depth of the model. This agrees with the conclusion obtained by Andersen [46] that the accuracy of the simulation results increases with the distance from the tunnel; in other words, the depth of the model is significant to obtain more accurate estimates.

Figure 14 shows the amplitude–frequency curve of the models with different heights. It can be observed that: (1) There is an obvious peak in the curve when the model height is 80 m, while the amplitude–frequency curves are monotonic in other models. This is because the vertical frequency becomes smaller with the increase of model depth, and the vertical frequencies are less than 1 Hz to most models. However, the applied frequencies are equal to or greater than 1 Hz, larger than the vertical frequency. Therefore, there are no obvious peak values of most models in the amplitude–frequency curve. (2) The frequencies corresponding to the maximum error of the amplitude–frequency curves are mainly in the low frequency (1–6 Hz). Moreover, the error of amplitude–frequency curve is less than 5% when the model height is 640 m, indicating that numerical results are stable.

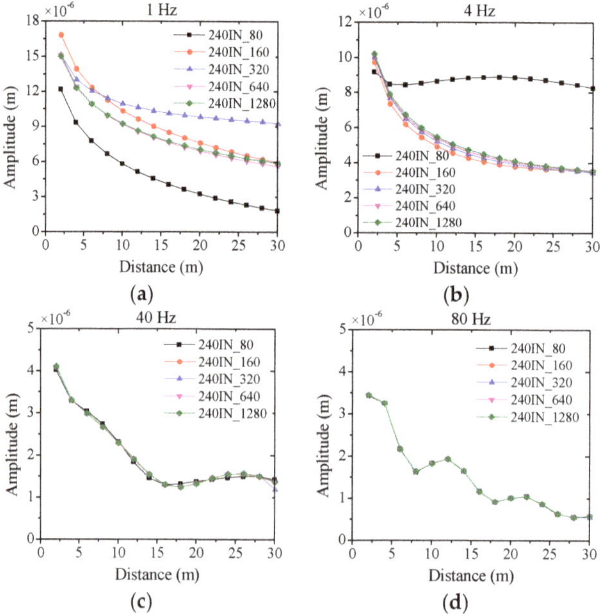

Figure 13. The attenuation curves with different model depths under the excitation of: (**a**) 1 Hz; (**b**) 4 Hz; (**c**) 40 Hz; and (**d**) 80 Hz.

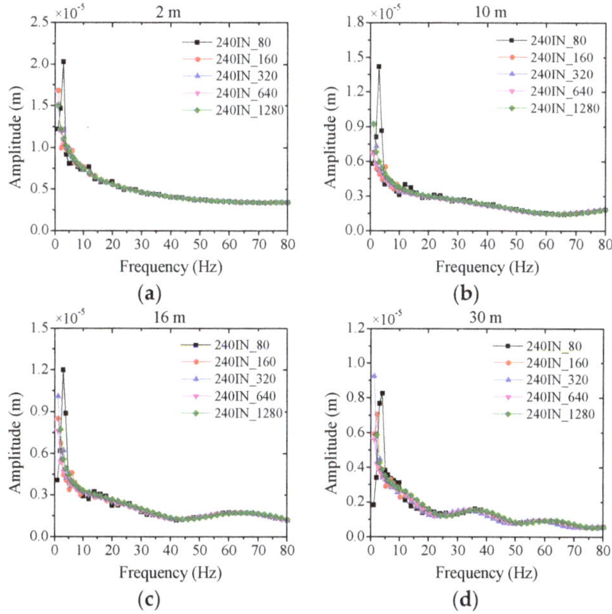

Figure 14. The amplitude–frequency curve with different model depths at the ground evaluation point of: (**a**) 2 m; (**b**) 10 m; (**c**) 16 m; and (**d**) 30 m.

4.2. The Effect of Model Lateral Sides with Infinite Element Boundary on the Simulation Results

Table 3 shows the numerical cases of models with lateral infinite element. The applied boundary is similar to the models in Section 4.1. In this section, the model height remains unchanged to study the influence of model width on the numerical results with the infinite element applied to the lateral side.

Table 3. Numerical cases of different model widths with lateral infinite element.

Case	1	2	3	4	5
L × H (m × m)	80 × 80	160 × 80	240 × 80	320 × 80	480 × 80
Expression	80IN_80	160IN_80	240IN_80	320IN_80	480IN_80

The numerical simulation shows that the solution is stable with the model width 1280 m, based on the discussion in Section 2.2. Therefore, the result of model 1280_80 is used as a benchmark to be compared with other models' results in this section.

The attenuation curves of the infinite element model with different widths under the excitation of vibration frequencies are shown in Figure 15. It can be seen that: (1) There is a certain requirement on the dimensions of the finite element area with the infinite element boundary applying to the lateral side, since smaller finite element area generates greater calculation error. While with the increasing of width in the finite element area, the calculation results tend to be stable gradually. (2) The simulation errors of finite element model 160_80 and model 160IN_80 are 60.78% and 13.88%, respectively, indicating that the maximum error is significantly reduced with the infinite boundary. The finite element model notes that infinite element boundary is not applied to the model. The relative error is the simulation results comparison with that in the benchmark model 1280_80.

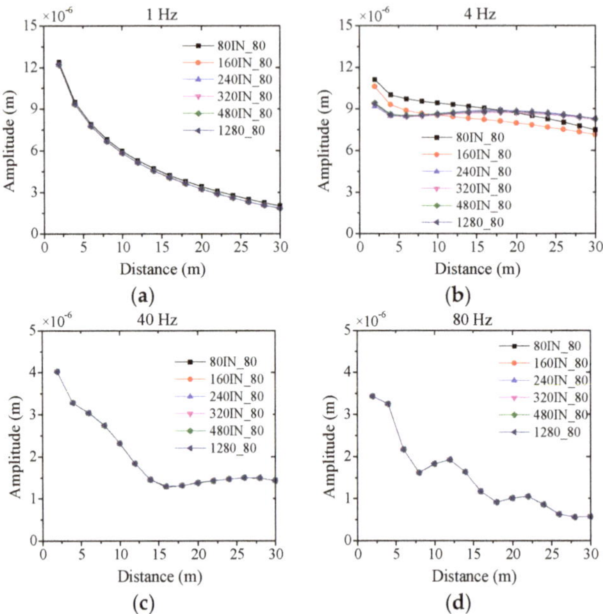

Figure 15. The attenuation curves of models with lateral infinite element boundary under the excitation of: (**a**) 1 Hz; (**b**) 4 Hz; (**c**) 40 Hz; and (**d**) 80 Hz.

Figure 16 shows the amplitude–frequency curve of the models with different widths and the infinite element boundary is applied to the lateral side in the simulation models. It can be observed that the errors of the infinite element model with 480 m width is less than 5%, and when the model width is 320 m, its corresponding error is 10%. The widths of the finite element model are 960 m and 640 m, respectively, to obtain the same accuracy. Therefore, the horizontal dimension of the finite element area can be reduced 50% effectively by means of the infinite element boundary, and the calculation accuracy remains unchanged. Moreover, the computational time can be saved due to the decreasing number of element units, demonstrating the higher performance of the combined finite element–infinite element boundary method.

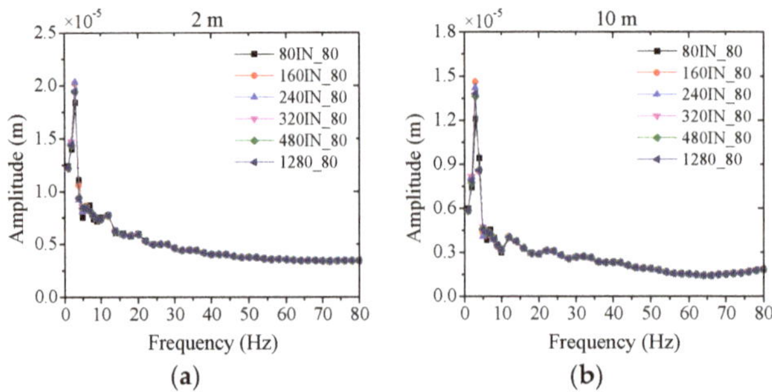

Figure 16. *Cont.*

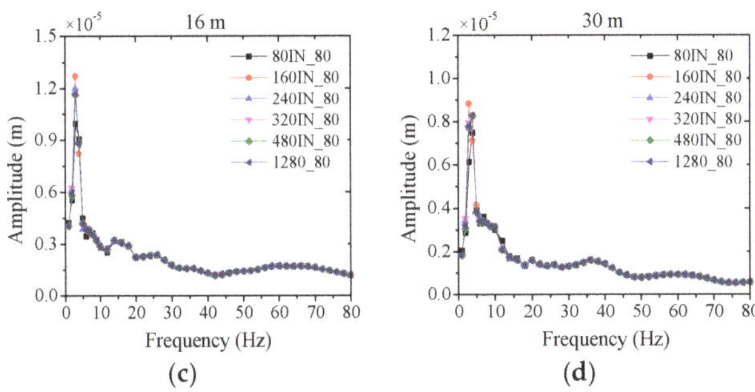

Figure 16. The amplitude–frequency curves of models with lateral infinite element boundary at the ground evaluation point of: (**a**) 2 m; (**b**) 10 m; (**c**) 16 m; and (**d**) 30 m.

4.3. The Effect of Model Base with Infinite Element Boundary on the Simulation Results

Table 4 shows the numerical cases with different model depths. The model can be expressed as L IN_H IN. The infinite element is applied to the right side and the base of the model. In these cases, the model width remains unchanged to study the influence of model depth on the numerical results with the infinite element. The numerical simulation shows that the solution is stable when the model's height is 1280 m based on the discussion in Section 4.1. Therefore, the result of model 240IN_1280 is used as a benchmark to be compared with other model results in this section.

Table 4. Numerical cases of different model depths with base infinite element.

Case	1	2	3
L × H (m × m)	240 × 80	240 × 160	240 × 240
Expression	240IN_80IN	240IN _160IN	240IN_240IN

The attenuation curves of the infinite element models with different heights under the excitation of vibration frequencies are shown in Figure 17, from which it can be concluded that there is a relatively large difference between the calculated curve and the stable solution under the low frequency excitation when the model depth is small. However, the results are close to the stable solution under the high frequency excitation, such as 40 Hz. Therefore, larger height and higher frequency load generally generate smaller simulation errors.

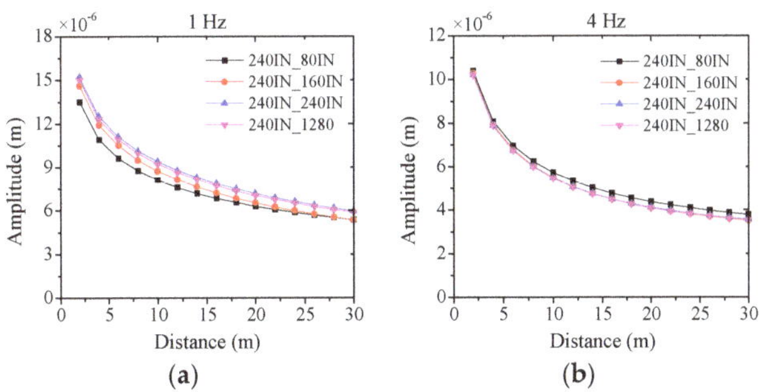

Figure 17. *Cont.*

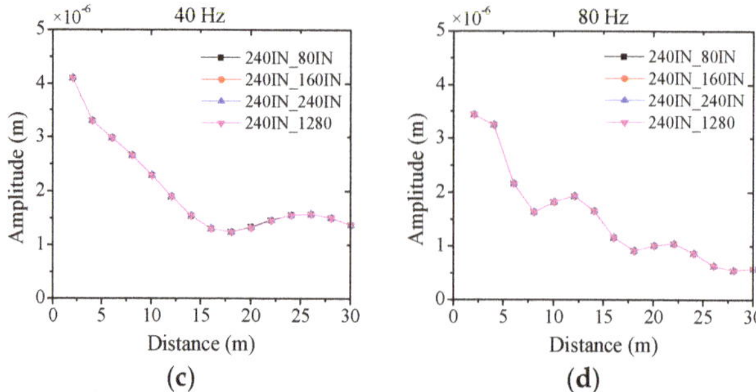

Figure 17. The attenuation curves of models with base infinite element boundary under the excitation of: (**a**) 1 Hz; (**b**) 4 Hz; (**c**) 40 Hz; and (**d**) 80 Hz.

Figure 18 shows the amplitude–frequency curve of the models with different heights as infinite element boundary is applied to the lateral side and base of the model. It can be observed that the results of amplitude–frequency in the infinite element models are much closer to the stable solution compared with that in Figure 14 in Section 4.1. The reason is that the fundamental frequency of the infinite models decreases, close to that of benchmark 240IN_1280. In addition, from the energy point of view, the reflection energy becomes weaker and close to the half-space infinite body after applying the infinite-element boundary. The near-source induced waves usually contain high energy, which may cause greater vibration [44].

Moreover, the model 240IN_640 in Section 4.1 can be replaced by the model 240IN_240IN in this section to achieve the same calculation accuracy, demonstrating that the height of finite element model can be reduced 62.5% after applying the infinite element boundary to the model base. Meanwhile, computational time is saved effectively, revealing the high performance of the coupled finite element–infinite element boundary method in decreasing the models' dimensions once more.

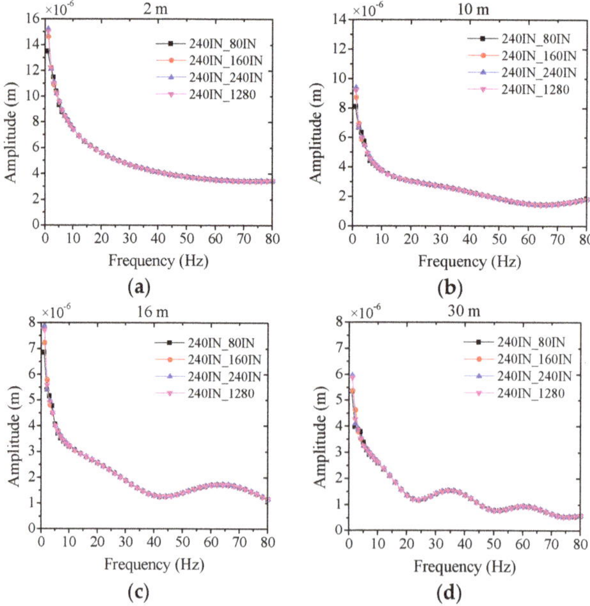

Figure 18. The amplitude–frequency curves of models with base infinite element boundary at the ground evaluation point of: (**a**) 2 m; (**b**) 10 m; (**c**) 16 m; and (**d**) 30 m.

5. Conclusions

In this study, a coupled finite element–infinite element boundary model based on the basic principles of the infinite element method was performed by the harmonic analysis based on the commercial software ABAQUS to address the issue of subway induced vibration, a major environmental concern in urban areas. In addition, the finite element method in the harmonic analysis was verified by comparing it with the thin layer method. Moreover, an interface program was developed to automatically read the simulation result files (dat file) in the ABAQUS harmonic analysis, and then the data were put into the frequency domain calculation on the MATLAB platform. Besides, a measurement of a practical engineering was performed on a track bed-tunnel lining-surrounding formation located on Line 2, Shanghai Metro, and the corresponding simulation model was established based on the introduced method. In this paper, the numerical results of vibration level and spectrum characteristic are discussed. Moreover, a simplified prediction formula of ground vibration level is proposed based on the measurement results. Finally, parametric study was conducted in an elastic half-space simulation to investigate the influence of model widths and depths with infinite element boundary on the numerical results, demonstrating the effectiveness and high performance of the coupled finite element–infinite element boundary model in the harmonic analysis based on ABAQUS. The following are the conclusions drawn from this study, which may provide insight towards the rational models of subway induced vibration.

1. The coupled simulation method is applicable for the practical engineering projects of subway induced vibration on the soft site in Shanghai, and the proposed formula for the vibration levels is reasonable for the prediction in the subway induced vibration.
2. There exists a vibration amplifying zone a certain distance from the tunnel under high frequency loads due to wave propagation and reflection, and the amplification has relationship with the spectrum of the applied excitation accelerations.
3. The finite element model dimensions can be reduced 50% and 66.67% effectively in the widths and depth, respectively, with the coupled method, applying the infinite element boundary to the lateral and base sides of the finite element model.
4. The model dimensions have great influence on the numerical results as the excited frequency is close to the model's vertical fundamental frequency, and the depth of the model should be designed seriously to reduce the numerical errors.

Author Contributions: This study is the result of full collaboration and therefore the authors accept full responsibility. J.Y. wrote the paper and developed the interface programs in Section 2; P.L. performed the field measurement of a practical engineering and established the simulation model in Section 3; and Z.L. conducted the parametric study in an elastic half-space simulation in Section 4.

Funding: This study was supported by the National Natural Science Foundation of China (Grant No. 51478355) and by the Ministry of Science and Technology of China (Grant No. SLDRCE14-B-22).

Conflicts of Interest: The authors declare no conflict of interest.

References

1. Xia, H.; Cao, Y.M. Problem of railway traffic induced vibrations of environments. *J. Railw. Sci. Eng.* **2004**, *1*, 44–51. (In Chinese)
2. Gupta, S.; Degrande, G.; Lombaert, G. Experimental validation of a numerical model for subway induced vibrations. *J. Sound Vib.* **2009**, *321*, 786–812. [CrossRef]
3. Gupta, S.; Liu, W.F.; Degrande, G.; Lombaert, G.; Liu, W.N. Prediction of vibrations induced by underground railway traffic in Beijing. *J. Sound Vib.* **2008**, *310*, 608–630. [CrossRef]
4. Lopez-Mendoza, D.; Romero, A.; Connolly, D.P.; Galvin, P. Scoping assessment of building vibration induced by railway traffic. *Soil Dyn. Earthq. Eng.* **2017**, *93*, 147–161. [CrossRef]
5. Zhong, H.; Qu, D.; Lin, G.; Li, H.J. Application of sensor networks to the measurement of subway-induced ground-borne vibration near the station. *Int. J. Distrib. Sens.* **2014**. [CrossRef]

6. Sun, K.; Zhang, W.; Ding, H.; Kim, R.E.; Spencer, B.F. Autonomous evaluation of ambient vibration of underground spaces induced by adjacent subway trains using high-sensitivity wireless smart sensors. *Smart Struct. Syst.* **2017**, *19*, 1–10. [CrossRef]

7. Takemiya, H. Analysis of wave field from high-speed train on viaduct at shallow/deep soft grounds. *J. Sound Vib.* **2008**, *310*, 631–649. [CrossRef]

8. Liu, C.; Wei, J.H.; Zhang, Z.X.; Liang, J.S.; Ren, T.Q.; Xu, H.Q. Design and evaluation of a remote measurement system for the online monitoring of rail vibration signals. *Proc. Mech. Eng. Part F* **2016**, *230*, 724–733.

9. Li, X.H.; Long, Y.; Ji, C.; Zhong, M.S.; Zhao, H.B. Study on the vibration effect on operation subway induced by blasting of an adjacent cross tunnel and the reducing vibration techniques. *J. Vibroeng.* **2013**, *15*, 1454–1462.

10. Ling, X.Z.; Chen, S.J.; Zhu, Z.Y.; Zhang, F.; Wang, L.N.; Zou, Z.Y. Field monitoring on the train-induced vibration responses of track structure in the Beiluhe permafrost region along Qinghai-Tibet railway in China. *Cold Reg. Sci. Technol.* **2010**, *60*, 75–83. [CrossRef]

11. Kouroussis, G.; Florentin, J.; Verlinden, O. Ground vibration induced by intercity/inter region trains: A numerical prediction based on the multibody/finite element modeling approach. *J. Vib. Control* **2016**, *22*, 4192–4210. [CrossRef]

12. Kaewunruen, S.; Remennikov, A.M. Sensitivity analysis of free vibration characteristics of an in situ railway concrete sleeper to variations of rail pad parameters. *J. Sound Vib.* **2006**, *298*, 453–461. [CrossRef]

13. Connolly, D.; Giannopoulos, A.; Fan, W.; Woodward, P.K.; Forde, M.C. Optimising low acoustic impedance back-fill material wave barrier dimensions to shield structures from ground borne high speed rail vibrations. *Constr. Build. Mater.* **2013**, *44*, 557–564. [CrossRef]

14. Kaewunruen, S.; Remennikov, A.M. Current state of practice in railway track vibration isolation: An Australian overview. *Aust. J. Civil Eng.* **2016**, *14*, 63–71. [CrossRef]

15. Indraratna, B.; Sun, Q.D.; Heitor, A.; Grant, J. Performance of rubber tire-confined capping layer under cyclic loading for railroad conditions. *J. Mater. Civil Eng.* **2018**, *30*. [CrossRef]

16. Navaratnarajahl, S.K.; Indraratna, B. Use of rubber mats to improve the deformation and degradation behavior of rail ballast under cyclic loading. *J. Geotech. Geoenviron.* **2017**, *143*. [CrossRef]

17. Hussaini, S.K.K.; Indraratna, B.; Vinod, J.S. A laboratory investigation to assess the functioning of railway ballast with and without geogrids. *Transp. Geotech.* **2016**, *6*, 45–54. [CrossRef]

18. Kaewunruen, S.; Remennikov, A.M. Dynamic crack propagations in prestressed concrete sleepers in railway track systems subjected to severe impact loads. *J. Struct. Eng* **2010**, *136*, 749–754. [CrossRef]

19. Remennikov, A.M.; Kaewunruen, S. Experimental load rating of aged railway concrete sleepers. *Eng. Struct.* **2014**, *76*, 147–162. [CrossRef]

20. Lu, Z.; Wang, Z.X.; Zhou, Y.; Lu, X.L. Nonlinear dissipative devices in structural vibration control: A review. *J. Sound Vib.* **2018**, *6*, 18–49. [CrossRef]

21. Dai, K.S.; Wang, J.Z.; Mao, R.F.; Lu, Z.; Chen, S.E. Experimental investigation on dynamic characterization and seismic control performance of a TLPD system. *Struct. Des. Tall Spec.* **2017**, *26*, e1350. [CrossRef]

22. Lu, Z.; Huang, B.; Zhang, Q.; Lu, X.L. Experimental and analytical study on vibration control effects of eddy-current tuned mass dampers under seismic excitations. *J. Sound Vib.* **2018**, *421*, 153–165. [CrossRef]

23. Lu, Z.; Chen, X.Y.; Li, X.W.; Li, P.Z. Optimization and application of multiple tuned mass dampers in the vibration control of pedestrian bridges. *Struct. Eng. Mech.* **2017**, *62*, 55–64. [CrossRef]

24. Lu, Z.; Chen, X.Y.; Zhang, D.C.; Dai, K.S. Experimental and analytical study on the performance of particle tuned mass dampers under seismic excitation. *Earthq. Eng. Struct. D* **2017**, *46*, 697–714. [CrossRef]

25. Lu, Z.; Chen, X.Y.; Zhou, Y. An equivalent method for optimization of particle tuned mass damper based on experimental parametric study. *J. Sound Vib.* **2018**, *419*, 571–584. [CrossRef]

26. Lu, X.L.; Liu, Z.P.; Lu, Z. Optimization design and experimental verification of track nonlinear energy sink for vibration control under seismic excitation. *Struct. Control Health Monit.* **2017**, *24*, e2033. [CrossRef]

27. Lu, Z.; Yang, Y.L.; Lu, X.L.; Liu, C.Q. Preliminary study on the damping effect of a lateral damping buffer under a debris flow load. *Appl. Sci.* **2017**, *7*, 201. [CrossRef]

28. Lu, Z.; Wang, Z.X.; Masri, S.F.; Lu, X.L. Particle Impact Dampers: Past, Present, and Future. *Struct. Control Health Monit.* **2018**, *25*, e2058. [CrossRef]

29. Lu, Z.; Lu, X.L.; Masri, S.F. Studies of the performance of particle dampers under dynamic loads. *J. Sound Vib.* **2010**, *329*, 5415–5433. [CrossRef]

30. Lu, Z.; Huang, B.; Zhou, Y. Theoretical study and experimental validation on the energy dissipation mechanism of particle dampers. *Struct. Control Health Monit.* **2018**, *25*, e2125. [CrossRef]

31. Gupta, S.; Hussein, M.F.M.; Degrande, G.; Hunt, H.E.M.; Clouteau, D. A comparison of two numerical models for the prediction of vibrations from underground railway traffic. *Soil Dyn. Earthq. Eng.* **2007**, *27*, 608–624. [CrossRef]

32. Sun, X.J.; Liu, W.N.; Zhai, H. Study on low-frequency vibration isolation performance of steel spring floating slab track. *Environ. Vib. Predict. Monit. Mitig. Eval.* **2009**, *1–2*, 429–434.

33. Galvin, P.; Francois, S.; Schevenels, M.; Bongini, E.; Degrande, G.; Lombaert, G. A 2.5D coupled FE-BE model for the prediction of railway induced vibrations. *Soil Dyn. Earthq. Eng.* **2010**, *30*, 1500–1512. [CrossRef]

34. Wolf, S. Potential low frequency ground vibration (<6.3 Hz) impacts from underground LRT operations. *J. Sound Vib.* **2003**, *267*, 651–661.

35. Hung, H.H.; Yang, Y.B. Analysis of ground vibrations due to underground trains by 2.5D finite/infinite element approach. *Earthq. Eng. Eng. Vib.* **2010**, *9*, 327–335.

36. Costa, P.A.; Calcada, R.; Cardoso, A.S. Track-ground vibrations induced by railway traffic: In-situ measurements and validation of a 2.5D FEM-BEM model. *Soil Dyn. Earthq. Eng.* **2012**, *32*, 111–128. [CrossRef]

37. Yuan, Z.H.; Bostrom, A.; Cai, Y.Q. Benchmark solution for vibrations from a moving point source in a tunnel embedded in a half-space. *J. Sound Vib.* **2017**, *387*, 177–193. [CrossRef]

38. Lopes, P.; Costa, P.A.; Ferraz, M.; Calcada, R.; Cardoso, A. Numerical modeling of vibrations induced by railway traffic in tunnels: From the source to the nearby buildings. *Soil Dyn. Earthq. Eng.* **2014**, *61–62*, 269–285. [CrossRef]

39. Chiacchiari, L.; Loprencipe, G. Measurement methods and analysis tools for rail irregularities: A case study for urban tram track. *J. Modern Transp.* **2015**, *23*, 137–147. [CrossRef]

40. Kaynia, A.M.; Madshus, C.; Zackrisson, P. Ground vibration from high-speed trains: Prediction and countermeasure. *J. Geotech. Geoenviron.* **2000**, *126*, 531–537. [CrossRef]

41. Huang, H.H.; Chen, G.H.; Yang, Y.B. Effect of railway roughness on soil vibrations due to moving trains by 2.5D finite/infinite element approach. *Eng. Struct.* **2013**, *57*, 254–266. [CrossRef]

42. Jiang, T.; Yue, J.Y. Analysis of environmental vibration induced by subway in Shanghai and Chengdu by using thin layer method. In Proceedings of the 5th International Symposium on Environmental Vibration, Chengdu, China, 20–22 October 2011; pp. 113–119.

43. Zou, C.; Wang, Y.M.; Guo, J.X. Measurement of ground and nearby building vibration and noise induced by trains in a metro depot. *Sci. Total Environ.* **2015**, *536*, 761–773. [CrossRef] [PubMed]

44. Ju, S.H. Three-dimensional analyses of wave barriers for reduction of train-induced vibrations. *J. Geotech. Geoenviron.* **2004**, *130*, 740–748. [CrossRef]

45. Sanayei, M.; Maurya, P.; Moore, J.A. Measurement of building foundation and ground-borne vibrations due to surface trains and subways. *Eng. Struct.* **2013**, *53*, 102–111. [CrossRef]

46. Andersen, L.; Jones, C.J.C. Coupled boundary and finite element analysis of vibration from railway tunnels—A comparison of two and three-dimensional models. *J. Sound Vib.* **2006**, *293*, 611–625. [CrossRef]

 *sustainability*

Article

Studies on Energy Dissipation Mechanism of an Innovative Viscous Damper Filled with Oil and Silt

Zheng Lu [1,2] , Junzuo Li [1] and Chuanguo Jia [3,4,*]

[1]   Research Institute of Structural Engineering and Disaster Reduction, Tongji University,
      Shanghai 200092, China; luzheng111@tongji.edu.cn (Z.L.); 1732580@tongji.edu.cn (J.L.)
[2]   State Key Laboratory of Disaster Reduction in Civil Engineering, Tongji University, Shanghai 200092, China
[3]   Key Laboratory of New Technology for Construction of Cities in Mountain Area, Chongqing University,
      Ministry of Education, Chongqing 400045, China
[4]   School of Civil Engineering, Chongqing University, Chongqing 400045, China
*   Correspondence: jiachuanguo@cqu.edu.cn; Tel.: +86-21-6598-2668; Fax: +86-21-6598-2224

Received: 20 March 2018; Accepted: 22 May 2018; Published: 29 May 2018

Abstract: To improve seismic performance of a traditional viscous damper, an innovative viscous damper is proposed, in which the conventional damping medium is replaced by a mixture of oil and silt. This new medium is expected to increase damping force and solve the problem that arises out of the fact that the energy dissipation capacity of a conventional fluid viscous damper is low for small displacements. Firstly, the design concept and device configurations are introduced, then a cyclic loading test is applied to investigate the damper's energy dissipation mechanism in different conditions. Experimental results show that the damper exhibits displacement-dependent characteristics for small displacements, indicating that the silt has changed its damping mechanism. Furthermore, the effect of multiple test parameters on the damping force is analyzed, showcasing that an increase in silt content can visibly increase the damping force. According to experimental data, fitting models of the damping force are obtained and verified, thus promoting further engineering applications.

Keywords: viscous damper; hybrid damper; seismic performance; cyclic loading test; silt

---

1. Introduction

The use of passive, semi-active, and active dampers to dissipate seismic energy is an effective approach in the creation of earthquake-resilient structures [1–4]. Such energy dissipation devices reduce structural seismic response and protect the structure from severe damage. As a type of passive damper, viscous dampers were firstly used as shock absorbers in mechanical engineering [5]. Makris and his group then began studying viscous dampers in civil engineering applications [6]. A fractional derivative Maxwell model was proposed for viscous dampers, which were used for vibration control and seismic isolation of buildings [6]. Fluid viscous dampers fabricated by Taylor Devices have been applied to hundreds of bridges and buildings, displaying satisfactory performance [7]. To further improve performance, semi-active viscous dampers, such as variable orifice dampers and magnetorheological (MR) dampers, were investigated [8,9]. These semi-active dampers can adjust damping forces through variable ways, such as changing the area of the orifice or changing the state of the damping medium to satisfy different excitations. Taylor and his group applied fluid viscous dampers in a high-rise structure in order to suppress anticipated wind-induced accelerations [10]. Viscous dampers have been proven to be useful to reduce structural vibration under both wind and earthquake excitations [10,11]. In addition, other scholars have also investigated performance tests, analytical models, design methods, and other aspects of viscous dampers [12–15]. Constantinou and Symans proved that viscous dampers could reduce both inter-story drifts and shear forces of a structure [12]. A research by Taylor and Constantinou presented the test methodology, procedure and results of full-scale viscous dampers with high-output forces, which was useful for further research [14]. Furthermore, many scholars have focused on the optimum design of viscous dampers for high-rise buildings [15]. The advantages of viscous dampers include their ability to not contribute to additional stiffness of a structure and great energy dissipation capacity. However, shortcomings do exist.

Viscous dampers are velocity-dependent. Compared to velocity-independent dampers such as friction dampers [16,17], viscous dampers have much lower energy dissipation capacity for small displacements [18]. In engineering applications, the performance of the damper is limited when structural displacements are not large enough to activate them [19]. Hence, some displacement amplification methods were studied. Berton and Bolander designed an amplification device based on a gear-type mechanism and verified the device's function through laboratory tests [20]. In addition, Ribakov et al. suggested magnification of the effect of viscous dampers by lever arms [21]. However, these methods make a structure more complicated and increase manufacturing costs, proving unfavorable to a wide range of engineering applications.

A multiple damping mechanism is an effective way to reduce the primary structure's response [22,23]. Tsai et al. combined velocity-dependent and velocity-independent devices in one frame, minimizing the shortcomings of individual dampers and effectively reducing structural response [24]. Marko et al. studied the seismic response of a structure equipped with multiple types of dampers [25]. Furthermore, some dampers with hybrid energy consumption mechanisms were designed to enhance damping performance [26–30]. Lee and his group developed a new hybrid energy dissipation device, combining a steel slit damper and rotational friction dampers [27]. Lu studied the performance of particle dampers that had several energy dissipation mechanisms [29]. Inspired by these experiences, a new type of viscous damper with a mixed damping medium consisting of silt and silicone oil is proposed in this paper.

As a very common engineering material, the dynamic properties of soil have been studied by scholars, who found that the deformation process of soil also dissipates energy [31–33]. The elastic-plastic model, which is displacement-dependent, is usually applied as the constitutive model of the silt. This research attempts to enrich the damping mechanism of the viscous damper by adding silt into the conventional damping medium. The high pressure in the cylinder of the damper can force the saturated silt to deform and dissipate energy. In addition, the connection between particles of saturated silt is weak, while its fluidity and viscosity is strong [34,35]. Therefore, the mixed medium has enough fluidity, and viscosity is increased to enhance the damping force.

This paper investigated this new type of viscous damper. A corresponding performance test was carried out to study its energy dissipation mechanism and a simplified damping force model was proposed (based on experimental results) to facilitate further engineering applications.

2. Conceptual Design of Innovative Viscous Damper

2.1. Viscous Fluid Design

Methyl silicone oil is a commonly used energy dissipation medium for viscous dampers. Table 1 lists its physical properties. It has excellent electrical insulation tolerance, temperature tolerance, and high compression modulus. It is hydrophobic and chemically inert, with small surface tension and viscosity-temperature coefficient.

Table 1. Physical properties of methyl silicone oil.

Density/kg/m^3	970
Kinematic viscosity/cst (25 °C)	1000 ± 50
Dynamic viscosity/kg/(m·s) (25 °C)	0.97
Flash point/°C	300
Freezing point/°C	−50

The silt applied in the test is shown in Figure 1a. Through moisture content test and sieve test, the specimen can be classified as low-plasticity silt, which is homogeneous with very little impurities. The particles are very small, like powder. Furthermore, the moisture content is 0.906%, which is extremely low. The plasticity index $I_p = 9.6$. On mixing the silt with methyl silicone oil, it was found that the mixed medium was fluidic after stirring and had great viscosity.

Firstly, the silt increases the viscosity of the damping medium and the maximum damping force. Furthermore, special dynamic characteristics change the original energy dissipation property of the viscous damper by bringing in a multiple energy dissipation mechanism. Seismic energy is not only dissipated by viscous damping, but also by the deformation of the silt particles and the friction between the particles and the cylinder. The reciprocating motion of the piston forms a relative movement between the silt and the silicone oil. At the same time, the silt is in a suspended state, where contact between the particles is rare. Hence, the fluidity of the mixture is enhanced. In addition, the excess pore pressure reduces the effective shear stress between the silt particles, which increases fluidity as well.

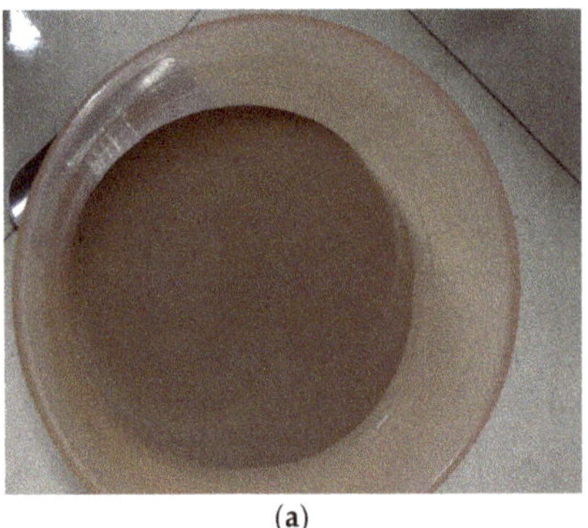

(a)

(b)

Figure 1. Viscous fluid composition: (**a**) silt; (**b**) mixture of methyl silicone oil and silt.

2.2. Structure of the Dampers

One type of damper (Damper 1) was designed and fabricated for a pre-test (based on a standard hydraulic cylinder), shown in Figure 2a. Two pores with diameter of 3 mm were punched on the piston. Then, by estimating the output force of the guide rod and the pressure on the piston plate, another type of damper (Damper 2) was designed to be tested by a different loading mechanism, which had a much higher loading capacity, shown in Figure 2b. Damper 2 had four pores with diameters of 5 mm.

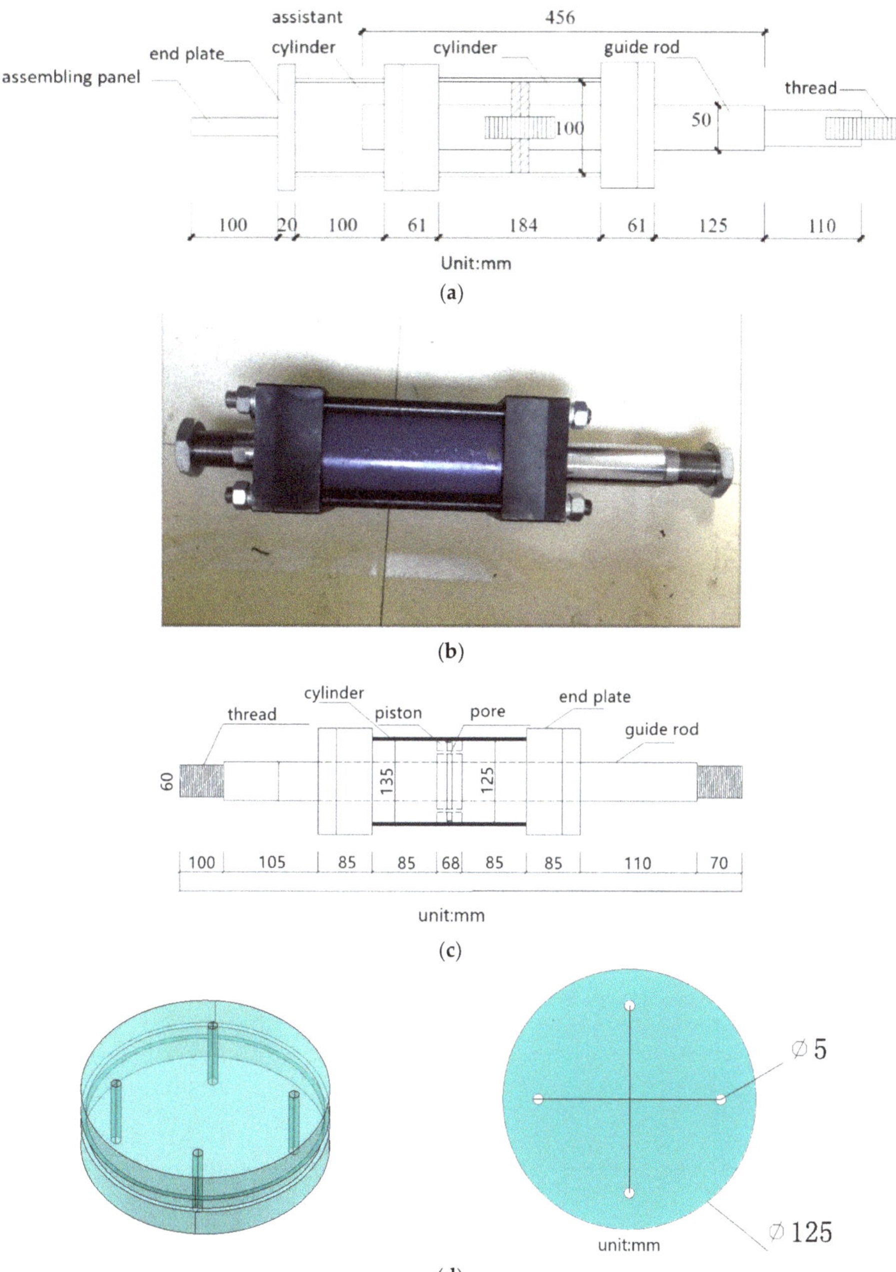

Figure 2. Structure of the dampers: (**a**) dimensions of Damper 1; (**b**) photo of Damper 2; (**c**) dimension of Damper 2; (**d**) piston of Damper 2.

2.3. The Sustainable Design

As an easily obtainable material, silt not only enhances the seismic performance of the damper, but also offers sustainability. Fluid viscous dampers are unable to exhibit great performance for small displacements, and adding silt to the damping medium solves this problem. The mixture of silt and silicone oil can bring displacement-dependent features to the damper, which will be discussed in later sections. Compared to other solutions (such as using lever arms to magnify the displacement or installing other types of dampers), the proposed damper possesses the desired properties without the added costs of construction materials. In addition, when compared to semi-active or active dampers such as MR dampers, the proposed damper can work without additional energy. Hence, the addition of silt is an economical approach to enhance the performance of conventional viscous dampers.

3. Experimental Study

The cyclic loading test is a pseudo-static test and is one of the most important experimental methods to study seismic performance of structural systems or members. These tests aim to obtain properties such as energy dissipation capacity, hysteretic property and ductility, which are vital for earthquake-resilient design.

3.1. Pre-Test

Harmonic excitations (sine wave) with displacement amplitudes and loading frequencies equal to 20 mm and 0.1 Hz were applied for preliminary study of the damper. The loading system in this pre-test had a capacity of 10 kN force. The loading procedure was controlled by displacement. Furthermore, Damper 1 was tested in two test conditions and each condition included 10 cycles. The damping medium of Condition#1 was silicone oil while that of Condition#2 was silicone oil with 30% silt (mass ratio).

The hysteretic curves of the two conditions are shown in Figure 3. The shapes of the two curves were similar. The maximum damping force of Condition#1 and Condition#2 was 0.79 kN and 1.18 kN, respectively. The silt increased the damping force by 49.4%.

In order to test this type of damper in more realistic operating conditions (large damping forces in real engineering projects), a formal large-scale experiment was planned. It sought to test dampers for small displacements (3 mm, 5 mm). Although the authors tried to test dampers without silt, the measured damping forces for small displacements (3 mm) were less than 2 kN because conventional viscous dampers could not yield correspondingly high damping forces for small displacements [7]. Considering that the loading system possessed a high capacity of 100 kN force, dampers without silt were not suitable to be tested in the formal experiment.

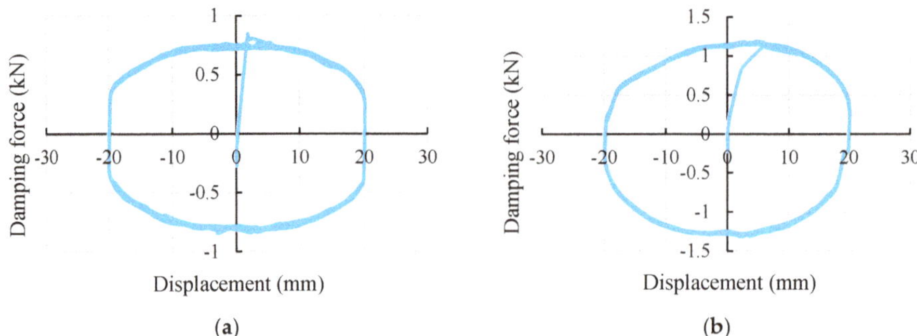

Figure 3. Hysteretic curves of the dampers in pre-test: (**a**) Condition#1 (without silt); (**b**) Condition#2 (with silt).

3.2. Test Setup of the Formal Experiment

The loading procedure was consistent with the pre-test. The test variables were silt content (mass ratio), loading displacement, and loading frequency.

Damper 2 was studied in the formal experiment and test conditions were divided into three groups with the silt mass ratio being 30%, 40%, and 50%, respectively. Orthogonal test in multiple loading conditions was applied to each group. The amplitude of the loading displacement varied from 3 mm to 20 mm and the loading frequency varied from 0.5 Hz to 2.0 Hz. There was a total of 33 test conditions and each condition included 15 cycles. The loading displacement and the damping force were recorded during each loading cycle to examine the energy dissipation capacity of the damper under different excitations.

The MTS electro-hydraulic servo system was used in the formal experiment. As shown in Figure 4, the two blue plates were steel backing plates, connected to the reaction wall and the floor, respectively. The yellow part was the actuator; the green part was the damper, fixed on an I-shape steel by bolts; the I-shape steel was also connected to the backing plate by bolts; and the steel backing plate was secured to the floor using pre-tensioned bolts to minimize slips.

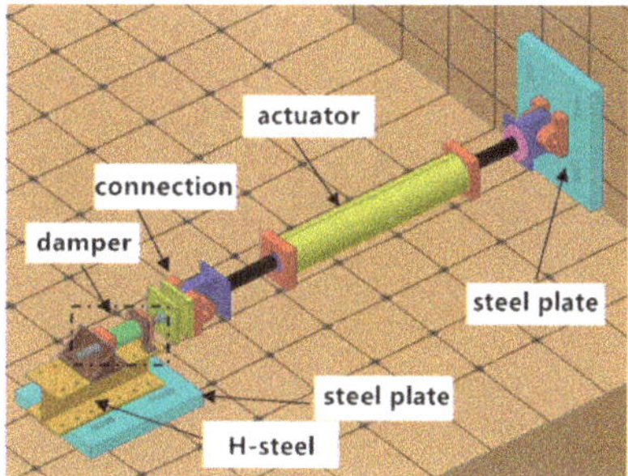

Figure 4. Test setup.

3.3. Experimental Results

3.3.1. Hysteretic Curves

Using the loading displacement as abscissa against the damping force as ordinate, the hysteretic curves under each test condition were drawn. The hysteretic curves of the three groups with different silt mass ratios were similar in how curves changed with displacement amplitude and loading frequency. Curves with silt mass ratio of 30% were presented as a typical case in Figure 5.

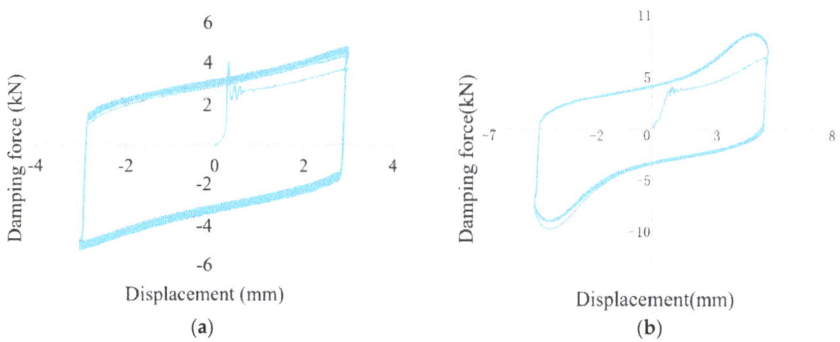

Figure 5. *Cont.*

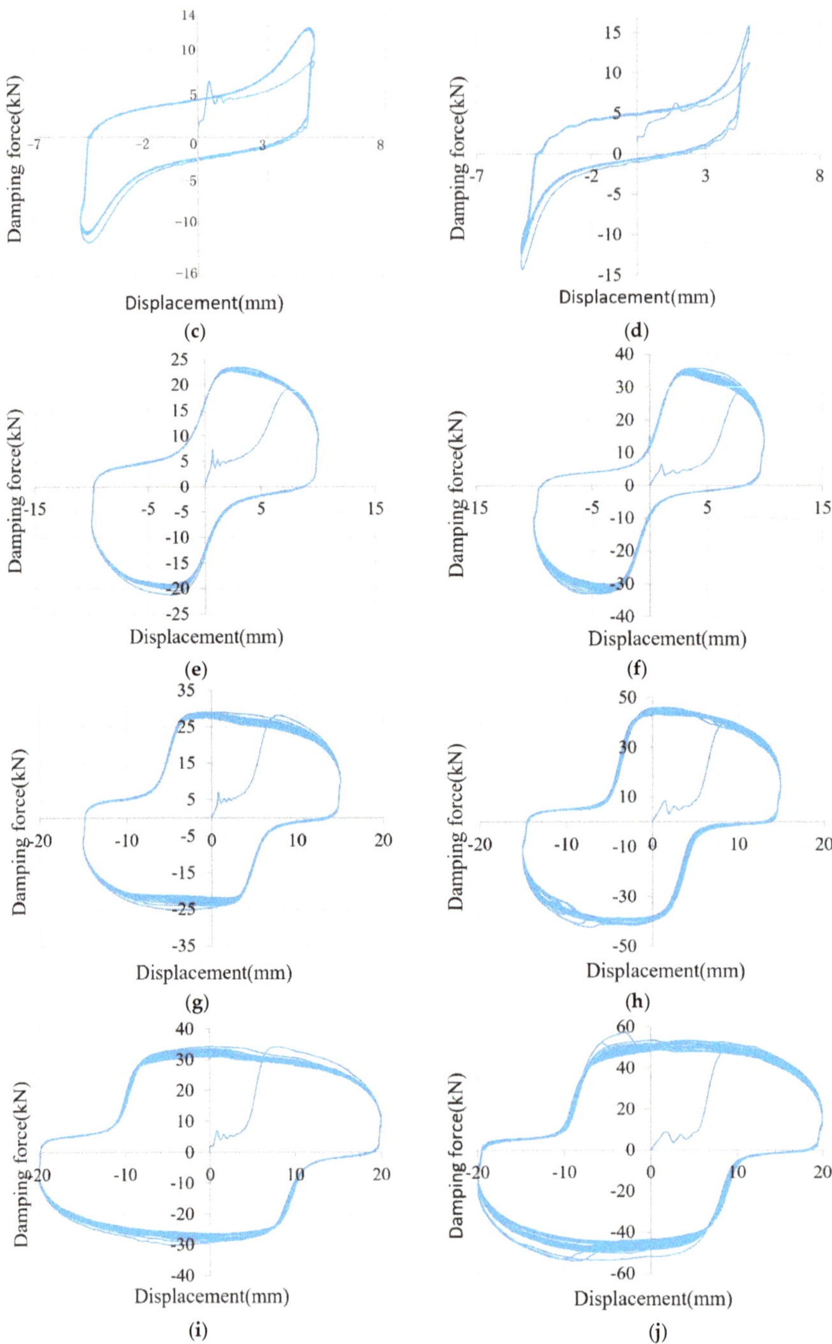

Figure 5. Hysteretic curves (silt mass ratio = 30%) (displacement amplitude, loading frequency): (**a**) 3 mm, 0.5 Hz; (**b**) 5 mm, 0.5 Hz; (**c**) 5 mm, 1.0 Hz; (**d**) 5 mm, 2.0 Hz; (**e**) 10 mm, 0.5 Hz; (**f**) 10 mm, 1.0 Hz; (**g**) 15 mm, 0.5 Hz; (**h**) 15 mm, 1.0 Hz; (**i**) 20 mm, 0.5 Hz; (**j**) 20 mm, 1.0 Hz.

By observing the changing process of the curves, it can be seen that when the amplitude of loading displacement was small, the curve was plump and the shape was similar to a parallelogram. With increase in amplitude, the curve became more rounded; then, despite the pinching phenomenon, the shape was closer to an oval. In conclusion, when the amplitude of loading displacement was large, the damper's energy dissipation model was close to the Kelvin model of the conventional viscous damper.

The difference between this hysteretic characteristic and that of a conventional viscous damper was that the damping force included not only a velocity-dependent viscous force but also a displacement-dependent hysteretic force. The displacement-dependent force was formed by the effective stress of the piston on the silt particles, which was related to the constitutive model of the silt. When the displacement was small, the energy was consumed by the elastic-plastic deformations of the silt. Hence, the hysteretic curve was close to a parallelogram, which was consistent with the elastic-plastic model of the silt. When the displacement was large, the damping force increased considerably as the silt increased the viscosity of the medium. Energy consumption also gradually showed velocity-dependent features and the damping force-displacement curve became an oval. Although the hysteretic curve under large displacement had a certain degree of pinching, the pattern of the change was still consistent with the expected result (that the energy dissipation mechanism was displacement-dependent for small displacement and velocity-dependent for large displacement). The pinching phenomenon was probably because the viscous medium was not dense enough, leading to the existence of porosity in the cylinder. In addition, it was also probably due to the plastic deformation ability of the medium. When the piston moved to one side, the viscous medium was unable to flow quickly through the pores to the other side of the cylinder. Hence, the other side was not fully filled by the damping medium.

3.3.2. Effect of Silt Content

As silt was the core difference between the proposed innovative viscous damper and the conventional one, the effect of the silt content on hysteretic curves was studied, as shown in Figure 6. Despite pinching, the curves were similar and almost oval. The damping force was considerably increased with an increase in silt content (mass ratio). Furthermore, the silt also made the curves plumper. However, it was also observed that the silt intensified the pinching phenomenon to a certain extent.

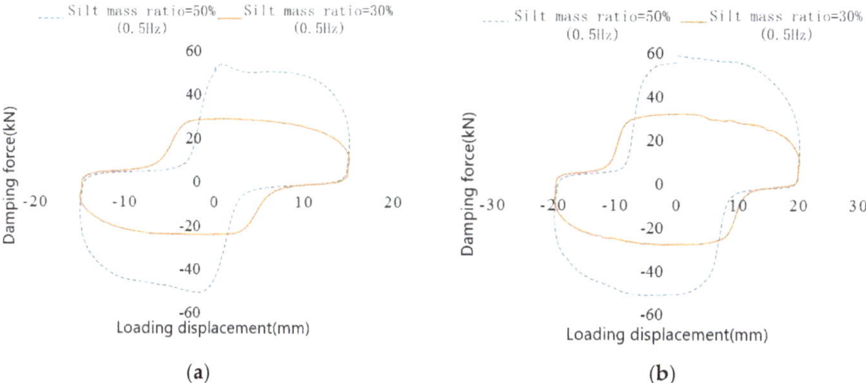

Figure 6. Comparison of the hysteretic curves with different silt contents: (**a**) displacement amplitude = 15 mm; (**b**) displacement amplitude = 20 mm.

The relationship between the silt content and the maximum damper force is shown in Figure 7. When the displacement amplitude was 5 mm, increasing the silt content did not visibly change the maximum damping force. When the amplitude was larger, the effect of the silt was more notable. This is because when the displacement was small, the main contribution of the silt was to dissipate energy by plastic deformation. This part of energy was small when compared to the total consumed energy, hence the damping force did not greatly increase. As the loading displacement continued to increase, the damping force gradually exhibited velocity-dependent characteristics. Since the silt considerably increased the viscosity of the medium, the growth of the damping force in large displacement positively correlated with an increase in silt content.

However, when the amplitude of displacement was 15 mm, increasing silt content did not prove to be effective when the mass ratio exceeded 40%, as too much silt weakened the fluidity of the medium, preventing it from passing thorough the pores.

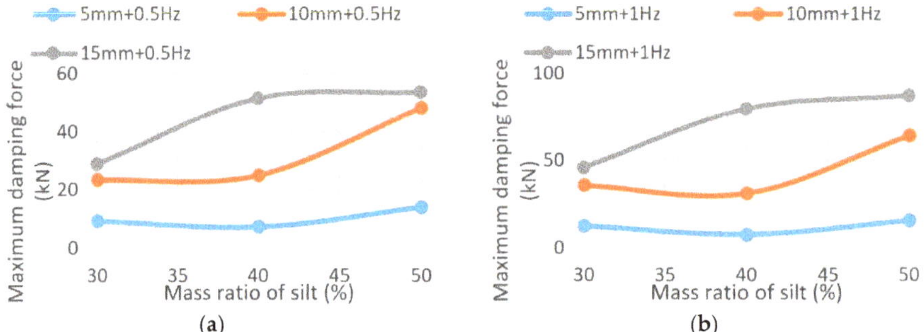

Figure 7. Relationship between the maximum damping force and the mass ratio of silt: (**a**) 0.5 Hz; (**b**) 1.0 Hz.

3.3.3. Effect of Displacement Amplitude and Loading Frequency

The effect of displacement amplitude and loading frequency on the maximum damping force is presented in Figures 8 and 9. It can be seen that, in general, the maximum damping force (Fmax) positively relates to displacement amplitude and loading frequency. However, it can still be seen that increasing loading frequency does not change Fmax visibly under small displacement amplitude, especially when the mass ratio of silt is large (> 40%). This phenomenon is consistent with the previous conclusion that the damper is displacement-dependent for small displacements.

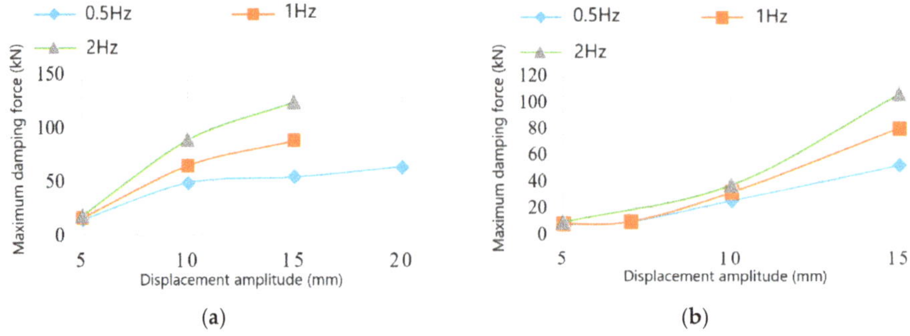

Figure 8. *Cont.*

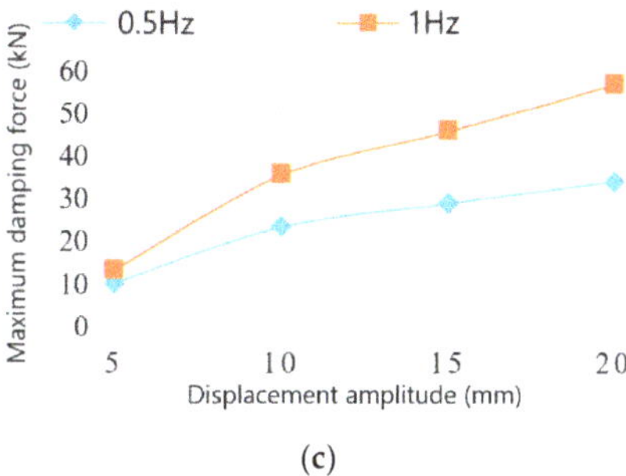

(**c**)

Figure 8. Relationship between the maximum damping force and the displacement amplitude: (**a**) mass ratio of silt = 50%; (**b**) mass ratio of silt = 40%; (**c**) mass ratio of silt = 30%.

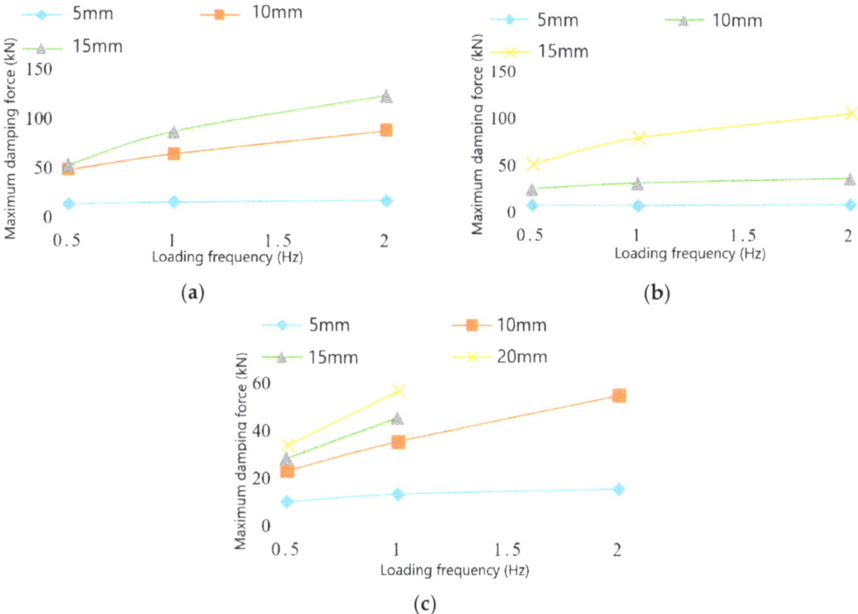

Figure 9. Relationship between the maximum damping force and the loading frequency: (**a**) mass ratio of silt = 50%; (**b**) mass ratio of silt = 40%; (**c**) mass ratio of silt = 30%.

4. Simulation Results and Discussion

The damping force models of the damper under small and large displacements can be obtained using the least square method.

As discussed earlier, under small displacement, the silt can help improve the performance of the damper by providing displacement-dependent features. The new damper can offer better control of the vibration of the structure during minor earthquakes, proving to be meaningful for structural comfort design. Taking the test condition that loading displacement and frequency equals to 3 mm and 0.5 Hz as a sample, the relationship between damping force and displacement can be fitted as a bilinear model, as shown in Figure 10, where K_0 and K_d represent the stiffness before and after yielding, respectively. Part of the experimental data is marked in Figure 10.

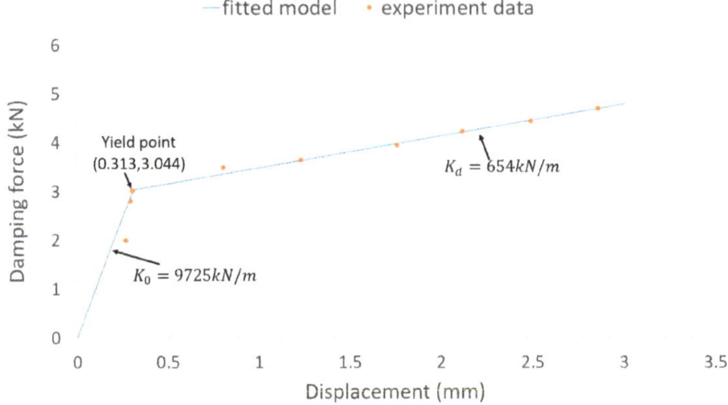

Figure 10. Fitted displacement-dependent model under a small displacement (3 mm).

Structures suffer large displacements during major earthquakes, and dampers are hence designed to protect them from severe damage or collapse. In this case, the bearing capacity of the damper is critical. As discussed earlier, the damper shows velocity-dependent features under large displacements. The data of test condition that loading displacement and frequency equal to 20 mm and 0.5 Hz can be fitted as below:

$$F = 3.008v^{0.5700} \left(F/\text{kN},\ v/\text{mm·s}^{-1} \right) \tag{1}$$

Velocity index is 0.57, in the commonly used value range (0.3~1.95) [7]. The data of all test conditions with silt content of 30% are taken to verify the rationality of Equation (1). Part of the typical data in the first and third quadrants of the hysteretic curves is shown in Figure 11. The difference between the experimental results and the fitted value is within 15%, indicating that the model can accurately predict the damping force.

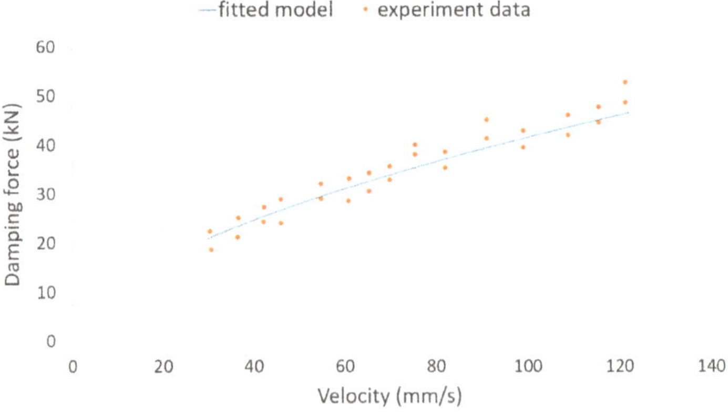

Figure 11. Fitted velocity-dependent model under large displacements.

5. Conclusions

This paper investigates a new type of viscous damper that adds silt to the damping medium to improve its energy dissipation performance. The energy dissipation mechanism of this damper is studied through a cyclic loading experiment. The following conclusions are drawn:

(1) Silt can considerably increase damping force by enhancing the viscosity of the damping medium.

(2) Silt can enrich the energy dissipation mechanism of the damper. Experimental results show that the damper is displacement-dependent with a plump hysteretic curve for small displacements. This displacement-dependent feature can improve the energy consumption ability of fluid viscous dampers for small displacements. For large displacements, the damper is still velocity-dependent.

(3) Silt content has a great effect on the maximum damping force. Unless displacement is very small (5 mm), increasing the content of silt can visibly increase the damping force. However, the relationship between the damping forces and the silt content is complex, and not always positive. Further studies engaging more test groups are required to determine the optimum mass ratio of silt.

(4) For small displacements, the displacement-dependent damping force can be predicted by a bilinear model. For large displacements, the damping force is velocity-dependent and can be reasonably fitted.

Author Contributions: Zheng Lu wrote the paper and directed the study. Junzuo Li analyzed the data and revised the paper. Chuanguo Jia conceived the idea, carried out the experiment, provided valuable discussions and revised the paper. He took responsibility of the corresponding work.

Acknowledgments: The authors gratefully acknowledge the financial supports from the National Natural Science Foundation of China (No. 51408080).

References

1. Takewaki, I.; Fujita, K.; Yamamoto, K.; Takabatake, H. Smart passive damper control for greater building earthquake resilience in sustainable cities. *Sustain. Cities Soc.* **2011**, *1*, 3–15. [CrossRef]

2. Housner, G.W.; Bergman, L.A.; Caughey, T.K.; Chassiakos, A.G.; Claus, R.O.; Masri, S.F.; Skelton, R.E.; Soong, T.T.; Spencer, B.F.; Yao, J.T. Structural Control: Past, Present, and Future. *J. Eng. Mech.* **1997**, *123*, 897–971. [CrossRef]

3. Lu, Z.; Wang, Z.; Masri, S.F.; Lu, X. Particle impact dampers: Past, present, and future. *Struct. Control Health Monit.* **2018**, *25*, e2058. [CrossRef]

4. Lu, X.; Liu, Z.; Lu, Z. Optimization design and experimental verification of track nonlinear energy sink for vibration control under seismic excitation. *Struct. Control Health Monit.* **2017**, *24*, e2033. [CrossRef]

5. Segel, L.; Lang, H.H. The Mechanics of Automotive Hydraulic Dampers at High Stroking Frequencies. *Veh. Syst. Dyn.* **1982**, *10*, 82–85. [CrossRef]

6. Makris, N. *Viscous Dampers: Testing Modeling and Application in Vibration and Seismic Isolation*; Technical Report NCEER-90-0028; State University at Buffalo: Buffalo, NY, USA, 1990.

7. Lee, D.; Taylor, D.P. Viscous damper development and future trends. *Struct. Des. Tall Spec. Build.* **2010**, *10*, 311–320. [CrossRef]

8. Hundal, M.S. Impact absorber with two-stage, variable area orifice hydraulic damper. *J. Sound Vib.* **1977**, *50*, 195–202. [CrossRef]

9. Dyke, S.J.; Spencer, B.F., Jr.; Sain, M.K.; Carlson, J.D. An experimental study of MR dampers for seismic protection. *Smart Mater. Struct.* **1998**, *7*, 693–703. [CrossRef]

10. Mcnamara, R.J.; Taylor, D.P. Fluid viscous dampers for high-rise buildings. *Struct. Des. Tall Spec. Build.* **2003**, *12*, 145–154. [CrossRef]

11. Lu, Z.; Wang, Z.X.; Zhou, Y.; Lu, X.L. Nonlinear dissipative devices in structural vibration control: A review. *J. Sound Vib.* **2018**, *423*, 18–49. [CrossRef]

12. Constantinou, M.C.; Symans, M.D. *Experimental and Analytical Investigation of Seismic Response of Structures with Supplemental Fluid Viscous Dampers*; Technical Report; National Center for Earthquake Engineering Research: Buffalo, NY, USA, 1992.

13. Makris, N.; Constantinou, M.C. Fractional-Derivative Maxwell Model for Viscous Dampers. *J. Struct. Eng.* **1991**, *117*, 2708–2724. [CrossRef]

14. Taylor, D.P.; Constantinou, M.C. Testing Procedures for High Output Fluid Viscous Dampers Used in Building and Bridge Structures to Dissipate Seismic Energy. *Shock Vib.* **1995**, *2*, 373–381. [CrossRef]
15. Uetani, K.; Tsuji, M.; Takewaki, I. Application of an optimum design method to practical building frames with viscous dampers and hysteretic dampers. *Eng. Struct.* **2003**, *25*, 579–592. [CrossRef]
16. López, I.; Busturia, J.M.; Nijmeijer, H. Energy dissipation of a friction damper. *J. Sound Vib.* **2004**, *278*, 539–561. [CrossRef]
17. Xu, Y.L.; Qu, W.L.; Chen, Z.H. Control of Wind-Excited Truss Tower Using Semiactive Friction Damper. *J. Struct. Eng.* **2001**, *127*, 861–868. [CrossRef]
18. Narkhede, D.I.; Sinha, R. Behavior of nonlinear fluid viscous dampers for control of shock vibrations. *J. Sound Vib.* **2014**, *333*, 80–98. [CrossRef]
19. Abbas, H. *A Methodology for Design of Viscoelastic Dampers in Earthquake-Resistant Structures*; Earthquake Engineering Research Center, University of California: Santa Barbara, CA, USA, 1993.
20. Berton, S.; Bolander, J.E. Amplification System for Supplemental Damping Devices in Seismic Applications. *J. Struct. Eng.* **2005**, *131*, 979–983. [CrossRef]
21. Ribakov, Y.; Reinhorn, A.M.; Asce, F. Design of amplified structural damping using optimal considerations. *J. Struct. Eng.* **2003**, *129*, 1422–1427. [CrossRef]
22. Dai, K.; Wang, J.; Mao, R.; Lu, Z.; Chen, S. Experimental investigation on dynamic characterization and seismic control performance of a TLPD system. *Struct. Des. Tall Spec. Build.* **2017**, *26*, e1350. [CrossRef]
23. Lu, Z.; Yang, Y.; Lu, X.; Liu, C. Preliminary Study on the Damping Effect of a Lateral Damping Buffer under a Debris Flow Load. *Appl. Sci.* **2017**, *7*, 201. [CrossRef]
24. Tsai, C.S.; Chen, K.C.; Chen, C.S. *Seismic Resistibility of High-Rise Buildings with Combined Velocity-Dependent and Velocity-Independent Devices*; ASME: New York, NY, USA, 1998; pp. 103–110.
25. Julius, M.; David, T.; Nimal, P. Influence of damping systems on building structures subject to seismic effects. *Eng. Struct.* **2004**, *26*, 1939–1956.
26. Lu, Z.; Lu, X.; Masri, S.F. Studies of the performance of particle dampers under dynamic loads. *J. Sound Vib.* **2010**, *329*, 5415–5433. [CrossRef]
27. Lee, J.; Kang, H.; Kim, J. Seismic performance of steel plate slit-friction hybrid dampers. *J. Constr. Steel Res.* **2017**, *136*, 128–139. [CrossRef]
28. Kim, J.; Shin, H. Seismic loss assessment of a structure retrofitted with slit-friction hybrid dampers. *Eng. Struct.* **2017**, *130*, 336–350. [CrossRef]
29. Lu, Z.; Huang, B.; Zhou, Y. Theoretical study and experimental validation on the energy dissipation mechanism of particle dampers. *Struct. Control Health Monit.* **2018**, *25*, e2125. [CrossRef]
30. Lu, Z.; Chen, X.Y.; Zhang, D.C.; Dai, K.S. Experimental and analytical study on the performance of particle tuned mass dampers under seismic excitation. *Earthq. Eng. Struct. Dyn.* **2017**, *46*, 697–714. [CrossRef]
31. Hardin, B.O.; Drnevich, V.P. Shear modulus and damping in soil: Measurement and parameter effects. *ASCE Soil Mech. Found. Div. J.* **1972**, *98*, 603–624.
32. Hardin, B.O.; Drnevich, V.P. Shear Modulus and Damping in Soils: Design Equations and Curves. *J. Soil Mech. Found. Div.* **1972**, *98*, 667–692.
33. Crouse, C.B.; Mcguire, J. Energy dissipation in soil-structure interaction. *Earthq. Spectra* **2001**, *17*, 235–259. [CrossRef]
34. Kaare, H.; Rune, D.; Sandbakken, G. Strength of Undisturbed versus Reconstituted Silt and Silty Sand Specimens. *J. Geotech. Geoenviron. Eng.* **2000**, *126*, 606–617.
35. Iravanai, S. Geotechnical Characteristics of Penticton Silt. Ph.D. Thesis, University of Alberta, Edmonton, AB, Canada, 1999.

 *sustainability*

Article

Characteristics of Corporate Contributions to the Recovery of Regional Society from the Great East Japan Earthquake Disaster

Rui Fukumoto [1,*] **, Yuji Genda** [2] **and Mikiko Ishikawa** [3]

[1] School of Engineering, Department of Urban Engineering, The University of Tokyo, 7-3-1 Hongo, Bunkyo-ku 113-8656, Japan

[2] Institute of Social Science, The University of Tokyo, 7-3-1 Hongo, Bunkyo-ku 113-0033, Japan; genda@iss.u-tokyo.ac.jp

[3] Faculty of Science and Engineering, Chuo University, 1-13-27 Kasuga, Bunkyo-ku 112-8551, Japan; ishikawa.27w@g.chuo-u.ac.jp

* Correspondence: wingbase@gmail.com

Received: 27 March 2018; Accepted: 20 May 2018; Published: 24 May 2018

check for
updates

Abstract: Municipalities in areas along the northeast coast of Japan were severely affected by the Great East Japan Earthquake. It was difficult for these municipalities to provide support to all devastated areas. It is important for communities in devastated areas to be resilient in order to autonomously and efficiently recover from natural disasters. This study focused on corporations, since they have various resources that can support disaster recovery. A postal questionnaire was sent to 1,020 corporations that included various industry types and small corporations located in Iwanuma and Natori, which were damaged by the Great East Japan Earthquake. The response rate was 39.22%. We analyzed the data using a logistic regression model. The study findings are as follows: (1) a total of 32.75% of corporations provided support for recovery after the disaster; (2) the ratio of corporations that provided actual support was lower than that of those that only had awareness of contributions; (3) the strongest characteristic was having not only awareness but also the opportunity to conduct support activities before the occurrence of disasters to enhance the efficient recovery of regional society; and (4) the characteristics of support differed according to industry type, location, and number of employees under certain conditions.

Keywords: corporation; resilience; disaster; recovery; Great East Japan Earthquake

---

1. Introduction

Following large-scale disasters such as the Great East Japan Earthquake, it is difficult for municipalities to provide support to all devastated areas. Communities in devastated areas need to have resilient routine disaster response structures in place during non-disaster times (hereinafter referred to as "ordinary times") in order to conduct efficient recovery, reconstruction, and revitalization. When considering such structures in devastated areas, it can be seen that corporations are one of the main actors that can support regional society by providing various resources such as human resources and financial support for recovery and disaster mitigation [1,2]. Regarding the efficient recovery, reconstruction, and revitalization of regional society after the Great East Japan Earthquake, many previous empirical studies have reported and focused on residents as the main actors [3–6]; however, only a few empirical studies have targeted corporations as disaster recovery actors. Most discussions about corporate activities in large-scale disasters have been focused on the recovery of the corporation itself or on business continuity planning (BCP) [7–9], corporate social responsibility (CSR) [1,10], capacity evaluation of corporations in disasters [11], or changes in employment or labor conditions [12]. Regarding corporate contributions to the recovery of regional societies after disasters, several studies focused on corporations' awareness of the "possibility of providing support" [13–17]. Previous empirical studies have identified "facts regarding corporate contributions" through the cases of the Hanshin–Awaji Earthquake disaster [18], the Niigata-ken Chuetsu-oki Earthquake [19], and Hurricane Katrina [1,19]. These studies targeted relatively large corporations that were listed on the Tokyo Stock Exchange, or were among the 1% Club of the Japan Business Federation or the most profitable corporations in the United States [1]; small and local corporations were not included. There is remarkably little research on disaster support by local corporations located in devastated areas right after disasters. Regarding corporations' contributions to disaster recovery, one study clarified the actual support to local communities provided by corporations located in devastated areas [20], although it only focused on local constructors after the Hanshin–Awaji Earthquake disaster.

While there is research related to the BCPs of local restaurants in devastated areas [9], there are few studies regarding the role of small corporations after the Great East Japan Earthquake. Through the abovementioned previous studies, it is clear that with regard to corporate contributions to recovery, reconstruction and revitalization after the Great East Japan Earthquake, no study has targeted corporations across various industry types, including small corporations whose headquarters are located in devastated areas. Moreover, no study has examined the gap between contributing awareness and the actual support provided by corporations. It is important to identify the characteristics of corporations that contribute to the recovery after disasters in order to formulate a regional disaster prevention plan and determine the characteristics of a regional society that can efficiently recover after disasters. Therefore, this study aims to understand the actual conditions of corporate contributions to the recovery of regional society after the Great East Japan Earthquake, and to identify the characteristics of the corporations that provided support.

2. Literature Review

In order to consider corporate contributions to recovery plans or local disaster prevention planning, it is necessary to identify the following: (1) the number of corporations that provided some support for the recovery of regional society; (2) the characteristics of those corporations; (3) the geographical spread or agglomeration of the corporations; and (4) the possibility of providing support by corporations in different situations. In this study, we focused on points 1 and 2 above as basic information for planning purposes. In this section, we review the current research in order to clarify which corporation variables are important to the planning process.

There are many basic theories related to corporations, focusing on economy [21], contract [22], behavior [23], evolution [24], and growth [25]. In this study, we focused on the theory of corporate behavior, since providing support is one of the behaviors of corporations. Since corporate behavior traditionally focuses on profit maximization, in our research, in order to understand this change, we focused on the corporate decision maker as a representative of the corporation. This approach is supported by Cyert and March [23].

Previous studies have shown a relationship between size, profitability, industry type, shareholder relation, financial factors and ownership structure among UK corporations and charitable donations [26–28].

Regarding corporate philanthropic disaster responses, relationships were found between response behavior and corporate size, profitability, geography, available cash, and leverage after the Sichuan earthquake in China in 2008 [29]. Murosaki and Iwami [18] examined the contents and triggers of support provided by 378 corporations after the Hanshin–Awaji Earthquake in Japan in 1995. Ohnishi [20] examined support contents and triggers, as well as corporation size, location, community relationship, and other local activities of 218 local construction corporations or carpenters after the Hanshin–Awaji Earthquake to identify the role of corporations in disasters. Toyota and Shoji [19] examined the provided support and industry type in order to identify the quantitative trends of the corporate contributions of 610 corporations during the Niigata-ken Chuetsu Earthquake in Japan in 2004, and of 345 corporations during Hurricane Katrina in the US in 2005.

We summarized the corporate variables in Table 1 to better organize the current research. We selected size, industry type, financial situation, community relationship, geographical location, and other local activities as important variables to illustrate recovery or local disaster prevention planning. In the next section, we examine actual conditions using results from an interview survey to formulate hypotheses.

Table 1. Summary of corporate variables.

Target	N	Corporate Variables	Author
Listed corporation (UK)	100	Leverage, Size, Profitability, and Ownership structure	Adams and Hardwick (1998)
Listed corporation (UK)	400	Size, Industry type, and Location	Brammer and Millington (2004)
Large corporation (UK)	550	Industry type, Size, Shareholder relations, and Financial factors	Brammer and Millington (2005)
Listed corporation (CN)	686	Corporate size, Profitability, Geography, Cash resource available, and Leverage	Zhang et al. (2009)
Fortune 500 or Listed corporation (US)	442	Industry type, Size, Profitability, Advertising intensity, R&D intensity, Leverage, and Local presence	Muller and Kräussl (2010)
Listed or Insurance corporation (JP)	378	Damage from disaster, Contents of provided supports, and Future earthquake countermeasures	Murosaki and Iwami (1995)
Local construction corporations or Carpenters (JP)	218	Size, Location, Contents of provided supports, Trigger of supports, Community relations, and Local activities	Ohnishi (1997)
1% Club of Japan Business Federation (JP) Corporations listed on National Donor Responses to Hurricane Katrina (US)	610 345	Contents of provided supports, Industry type	Toyota and Shoji (2007)

3. Case Study

This study is based on the assumption that corporations located in devastated areas provided support for the recovery of the regional society after the Great East Japan Earthquake. The interview survey conducted first aimed to understand the context of corporate contributions to recovery; after the survey was conducted, hypotheses were formulated to explain the trends.

3.1. Interview Survey

3.1.1. Outline of the Interview Survey

This interview survey was conducted with the presidents or directors of the disaster prevention departments of corporations located in Iwanuma, which was selected as the study area. This area has been frequently used to interpret the resilience of regional society [3–5] and social capital or health [30–35]. The implementation period was from April 2012 to December 2014. The interviews were conducted to gain an understanding of the actual situation of the support provided by the corporations. From the survey results, we formulated four hypotheses to identify the characteristics of corporations' disaster recovery support. With assistance from the Iwanuma City Society of Commerce and Industry, we selected six corporations located in devastated areas and six corporations from other areas. The survey method involved semi-structured interviews with questionnaires. An outline of the survey is shown in Table 2. We interviewed the corporations regarding disaster incidents, their thoughts, their actions, and the reasons for their actions from the time of their establishment to the current period, and organized the data in chronological order. Based on this context, we also interviewed the participants regarding the details of the provided support and the trigger of the support from right after the disaster until December 2011.

Table 2. Outline of the interview survey.

Targets	12 corporations located in Iwanuma
Interviewees	Presidents or directors of disaster prevention departments of each corporation
Implementation period	From April 2012 to December 2014
Semi-structural interviews (with questionnaires)	Part 1: History of the corporation and the context of the interviewee (with a questionnaire including fundamental characteristics of the corporations such as industry type, number of employees, and annual sales) Part 2: Business continuity (with a questionnaire including damage caused by the disaster, such as damage amount and content of damages) Part 3: Support for recovery of the regional society (content of the support, support recipients, and trigger for support)

3.1.2. Ethics

Since the Great East Japan Earthquake, the authors have taken part in support activities and planning for the recovery and revitalization of devastated areas, and formed relationships with local residents, presidents of local corporations, and local public officers. One of the authors has fundamental knowledge of corporate management acquired from seven years of experience managing his own corporations; this author conducted the interview survey, which was carefully administered to each corporation several times and across several years starting in April 2012. We never forced the interviewees to share their story regarding their experience of anxiety and fear related to the disaster. We also created a chronology of the data collected from the corporations based on the survey results and gave them feedback after the interview. We obtained their consent to publish the relevant parts of this research for academic purposes without revealing the names and year of establishment of each corporation.

3.2. Understanding the Situation of Corporate Contributions to the Recovery of Regional Society

Table 3 shows a summary of corporate contributions to the recovery of regional society based on the interview survey results. Among the 12 corporations, five provided support for the recovery of regional society.

Table 3. Summary of the interview survey.

Items	Target Corporations											
	A	B	C	D	E	F	G	H	I	J	K	L
Industry type	CONS	MANU	MANU	CONS	ELEC	AGRI	MANU	CONS	ELEC	MINI	CONS	CONS
Number of employees	70	240	98	5	7	7	15	30	16	26	4	6
Awareness	HAA	HAN	HAN	HAA	HAA	HAA	HAN	HAN	HAN	HAA	NC	HAA
Distance from coastline	2.68	2.24	1.58	2.18	2.29	2.49	5.89	6.94	7.37	10.98	6.90	8.03
Damage/sales	4.00%	8.10%	26.00%	27.50%	40.60%	300.00%	2.50%	2.50%	10.70%	0.40%	4.90%	4.20%
Damage from tsunami	+	+	+	+	+	+	-	-	-	-	-	-
Some support	+	-	-	+	+	+	-	-	-	-	-	+
Funding support	-	-	-	-	-	-	-	-	-	-	-	-
Daily commodities support	+	-	-	-	-	-	-	-	-	-	-	+
Equipment and materials support	+	-	-	+	+	+	-	-	-	-	-	+
Human resources support	+	-	-	+	+	+	-	-	-	-	-	+
Information support	+	-	-	+	-	+	-	-	-	-	-	+
Temporary use of lands and buildings support	+	-	-	-	-	-	-	-	-	-	-	-

Abbreviations: CONS: Construction, MANU: Manufacturing, ACCO: Accommodation, Eating, and Drinking Services, MINI: Mining and Quarrying of Stone, ELEC: Electricity, Gas, Heat, and Water Supply, AGRI: Agriculture, Forestry, and Fisheries, REAL: Real Estate and Goods Rental and Leasing, HAA: Having an awareness of contributions to the recovery of regional society, and having already conducted related activities before the occurrence of the disaster, HAN: Having an awareness of contributions to the recovery of regional society, but having conducted no related activities before the occurrence of the disaster, NC: Not having the capacity to contribute to the recovery of regional society before the disaster.

We found the following categories of support: "Daily commodities", "Equipment and materials", "Human resources", "Useful information", and "Temporary use of land and buildings".

In this study, "Some support" was defined as the provision of one or more of the following: "Funding", "Daily commodities", "Equipment and materials", "Human resources", "Useful information", and "Temporary use of land and buildings".

Table 4 shows the descriptions of support for recovery provided by the corporations.

Table 4. Descriptions of provided support based on the results of the interview survey.

Support Type	Provider	Destination	Description
Funding	None	None	This type of support was not found in the interview survey. However, corporation I provided heavy machines, fuel, and funds for corporation E in the same industry that provided support to the regional society.
Daily Commodities	A	Neighborhood Association	Provided materials that were stocked by them during ordinary times to the local evacuation center of the neighborhood association.
Daily Commodities	L	Elementary School	Provided materials that were collected by the corporation right after the disaster to elementary schools.
Equipment and Materials	A	Neighborhood Association	Removed mud and rubble piled up on roads using heavy machines for the neighborhood association. They could provide this support for only three days after the disaster. Starting on the fourth day after the disaster, they declined personal requests because they received official business orders from the local government and found it necessary to devote their efforts to these orders.
Equipment and Materials	D	Neighborhood Association	Provided construction tools and materials for removing mud and rubble. They could not provide heavy machines for construction because the machines were damaged by the tsunami.
Equipment and Materials	D	Mutual Fire Association	Provided construction tools to mutual fire association for finding missing persons.
Equipment and Materials	E	Neighborhood Association	Provided support in cooperation with other corporations and removed mud and rubble piled up on roads and near buildings using heavy machines. They borrowed heavy machines and fuel from unions in the same industry.
Equipment and Materials	E	Elementary School	Removed mud from elementary school grounds using heavy machines.
Equipment and Materials	F	Neighborhood Association	Provided construction tools and materials for removing mud and rubble. They could not provide heavy machines for construction because the machines were damaged by the tsunami.
Equipment and Materials	F	Kindergarten	Removed mud from kindergarten grounds using heavy machines.
Equipment and Materials	L	Neighborhood Association	Corporation L provided materials and tools for repairing the floors of houses located in areas devastated by the tsunami.
Equipment and Materials	L	Medical Worker	Provided paper and fuel to local medical workers for medical activities in devastated areas.
Human Resources	A	Neighborhood Association	Provided human resources to the neighborhood association for removing mud and rubble.
Human Resources	D	Neighborhood Association	Provided human resources to the neighborhood association to remove mud and rubble.
Human Resources	E	Neighborhood Association	Provided human resources to the neighborhood association to remove mud and rubble.
Human Resources	F	Kindergarten	Removed mud from kindergarten grounds using construction tools.
Human Resources	L	Neighborhood Association	Provided materials that were stocked by them during ordinary times to the local evacuation center of the neighborhood association.
Useful Information	A	Iwanuma City	Provided information regarding the condition of recovery in Iwanuma Rinku Industrial Park to Iwanuma City, the neighborhood association, and police officers.
Useful Information	A	Neighborhood Association	Provided information regarding the condition of recovery in Iwanuma Rinku Industrial Park to Iwanuma City, the neighborhood association, and police officers.
Useful Information	A	Police officer	Provided information regarding the condition of recovery in Iwanuma Rinku Industrial Park to Iwanuma City, the neighborhood association, and police officers.
Useful Information	D	Elementary School	Provided information regarding the condition of recovery in the regional society and information regarding the safety of elementary school children.
Useful Information	F	Kindergarten	Provided information regarding the condition of recovery in the regional society and land for the shifting of a kindergarten.
Useful Information	L	Medical Worker	Provided information regarding the condition of recovery in the regional society and about people with medical needs to medical workers.
Temporary use of lands and buildings	A	Mutual Fire Association	Provided permission for temporary use of lands and buildings for rescue activities.
Temporary use of lands and buildings	A	Police officer	Provided permission for temporary use of lands for activities for the safety of the regional society.

3.2.1. Daily Commodities Support

Based on the interview survey, it was found that two corporations provided support in the form of daily commodities. Corporation A provided water and food provisions, which were stocked by them during ordinary times, to the local evacuation center of the neighborhood association. Corporation L provided water and food provisions, blankets, and towels, which were collected by the corporation right after the disaster, to an elementary school. Both corporations had a relationship with their support recipients prior to the disaster. They said, "we saw a terrible scene that we had never seen before. We provided support to people affected by the disaster because we could not predict our corporation's future, such as when it would be possible to resume business".

3.2.2. Equipment and Materials Support

We found that five corporations provided support in the form of equipment and materials.

Corporation A helped restore a road in the nearby region, which was essential for residents, using their equipment for three days after the disaster. Starting on the fourth day, they could not conduct this support activity because the corporation received a formal business request from municipalities regarding their support, so they could not devote time to informal support.

Corporation E cooperated with corporations D and F and removed the mud and rubble piled up on roads and around buildings using heavy machines. Through their connections with unions in the same industry, they borrowed heavy machines, fuel, and tools for construction. They had participated in activities of the residential association during ordinary times, so the land to dump the removed mud and rubble was provided by the board members of the neighborhood association. Corporation E said that "it was necessary to restore roads in the regional area in order to continue our business and to receive recovery support from the municipality and government". Corporations D and F said that they could not think about anything else apart from providing support because the disaster happened in their home town. They worked to solve the problem by cooperating with other corporations and procuring necessary resources together.

Corporation L provided materials and tools for the repair of the floors of houses located in areas devastated by the tsunami. The trigger for providing the support was that they received a request for construction materials from a builder who lived in an area that was completely destroyed by the tsunami. Corporation L also provided paper and fuel to doctors, nurses, pharmacies, and clinical psychologists who were engaged in medical treatment in devastated areas. The trigger for providing this support was a referral from the staff of the Social Welfare Council.

3.2.3. Human Resources Support

We found that five corporations provided support in the form of human resources. Corporations A, D, and E provided resources to the neighborhood association to remove mud and rubble. Corporation E also provided resources to a kindergarten to remove mud and rubble using construction tools. Corporation L provided resources to the neighborhood association to repair the floors of houses damaged by the tsunami.

3.2.4. Useful Information Support

Four corporations provided support in the form of useful information.

Corporation A provided local information (conditions of damage, operation status, and employee status of Iwanuma Rinku Industrial Park) to residents and supporters who assisted in the rescue work from the outside cities. The trigger for this support was a request from the residents after the disaster.

Corporation D provided information regarding local conditions and about the safety of children who lived in the region to an elementary school. The trigger of this support was that the elementary school was one of the local evacuation centers.

Corporation F provided information regarding land for the possible shifting of the local kindergarten. This corporation already had a relationship with the staff of the kindergarten and received a request to provide information after the disaster. Subsequently, Corporation F collected information from some real estate corporations, and gave the kindergarten staff this necessary information.

Corporation L provided information on local conditions and about people with medical needs to medical workers. The trigger of this support was that they were introduced to these workers by the Social Welfare Council.

3.2.5. Temporary Use of Land and Buildings Support

One corporation provided support in the form of granting permission for the temporary use of land and buildings.

Corporation A provided permission for the temporary use of land and buildings to a mutual fire association and police officers for their activities.

3.2.6. Funding Support

None of the corporations interviewed in this survey provided funding support. However, Corporation I provided heavy machines, fuel, and funds for Corporation E in the same industry, which in turn provided support for the recovery of regional society. In general, as seen with the Japan Platform [36] or the 1% Club of the Japan business federation [37], there were many cases where corporations provided funding support as donations. Therefore, it is possible that corporations that were located in the devastated areas also provided funding support to regional society.

3.3. Understanding the Characteristics of the Corporations That Provided Support for the Recovery of Regional Society

Based on the results of the interview survey and literature review, the characteristics of the corporations that provided recovery support were extracted in order to formulate hypotheses.

3.3.1. Damages Caused by the Tsunami

Based on the literature review, we decided that the most important corporation variables were financial situation and geographical location. In this study, "damages caused by the tsunami" was a comprehensive variable. Four of the five corporations that provided support for the recovery of regional society were damaged by the tsunami, which caused great destruction. Nonetheless, corporations damaged by the 2011 tsunami provided support to regional society. Therefore, it is possible that one characteristic that influences corporations to provide recovery support is damage caused by the tsunami.

3.3.2. Number of Employees

Based on the literature review, we decided that size was also an important variable. In this study, "Number of employees" was adopted as the indicator to measure size. The survey results indicated that four of the five corporations that provided recovery support had a small number of employees. In cases where the entire region suffered from the tsunami, it was necessary to restore roads in cooperation with residents. For this reason, small-scale corporations tend to provide recovery support, and there is a possibility that the number of employees in a corporation is one characteristic that leads to recovery support.

3.3.3. Awareness of Corporations to Contribute to the Recovery of Regional Society

Based on the literature review, we selected community relations and local activities as important variables. In this study, "Awareness of corporations to contribute to the recovery of regional society" was adopted as an indicator to describe community relations and local activities. In general, corporations that would like to contribute to the recovery of regional society after disasters can easily provide that support; five of the eleven corporations that wanted to contribute provided recovery support. In addition, five of the seven corporations that had an awareness of contributions and conducted practical activities provided recovery support. Thus, it is possible that awareness and the carrying out of related activities during ordinary times is correlated with recovery support.

3.3.4. Industry Type

We found five industry types that provided support for the recovery of regional society. After the Great Hanshin–Awaji Earthquake, local construction corporations provided recovery support in the form of "Rescue from collapsed houses", "Providing equipment and materials", and offering "Human resources" [20]. In addition, a survey conducted on the participation and contribution of corporations in Sendai, Ota Ward, Shizuoka, Nagasaki, and Kobe reported different results for the support provided by each industry type [13]. Therefore, there is a possibility that a corporation's industry type is a relevant characteristic.

3.4. Formulating Hypotheses

Based on the characteristics discussed in the previous section, the following hypotheses were formulated. Hypothesis 1 was formulated based on "damages caused by the tsunami".

Hypothesis 1. *No significant difference in damage exists between corporations responding and those not responding to the disaster.*

Most corporations examined in this research are located on alluvial flats. Therefore, it is possible to estimate the damage they suffered based on the distance between their location and the nearest coastal line. To examine this hypothesis, it was necessary to investigate the differences between the recovery support provided by corporations in devastated areas and that provided by those in other areas.

Hypothesis 2 was formulated based on "number of employees".

Hypothesis 2. *No significant difference exists in the number of employees between corporations responding and those not responding to the disaster.*

For this, it was necessary to investigate the differences between small-scale, medium-scale, and large-scale corporations in terms of their contributions to recovery.

Hypothesis 3 was formulated based on "awareness of contributions to the recovery of regional society".

Hypothesis 3. *Corporations that have an awareness of contributions to recovery support and had already conducted related activities before the disaster are more likely to provide support for the recovery of regional society than are other corporations.*

To test this hypothesis, it was necessary to investigate the differences in recovery support provided by different corporations, corporations' awareness of contributions to recovery and related activities conducted before the disaster, and their capacities to contribute to recovery.

Hypothesis 4 was formulated based on "corporations' industry type".

Hypothesis 4. *The content of support differs according to industry types.*

For this, it was necessary to investigate the differences in recovery support between corporations with different industry types or based on the location of their main customers.

4. Research Design and Method

This research involved understanding the circumstances of support provided by corporations for the recovery of regional society after disasters, and identifying the specific characteristics of such corporations that provide recovery support.

4.1. Study Area

This study selected Iwanuma as the study area. In addition, Natori was included as a study area because Sendai Airport is located near the boundary between Natori and Iwanuma, and there was a large industrial zone near the Sendai Airport.

The industrial zone of the Iwanuma Rinku Industrial Park was adequate for this study to investigate the diverse characteristics of corporations, including small-scale and large-scale corporations in terms of the number of employees. Figure 1 shows the location of corporations on the land use map of the study site. Table 5 shows an overview of the damages from the Great East Japan Earthquake [38,39].

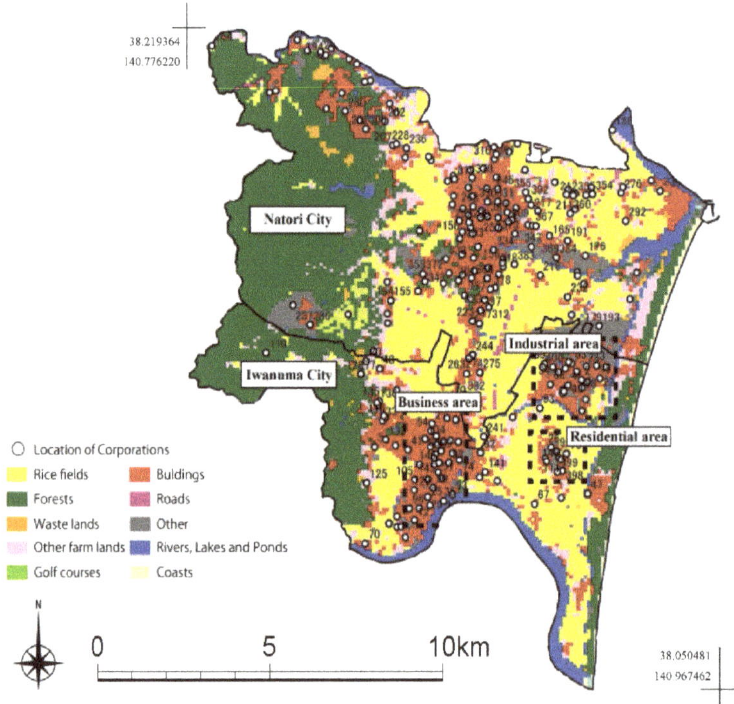

Figure 1. Location of corporations on the land use map in Iwanuma and Natori.

Table 5. Overview of damages from the disaster.

Item	Iwanuma City	Natori City
Dead or missing	187 people	923 people
Completely destroyed	736 buildings	2801 buildings
Large-scale partial destruction	509 buildings	219 buildings
Partial destruction	1097 buildings	910 buildings
Partial damage	3086 buildings	10,061 buildings

4.2. Data Collection

4.2.1. Questionnaire Survey

This survey collected data in order to understand the actual situation of recovery support and to test the hypotheses discussed in the previous section. A mail survey was administered to

1020 corporations from January to August 2015, whose headquarters were located in Iwanuma or Natori. Table 6 shows the outline of this survey. These corporations were chosen from a corporate database of Teikoku Databank, Ltd.; those categorized under "political, economic, and cultural group" and "local government" were not included in the questionnaire survey. The database is a corporate database of 1020 corporations corresponding to 36.82% of all 2941 corporations in the target area.

Table 6. Outline of the questionnaire survey.

Targets	1020 corporations located in Iwanuma or Natori
Survey method	Mail survey questionnaire
Implementation period	From January to August 2015
Response rate	39.22%

4.2.2. Description of the Questionnaire

The questionnaire items were categorized as shown in Table 7.

Table 7. Description of the questionnaire.

Categories	Questionnaire Items	Factors
Provided Support	Support recipients in regional society	"Neighborhood Association", "School (elementary, junior high, high), kindergarten, etc"., "Hospital/Medical worker", "Police station/Police officer", "Fire Station/Mutual fire association", "Iwanuma/Natori"
	Content of support	"Funding", "Daily commodities such as water and food", "Equipment and materials", "Human resources", "Useful information", "Temporary use of lands and buildings"
	Relationship between the corporation and support recipients	"Relationship before the disaster"
		"No relationship before the disaster and contact made only after the disaster"
		"No relationship before the disaster and were introduced by others after the disaster"
		"Other"
	Contents of the relationship	"Have not conducted any particular activities"
		"Have done some business"
		"Have done some community activities"
		"Have done some exchange information"
		"Have done other activities"
Damages caused by the disaster	Approximate value of the damages up to September 2011 from the occurrence of the disaster	Numerical answer
	Details of the damage	"Equipment damage", "Inventory damage", "Surging personnel costs", "Surging outsourcing costs", "Decrease in customers", "Decrease in sales", "Surging purchases", "Increase in borrowing", "Other".
	Shutdown period	Numerical answer
Awareness	Awareness of corporations to contribute to the recovery of regional society	"Having awareness of contributions to regional society, and having already conducted related activities"
		"Having awareness of contributions to regional society, but having conducted no related activities earlier"
		"Not having the capacity to contribute"
		"Having never thought about making contributions"
Fundamental characteristics of corporations	Industry type	"Medical, health care, and welfare", "Transport and postal services", "Wholesale and retail trade", "Scientific research, professional, and technical services", "Education and learning support", "Construction", "Finance and insurance", "Mining and quarrying of stone", "Accommodation, eating, and drinking services", "Information and communications", "Living-related and personal services and amusement services", "Manufacturing", "Electricity, gas, heat, and water supply", "Agriculture, forestry, and fisheries", "Real estate and goods rental and leasing", "Miscellaneous business services, n.e.c".
	Number of employees	Numerical answer
	Annual sales	Numerical answer
	Business contents	Descriptive answer
	Key business partner	Descriptive answer
	Location	Descriptive answer
	Contact information	Descriptive answer

Provided Support for the Recovery of Regional Society after the Disaster

In order to understand the kind of support provided by the target corporations after the Great East Japan Earthquake, we asked questions about support recipients, the content of support, and the relationship between the corporation and support recipients before the disaster. "Support recipients" included "individuals", "corporations of the same industry", "corporations of other industries", "neighborhood association", "school (elementary, junior high, high), kindergarten, etc"., "hospital/medical worker", "police station/police officers", "fire station/mutual fire association", "NPO/voluntary organization", "Social Welfare Council/public interest group", "university/research institution", "economic organizations such as Chamber of Commerce", "enterprise association and exchange organization", "Iwanuma or Natori", "Miyagi prefecture", and "government".

Among these recipients, "neighborhood association", "school (elementary, junior high, high), kindergarten, etc"., "hospital/medical worker", "police station/police officer", "fire station/mutual fire association", and "Iwanuma or Natori" were defined as the regional society for this research. The content of the support was divided into "funding support", "everyday commodities such as water and food", "equipment and material support", "human resources", "useful information", and "temporary use of land and buildings".

Regarding the relationship between the corporation and support recipients before the disaster, the questionnaire items included "a relationship before the disaster", "no relationship before the disaster and contact made only after the disaster", "no relationship before the disaster and were introduced by others after the disaster", and "other". For the contents of the relationship, the questionnaire items also included "have not conducted any particular activities", "have done some business", "have done some community activities", "have done some exchange information" and "have done other activities".

Damages Caused by the Tsunami

Regarding the damages suffered by the corporations due to the disaster, "approximate value of the damages up to September 2011 from the occurrence of the disaster" and "details of the damages" were investigated. The factors included in this questionnaire item were "equipment damage", "inventory damage", "surging personnel costs", "surging outsourcing costs", "decrease in customers", "decrease in sales", "surging purchases", "increase in borrowing", and "other", with the option to select multiple factors. In addition, changes in "sales" and "employees" based on the latest accounting period and the accounting period before the disaster were investigated. The existence of a "shutdown period" was also investigated.

Awareness of Contributions and Support Activities for the Recovery of Regional Society

In order to determine target corporations' awareness of contributions before the disaster, we included and investigated the following factors in this questionnaire item: "having awareness of contributions to regional society, and having already conducted related activities before the disaster occurred", "having awareness of contributions to regional society, but having conducted no related activities before the disaster occurred", "not having the capacity to contribute", and "having never thought about making contributions".

Fundamental Characteristics of Corporations

As fundamental characteristics before and after the disaster, "number of employees", "annual sales", "industry type", "business contents", "major business partner", "location", and "contact information" were investigated.

4.2.3. Information on the Locational Attributes of the Corporations

Regarding the locational attributes of the corporations, "distance between the corporation and the coastline" and "land use of the region around the corporation" were calculated based on the location information of the corporation. For calculating the "distance between the corporation and the coastline", the National Land Numeral Information—Coastline data (FY 2006) [40] was used. Regarding information of land use, using land use subdivision mesh data (FY 2009) [40], a circular buffer with a radius of 500 m centered on the location information of the corporation was created, and the land use ratio in the buffer was calculated.

4.2.4. Ethics

We obtained the consent of the survey participants in advance through an informed consent statement [41] that included the following: (1) a brief description of the purpose and procedure of the research, including the expected duration of the study; (2) a statement regarding discomfort associated with participation by being reminded of the terrible experiences of the disaster; (3) a guarantee of the anonymity and confidentiality of the collected data; (4) the identity of the corresponding researcher and the location of information about participants' rights or questions regarding the study; (5) a statement that participation is completely voluntary and can be terminated at any time without penalty; (6) a statement describing alternative procedures that may be used; (7) a statement regarding providing a ballpoint pen as compensation to participants; and (8) an offer to provide a summary of the findings.

4.3. Logistic Regression Model

The analysis was carried out using a logistic regression model to extract explanatory variables with a strong explanatory power for response variables. There were two values of response variables in this study (provided support or did not provide support), translated into the logit as $-\infty$ $+\infty$ values, which were adopted to comprehend their relationship with the explanatory variables. In general, the logistic regression model can be expressed using the following formula.

$$\log\left(\frac{p(x)}{1 - p(x)}\right) = b_0 + b_1 x_1 + b_2 x_2 + ... + b_p x_p \qquad (1)$$

The provability of the provision of each support by corporation is $p(x)$ and the provability of the non-provision of each support by corporation is $1 - p(x)$ in this study. The fixed value (constant) is b0, the regression coefficient of the xp is bp, and each explanatory variable is presented as x.

In the selection of explanatory variables, multiple sets of data largely influential on the response variables were selected as explanatory variables from the pre-processing data other than the response variables, using the stepwise method. The variance inflation factor (VIF) was also calculated to determine multicollinearity in each analysis. The results of the VIF are shown as notes in corresponding Tables. The sets of explanatory variables in each model were optimized by AIC (Akaike's information criterion) to start with the full set of explanatory variables. In addition, the Wald confidence interval was used to investigate whether 0 is or is not included in the reliable section. The statistical analysis software R (version 3.3.2) was used for the analysis.

4.3.1. Response Variables

The response variables were the presence or absence of contributions for recovery of regional society and included the following: "some support", "funding", "daily commodities such as water and food", "equipment and materials", "human resources", "useful information", and "temporary use of lands and buildings". In this research, we analyzed seven cases that had sets of each response variable and each explanatory variable.

4.3.2. Explanatory Variables

The explanatory variables in the analysis were "number of employees (dummy)", "the ratio of damage amount to annual sales [%]", "industry type (dummy)", "location of main customers (dummy)", "awareness of contributions to the recovery of regional society before the disaster (dummy)", "damages caused by the tsunami based on distance from coastal line (dummy)", and "ratio of the surrounding land use [%]".

5. Results

5.1. Results of the Questionnaire Survey

The results of the questionnaire survey are shown in Table 8. The questionnaires were mailed to 1020 corporations located in Iwanuma or Natori. The response rate was 39.22% (400 corporations) by 31 August 2015, which was better than initially assumed. This was possibly due to two reasons. First, most corporations had already recovered from damages because four years had passed since the disaster. Hence, they were able to respond to the mail. Second, since the disaster, the authors had been engaged in revitalization activities and support planning in the devastated area, and they had thus formed a relationship with the residents.

Table 8. Descriptive statistics based on the results of the questionnaire survey.

Variable	N	Mean	Median	SD	Min	Max
Response variable						
Provided some support for the recovery of regional society (dummy)	131	-	-	-	0	1
Provided funding support (dummy)	18	-	-	-	0	1
Provided daily commodities support (dummy)	51	-	-	-	0	1
Provided equipment and materials support (dummy)	60	-	-	-	0	1
Provided human resources support (dummy)	78	-	-	-	0	1
Provided useful information support (dummy)	20	-	-	-	0	1
Provided temporary use of lands or buildings support (dummy)	15	-	-	-	0	1
Explanatory variable						
Number of employees	400	32.2	7	169.3	0	3049
0 (dummy)	16	-	-	-	0	1
1–7 (dummy)	304	-	-	-	0	1
8–20 (dummy)						
21–100 (dummy)	64	-	-	-	0	1
>101 (dummy)	16	-	-	-	0	1
The ratio of damage amount to annual sales [%]	337	36.8	29.4	149.1	0.0	1500.0
Industry type						
Medical, Health Care, and Welfare (dummy)	29	-		-	0	1
Transport and Postal Services (dummy)	24	-		-	0	1
Wholesale and Retail Trade (dummy)	78	-		-	0	1
Scientific Research, Professional, and Technical Services (dummy)	3	-		-	0	1
Education and Learning Support (dummy)	4	-		-	0	1
Construction (dummy)	120	-		-	0	1
Mining and Quarrying of Stone (dummy)	1	-		-	0	1
Accommodation, Eating, and Drinking Services (dummy)	8	-		-	0	1
Information and Communications (dummy)	4	-		-	0	1
Living-related and Personal Services and Amusement Services (dummy)	11	-		-	0	1
Manufacturing (dummy)	42	-		-	0	1
Electricity, Gas, Heat, and Water Supply (dummy)	10	-		-	0	1
Agriculture, Forestry, and Fisheries (dummy)	2	-		-	0	1
Real Estate and Goods Rental and Leasing (dummy)	20	-		-	0	1
Miscellaneous Business Services, n.e.c. (dummy)	44	-		-	0	1
Location of main customers						
Inside the same city limits (dummy)	157	-		-	0	1
Outside the city limits and inside Miyagi Pref. limits (dummy)	227	-		-	0	1
Outside Miyagi Pref. limits (dummy)	127	-		-	0	1
Awareness						
Consideration toward the recovery of regional society before the disaster						
Having awareness of contributions to the recovery of regional society, and having already conducted related activities (dummy)	117	-	-	-	0	1
Having awareness of contributions to the recovery of regional society, but having conducted no related activities earlier (dummy)	112	-	-	-	0	1
Not having the capacity to contribute (dummy)	139	-	-	-	0	1
Having never thought about making contributions (dummy)	9	-	-	-	0	1
Geographical location						
Distance from coastal line [km]	389	7.40	7.31	3.24	1.12	17.45
Districts without tsunami damage far from the devastated area (dummy)	42	-	-	-	0	1
Districts without tsunami damage near the devastated area (dummy)	261	-	-	-	0	1
Partially destroyed or flooded area (dummy)	44	-	-	-	0	1
Large-scale partially destroyed area (dummy)	37	-	-	-	0	1
Completely destroyed area (dummy)	5	-	-	-	0	1

Table 8. *Cont.*

Variable	N	Mean	Median	SD	Min	Max
Surrounding land use						
Rice field area [%]	389	19.12	-	25.82	0	100
Other farm area [%]	389	5.36	-	11.07	0	100
Forest area [%]	389	4.87	-	12.51	0	86.27
Waste land area [%]	389	0.48	-	3.01	0	33.96
Buildings area [%]	389	34.83	-	34.66	0	100
Roads area [%]	389	0.57	-	1.97	0	18.75
Golf area [%]	389	0.20	-	3.13	0	61.54
Rivers, lakes, and ponds area [%]	389	3.10	-	11.02	0	50

5.1.1. Industry Types and Response Rates

Regarding the number of respondents by industry type, the number exceeded 40 (10%) for "Construction" (120 corporations), "Wholesale and Retail Trade" (78 corporations), "Manufacturing" (42 corporations), and "Miscellaneous Business Services, n.e.c". (40 corporations).

Figure 2 shows the response rates for each industry type of the 1123 corporations, including 103 corporations that were closed down between March 2011 and December 2014. The response rates for most industry types were over 30.0%. It should be noted that the response rate of "Finance and Insurance" was 0%.

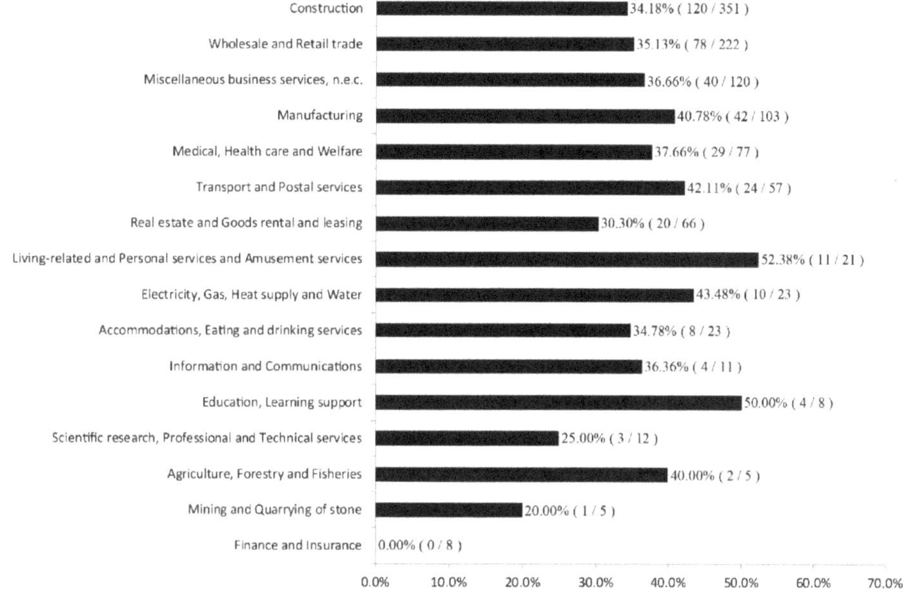

Figure 2. Response rate of each industry.

Figure 3 shows the location of the major business partners. The results were "within the same city limits" (39.50%), "outside the city and within Miyagi Pref. limits" (55.50%), and "outside of Miyagi Pref. limits" (29.75%).

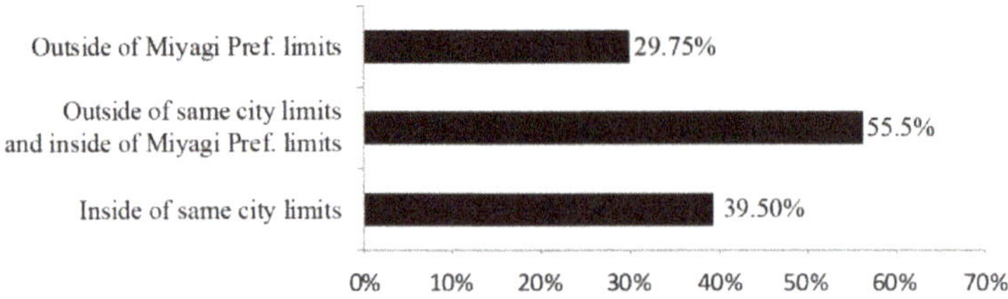

Figure 3. Location of the major business partners.

5.1.2. Number of Employees and Annual Sales

Figure 4 shows the classification of the number of employees. The average number of employees was 32.23, with a median of 7, maximum of 3049, and minimum of 0. Corporations with more than 100 employees are defined as large-scale corporations in Japan; those with more than 20 but fewer than 100 are defined as medium-scale corporations; and those with fewer than 20 employees are defined as small-scale corporations. There is no definition of "micro corporation" in Japan. In this study, corporations with seven or less employees were defined as micro corporations. Therefore, the target corporations in this survey were classified into five categories according to the number of employees: "the number of employees is 0" (4.00%), "the number of employees is more than 1 but fewer than 7, micro corporation" (46.50%), "the number of employees is more than 8 but fewer than 20, small corporation" (29.50%), "the number of employees is more than 21 but fewer than 100, medium corporation" (16.00%), and "more than 100 employees, large corporation" (4.00%).

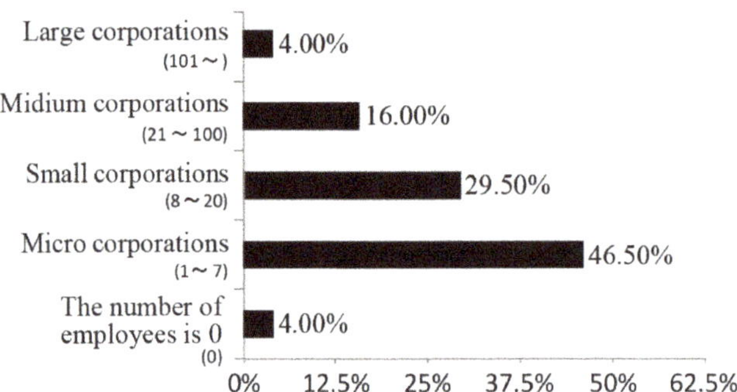

Figure 4. Classification of the number of employees.

Figure 5 shows the classification of annual sales. The average annual sales were 1613.47 million yen with a median of 120 million yen, maximum of 155,381 million yen, and minimum of 0.18 million yen. Considering annual sales, small corporations were included in the target corporations, since the sales of most corporations were less than 500 million JPY.

Based on the number of employees and annual sales, it was confirmed that there were micro, small, medium, and large corporations among the target corporations. The average number of employees in the corporations in Iwanuma and Natori based on the Economic Census for Business Activity in Japan (FY 2009) [42] was about 11, and among the target corporations in this survey, the proportion of small corporations was relatively higher, since the median in this survey was less than that in the census. According to the census, the ratio of large corporations in the target area was 1.10% and the ratio of medium corporations was 10.12%. The ratio of large corporations in our survey was 4.00% and the ratio of medium corporations was 16.00%. Therefore, the data of the survey are comparable to those of the target area.

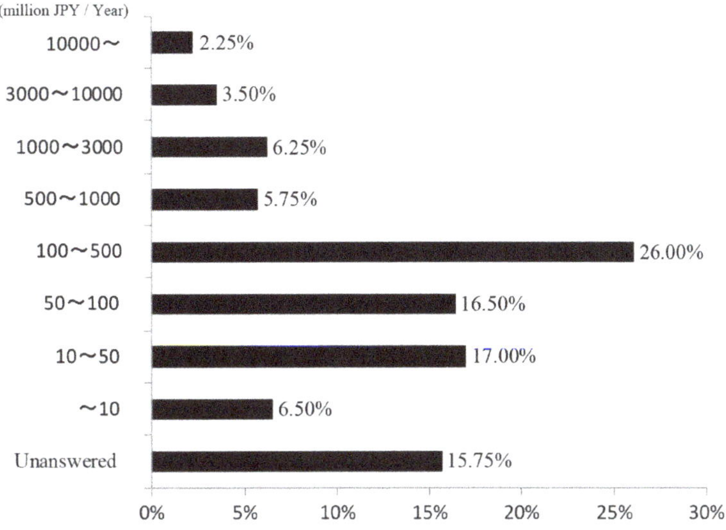

Figure 5. Annual sales of the target corporations.

5.1.3. Damage Caused by the Disaster

Figure 6a shows the classification of the ratio of damage amount to annual sales. The average was 36.83%, with a median of 29.38%, maximum of 1500.00%, and minimum of 0.00%. It was found that most corporations were damaged by the disaster, since the rate of corporations with no damage was only 6.25%.

Figure 6b shows the types of damages: "equipment damage" (59.25%), "decrease in sales" (45.75%), "decrease in customers" (29.75%), "inventory damage" (26.50%), "increase in borrowing" (16.00%), "surging purchases" (13.75%), "surging outsourcing costs" (13.25%), and "surging personnel costs (9.50%)". Further, 16.80% of corporations had to temporarily discontinue their business after the Great East Japan Earthquake.

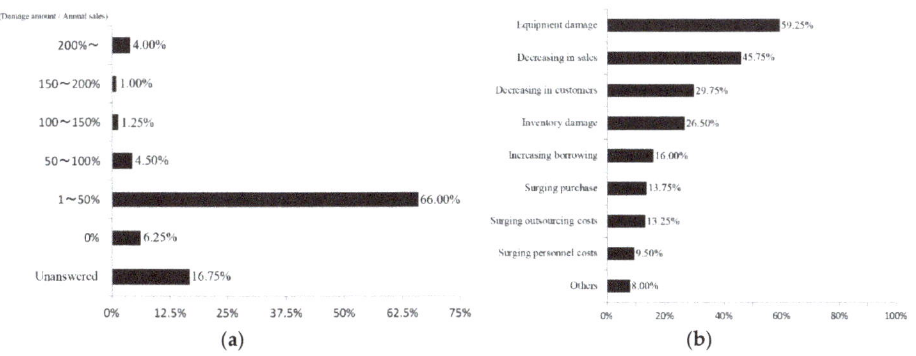

Figure 6. (**a**) Ratio of the damage amount to annual sales; (**b**) Proportion of the type of damages.

5.1.4. Awareness of Contributions

Figure 7 shows the corporations' awareness of contributions to the recovery of regional society before the disaster. The results were "having awareness of contributions, and having already conducted related activities before the disaster (HAA)" (29.25%), "having awareness of contributions, but having conducted no related activities before the disaster (HAN)" (28.00%), "not having the capacity to contribute before the disaster (NHC)" (35.25%), "having never thought about making contributions before the disaster (NTC)" (2.25%).

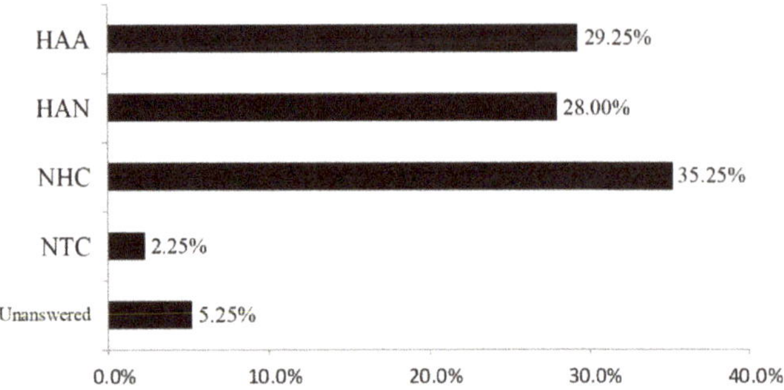

Figure 7. Corporations' awareness of contributions to the recovery of the regional society.

5.1.5. Location Attributes

The average distance from the coastline was 7.40 km, with a median of 7.31 km, a longest value of 17.45 km, and a shortest value of 1.12 km. Figure 8 shows the classification of damages caused by the tsunami based on the distance between the location of each corporation and the coastline. The results were as follows: the ratio of corporations located in the completely destroyed area (<1.50 km) was 1.25%, that of those located in the large-scale partially destroyed area (1.50 to 3.00 km) was 9.25%, that of those located in the partially destroyed or flooded areas (3.00 to 5.50 km) was 11.0%, that of those located in districts without tsunami damage near the devastated area (5.50 to 11.00 km) was 65.25%, that of those located in districts without tsunami damage far from the devastated area (>11.00 km) was 10.50%, and that of corporations with unknown locations was 2.75%. Many corporations were located in districts without tsunami damage near the devastated area, since Iwanuma station and Natori station are the main railway stations of both cities.

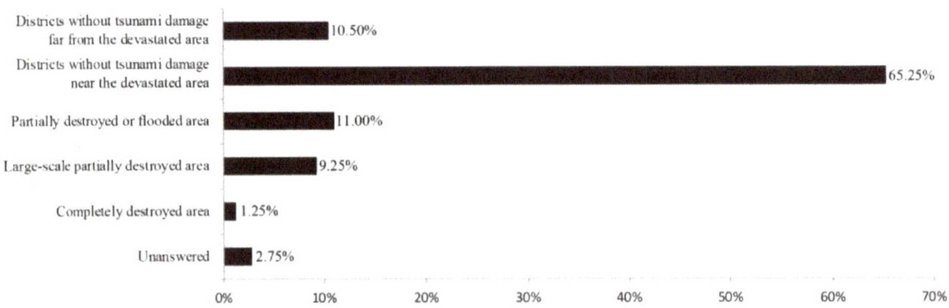

Figure 8. The classification of damages caused by the tsunami.

Regarding the land use ratio of the area surrounding the corporations, the results were as follows: "rice fields" (19.12%), "other agricultural lands" (5.36%), "forests" (4.87%), "wastelands" (0.48%), "buildings" (34.83%), "roads" (0.57%), "golf course area" (0.20%), and "river, lake, and pond area" (3.10%). As shown in Figure 1, many corporations were located in the commercial area near railway stations. In the tsunami-affected area, most corporations were located in the industrial park or the eastern residential area.

5.1.6. Provided Support for the Recovery of Regional Society

Figure 9 shows the ratio of corporations that made contributions. The corporations that provided some support to recipients represented 53.50% of the total. The ratio of corporations that provided support for the recovery of regional society was 32.75%.

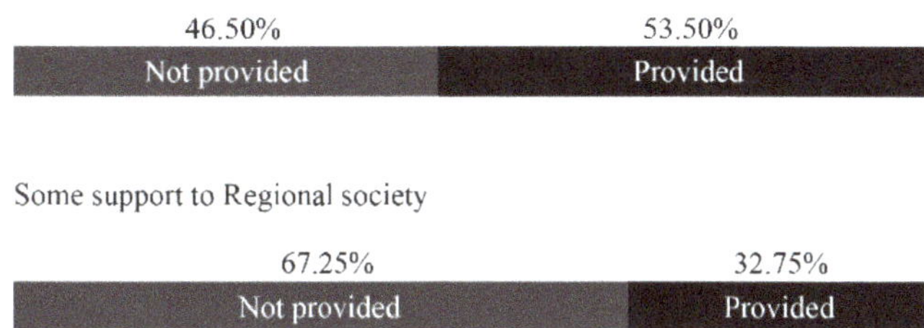

Figure 9. The ratio of corporations that provided some type of support.

Figure 10 shows the content of support provided for the recovery of regional society. The results are as follows: "human resources" (59.54%), "equipment and materials" (45.80%), "daily commodities such as water and food" (38.93%), "useful information" (15.27%), "funding support" (13.7%), and "temporary use of land and buildings" (11.45%).

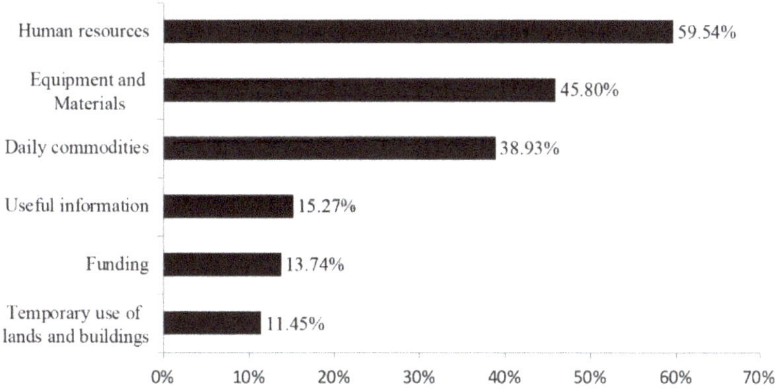

Figure 10. The content of the provided support.

Figure 11 shows the results of the questionnaire item regarding the relationship between the corporations and support recipients. The ratio of corporations that had a relationship with the support recipients before the disaster was 74.50%.

Figure 11. The relationship between the corporations and support recipients.

Figure 12 shows the results of the questionnaire item regarding the contents of the relationships. The ratio of "have not conducted any particular activities" was 50.40%.

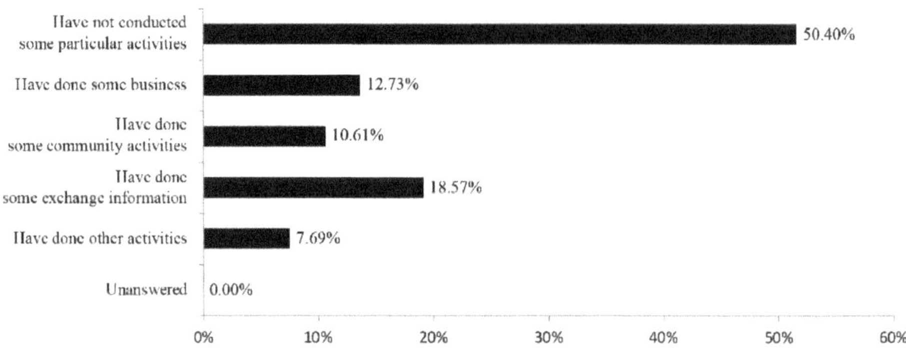

Figure 12. Contents of the relationships.

Considering the characteristics of corporations discussed in the previous section, the following parts discuss the results regarding the ratio of corporations that provided each type of support for the recovery of regional society.

5.1.7. Differences between Groups Categorized by Damages Caused by the Tsunami

Figure 13 shows the rate of recovery support provided by corporations classified according to the damages caused by the tsunami. First, the response rates in each location were as follows: "completely destroyed area" (35.71%), "large-scale partially destroyed area" (56.92%), "partially destroyed or flooded area" (57.14%), "districts without tsunami damage near the devastated area" (36.97%), and "districts without tsunami damage far from the devastated area" (33.54%).

Damages caused by the tsunami (Distance from coastal line)	N Sample		Some Support	Funding	Daily Commodity	Equipment and Materials	Human Resources	Useful Information	Temporaly use of Lands and Buildings
Districts without tsunami damage far from the devastated area (11.00 km <)	53	→	30.19%	5.66%	13.21%	15.09%	7.55%	1.89%	1.89%
Districts without tsunami damage near the devastated area (5.50 to 11.00 km)	261	→	34.10%	5.75%	13.03%	13.79%	23.37%	4.60%	2.30%
Partially destroyed or flooded (3.00 to 5.50 km)	44	→	31.82%	0.00%	11.36%	15.91%	15.91%	6.82%	4.55%
Large-scale partially destroyed (1.50 to 3.00 km)	37	→	24.32%	0.00%	10.81%	16.22%	13.51%	10.81%	13.51%
Completely destroyed (< 1.50 km)	5	→	0.00%	0.00%	0.00%	0.00%	0.00%	0.00%	0.00%

Figure 13. The ratio of corporate contribution by damage caused by the tsunami.

Next, the percentage of corporations that provided some support is as follows: "completely destroyed area (0.00%)", "large-scale partially destroyed area" (24.32%), "partially destroyed or flooded area" (31.82%), "districts without tsunami damage near the devastated area" (34.10%), and "districts without tsunami damage far from the devastated area" (30.19%).

The percentage of corporations based on the type of support provided is as follows: regarding "funding support", the percentage of corporations that provided this support was 0.00% in the devastated areas. Regarding "daily commodities support", the percentage of corporations that provided this support in each district ranged from 10.81% to 13.21%, and it was found that there were no significant differences based on damages. Regarding "equipment and materials support", the percentage of corporations that provided this support in each group based on this classification ranged from 13.79% to 16.22%, and no significant differences were found between the groups. Regarding "human resources support", the percentage of corporations that provided this support in the districts without tsunami damage near the devastated area was 23.37%, higher than that in the other districts. Regarding "useful information support", as the distance from the coastline decreased, the rate of corporations that provided this support increased. Regarding "temporary use of land and buildings support", as the distance from the coastline decreased, the rate of corporations that provided this type of support increased.

5.1.8. Differences between Groups Categorized by the Number of Employees

Figure 14 shows the results of recovery support provided by corporations classified according to the number of employees. The percentage of corporations in each group that provided some support is as follows: 0 (25.00%), 1–7 (33.87%), 8–20 (29.66%), 21–100 (35.94%), and more than 101 (37.50%).

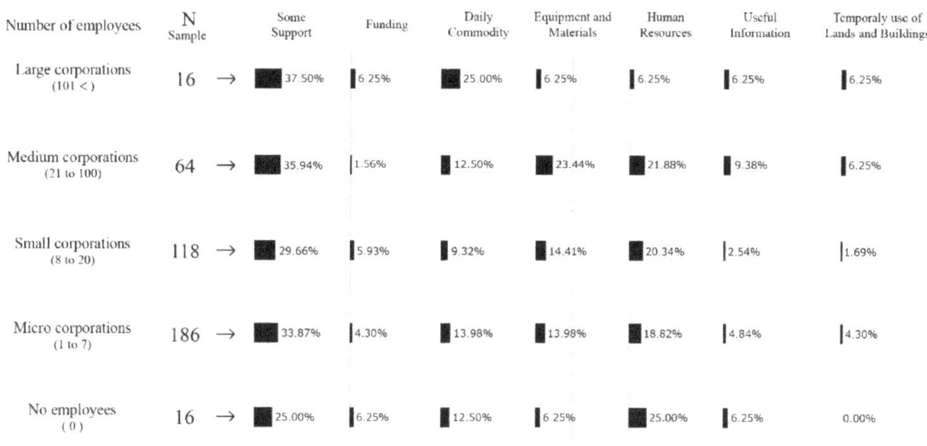

Figure 14. The ratio of corporate contribution by number of employees.

The percentage of corporations that provided each type of support is as follows: regarding "funding support", the rate of corporations that provided this support in each group ranged from 1.56–6.25%, and it was found that there were no significant differences based on the size of the corporation. Regarding "daily commodities support", the rate of corporations that provided this support in each district based on this classification ranged from 9.32% to 25.00%, and the rate of corporations that provided this support was higher in the "over 101 employees" group than in the other groups. The rate of corporations in this group that provided this support was also higher than the rate of corporations from this group that provided other types of support. Regarding "equipment and materials support", the rate of corporations that provided this support in each group ranged from 6.25% to 23.44%, and the rate of corporations that provided this support in the "21–100 employees" group was higher than in the other groups. Regarding "human resources support", the rate of corporations in each group that provided this support ranged from 6.25% to 25.0%, and the rate of corporations in the "over 101 employees" group was lower than in the other groups. Regarding "useful information support", the rate of corporations that provided this support in each group ranged from 2.54% to 9.38%, and no significant differences between the groups were found. Regarding "temporary use of land and buildings support", the rate of corporations in each group that provided this support ranged from 0.00% to 6.25%, and the rate of corporations that provided this support in the "0 employee" group was 0.00%.

5.1.9. Differences between Groups Classified by Their Awareness of Contributions

Figure 15 shows the rate of corporations that provided each support based on the classification by corporations' awareness. The results are as follows: "having awareness of contributions to the recovery of regional society, and having already conducted related activities (HAA)" (55.56%), "having awareness of contributions to the recovery of regional society, but having conducted no related activities earlier (HAN)" (24.11%), "not having the capacity to contribute (NHC)" (20.14%), and "having never thought about making contributions (NTC)" (33.33%). The ratio of corporations that have an awareness of contributions was 57.25% (229 corporations). However, among these corporations, the ratio of those that also provided support was 40.17% (92 corporations). The number of corporations in the "having never thought about making contributions" group was fewer than that in the other groups. The rate of corporations that provided support in the "having awareness of contributions to the recovery of regional society, and having already conducted related activities" group was higher than that in the other two groups for all types of support.

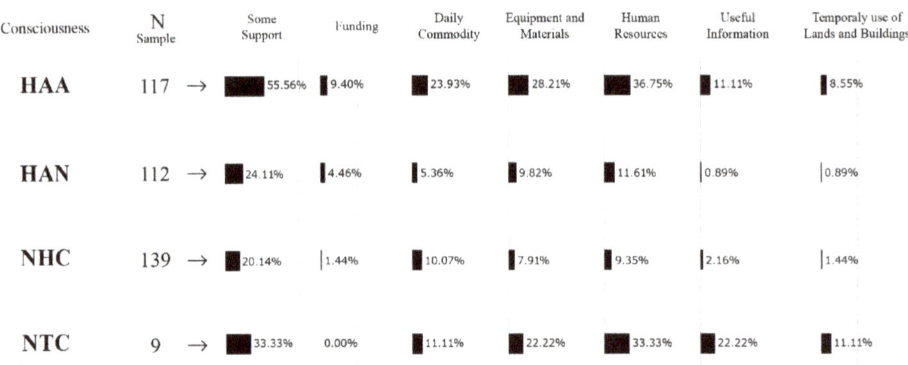

Consciousness	N Sample	Some Support	Funding	Daily Commodity	Equipment and Materials	Human Resources	Useful Information	Temporaly use of Lands and Buildings
HAA	117 →	55.56%	9.40%	23.93%	28.21%	36.75%	11.11%	8.55%
HAN	112 →	24.11%	4.46%	5.36%	9.82%	11.61%	0.89%	0.89%
NHC	139 →	20.14%	1.44%	10.07%	7.91%	9.35%	2.16%	1.44%
NTC	9 →	33.33%	0.00%	11.11%	22.22%	33.33%	22.22%	11.11%

Figure 15. The ratio of corporate contribution by corporations' awareness.

Differences between groups according to industry type

Figure 16 shows the rate of corporations in each industry that provided each type of support. The response rate of each industry type is shown in Figure 2. The rate of corporations that provided any support in each industry type is as follows: "Construction" (40.83%), "Wholesale and Retail Trade" (34.62%), "Miscellaneous Business Services, n.e.c". (29.55%), "Manufacturing" (23.81%), "Medical, Health Care, and Welfare" (31.03%), "Transport and Postal Services" (12.50%), "Real Estate and Goods Rental and Leasing" (35.00%), "Living-related and Personal Services and Amusement Services" (36.36%), "Electricity, Gas, Heat, and Water Supply" (30.00%), "Accommodation, Eating, and Drinking Services" (50.00%), "Information and Communications" (25.00%), "Education, Learning Support" (0.00%), "Scientific Research, Professional, and Technical Services" (33.33%), "Agriculture, Forestry, and Fisheries" (50.00%), and "Mining and Quarrying of Stone" (0.00%).

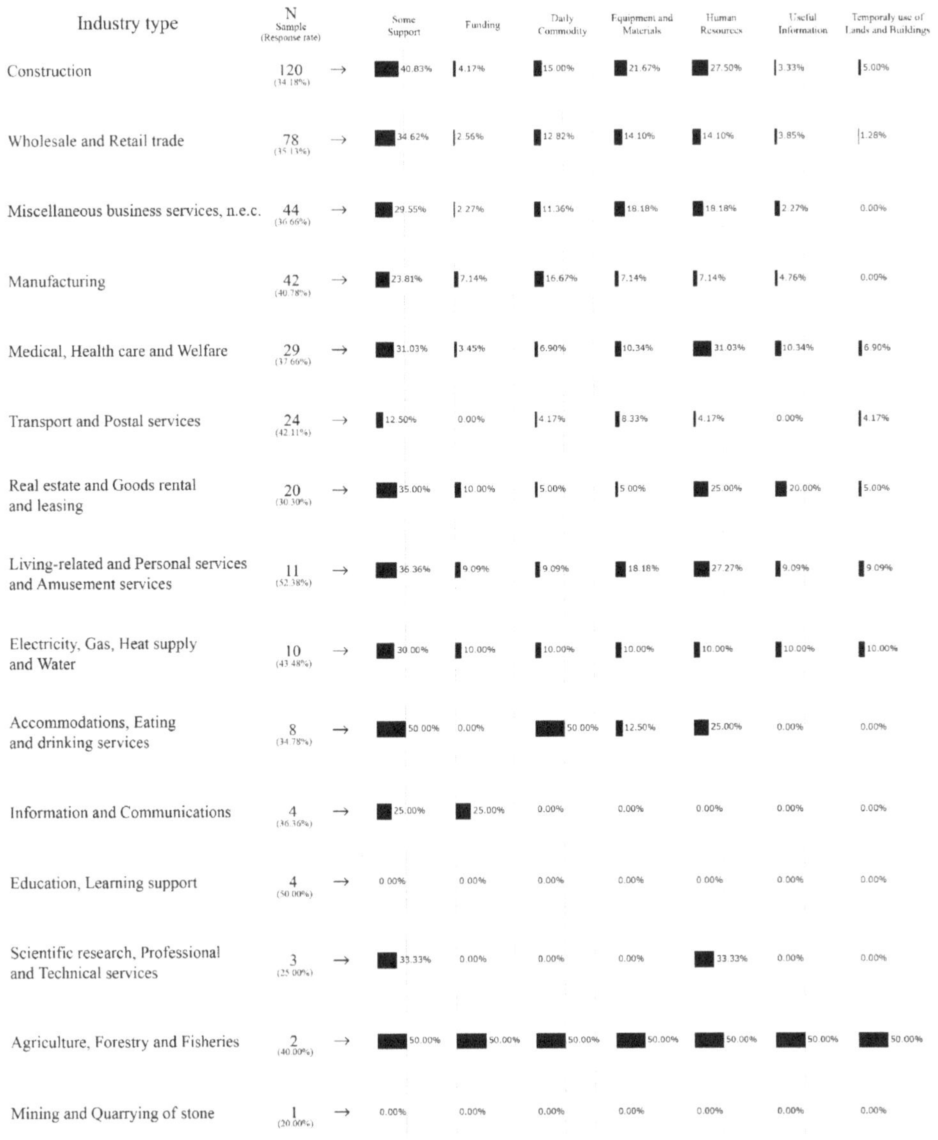

Industry type	N Sample (Response rate)		Some Support	Funding	Daily Commodity	Equipment and Materials	Human Resources	Useful Information	Temporaly use of Lands and Buildings
Construction	120 (34.18%)	→	40.83%	4.17%	15.00%	21.67%	27.50%	3.33%	5.00%
Wholesale and Retail trade	78 (35.13%)	→	34.62%	2.56%	12.82%	14.10%	14.10%	3.85%	1.28%
Miscellaneous business services, n.e.c.	44 (36.66%)	→	29.55%	2.27%	11.36%	18.18%	18.18%	2.27%	0.00%
Manufacturing	42 (40.78%)	→	23.81%	7.14%	16.67%	7.14%	7.14%	4.76%	0.00%
Medical, Health care and Welfare	29 (37.66%)	→	31.03%	3.45%	6.90%	10.34%	31.03%	10.34%	6.90%
Transport and Postal services	24 (42.11%)	→	12.50%	0.00%	4.17%	8.33%	4.17%	0.00%	4.17%
Real estate and Goods rental and leasing	20 (30.30%)	→	35.00%	10.00%	5.00%	5.00%	25.00%	20.00%	5.00%
Living-related and Personal services and Amusement services	11 (52.38%)	→	36.36%	9.09%	9.09%	18.18%	27.27%	9.09%	9.09%
Electricity, Gas, Heat supply and Water	10 (43.48%)	→	30.00%	10.00%	10.00%	10.00%	10.00%	10.00%	10.00%
Accommodations, Eating and drinking services	8 (34.78%)	→	50.00%	0.00%	50.00%	12.50%	25.00%	0.00%	0.00%
Information and Communications	4 (36.36%)	→	25.00%	25.00%	0.00%	0.00%	0.00%	0.00%	0.00%
Education, Learning support	4 (50.00%)	→	0.00%	0.00%	0.00%	0.00%	0.00%	0.00%	0.00%
Scientific research, Professional and Technical services	3 (25.00%)	→	33.33%	0.00%	0.00%	0.00%	33.33%	0.00%	0.00%
Agriculture, Forestry and Fisheries	2 (40.00%)	→	50.00%	50.00%	50.00%	50.00%	50.00%	50.00%	50.00%
Mining and Quarrying of stone	1 (20.00%)	→	0.00%	0.00%	0.00%	0.00%	0.00%	0.00%	0.00%

Figure 16. The ratio of corporate contribution by industry type.

The rates of corporations in each industry type whose response rate was over 30.00%, whose sample was over 1.00% of the target corporations (N > 11), and who conducted each type of support, are discussed below.

Regarding "funding support", the ratio of corporations that provided support in each industry type ranged from 0.00% to 10.00%, and no significant differences between industry type were found.

Regarding "daily commodities support", the ratio of corporations that provided support in each industry type ranged from 4.17% to 50.00%, and the ratio of corporations that provided support in the

"Accommodation, Eating, and Drinking Services" group was the highest in this survey. The ratios in the "Manufacturing" and "Construction" groups were also higher than those of the other groups.

Regarding "equipment and materials support", the ratio of corporations that provided this type of support ranged from 5.00 to 21.67%, and the ratio in the "Construction" group was the highest in this survey. The ratio in the "Living-related and Personal Services and Amusement Services" and "Miscellaneous Business Services, n.e.c". groups were also higher than that of the other groups.

Regarding "human resources support", the ratio of corporations that provided support ranged from 4.17% to 31.03%, and the ratio in the "Medical, Health Care, and Welfare" group was the highest in this survey. The ratio in the "Construction", "Living-related and Personal Services and Amusement Services", "Miscellaneous Business Services, n.e.c"., "Real Estate and Goods Rental and Leasing", and "Accommodation, Eating, and Drinking Services" groups were also higher than that of the other groups.

Regarding "useful information", the ratio of corporations that provided support ranged from 0.00% to 20.00%, and the ratio in the "Real Estate and Goods Rental and Leasing" group was the highest in this survey.

Regarding "temporary use of land and buildings", the ratio of corporations that provided this support ranged from 0.00% to 10.00%, and the ratio in the "Electricity, Gas, Heat, and Water Supply" group was the highest in this survey.

5.2. Results of Logistic Regression

In this section, we discuss the results of the logistic regression analysis based on the response variables (each type of support), as well as the explanatory variables for validating the hypotheses.

5.2.1. Parameter Estimation for "Some Support"

Table 9 shows the results of the parameter estimation for any support. The strongest positive relationship was found between providing "some support" and "having awareness of contributions to the recovery of regional society, and having already conducted related activities". There was a strong positive relationship between providing "some support" and "building area". In addition, there was also a positive relationship between providing "some support" and the location of main customers being "inside the same city limits".

Table 9. Results of the parameter estimation for "some support".

Explanatory Variable		Estimate	S.E	Z	Pr(>|z|)	Crude OR (95% CI)	adj. OR (95% CI)
Category	Value						
	(Intercept)	−2.041	0.282	−7.241	4.46×10^{-13} ***		
Number of employees	1–7	0.325	0.237	1.372	0.170	1.10 (0.72–1.67)	1.38 (0.87–2.20)
Awareness	Having awareness of contributions to the recovery of regional society, and having already conducted related activities	1.481	0.245	6.039	1.55×10^{-9} ***	4.11 (2.60–6.49)	4.4 (2.72–7.11)
Location of main customers	Inside the same city limits	0.471	0.234	2.017	0.044 *	1.70 (1.11–2.59)	1.60 (1.01–2.53)
Surrounding land use	Rice field area	0.579	0.461	1.256	0.209	1.25 (0.56–2.78)	1.78 (0.72–4.40)
	Building area	0.999	0.339	2.945	0.003 **	2.61 (1.43–4.75)	2.72 (1.40–5.28)

Null deviance: 505.92 on 399 degrees of freedom. Residual deviance: 451.35 on 394 degrees of freedom. AIC: 463.35. None of the VIF was over 10.0. Mean VIF of the model was 1.058 with a maximum of 1.074 and minimum of 1.019. *** 0 is not included in 99.9% Wald confidence interval. ** 0 is not included in 99% Wald confidence interval. * 0 is not included in 95% Wald confidence interval.

5.2.2. Parameter Estimation for Funding Support

Table 10 shows the results of the parameter estimation for "funding support". There was a strong positive relationship between providing "funding support" and "having awareness of contributions to the recovery of regional society, and having already conducted related activities".

Table 10. Results of parameter estimation for funding support.

Explanatory Variable		Estimate	S.E	Z	Pr(>|z|)	Crude OR (95% CI)	adj. OR (95% CI)
Category	Value						
	(Intercept)	−3.726	0.417	−8.936	$<2 \times 10^{-16}$ ***		
Number of employees	21–100	−1.484	1.056	−1.406	0.160	0.3 (0.04–2.28)	0.23 (0.03–1.80)
Awareness	Having awareness of contributions to the recovery of regional society, and having already conducted related activities	1.484	0.517	2.869	0.004 **	4.09 (1.55–10.83)	4.41 (1.60–12.16)
Industry type	Manufacturing	1.010	0.688	1.468	0.142	1.76 (0.49–6.35)	2.75 (0.71–10.57)
	Agriculture, Forestry, and Fisheries	2.242	1.457	1.538	0.124	22.41 (1.34–373.77)	9.41 (0.54–163.64)

Null deviance: 146.82 on 399 degrees of freedom. Residual deviance: 132.06 on 395 degrees of freedom. AIC: 142.06. None of the VIF was over 10.0. Mean VIF of the model was 1.037 with a maximum of 1.056 and minimum of 1.020. ** 0 is not included in 99% Wald confidence interval.

5.2.3. Parameter Estimation for Daily Commodities Support

Table 11 shows the results of the parameter estimation for "daily commodities support". The strongest positive relationship was found between providing "daily commodities support" and "having awareness of contributions to the recovery of regional society, and having already conducted related activities". There was also a strong positive relationship between providing "daily commodities support" and "Accommodation, Eating, and Drinking Services". Further, there were positive relationships between providing "daily commodities support" and the "Manufacturing" industry and the "location of main customers—inside the same city limits".

Table 11. Results of parameter estimation for daily commodities support.

Explanatory Variable		Estimate	S.E	z	Pr(>|z|)	Crude OR (95% CI)	adj. OR (95% CI)
Category	Value						
	(Intercept)	−3.322	0.398	-8.343	$<2 \times 10^{-16}$ ***		
Awareness	Having awareness of contributions to the recovery of regional society, and having already conducted related activities	1.378	0.322	4.279	1.88×10^{-5} ***	3.56 (1.95–6.49)	3.97 (2.11–7.46)
Industry type	Wholesale and Retail Trade	0.631	0.472	1.337	0.181	1.01 (0.48–2.11)	1.88 (0.74–4.75)
	Construction	0.603	0.409	1.473	0.141	1.20 (0.65–2.23)	1.83 (0.82–4.07)
	Accommodation, Eating, and Drinking Services	2.659	0.825	3.224	0.001 **	7.34 (1.78–30.34)	14.28 (2.84–71.92)
	Manufacturing	1.083	0.538	2.014	0.044 *	1.43 (0.60–3.41)	2.95 (1.03–8.47)
Location of main customers	Inside the same city limits	0.658	0.248	−2.012	0.044 *	1.87 (1.04–3.38)	1.93 (1.03–3.61)

Null deviance: 305.29 on 399 degrees of freedom. Residual deviance: 272.26 on 393 degrees of freedom. AIC: 286.26. None of the VIF was over 10.0. Mean VIF of the model was 1.257 with a maximum of 1.545 and minimum of 1.026. *** 0 is not included in 99.9% Wald confidence interval. ** 0 is not included in 99% Wald confidence interval. * 0 is not included in 95% Wald confidence interval.

5.2.4. Parameter Estimation for Equipment and Materials Support

Table 12 shows the results of the parameter estimation for "equipment and materials support". There was an extremely strong positive relationship between providing "equipment and materials support" and "having awareness of contributions to the recovery of regional society, and having already conducted related activities". There was also a strong positive relationship between providing "equipment and materials support" and the "construction" industry. Further, there were positive relationships between providing "equipment and materials support" and having 21–100 employees and "surrounding land use—rice field area".

Table 12. Results of parameter estimation for equipment and materials support.

Explanatory Variable		Estimate	S.E	Z	Pr(> |z|)	Crude OR (95% CI)	adj. OR (95% CI)
Category	Value						
	(Intercept)	−3.463	0.428	−8.088	6.05×10^{-16} ***		
Number of employees	21–100	0.758	0.375	2.024	0.043*	1.98 (1.03–3.82)	2.13 (1.02–4.45)
Awareness	Having awareness of contributions to the recovery of regional society, and having already conducted related activities	1.225	0.301	4.074	4.62×10^{-5} ***	3.72 (2.12–6.55)	3.4 (1.89–6.13)
Industry type	Wholesale and Retail Trade	0.655	0.454	1.443	0.149	0.91 (0.45–1.85)	1.93 (0.79–4.69)
	Construction	1.008	0.382	2.639	0.008 **	1.81 (1.03–3.17)	2.74 (1.30–5.79)
	Agriculture, Forestry, and Fisheries	1.907	1.481	1.288	0.198	5.75 (0.35–91.13)	6.73 (0.37–122.69)
	Miscellaneous Business Services, n.e.c.	0.702	0.508	1.382	0.167	1.30 (0.57–2.95)	2.02 (0.75–5.45)
Surrounding land use	Rice field area	1.166	0.577	2.022	0.043 *	2.35 (0.87–6.33)	3.21 (1.04–9.93)
	Building area	0.688	0.460	1.496	0.135	1.86 (0.86–4.0)	1.99 (0.81–4.9)

Null deviance: 338.17 on 399 degrees of freedom. Residual deviance: 300.85 on 391 degrees of freedom. AIC: 318.85. None of the VIF was over 10.0. Mean VIF of the model was 1.235 with a maximum of 1.610 and minimum of 1.030. *** 0 is not included in 99.9% Wald confidence interval. ** 0 is not included in 99% Wald confidence interval. * 0 is not included in 95% Wald confidence interval.

5.2.5. Parameter Estimation for Human Resources Support

Table 13 shows the results of the parameter estimation for "human resources support". There was an extremely strong positive relationship between providing "human resources support" and "having awareness of contributions to the recovery of regional society, and having already conducted related activities". There was also a strong positive relationship between providing "human resources support" and "surrounding land use—buildings area". Further, there were positive relationships between providing "human resources support" and "located in districts without tsunami damage near the devastated area" and "location of main customers—inside the same city limits".

Table 13. Results of parameter estimation for human resources support.

Explanatory Variable		Estimate	S.E	z	Pr(>|z|)	Crude OR (95% CI)	adj. OR (95% CI)
Category	Value						
	(Intercept)	−3.208	0.368	−8.729	$<2 \times 10^{-16}$ ***		
Awareness	Having awareness of contributions to the recovery of regional society, and having already conducted related activities	1.452	0.276	5.270	1.37×10^{-7} ***	4.12 (2.46–6.90)	4.27 (2.49–7.33)
Distance from coastline	Districts without tsunami damage near the devastated area	0.651	0.324	2.013	0.044 *	2.19 (1.22–3.92)	1.92 (1.02–3.62)
Location of main customers	Inside the same city limits	0.659	0.275	2.402	0.016 *	1.94 (1.18–3.20)	1.93 (1.13–3.31)
Surrounding land use	Building area	1.199	0.383	3.133	0.002 **	4.22 (2.10–8.46)	3.32 (1.57–7.02)

Null deviance: 394.71 on 399 degrees of freedom. Residual deviance: 340.90 on 395 degrees of freedom. AIC: 350.90. None of the VIF was over 10.0. Mean VIF of the model was 1.032 with a maximum of 1.053 and minimum of 1.012. *** 0 is not included in 99.9% Wald confidence interval. ** 0 is not included in 99% Wald confidence interval. * 0 is not included in 95% Wald confidence interval.

5.2.6. Parameter Estimation for Useful Information Support

Table 14 shows the results of the parameter estimation for "useful information support". There was an extremely strong positive relationship between providing "useful information supports" and the "Real Estate and Goods Rental and Leasing" industry. There was also a strong positive relationship between providing "useful information support" and "having awareness of contributions to the recovery of regional society, and having already conducted related activities". In addition, there were positive relationships between providing "useful information support" and the "Agriculture, Forestry, and Fisheries" industry and "surrounding land use—building area".

Table 14. Results of parameter estimation for useful information support.

Explanatory Variable		Estimate	S.E	z	Pr(>|z|)	Crude OR (95% CI)	adj. OR (95% CI)
Category	Value						
	(Intercept)	−4.804	0.563	−8.538	$< 2 \times 10^{-16}$ ***		
Number of employees	21–100	0.918	0.573	1.602	0.109	2.38 (0.88–6.44)	2.50 (0.81–7.70)
Awareness	Having awareness of contributions to the recovery of regional society, and having already conducted related activities	1.495	0.524	2.854	0.004 **	4.93 (1.91–12.69)	4.46 (1.60–12.44)
Industry type	Agriculture, Forestry, and Fisheries	2.976	1.504	1.978	0.048 *	19.95 (1.20–331.27)	19.61 (1.03–374.13)
	Real Estate and Goods Rental and Leasing	2.456	0.725	3.390	7×10^{-4} ***	5.69 (1.71–18.97)	11.66 (2.82–48.27)
Surrounding land use	Buildings area	1.507	0.682	2.210	0.027 *	4.85 (1.41–16.71)	4.51 (1.19–17.16)

Null deviance: 158.81 on 399 degrees of freedom. Residual deviance: 128.85 on 394 degrees of freedom. AIC: 140.85. None of the VIF was over 10.0. Mean VIF of the model was 1.100 with a maximum of 1.142 and minimum of 1.066. *** 0 is not included in 99.9% Wald confidence interval. ** 0 is not included in 99% Wald confidence interval. * 0 is not included in 95% Wald confidence interval.

5.2.7. Parameter Estimation for Temporary Use of Land and Buildings

Table 15 shows the results of the parameter estimation for "temporary use of land and buildings support". There was an extremely strong positive relationship between providing "temporary use of land and buildings support" and "having awareness of contributions to the recovery of regional society, and having already conducted related activities". There was also a positive relationship between providing "temporary use of land and buildings support" and "located in large-scale partially destroyed area".

Table 15. Results of parameter estimation for temporary use of land and buildings.

Explanatory Variable		Estimate	S.E	Z	Pr(>|z|)	Crude OR (95% CI)	adj. OR (95% CI)
Category	Value						
	(Intercept)	−4.596	0.569	−8.080	6.45×10^{-16} ***		
The ratio of damage amount to annual sales		0.0810	0.105	0.774	0.439	1.16 (0.96–1.39)	1.08 (0.88–1.33)
Awareness	Having awareness of contributions to the recovery of regional society, and having already conducted related activities	1.478	0.573	2.577	9.95×10^{-3} ***	5.20 (1.74–15.55)	4.38 (1.42–13.48)
Location of main customers	Inside the same city limits	0.750	0.555	1.352	0.176	2.38 (0.83–6.81)	2.12 (0.71–6.27)
Distance from coastline	Large-scale partially destroyed area	1.311	0.609	2.152	0.031*	5.52 (1.78–17.12)	3.71 (1.12–12.24)

Null deviance: 127.93 on 399 degrees of freedom. Residual deviance: 111.02 on 395 degrees of freedom. AIC: 121.02. None of the VIF was over 10.0. Mean VIF of the model was 1.026 with a maximum of 1.041 and minimum of 1.006. *** 0 is not included in 99.9% Wald confidence interval. * 0 is not included in 95% Wald confidence interval.

6. Discussion

6.1. Actual Condition of Corporate Contributions to the Recovery of Regional Society

This study selected 1,020 corporations whose headquarters are located in Iwanuma or Natori. We conducted a mail survey, the response rate of which was 39.22%, and it was identified that 32.75% of corporations had provided some type of support for the recovery of regional society after the Great East Japan Earthquake. The data suggest that the target area already had some structures for the autonomous and efficient recovery of regional society set up by the corporations during ordinary times, because the ratio of the corporations that already had a relationship with the support recipients before the occurrence of the disaster was 74.50%. Many of those relationships had "not conducted any particular activities". In other words, strong existing relationships such as business and community activities and information exchange triggered the provision of support, but weak relationships, such as only becoming acquainted with support recipients, also triggered the provision of support. Most previous studies on corporate contributions to the recovery of regional society have only focused on corporations' awareness of contributions [13–17]. In this study, the ratio of corporations that had such an awareness was 57.25%, whereas the ratio of those that actually provided support among corporations with awareness was 40.17%. In addition, among those that provided support, the ratio of corporations "having awareness of contributions to regional society, and having already conducted related activities" was higher than that of those that only had the awareness but had not conducted any related activities. Thus, the data show that it is necessary to have an awareness of contributions as well as opportunities to conduct related activities and become acquainted with support recipients before the occurrence of a disaster to enhance the resilience of regional society in order to achieve autonomous and efficient recovery after disaster.

According to the report of an investigation by the Disaster Management Cabinet Office Japan [17], regarding the questionnaire item "Do you have cooperative ties or relationships with regional society for the purpose of preparation against disaster risk?", it was found that the ratio of corporations that answered "having the opportunity to participate in activities of regional society, such as neighborhood association or independent anti-disaster organization" was 27.4%, and the ratio of corporations that answered "There was nothing that corresponded to that" was 56.6%. The results indicated that the majority of corporations do not have cooperative ties or relationships with regional society. Based on the results and our findings, we identified the problem that must be solved in order to make regional society resilient, namely how to enhance the "activities of corporations that do not have ties or relationships of cooperation with regional society" and "make connections between the corporations and regional society" during ordinary times. In order to resolve this issue, it is necessary to build a contact system and facilitate dialogue between corporations and the regional society through the building of relationships with support recipients, or through the development of not only various disaster-related activities such as drills and meetings but also other activities such as cleaning/beautification, crime prevention/patrols, local festivals/events, and so on.

Regarding corporate contributions to recovery after the Great East Japan Earthquake, previous studies have not documented that the targeted corporations are from various industry types and include small-scale corporations with headquarters in the devastated area. Thus, our results can contribute to the formulation of regional disaster prevention plans for large-scale disasters such as the Great East Japan Earthquake.

6.2. Characteristics of Corporate Contributions to the Recovery of Regional Society

Table 16 shows the summary of parameter estimation using the logistic regression model. It was found that there was one common characteristic and some other specific characteristics based on the analysis of the provided support. In this section, we discuss these characteristics. Our hypotheses were validated based on these characteristics.

Table 16. Summary of the characteristics of the corporations that provided support.

Type of Support	Awareness	Number of Employees	Industry Type	Location of Main Customers	Surrounding Land Use	Damage Caused by the Tsunami
Some of the support	HAA ***					
Funding	HAA **					
Daily commodities	HAA ***		ACCO *** MANU *	Inside the same city limits *		
Equipment and materials	HAA ***	21–100 *	CONS **		Rice field area *	
Human resources	HAA ***			Inside the same city limits *	Building area **	Districts without tsunami damage near the devastated area *
Useful information	HAA **		REAL ***		Building area *	
Temporary use of lands and buildings	HAA **					Large-scale partially destroyed area *

*** 0 is not included in 99.9% Wald confidence interval. ** 0 is not included in 99% Wald confidence interval. * 0 is not included in 95% Wald confidence interval.

First, the common characteristics influential in the provision of all types of support were the following: "having awareness of contributions to the recovery of regional society, and having already conducted related activities before the occurrence of disaster", "location of main customers—inside the same city limits", and "main surrounding land use—building area". However, the latter two characteristics were significantly influenced by "human resources support". Therefore, it was reasonable to exclude these two characteristics from the common characteristics.

Further, there were differences in individual characteristics between each support type. With the exception of awareness of contributions, these characteristics are described below by each support type.

Regarding providing "funding support", only the characteristic of "having awareness of contributions to the recovery of regional society and having already conducted related activities before the disaster" were found to be influential. Most previous studies reported that most corporations provided funding support [1,18,19]. In this study, however, there were not many corporations that were found to have provided funding support. The differences between the target corporations were considered as one reason for this. The target corporations in previous studies were large corporations that were not located in devastated areas, such as those in the 1% Club, or those listed on the Tokyo Stock Exchange. However, the target corporations in this study were not only large corporations but also medium corporations, small corporations, and micro corporations that were located in the devastated area. The objective of funding support was not investigated in this study, though. Therefore, it is necessary to obtain detailed information based on each funding support objective to identify more characteristics of corporations that provide this type of support.

Regarding providing "daily commodities support", the following were influential: "having the awareness of contributions to the recovery of regional society, and having already conducted related activities before the disaster", the "Accommodation, Eating, and Drinking Services" and "Manufacturing" industries, and "location of main customers—inside the same city limits". These characteristics were found to be reasonably influential. It was assumed that providing support was facilitated based on the condition of corporations because those in the "Accommodation, Eating, and Drinking Services" industry have a sufficient stock of daily commodities during ordinary times. Notably, corporations belonging to the food industries were also categorized under the "Manufacturing" industry. It was also assumed that "Manufacturing" had the same conditions as the "Accommodation, Eating, and Drinking Services" industry because it was reported by the Food Industry Affairs Bureau that, after the Great East Japan Earthquake, most staple foods were provided by corporations (85.9%) belonging to the "Manufacturing" industry [43]. In the case of the Hanshin–Awaji Earthquake Disaster, it was reported that the ratio of listed corporations that provided support was 62.3%; however, the industry types of the corporations were not identified [18]. In the case of the Niigata-ken Chuetsu-oki Earthquake, it was reported that the ratio of corporations that belonged to the 1% Club and provided support was 8.8% (household goods) and 21% (water or food provisions) [19]. It was also reported that the industry type of the corporations that provided water or food provision support was "Manufacturing" [19]. Regarding the provision of support of "water or food provisions", it was suggested that industry type was more influential than other characteristics such as the size or location of corporations. It was reported that the industry type of the corporations that provided household goods support were "Insurance", "Finance", and "Business Services", according to a previous study [19]. The industry type of "Accommodation, Eating, and Drinking Services", as a characteristic, was not included in the previous study. Thus, it was implied that the size or location of a corporation were influential characteristics. The industry types of "Insurance" and "Finance" were not identified because there were no respondents from these industry types in this study.

Regarding the provision of "equipment and materials support", the following were found to be influential: "having awareness of contributions to the recovery of regional society, and having already conducted related activities before the disaster", "number of employees—21–100", "Construction" industry, and "main surrounding land use—rice field areas". The "Construction" industry was included because of previous studies [13,19,20], and the interview surveys from this research indicate that one of the roles of construction industries was to provide support for the recovery of regional society through machinery. The "number of employees—21–100" indicates that this support was provided by medium-sized corporations in the study area. As shown in Figure 1, it was found that "main surrounding land use—rice field areas" indicates that the corporations were located in the eastern part of the study area, probably near the eastern residential or industrial area.

Regarding the provision of "human resources support", the following were found to be influential: "having awareness of contributions to the recovery of regional society, and having already conducted related activities before the disaster", "location of main customers—inside the same city limits", "main surrounding land use—building areas", and "location of the corporations—districts without tsunami damage near the devastated area". "Main surrounding land use—building areas" and "location of the corporations—districts without tsunami damage near the devastated area" indicate that the corporations were located near the Iwanuma or Natori railway stations. Therefore, it was reasonable to include these as influential characteristics because such corporations, located near stations that were not damaged by the tsunami, had the capacity to provide support after the disaster. On the other hand, industry types were not found to be influential characteristics in this study. Previous studies found that

"Construction", "Transport", and "Manufacturing" were the main industry characteristics with the possibility of providing such support [13]. Figure 15 shows that the ratio of corporations that belonged to the "Construction" industry and provided such support was 27.50%, and the ratios of corporations that belonged to the "Transport" and "Manufacturing" industry and provided such support were 4.17% and 7.14%. It was suggested that belonging to the "Construction" industry was more influential than belonging to the "Transport" or "Manufacturing" industry in this study. However, belonging to the "Construction" industry was not an identified characteristic based on logistic regression analysis.

Regarding providing "useful information support", the following were found to be influential: "having awareness of contributions to the recovery of regional society, and having already conducted related activities before the disaster", the "Real Estate and Goods Rental and Leasing" and "Agriculture, Forestry, and Fisheries" industries, and "main surrounding land use—building areas". However, the "Agriculture, Forestry, and Fisheries" industry was excluded from the characteristics because there were very few corporations belonging to this industry among the target corporations. The "Real Estate and Goods Rental and Leasing" industry was included since the interview survey indicated that the demand for information regarding land or buildings increased owing to the study area being designated as a disaster risk area due to the damage caused by the tsunami. The characteristics for providing this support were not found in previous studies.

Regarding providing "temporary use of lands and building support", the following characteristics were found to be influential: "having awareness of contributions to the recovery of regional society, and having already conducted related activities before the disaster" and "located in large-scale partially destroyed area". From the interview survey, it was found that most camps intended for rescue and recovery activities were located in a large-scale partially destroyed area. Thus, it was reasonable to include "located in large-scale partially destroyed area" as an influential characteristic for this support. In previous studies, it was found that "Construction" and "Manufacturing" were the characteristics with the most possibility of providing this support [13]. However, Figure 15 shows that the ratio of corporation that belonged to the "Construction" and "Manufacturing" industry and provided this support was 5.00% and 0.00%. It was suggested that "awareness" and "damage from the tsunami" were more influential as characteristics than industry type.

According to the above discussion, Hypothesis 1, *"No significant difference in damage exists between corporations responding and those not responding to the disaster"*, was correct except for the condition that the corporations provided "temporary use of lands and building support" and were located in a large-scale partially destroyed area.

As shown in Figure 13, the ratio of large-scale corporations that provided "daily commodities support" was higher than that of other corporations. Small-scale and micro-scale corporations provided this support to some extent. In short, corporations of all scales provided support; however, only the relationship between "medium-scale corporations" and providing "equipment and materials support" was identified as a characteristic based on statistical analysis.

According to the above discussion, Hypothesis 2, "No significant difference exists in number of employees between corporations responding and those not responding to the disaster", was correct except for the condition that the corporations provided "equipment and materials support" and the number of employees was 21–100.

Hypothesis 3, "Corporations that have an awareness of contributions to recovery support and had already conducted related activities before the disaster are more likely to provide support for the recovery of regional society than are other corporations", was also found to be correct because this characteristic has a strong relationship with all support types.

Hypothesis 4, *"The content of support differs according to industry types"*, was correct too, as it was found that, for example, "daily commodities support" was mostly provided by corporations belonging to the "Accommodation, Eating, and Drinking Services" and "Manufacturing" industries, "equipment and materials support" was mainly provided by those in the "Construction" industry, and "useful information support" was mostly provided by those in the "Real Estate and Goods Rental and Leasing" industry.

6.3. Limitations

This study discussed the characteristics of corporations that spontaneously contributed to the recovery of regional society. However, it has certain limitations, which are as follows.

First, the questionnaire survey targeted corporations that were still conducting business in January 2015; therefore, corporations that had closed their business between March 2011 and December 2014 were not included in this survey.

Second, we could not identify a "causal relationship" in this research, since the results (found by using a logistic regression model) used dummy variables that were translated from quantitative data.

Third, the study area was a provincial city developed from an agriculture-based society. The awareness of residents in the study area was relatively high, and they had strong connections; the area had a system for receiving support from the corporations. Hence, to help formulate disaster-management plans to deal with disasters in urban areas that do not have strong connections, in particular for inland earthquakes with an epicenter directly below big cities, it is necessary to conduct further studies focusing on the characteristics of urban areas.

Fourth, the data collected in this study were based on the observations and judgments of a single representative of the targeted corporations; data based on different individual circumstances or the situation of various people and associations related to the corporate contributions were not reflected in this study. The interview survey also showed that small-scale corporations cooperated with other corporations and associations; however, this study could not analyze this further. In order to examine the situation in more detail, it is necessary to investigate the support provided by local business networks or groups, the relationship between provided support and the geographical spread or agglomeration of corporations, case studies of some corporations that include employee activities, and the actual situation of corporate contributions to the recovery of regional society from the perspective of residents.

7. Conclusions

This study conducted a mail survey that was sent to 1020 corporations located in Iwanuma or Natori, Miyagi Prefecture, Japan, which were damaged due to the Great East Japan Earthquake and tsunami. Of the 1020 corporations, 400 responded to the mail survey. We drew two major conclusions based on the results.

1. In order to understand the actual situation of corporate contributions to the recovery of regional society, we identified the following by simple totaling and cross-totaling.

    - Among the target corporations, 32.75% provided some type of support for the recovery of regional society after the disaster.
    - Further, 74.50% of corporations that provided support had a relationship with the support recipients before the occurrence of the disaster. This refer not only to those corporations with strong relationships such as business, community activities, and information exchanges, but also to those with weak relationships such as maintaining acquaintances with support recipients; both types of relationships were indicated as triggers for providing support.
    - The contents of the support were as follows: "human resources" (59.54%), "equipment and materials" (45.80%), "daily commodities such as water and food" (38.93%), "useful information" (15.27%), "funding support" (13.74%), and "temporary use of land and buildings" (11.45%).
    - The proportion of corporations that have awareness of contributions was 57.25%, and of these, the proportion of those that actually provided some type of support was 40.17%.

2.   The following was identified through the logistic regression analysis and helped determine the characteristics of the corporations that provided support for the recovery of regional society.

- In order to providing any type of support, it was found that corporations need to have an awareness of their possible contributions, as well as opportunities to conduct related activities before the occurrence of a disaster to enhance the resilience of regional society in order to achieve efficient recovery.
- The influential characteristics for each type of support differed by industry type, location, and number of employees, under the following conditions.
- For "temporary use of land and buildings support", "location of the corporation—large-scale partially destroyed area" was an influential characteristic.
- For "equipment and materials support", the influential characteristics were "medium-scale corporations" and the "Construction" industry.
- For "daily commodities support", the influential characteristics were the "Accommodation, Eating, and Drinking Services" and "Manufacturing" industries.
- For "useful information support", the "Real Estate and Goods Rental and Leasing" industry was an influential characteristic.

These findings provide evidence for formulating disaster support policies and building the relationships and agreements between corporations and municipalities/residential associations needed for disaster response and recovery.

Further studies are needed to overcome this study's limitations, particularly the fourth one as discussed above. We especially want to identify the "the support provided by local business networks or groups" and "the actual situation of corporate contributions to the recovery of regional society from the perspective of residents" in future research.

Author Contributions: R.F. and M.I. conceived, designed and conducted the interview survey; R.F. and Y.G. conceived, designed and conducted the questionnaire survey; R.F. analyzed the data and wrote the paper.

Acknowledgments: The authors would like to thank Makoto Yokohari, Fumihiko Seta, U Hiroi, Toru Terada, Akiko Iida, Keisuke Sakamoto and Takeki Izumi for useful discussions. This research was supported by the Research Institute of Science and Technology for Society, Japan Science and Technology Agency (Category II Project, "Redevelopment of Tsunami Impacted Coastal Region to Save Life and to Implement Disaster Resilient Community", Mikiko ISHIKAWA, R&D Focus Area: Creating Community-based Robust and Resilient Society).

Conflicts of Interest: The authors declare no conflict of interest.

Abbreviations

CONS	Construction
MANU	Manufacturing
ACCO	Accommodation, Eating, and Drinking Services
MINI	Mining and Quarrying of Stone
ELEC	Electricity, Gas, Heat, and Water Supply
AGRI	Agriculture, Forestry, and Fisheries
REAL	Real Estate and Goods Rental and Leasing
HAA	Having awareness of contributions to the recovery of regional society, and having already conducted related activities before the occurrence of disaster
HAN	Having awareness of contributions to the recovery of regional society, but having conducted no related activities before the occurrence of disaster
NHC	Not having the capacity to contribute to the recovery of regional society before the disaster.
NTC	Having never thought about making contributions

References

1. Johnson, B.R.; Connolly, E.; Carter, T.S. Corporate social responsibility: The role of Fortune 100 companies in domestic and international natural disasters. *Corp. Soc. Responsib. Environ. Manag.* **2011**, *18*, 352–369. [CrossRef]
2. Hatakeyama, S.; Sakata, A.; Kawamoto, A.; Itoh, N.; Shiraki, W. Drawing Concept of Company BCP with Considering District Continuity and Propose of Measures for Reinforcing Disaster Resilience. *J. Jpn. Soc. Civ. Eng.* **2013**, *69*, 25–30. [CrossRef]
3. Ishikawa, M. A Study on Community-Based Reconstruction from Great East Japan Earthquake Disaster—A Case Study of Iwanuma City in Miyagi-Pref. *J. Disaster Res.* **2015**, *10*, 807–817. [CrossRef]
4. Sonoda, C.; Sakamoto, K.; Ishikawa, M. A Study about the Function and Influence of the Workshops in City Reconstruction Planning Process Through the Reconstruction of Iwanuma City, Miyagi Pref. from the Tohoku Earthquake. *J. City Plan. Inst. Jpn.* **2013**, *48*, 849–854.
5. Murakami, A.; Kumakura, E.; Ishikawa, M. Reconstruction of Coastal Villages Swept Away by Tsunami by 3D Digital Model. *J. Disaster Res.* **2015**, *10*, 818–829. [CrossRef]
6. Kariya, T.; Ubaura, M. A Study of the process of Community Participation in the Initial Stage of Reconstruction from Disaster, A case Study of Recovery Process from Tsunami Disaster in Ishinomaki City. *J. City Plan. Inst. Jpn.* **2013**, *48*, 837–842.
7. Olcott, G.; Oliver, N. Social Capital, Sensemaking, and Recovery: Japanese Companies and the 2011 Earthquake. *Calif. Manag. Rev.* **2014**, *56*, 5–22. [CrossRef]
8. Isouchi, C.; Mano, K.; Shiraki, W.; Inomo, H. A Proposal of District Continuity Intensification Through Supporting Drowing Up Business Continuity Plan (BCP) for Construction Companies. *J. Jpn. Soc. Civ. Eng.* **2011**, *67*, 59–64. [CrossRef]
9. Sakamoto, M.; Sato, S.; Abe, K.; Ogata, K.; Nakagawa, M.; Otsuka, T. Business Continuity by Small Business Owners -Business Continuity based on Disaster Experience Review by Ishinomaki Mebaekai. *J. Soc. Saf. Sci.* **2015**, *26*, 19–26.
10. Twigg, J. *Corporate Social Responsibility and Disaster Reduction: A Global Overview*; Benfield Greig Hazard Research Centre: London, UK, 2001.
11. Kaji, H.; Yamaki, T. Jishin ni taisuru Kigyou Bousairyoku Hyouka system: CMP hou no Kaihatsu (Evaluate system of Corporate Disaster Prevention against Earthquake Disaster: Development of CMP Method). *Yobou Jihou* **2004**, *219*, 40–48.
12. Genda, Y. Higashi Nihon Daishinsai ga Shigoto ni Ataeta Eikyou ni tsuite (Influence on Labor and Employment caused by the Great East Japan Earthquake Disaster). *Jpn. J. Labour Stud.* **2014**, *653*, 100–120.
13. Murosaki, Y. Minkan Kigyou no Bousai Taiou no Jittai (Trends in Corporate Disaster Prevention). In Proceedings of the Annual Conference of the Institute of Social Safety Science, Shizuoka, Japan, 15–16 May 1992; Volume 2, pp. 145–154.
14. Ogawa, Y.; Nagano, Y. Nen Hanshin Daishinsai wo Keiki tosuru Kigyou no Bousai Ishiki no Henka ni kansuru Kentou (Changes in Corporate Awareness of Disaster Prevention after Hanshin-Awaji Great Earthquake). In Proceedings of the Annual Conference of the Institute of Social Safety Science, Shizuoka, Japan, 16–18 November 1995; Volume 5, pp. 141–144.
15. Ito, M.; Kameno, T. Assistance Awareness of the Private Sector in Local Disaster Management: A Case Study of Coastal Areas in Oita. *Rep. City Plan. Inst. Jpn.* **2014**, *13*, 31–35.
16. Nakamura, J.; Harada, K. A Survey of Corporate Efforts for Regional Disaster Prevention Related to Corporate Social Responsibility (CSR). *J. Soc. Saf. Sci.* **2014**, *24*, 53–60.
17. Disaster Management Cabinet Office Japan. *Heisei 27 Nendo Kigyou no Jigyou Keizoku oyobi Bousai no Torikumi ni kansuru Jittai Cyousa (Actual Conditions Survey of Business Continuity and Activities for Disaster Prevention)*; Disaster Management Cabinet Office Japan: Koshigaya, Japan, 2015.
18. Murosaki, Y.; Iwami, T. Hanshin Awaji Daishinsai to Kigyou no Bousai Taiou (Corporate Disaster Prevention Activities after Hanshin Awaji Great Earthquake). In Proceedings of the Annual Conference of the Institute of Social Safety Science, Shizuoka, Japan, 16–18 November 1995; Volume 5, pp. 129–134.
19. Toyoya, A.; Shoji, G. Clarification of Corporate Contribution in a Regional Disaster: A Case Study of 2004 Niigata-ken Chuetsu Earthquake and 2005 Hurricane Katrina. *J. Soc. Saf. Sci.* **2007**, *9*, 9–19.

20. Ohnishi, K. A Study on the Role of Local Construction Companies and Carpenters in The Great Hanshin-Awaji Earthquake. In Proceedings of the Annual Conference of the Institute of Social Safety Science, Shizuoka, Japan, 7–9 November 1997; Volume 7, pp. 234–237.
21. Henderson, J.M.; Quandt, R.E. *Microeconomic Theory: A Mathematical Approach*; McGraw-Hill: New York, NY, USA, 1958.
22. Alcahian, A.A.; Demsetz, H. Production, Information Cost, and Economic Organization. *Am. Econ. Rev.* **1972**, *62*, 4.
23. Cyert, R.H.; March, J.G. *A Behavioral Theory of the Firm*; Prentice Hall: Englewood Cliffs, NJ, USA, 1963.
24. Nelson, R.; Winter, S. *An Evolutionary Theory of Economic Change*; Belknap: Louisville, KY, USA, 1982.
25. Penrose, E. *The Theory of Growth of the Firm*; Wiley: Hoboken, NJ, USA, 1995.
26. Adams, M.; Hardwick, P. An Analysis of Corporate Donations: United Kingdom Evidence. *J. Manag. Stud.* **1998**, *35*, 641–654. [CrossRef]
27. Brammer, S.; Millington, A. The Development of Corporate Charitable Contributions in the UK: A Stakeholder Analysis. *J. Manag. Stud.* **2004**, *41*, 1411–1434. [CrossRef]
28. Brammer, S.; Millington, A. Profit Maximization vs. Agency: An Analysis of Charitable Giving by UK Firms. *Camb. J. Econ.* **2005**, *29*, 517–534. [CrossRef]
29. Zhang, R.; Rezaee, Z.; Zhu, J. Corporate Philanthropic Disaster Response and Ownership Type: Evidence from Chinese Firms' Response to the Sichuan Earthquake. *J. Bus. Ethics* **2010**, *91*, 51–63. [CrossRef]
30. Koyama, S.; Aida, J.; Kawachi, I.; Kondo, N.; Subramanian, S.V.; Ito, K.; Kobashi, G.; Masuno, K.; Kondo, K.; Osaka, K. Social Support Improves Mental Health among the Victims Relocated to Temporary Housing following the Great East Japan Earthquake and Tsunami. *Tohoku J. Exp. Med.* **2014**, *234*, 241–247. [CrossRef] [PubMed]
31. Tsuboya, T.; Aida, J.; Hikichi, H.; Subramanian, S.V.; Kondo, K.; Osaka, K.; Kawachi, I. Predictors of depressive symptoms following the Great East Japan Earthquake: A Prospective study. *Soc. Sci. Med.* **2016**, *161*, 47–54. [CrossRef] [PubMed]
32. Hikichi, H.; Aida, J.; Tsuboya, T.; Kondo, K.; Kawachi, I. Can Community Social Cohesion Prevent Posttraumatic Stress Disorder in the Aftermath of a Disaster? A Natural Experiment from the 2011 Tohoku Earthquake and Tsunami. *Am. J. Epidemiol.* **2016**, *183*, 902–910. [CrossRef] [PubMed]
33. Sasaki, Y.; Aida, J.; Tsuji, T.; Miyaguni, Y.; Tani, Y.; Koyama, S.; Matsuyama, Y.; Sato, Y.; Tsuboya, T.; Nagamine, Y.; et al. Does Type of Residential Housing Matter for Depressive Symptoms in the Aftermath of a Disaster? Insights from the Great East Japan Earthquake and Tsunami. *Am. J. Epidemiol.* **2017**, *187*, 455–464. [CrossRef] [PubMed]
34. Aida, J.; Hikichi, H.; Matsuyama, Y.; Sato, Y.; Tsuboya, T.; Tabuchi, T.; Koyama, S.; Subramanian, S.V.; Kondo, K.; Osaka, K.; et al. Risk of mortality during and after the 2011 Great East Japan Earthquake and Tsunami among older coastal residents. *Sci. Rep.* **2017**, *7*, 16591. [CrossRef] [PubMed]
35. Hikichi, H.; Sawada, Y.; Tsuboya, T.; Aida, J.; Kondo, K.; Koyama, S.; Kawachi, I. Residential relocation and change in social capital: A natural experiment from the 2011 Great East Japan Earthquake and Tsunami. *Sci. Adv.* **2017**, *3*, e1700426. [CrossRef] [PubMed]
36. Japan Platform. Available online: http://www.japanplatform.org/E/ (accessed on 9 March 2018).
37. 1% club of Japan Business Federation. Available online: http://www.keidanren.or.jp/japanese/profile/1p-club/ (accessed on 9 March 2018). (In Japanese)
38. Iwanuma City. 2011.3.11 Higashi Nihon Daishinsai Iwanuma Shi no Kiroku (Historical Records of Iwanuma City after the Great East Japan Earthquake, 11 March 2011). 2014. Available online: https://www.pref.miyagi.jp/pdf/kiki/iwanuma2603.pdf (accessed on 9 March 2018).
39. Natori City. Higashi Nihon Daishinsai Natori Shi no Kiroku (Historical Records of Natori City after the Great East Japan Earthquake). 2014. Available online: http://www.city.natori.miyagi.jp/content/download/29726/173138/filenatorishi-kiroku-all.pdf (accessed on 9 March 2018).
40. National Land Numerical Information Download Service. Available online: http://nlftp.mlit.go.jp/ksj-e/index.html (accessed on 9 March 2018).
41. Neuman, W.L. *Social Research Methods: Qualitative and Quantitative Approaches*, 7th ed.; Pearson Education Limited: London, UK, 2014.

42. Statistics Bureau, Ministry of Internal Affairs and Communications. Economic Census for Business Activity in Japan. Available online: http://www.stat.go.jp/english/data/e-census/2012/index.htm (accessed on 9 March 2018).

43. Doi, K. Issue and Recommendation of Governmental Emergency Food Supply Based on the Great East Japan Earthquake Experience. *J. Jpn. Soc. Irrig. Drain. Rural Eng.* **2013**, *81*, 31–34.

MDPI

St. Alban-Anlage 66

4052 Basel

Switzerland

Tel. +41 61 683 77 34

Fax +41 61 302 89 18

www.mdpi.com

Sustainability Editorial Office

E-mail: sustainability@mdpi.com

www.mdpi.com/journal/sustainability